Proceedings of the 23rd International Conference on Industrial Engineering and Engineering Management 2016

Ershi Qi · Jiang Shen · Runliang Dou
Editors

Proceedings of the 23rd International Conference on Industrial Engineering and Engineering Management 2016

Theory and Application of Industrial Engineering

ATLANTIS
PRESS

Editors
Ershi Qi
College of Management and Economics
Tianjin University
Tianjin
China

Runliang Dou
College of Management and Economics
Tianjin University
Tianjin
China

Jiang Shen
College of Management and Economics
Tianjin University
Tianjin
China

ISBN 978-94-6239-254-0 ISBN 978-94-6239-255-7 (eBook)
DOI 10.2991/978-94-6239-255-7

Library of Congress Control Number: 2016960282

Printed on acid-free paper

Contents

Selection Of Port Enterprise Logistics Service Providers Based On The Combination Weighting-Grey Synthetic Decision-Making Method

Ke-jia CHEN*, Xuan-nan LI
School of Economics and Management, Fuzhou University, Fuzhou, China
(*e-mail: kjchen@fzu.edu.cn)

Abstract - **Selection of port enterprises logistics service providers is a complex multiple attribute decision-making problem. In order to improve the scientific nature of selection, a set of reasonable decision-making methods of logistics service providers selection are needed to establish. Accordingly, a combination weighting-grey synthetic decision-making method is proposed, which is used to solve the problem of providers selection. Firstly, this method combines the network analysis method with the entropy method to determine combined weights of evaluation indexes. Secondly, the improved center-point triangular whitenization weight function is introduced to cluster alternative plans and judge its corresponding grey classes, and then synthetic weighted decision-making vectors are used to solve grey synthetic decision-making coefficient vectors. Next, grey synthetic clustering decision-making coefficients are calculated to conduct a synthetic decision-making rank of alternative plans. Finally, an empirical study is conducted to verify the effectiveness of the proposed method by taking logistics service providers selection of a port enterprise.**

Keywords - **Combination weighting, grey synthetic decision-making, logistics service, providers selection**

I. INTRODUCTION

In the market environment, the reasonable choice of logistics service providers is one of the important problem that enterprises need to face. Due to the uncertainty of the evaluation process, it makes the problem appear the grey character, and causes it to become a complex multiple attribute decision-making problem [1]. At present, logistics service providers of port enterprises in our country are mainly small and medium sized enterprises. The overall strength and operation standardization degree of these companies are not high enough, they provide the quality of logistics service is also sometimes difficult to make port enterprises satisfaction. Therefore, the study of the selection strategy of port enterprises logistics service providers, has a certain practical significance to improve the operation efficiency and operation level of port enterprises.

To solve this problem, many scholars have done researches. Reference [2] introduced the compatibility of logistics service providers in the evaluation index system, through the network analysis method (ANP) for providers selection. Reference [3] introduced the grey clustering method in providers selection problem and established a decision-making model of grey clustering for providers evaluation. Reference [4] constructed the model of providers evaluation based on the extension AHP and the grey relational analysis method. Reference [5] constructed a double-layer planning model, respectively to the logistics cost and customer satisfaction as the optimization goal, and calculated the weight of each index by entropy method, finally using cloud adaptive genetic algorithm to solve the model. Reference [6] set up a double-layer planning model based on the logistics cost and service quality. In this foundation, the QFD model was reconstructed and the AHP was used to evaluate logistics service providers. Reference [7] used the ANP method to obtain corresponding weights based on the logistics strength, the cost price, the service quality and the cooperation risk of providers, then used the VIKOR to carry on the final sorting. Reference [8] combined the grey relational analysis and the TOPSIS method, then proposed a combination weighting GI-TOPSIS method was used to solve the logistics service providers selection problem. Reference [9] combined with the entropy method and the G1 method to evaluate providers. Reference [10] proposed using the rough set reduction theory to obtain the index system weight, and finally according to the TOPSIS method to calculate various logistics service providers to the close degree of the grey ideal solution to determine the appropriate provider. Reference [11] proposed a multiple objective decision-making model based on the grey clustering and the entropy weight method, which was used to select the best provider. Reference [12] combined the grey system theory with the ANP, and applied to evaluation and selection of logistics service providers. Reference [13] used the grey relational analysis method to choose providers, and through the change of weights to study the sensitivity of the change of the selection result. Reference [14] combined the entropy method with the principal component analysis method to solve weights, then the grey relational analysis method was used to select providers. In 2014, reference [15] improved the original center-point triangular whitenization weight function. In the new method, the class 1 was constructed to a whitenization weight function of lower measure, and the class s was constructed to a whitenization weight function of upper measure, through the improved center-point triangular whitenization weight function can effectively solve the extension problem of the range of each clustering index.

In summary, there are some shortcomings in selection of logistics service providers. For example, the index system cannot well reflect the interdependence relationship between indexes; the way of solving indexes weights too much rely on expert judgment and other subjective factors, the degree of attention to objective

E. Qi et al. (eds.), *Proceedings of the 23rd International Conference on Industrial Engineering and Engineering Management 2016*, DOI 10.2991/978-94-6239-255-7_1

factors is not high enough; the way of handling the grey character of problem is not appropriate and so on. In this paper, to determine providers evaluation indexes weights not only considers the expert judgment and other subjective factors, but also considers the objective information of evaluation indexes data and the grey character. Accordingly, a combination weighting-grey synthetic decision-making method is proposed, which is used to help enterprises solve the problem of selecting logistics service providers, through empirical research, to verify the effectiveness of the decision-making method.

II. The evaluation index system of logistics service providers of port enterprises

A. Construction of the evaluation index system

Based on the existing research results, the evaluation index system of selecting logistics service providers of port enterprises is established, as shown in Table I.

B. Analysis of reliability and validity of the evaluation index system

According to the constructed evaluation index system above, the questionnaire survey is used to evaluate the importance of each evaluation index.

1) Reliability Analysis

Through the reliability analysis, the overall reliability coefficient of the questionnaire is 0.738. The internal consistency and reliability of the questionnaire are good.

2) Validity Analysis

According to the result of the analysis of validity, evaluation indexes are grouped into first level indexes of five groups, as shown in Table I.

III. The combination weighting-grey synthetic decision-making method

For a logistics service, there are m alternative plans on the market. There are n evaluation indexes for the evaluation of alternative plans. Evaluation data of alternative plans $i(i=1,2,\dots,m)$ about evaluation indexes $j(j=1,2,\dots,n)$ are $x_{ij}(i=1,2,\dots,m;\ j=1,2,\dots,n)$. This paper contains a total of s grey class. $w_j=(w_1,w_2,\dots,w_n)$ represents the combined weight of each evaluation index. In this paper, decision steps of the combination weighting-grey synthetic decision-making method are shown as follows.

A. Combination weighting

1) Determination of subjective weights based on the ANP method

Network structures of ANP are mainly composed of two parts: the control layer and the network layer. Elements can be divided into three relations, which are external influence relations, internal influence relations and mutual influence relations.

TABLE I
THE EVALUATION INDEX SYSTEM AND EVALUATION DATA OF LOGISTICS SERVICE PROVIDERS

Evaluation index		Alternative logistics service providers				
First level indexes	Second level indexes	G_1	G_2	G_3	G_4	G_5
Service quality A	Customer satisfaction rate A_1 (%)	83	80	87	89	84
	Order completion rate A_2 (%)	90	87	94	93	88
	On-time delivery rate A_3 (%)	85	86	87	88	89
	Product integrity rate A_4 (%)	90	94	91	93	94
	Problem handling rate A_5 (%)	92	88	94	89	91
Business ability B	Business diversity B_1 (Hundred mark)	81	77	88	85	84
	Logistics network B_2 (Hundred mark)	80	78	90	86	84
	Operation management ability B_3 (Hundred mark)	82	80	89	92	83
	Risk coping ability B_4 (Hundred mark)	67	63	72	69	77
Logistics cost C	Transportation cost C_1 (Vehicle 16M)	9300	8900	8800	8700	9600
	Storage cost C_2 (Yuan/(m^2*day))	2.8	3.3	3.6	3.8	3.4
	Circulation processing cost C_3 (Yuan/m^3)	17	20	14	12	16
	Handling and carrying cost C_4 (Yuan/t*1)	16	16	12	10	13
Enterprise soft power D	Market share D_1 (%)	5	5	6	4	3
	Enterprise qualification D_2 (Hundred mark)	87	82	92	93	90
	Enterprise reputation D_3 (Hundred mark)	78	82	80	79	84
	Professional talent ratio D_4 (%)	8	5	11	12	6
Cooperation E	Cooperation success rate E_1 (%)	74	77	86	88	83
	Informatization degree E_2 (Hundred mark)	82	72	80	68	81
	Communication level E_3 (Hundred mark)	75	85	83	83	69

By solving the limit matrix, the subjective weight s_j of each evaluation index is obtained.

2) Determination of objective weights based on the entropy method

Because there are some differences in the dimension of each evaluation index, therefore, before using the entropy method to solve objective weights, it is necessary to standardize evaluation data x_{ij}. When evaluation indexes are positive or reverse type, the standard formula (1) and (2) are used to convert evaluation data into non dimensional evaluation data r_{ij}.

$$r_{ij} = \frac{x_{ij} - \min_i x_{ij}}{\max_i x_{ij} - \min_i x_{ij}} \quad j = 1,2,\cdots,n \qquad (1)$$

$$r_{ij} = \frac{\max_i x_{ij} - x_{ij}}{\max_i x_{ij} - \min_i x_{ij}} \quad j = 1,2,\cdots,n \qquad (2)$$

After standardized treatment, the objective weight o_j of each evaluation index is obtained through using the entropy method.

3) Calculation of combined weights

Combined weights of subjective weights and objective weights are calculated by the formula (3).

$$w_j = s_j \times o_j \Big/ \sum_{j=1}^{n} s_j \times o_j \quad j = 1,2,\cdots,n \qquad (3)$$

B. Grey synthetic decision-making

1) Determination of turning points or center points of evaluation indexes grey classes

The value range of hypothesis evaluation indexes j are between [a_j, b_j]. According to overall evaluation requirements, evaluation indexes are divided into a total of s grey classes. Next, turning points λ_j^1 、 λ_j^s of grey classes 1 and grey classes s and center points λ_j^2 、 λ_j^3 … λ_j^{s-1} of grey classes $k(k \in \{2,3,...,s-1\})$ are determined respectively.

2) Construction of corresponding grey classes whitenization weight functions

Grey classes 1 are constructed to a corresponding whitenization weight function $f_j^1[-,-,\lambda_j^1,\lambda_j^2]$ of lower measure; grey classes s are constructed to a corresponding whitenization weight function $f_j^s[\lambda_j^{s-1},\lambda_j^s,-,-]$ of upper measure; for grey classes $k(k \in \{2,3,...,s-1\})$, they are needed to connect the point $(\lambda_k,1)$ and the center point $(\lambda_j^{k-1},0)$ of the grey class $k-1$ (or the turning point $(\lambda_j^1,0)$ of grey class 1) as well as the center point $(\lambda_j^{k+1},0)$ of the grey class $k+1$ (or the turning point

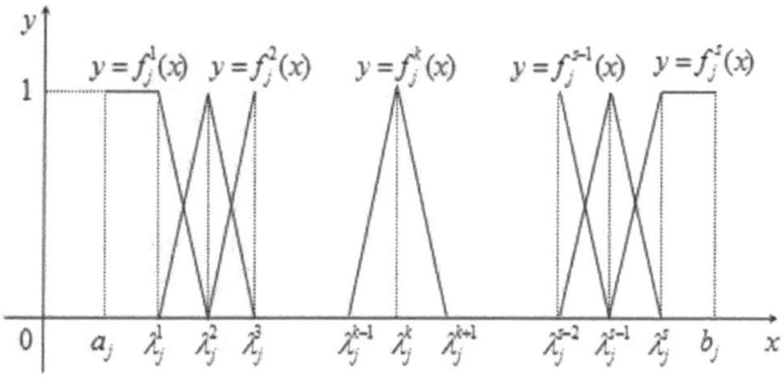

Fig.1. Schematic diagram of the improved center-point triangular whitenization weight function

$(\lambda_j^s,0)$ of grey class s). Thus, it can get triangular whitenization weight functions of evaluation indexes j for grey classes k. The function diagram shown in Figure 1.

3) The calculation of grey clustering coefficients

By the formula (4) we can calculate grey clustering coefficients σ_i^k of alternative plans $i(i = 1,2,...,m)$ for grey classes $k(k = 1,2,...,s)$.

$$\sigma_i^k = \sum_{j=1}^{n} f_j^k(x_{ij}) w_j \qquad (4)$$

$f_j^k(x_{ij})$ represent whitenization weight functions of evaluation indexes j for grey classes k, w_j represent the combined weight of each evaluation index.

4) The judgment of alternative plans corresponding grey classes

Order $\delta_i^k = \frac{\sigma_i^k}{\sum_{k=1}^{s} \sigma_i^k}$, called δ_i^k as the normalized grey clustering coefficients of alternative plans i for grey classes k. Obviously, $\delta_i^k (k = 1,2,...,s)$ can meet the condition of $\sum_{i=1}^{s} \delta_i^k = 1$. $\delta_i = (\delta_i^1, \delta_i^2, ..., \delta_i^s)(i = 1,2,...,m)$ are called as the normalized grey clustering coefficient vectors of alternative plans i. According to the value of the $\max_{1 \le k \le s} \{\delta_i^k\} = \delta_i^{k^*}$, can judge alternative plans corresponding grey classes.

5) The calculation of synthetic weighted decision-making vectors

Assuming the existing a total of s grey classes, $\eta_k = (\eta_k^1, \eta_k^2, ..., \eta_k^s)(k = 1,2,...,s)$ are called synthetic weighted decision-making vector of grey classes k. The kth component of these vectors is assigned to the value s, and the kth component is took as the midpoint. When assigning a value to its left and right components, descending order, it indicates that the kth component is the largest contribution for alternative plans belonging grey classes k, so give it the greatest weight s. The calculation method is as follows:

$$\eta_k = \frac{(s-k+1, s-k+2, ..., s-1, s, s-1, ..., k)}{\frac{s(s+1)}{2} + [(k-1)s - k(k-1)]} \quad (5)$$

6) The calculation of grey synthetic decision-making coefficient vectors

By the formula (6) can calculate initial grey synthetic decision-making coefficients of alternative plans i for grey classes k, and $\psi_i' = (\psi_i^{1'}, \psi_i^{2'}, ..., \psi_i^{s'})(i = 1,2,...,m)$ are called initial grey synthetic decision-making coefficient vectors of alternative plans i.

$$\psi_i^{k'} = \eta_k \delta_i^T \quad (6)$$

After the normalization process, we can get $\psi_i = (\psi_i^1, \psi_i^2, ..., \psi_i^s)(i = 1,2,...,m)$ which are called grey synthetic decision-making coefficient vectors of alternative plans i.

7) The calculation of grey synthetic clustering decision-making coefficients

The weight vector of grey synthetic clustering decision-making coefficients is $\gamma = (1,2,..., s-1, s)^T$, by the formula (7) can calculate grey synthetic clustering decision-making coefficients π_i of alternative plans i.

$$\pi_i = \psi_i \cdot \gamma = \sum_{k=1}^{s} k \cdot \psi_i^k \, (i = 1,2,...,m) \quad (7)$$

Because $0 \le \psi_i^k \le 1$, therefore, the maximum value of grey synthetic clustering decision-making coefficient $\pi_i = \psi_i \cdot \gamma (i = 1,2,...,m)$ of alternative plans i is s, and the minimum value is 1. We can draw a conclusion: $1 \le \pi_i \le s$. At the same time, the decision-making coefficient matrix can be obtained as follows:

$$\pi_i = \begin{bmatrix} \pi_1 \\ \pi_2 \\ \cdots \\ \pi_m \end{bmatrix} = \begin{bmatrix} \psi_1^1 & \psi_1^2 & \cdots & \psi_1^s \\ \psi_2^1 & \psi_2^2 & \cdots & \psi_2^s \\ \cdots & \cdots & \cdots & \cdots \\ \psi_m^1 & \psi_m^2 & \cdots & \psi_m^s \end{bmatrix} \cdot (1,2,..., s-1, s)^T$$

8) The rank of the synthetic decision-making

By comparing the size of grey synthetic clustering decision-making coefficients of alternative plans i in the same grey classes k, we can conduct a synthetic decision-making rank.

IV. Empirical study

A port enterprise mainly produce various concrete block and concrete prefabricated components. These

TABLE II
EVALUATION INDEX COMBINATION WEIGHTS AND GREY CLASSES CLASSIFICATION OF LOGISTICS SERVICE PROVIDERS

Evaluation index			Turning points or center points of evaluation indexes grey classes					
First level indexes	Second level indexes	Combination weights	Lower limit	Poor	Medium	Good	Excellent	Upper limit
Service quality A	Customer satisfaction rate A_1 (%)	0.005	70	75	80	85	90	100
	Order completion rate A_2 (%)	0.020	70	80	85	90	95	100
	On-time delivery rate A_3 (%)	0.024	70	80	85	90	95	100
	Product integrity rate A_4 (%)	0.015	70	80	85	90	95	100
	Problem handling rate A_5 (%)	0.020	70	80	85	90	95	100
Business ability B	Business diversity B_1 (Hundred mark)	0.049	60	65	75	80	90	100
	Logistics network B_2 (Hundred mark)	0.087	60	70	75	85	90	100
	Operation management ability B_3 (Hundred mark)	0.174	60	65	75	85	90	100
	Risk coping ability B_4 (Hundred mark)	0.109	55	60	65	75	85	100
Logistics cost C	Transportation cost C_1 (Vehicle 16M)	0.034	10000	9800	9400	9000	8600	8000
	Storage cost C_2 (Yuan/(m^2*day))	0.025	8	4	3.5	3	2.5	2
	Circulation processing cost C_3 (Yuan/m^3)	0.011	30	25	19	15	11	10
	Handling and carrying cost C_4 (Yuan/t*1)	0.018	25	22	18	15	10	8
Enterprise soft power D	Market share D_1 (%)	0.036	0	1	3	5	7	10
	Enterprise qualification D_2 (Hundred mark)	0.170	70	75	80	85	90	100
	Enterprise reputation D_3 (Hundred mark)	0.059	60	65	75	85	90	100
	Professional talent ratio D_4 (%)	0.050	1	4	7	10	13	15
Cooperation E	Cooperation success rate E_1 (%)	0.018	60	70	75	85	90	100
	Informatization degree E_2 (Hundred mark)	0.053	60	65	70	75	80	100
	Communication level E_3 (Hundred mark)	0.023	55	60	70	80	85	100

TABLE III
GREY CLUSTERING COEFFICIENTS OF PROVIDERS

Grey clustering coefficients	Poor	Medium	Good	Excellent
G_1	0.004	0.346	0.506*	0.144
G_2	0.079	0.499*	0.378	0.044
G_3	0.005	0.097	0.290	0.608*
G_4	0.036	0.174	0.269	0.521*
G_5	0.036	0.197	0.480*	0.287

products are widely used in port and waterway engineering, buildings, roads and other buildings. Due to the limitation of the allocation of funds, the enterprise decide to choose logistics service providers to help them deal with the logistics business. Selection and evaluation of alternative providers are based on the combination weighting-grey synthetic decision-making method. Evaluation data of each alternative provider are shown in Table I.

A. Evaluation indexes combination weighting of port enterprise logistics service providers selection

By using the combination weighting method, combined weights of evaluation indexes of logistics service providers are obtained and results are shown in Table II.

B. Grey synthetic decision-making of port enterprise logistics service providers selection

1) Determination of turning points or center points of evaluation indexes grey classes

According to evaluation requirements, evaluation indexes of logistics service providers are divided into 4 grey classes: poor, medium, good and excellent, as shown in Table II.

2) Construction of corresponding grey classes whitenization weight functions

Taking the customer satisfaction rate as an example, the corresponding triangular whitenization weight function of lower measure $f_j^1[-,-,75,80]$ is constructed for the grey class 1; the corresponding triangular whitenization weight function of upper measure $f_j^4[85,90,-,-]$ is constructed for the grey class 4; the corresponding triangular whitenization weight functions $f_j^2[75,80,-,85]$ and $f_j^3[80,85,-,90]$ are constructed for the grey class 2 and the grey class 3.

3) The calculation of grey clustering coefficients

Grey clustering coefficients of each alternative provider are calculated and results are summarized as shown in Table III.

TABLE IV
PROVIDERS CORRESPONDING GREY CLASSES

G_1	G_2	G_3	G_4	G_5
Good	Medium	Excellent	Excellent	Good

TABLE V
GREY SYNTHETIC CLUSTERING DECISION-MAKING COEFFICIENTS OF PROVIDERS

	G_1	G_2	G_3	G_4	G_5
π_i	2.593	2.461	2.840	2.760	2.673

4) The judgment of alternative providers corresponding grey classes

By observing the maximum value of grey clustering coefficients of each alternative provider, it can be judged that providers belong to grey classes and results are shown in Table IV.

5) The calculation of synthetic weighted decision-making vectors

This paper involves four kinds of grey classes, which can be used to calculate synthetic weighted decision-making vectors of grey classes k:

$$\eta_1 = \frac{1}{10}(4,3,2,1); \eta_2 = \frac{1}{12}(3,4,3,2);$$

$$\eta_3 = \frac{1}{12}(2,3,4,3); \eta_4 = \frac{1}{10}(1,2,3,4)$$

6) The calculation of grey synthetic decision-making coefficient vectors

Calculate initial grey synthetic decision-making coefficient vectors of alternative providers i for grey classes k, after the normalization process, the grey synthetic decision-making coefficient vectors matrix can be obtained:

$$\psi_i = \begin{bmatrix} 0.209 & 0.252 & 0.276 & 0.263 \\ 0.246 & 0.271 & 0.259 & 0.224 \\ 0.153 & 0.211 & 0.279 & 0.357 \\ 0.174 & 0.223 & 0.272 & 0.331 \\ 0.192 & 0.236 & 0.279 & 0.293 \end{bmatrix}$$

7) The calculation of grey synthetic clustering decision-making coefficients

The grey synthetic clustering decision-making coefficient of each alternative provider is obtained, as shown in Table V.

8) The rank of the synthetic decision-making

According to results in Table IV and Table V: G_2 belongs to the medium grey class; G_1 and G_5 belongs to the good grey class, because of 2.673>2.593, so that G_5 is better than G_1; G_3 and G_4 belongs to the excellent grey class, because of 2.840>2.760, so that G_3 is better than G_4. The final ranking for alternative providers is: $G_3 > G_4 > G_5 > G_1 > G_2$. This result show that the provider G_3 has the largest grey synthetic clustering decision-making coefficient and the provider G_4 ranks second; Seeing from another angle, the provider G_2 is relatively poor. Therefore, the port enterprise should choose the provider G_3 as its partner, to help them deal with the logistics business.

V. CONCLUSION

Selection of logistics service providers is an important problem that enterprises must consider in supply chain management.Through the analysis of reliability and validity of the questionnaire, this paper determines the provider evaluation index system. Then, the network analysis method and the entropy method are combined to determine combined weights, the improved center-point triangular whitenization weight function is used to cluster alternative plans. Finally, grey synthetic clustering decision-making coefficients are calculated to make the synthetic decision-making rank of alternative plans. The validity of this method is proved through empirical research.

ACKNOWLEDGMENT

This work is financially supported by National Natural Science Foundation of China under the project of 71401039 and Program for New Century Excellent Talents in University under the project of NCET-11-0903.

REFERENCES

[1] V. Thakur, R. Anbanandam, "Supplier selection using grey theory: a case study from Indian banking industry," *Journal of Enterprise Information Management*, vol. 28, no. 6, pp. 769-787, 2015.

[2] S. Jharkharia, R. Shankar, "Selection of logistics service provider: an analytic network process (ANP) approach," *Omega*, vol. 35, no. 3, pp. 274-289, 2007.

[3] X. Ping, Z. Qi-shan, "Decision-making model of grey clusterng for supplier selection," *Journal of Harbin Institute of Technology*, vol. 39, no. 12, pp. 2006-2008, 2007.

[4] Z. Ping, Y. Yi-nan, "Supplier selection based on EAHP/GRAP," (in Chinese) *Systems Engineering - Theory & Practice*, vol. 29, no. 3, pp. 69-75, 2009.

[5] W. Xu, S. Lin, G. Xian-long, "Bi-level programming model and cloud genetic algorithm for logistics service supplier selection," (in Chinese) *Application Research of Computers*, vol. 28, no. 8, pp. 2871-2875, 2011.

[6] C. Hu, "Research on functional logistics provider selection based on QFD and bi-level programming modeling," (in Chinese) *Computer Engineering and Applications*, vol. 48, no. 27, pp. 200-205, 2012.

[7] W. Yan, X. Rui-rui, "The two-stage hybrid decision research of logistics outsouring partners selection: using the DEMATEL-ANP-VIKOR," (in Chinese) *China Business and Market*, vol. 26, no. 8, pp. 30-36, 2012.

[8] C. Ke-jia, Y. Xian-kang, Y. Li-xi, "GI-TOPSIS based on combinational weight determination and its application to selection of reverse logistics service providers," *The Journal of Grey System*, vol. 25, no. 3, pp. 16-33, 2013.

[9] L. Qiu-sheng, W. Xiu-zhu, H. Yun-zhang, "Research on third party reverse logistics provider evaluation based on entropy," (in Chinese) *Science and Technology Management Research*, no. 10, pp. 179-182, 2013.

[10] L. Xiao-li, "Rough set - based grey - topsis approach to the third party reverse logistic vendor selection," (in Chinese) *Science and Technology Management Research*, no. 14, pp. 67-71, 2013.

[11] Z. Ye-wang, "The research on supplier selection model of the third party logistics based on grey clustering," *International Journal of Modeling and Optimization*, vol. 4, no. 6, pp. 489-492, 2014.

[12] W. Xing-hua, Y. Xun, L. Si-huan, "Research on the selection of low carbon logistics service providers based on GST-ANP model," (in Chinese) *Science and Technology Management Research*, no. 20, pp. 253-258, 2015.

[13] R. Rajesh, V. Ravi, "Supplier selection in resilient supply chains: a grey relational analysis approach," *Journal of Cleaner Production*, vol. 86, pp. 343-359, 2015.

[14] P. Pitchipoo, P. Venkumar, S. Rajakarunakaran, "Grey decision model for supplier evaluation and selection in process industry: a comparative perspective," *The International Journal of Advanced Manufacturing Technology*, vol. 76, pp. 2059-2069, 2015.

[15] L. Si-feng, F. Zhi-geng, Y. Ying-jie, "Two stages decision model with grey synthetic measure and a betterment of triangular whitenization weight function," (in Chinese) *Control and Decision*, vol. 29, no. 7, pp. 1232-1238, 2014.

Application of evaluation of aircraft material demand forecasting method and mining of association rules

Jun WANG[1,*], Li-geng WANG[1], Wei WEI[1]
[1] School of Economics and Management, Beihang University, Beijing, PR China
(king.wang@buaa.edu.cn*, mailwangligeng@163.com, dw123dw456@hotmail.com)

Abstract - **The support probability of aviation materials is a crucial part in the process of normal operation of airlines. But the higher support probability will inevitably lead to the increase of inventory cost of aviation materials, and restrict airlines to improve their efficiency. Hence it is of great significance for airlines reducing the material cost on the premise of normal operation to predict the material requirements accurately based on reasonable models. This paper summarizes a series of prediction models of aviation material requirements, and applies the grey comprehensive correlation degree to evaluate the models. On this basis, the method of association rules is used to discover the association relationships between the types of aviation materials and the prediction models.**

Keywords - **Association rule, aviation material prediction, grey comprehensive correlation degree, model recommendation**

I. INTRODUCTION

With the rapid development of China's economy, the aviation industry of China also has made great progress. However, with the rapid development, airlines are also facing with a series of questions. Especially, supply chain and security of aviation materials have been the focus of attention of each airline. The cost of aviation materials is an important part of aircraft operating costs of aviation transport enterprises, which ranks only second to the cost of purchase of aircraft and aviation fuel. Thus, to reduce aviation materials inventory cost of airlines effectively, it is crucial to predict the quantity demanded of aviation materials scientifically and reasonably.

At present, the study of methods of aviation materials demand forecasting has been paid highly attention by academia and application fields. Regattieri et al. applied the AW/M model to predict the fluctuant demands of spare parts of aviation materials [1]. Johnston et al. evaluated the optimal prediction parameters through analyzing the cycle length from 2 to 12, and built moving average model (MA(i)) to achieve prediction [2]. Aslam et al. put forward the prediction model based on exponentially weighted moving average [3]. Croston raised a method called Croston-CR, which overcame the shortage of exponential smoothing method, respectively evaluated the quantity demanded and the frequency of demand and prevented the shortage of inventory by adjusting the safety stock [4]. On the basis of Croston, Segerstedt computed the quantity of orders and order lead time [5]. Willemain et al. applied Croston's method to predict the intermittent demand for spare parts, which can bring tangible benefits for enterprises [6]. Godfrey carried out medicine research with linear regression model [7]. Isobe et al. applied linear regression model to research in astronomy [8]. Liu et al. used linear regression model to analyze near infrared spectrum [9]. And Figura applied it to predict underground water temperature [10]. Ramos et al. used ARIMA model to forecast consumer retail sales [11]. Li & Wang developed an automatic auto regressive integrated moving average modeling based data aggregation scheme in wireless sensor networks [12]. Chang et al. predicted the Internet users and the industry revenue of online games by using grey theory [13]. Ou applied GM(1,1) model to forecast the agricultural output [14]. Zhou et al. utilized GM(1,1) to predict the output of the fuel [15]. Yang & Shieh used SVR to predict consumers' affective responses [16].

The previous related research made outstanding contributions to improve the efficiency of demand forecasting. But in the practical application, the prediction methods suited to the different materials differ from one another. This paper put forward the prediction methods based on multi model and use grey comprehensive correlation degree to evaluate the predictive effect. And we mine the association pattern between material types and prediction models, so that the applicable prediction models can be given based on the association rules after airlines accumulating a certain amount of prediction record.

The paper is organized as follows: The second section mainly introduces the process of prediction and recommendation. In the section III, we make a case analysis. Finally, we present our conclusions in Section IV.

II. THE ANALYSIS AND RECOMMENDATION OF ALTERNATIVE PREDICTION MODELS

The statistical characteristics of different types of demand data of aviation materials are not the same. Because there are a great variety of aviation materials, it is difficult for each aviation material requirement to build and verify mathematical model respectively. To solve the above problems, eight prediction models are introduced to predict the aviation material requirements, which are Poisson distribution model, linear regression model, auto regression (AR) model, integrated autoregressive moving average (ARIMA) model, automatic arima model, holt-winters model, grey prediction model GM(1,1) and support vector machine regression (SVR) model. The TABLE I shows the description of the alternative prediction models.

E. Qi et al. (eds.), *Proceedings of the 23rd International Conference on Industrial Engineering and Engineering Management 2016,* DOI 10.2991/978-94-6239-255-7_2

In this paper, the prediction and recommendation of material demand is divided into five steps:

(1) The preprocessing of data, including the outbound quantity, flight hours and flight frequency of aviation material;

(2) Multi model prediction. We apply eight kinds of prediction models to predict the demand of aviation material.

(3) Calculate the grey comprehensive correlation degree. Grey comprehensive correlation degree are used to evaluate the effect of prediction models.

(4) Mine the association rules. We apply apriori algorithm to mine the association pattern between material types and prediction models.

(5) Model recommendation. The optimal models are given based on the association rules mined by step (4) according to different aviation material.

III. RESULTS FOR CASE STUDY

In order to verify the multi prediction models and the model recommendation methods, this paper selected the monthly data of the material A of an airline company from January 2000 to July 2014, including the consumption quantity of the material A, the number of flight hours and the landing times. The detailed calculation and analysis results will be presented in the next part of this paper.

TABLE I
THE ALTERNATIVE PREDICTION MODELS

Name	Model description
Poisson distribution model	$p_k(t)=\frac{(\lambda_t)^k}{k!}e^{-\lambda_t}, k\in Z^+, t>0$ (1) $p_k(t)$ is the probability that the quantity demand of the aviation material is k within the time (0, t).
Linear regression model	Y_i is used to reflect the outbound quantity of aviation material, and X_i is the flight frequency or flight hours of aviation material. $Y_i = a + bX_i + \varepsilon_i$ (2)
Auto regression(AR) model	It is built according to the outbound quantity Y of aviation material. $Y(t)=\varepsilon(t)+\sum_{i}^{p}\varphi_i Y(t-i)$ (3)
ARIMA(p,d,q)	$X(t)=\beta_0+\sum_{i=1}^{q}\beta_i X(t-i)-\varepsilon(t)-\sum_{j=1}^{p}\alpha_j\varepsilon(t-j)$ (4)
Automatic arima model	There are tens of thousands of aviation materials in aviation enterprises. It is impossible for every time series data to arrange the staff to build model. Thus, it is necessary to introduce the automatic arima model, which can automatically identify the pattern of time series and estimate the parameters of models.
Holt-winters model	$a_i=\alpha(y_i-c_{i-p})+(1-\alpha)(a_{i-1}+b_{i-1})$ (5) $b_i=\beta(a_i-a_{i-1})+(1-\beta)b_{i-1}$ (6) $c_i=\gamma(y_i-a_{i-1}-b_{i-1})+(1-\gamma)c_{i-p}$ (7)
Grey prediction model GM(1,1)	$\frac{dx^{(1)}}{dt}+ax^{(1)}=u$ (8) $x^{(1)}$ is the cumulative sequence in term of the historical outbound quantity of aviation material. t is the time. a,u are development grey number and Internal control grey number.
Support vector regression (SVR) model.	$f(x)=w^t\phi(x)+b,\ w\in R^n, b\in R$ (9) The $f(x)$ reflects the historical outbound quantity of aviation material. x is flight hours or flight frequency. $\phi(x)$ is the nonlinear function.

A. The results of the prediction models

The paper selected the monthly data of the material A of an airline company from January 2000 to July 2014. Because the parameters of the model need a certain amount of data, the first 12 months of the data are only used to estimate of the model parameters, not to forecast the demand. Fig. 1 to Fig. 9 show the actual monthly outbound quantity of material A and eight kinds of prediction models from January 2001 to July 2014.

From the comparison of the actual demand with the forecasted results, among the eight models, the effect of grey prediction model GM(1,1) is the worst. Based on the analysis of the application of GM(1,1) in the grey forecasting model, it is generally believed that the necessary condition for obtaining high precision with GM(1, 1) model is that the equal time interval, the non negative and the monotonicity. Although the demand of A is satisfied with the equal time interval and the non negative, it is not satisfied with the monotone. So the prediction accuracy of the grey model GM(1,1) is poor. In contrast, the time series methods and the support vector machine regression model have better prediction effect.

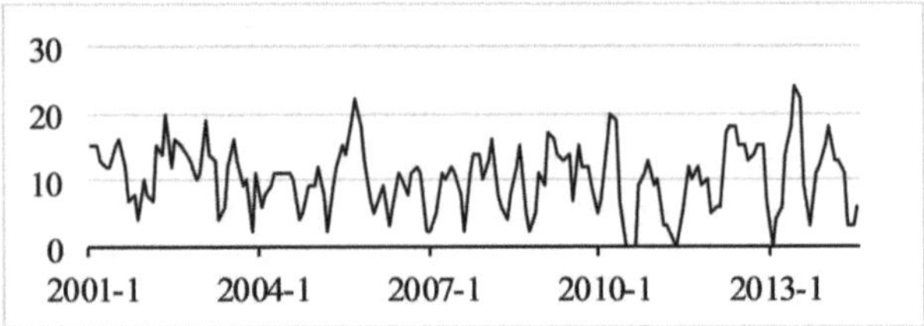

Fig. 1. Actual demand.

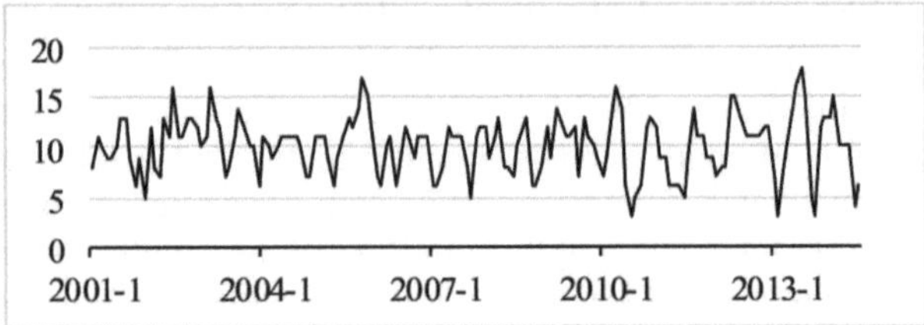

Fig. 2. Result of AR model.

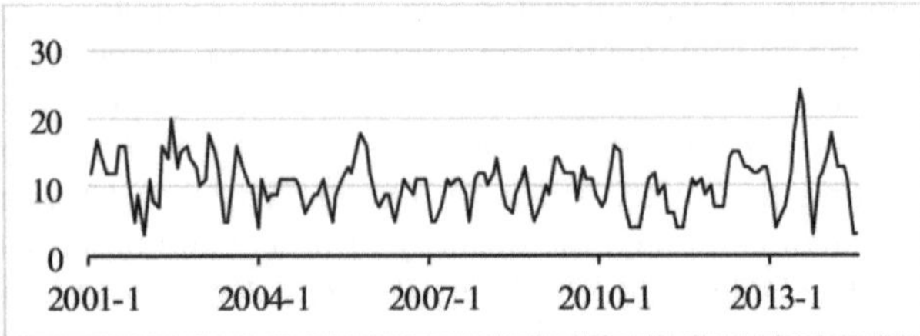

Fig. 3. Result of ARIMA model.

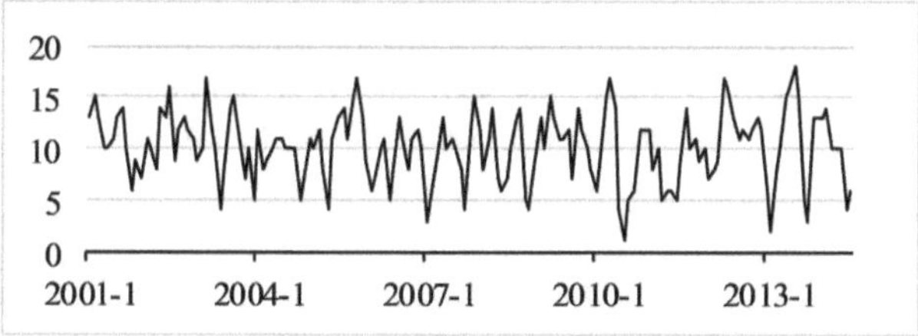

Fig. 4. Result of Automatic arima model.

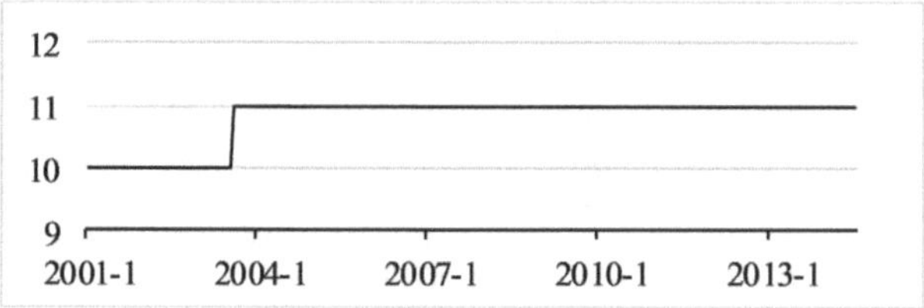

Fig. 5. Result of GM(1,1) model.

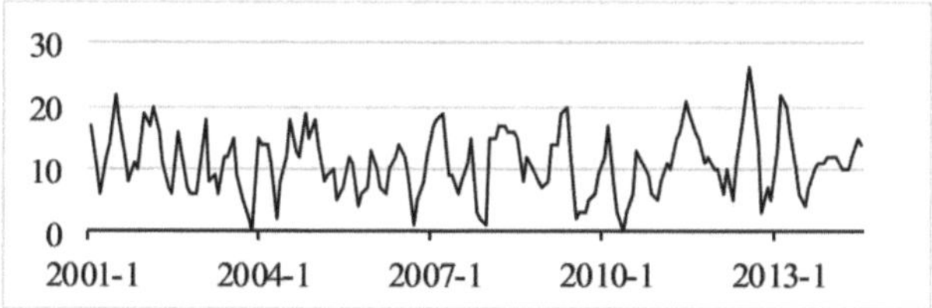

Fig. 6. Result of holt-winters model.

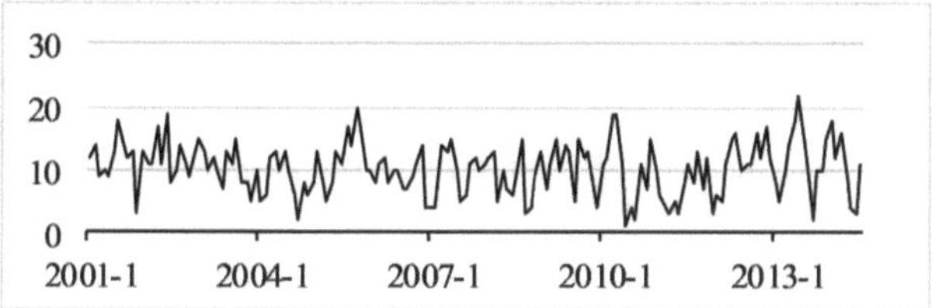

Fig. 7. Result of linear regression model.

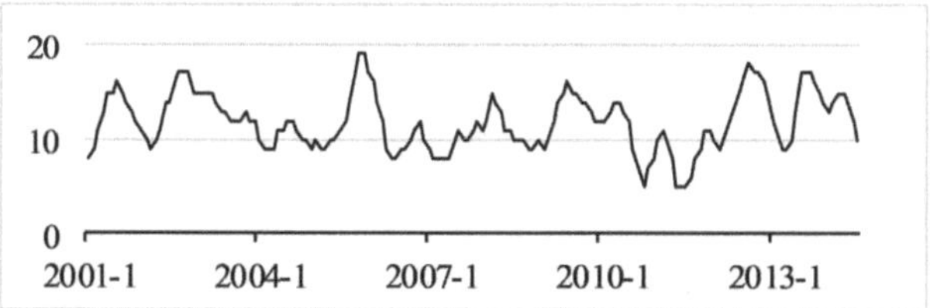

Fig. 8. Result of Poisson model.

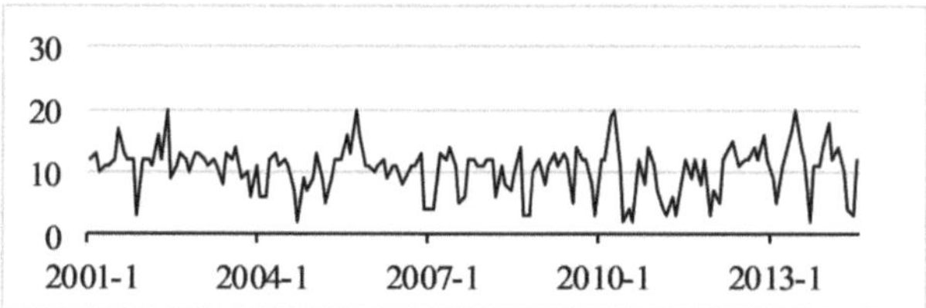

Fig. 9. Result of SVR model.

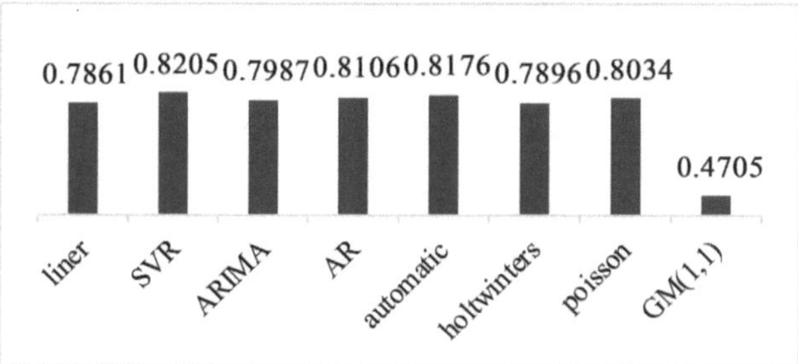

Fig. 10. Grey comprehensive correlation degree-December, 2011

B. Evaluation results of grey comprehensive correlation degree

In this paper, the grey comprehensive correlation degree is used to evaluate the prediction models. Taking the calculation process of the grey comprehensive correlation degree in December 2011 for example, the specific analysis process is as follows.

(1) Select the actual demand of the material *A* and the results of eight prediction models.

(2) Set the weight of grey absolute correlation degree and grey relative correlation degree to 0.5.

(3) Get the grey comprehensive correlation degree, as shown in the Fig.10.

According to the order of the grey comprehensive correlation degree from large to small, it shows that the prediction effect of the support vector machine regression model outperform other forecasting models in December 2011. The predictive effect of Automatic ARIMA model is worse than support vector machine regression model, but it is better than other models. The effect of grey forecasting model GM(1,1) is the most unsatisfactory, which is consistent with the conclusion of the last section.

(4) Obtain the forecast record according to the threshold of the grey correlation degree

According to the characteristics of aircraft demand and the requirement of aircraft management, set the threshold value of the grey comprehensive correlation degree to 0.8. Set the prediction model as an alternative model, whose grey comprehensive correlation degree is more than 0.8. The forecast record in December 2011 is as follows:

$$\{A, SVR, AR, Automatic - ARIMA, Poisson\} \tag{10}$$

C. The association rules between the aircraft material types and prediction models

According to the relationship between material types and alternative prediction models, the confidence level is set to 0.8, and the support level is set to 0.5. The following association rules can be obtained by mining the forecast records from February 2001 to July 2014 which contains 162 monthly forecast records.

According to the association rules obtained from the TABLE II, it can be known that the applicability of Automatic ARIMA model is higher than other forecast models for material A, and the second is the support vector machine regression model. The enterprise will be able to directly select the Automatic ARIMA model to predict the future demand of material A, which can greatly improve the efficiency of the demand prediction.

IV. CONCLUSION

In this paper, we analyzed the problems of aircraft material demand forecasting, and summarized the 8 kinds

TABLE II
ASSOCIATION RULES

NO	Association rules	Support	Confidence
1	{A} => { Automatic ARIMA}	0.74	0.92
2	{A} => { SVR}	0.55	0.85

of demand forecasting models. Then, the grey comprehensive correlation degree between the predicted results of each model and the actual consumption is calculated. According to the threshold value of the grey comprehensive correlation degree, the forecast model was selected, and the forecast record was obtained. Finally, the apriori algorithm is introduced in the paper, and the association rules between the aviation material type and the prediction models were obtained by mining the forecast records. In the future, the optimal forecasting model can be directly given according to the association rules and the type of aviation material, which greatly improves the efficiency of the demand forecasting.

In this paper, the grey comprehensive correlation degree is used to evaluate the eight prediction models. Although it makes up for some deficiencies of absolute degree of grey incidence and the relative degree of incidence, there is not a reliable basis about how to determine the weight between the grey absolute correlation and the relative, which is mainly adjusted according to the experience and historical data. In future studies, we will focus on the weight, so that it can be determined more scientific standards.

ACKNOWLEDGMENT

The work described in this paper was supported by the National Natural Science Foundation of China under Grant No. 71271018 and No. 71531001.

REFERENCES

[1] A. Regattieri, M. Gamberi, R. Gamberini and R. Manzini, "Managing lumpy demand for aircraft spare parts", *Journal of Air Transport Management*, vol. 11, no. 6, pp. 426-431, Nov. 2005.

[2] F. R. Johnston, J. E. Boylan, E. Shale and M. Meadows, "A robust forecasting system, based on the combination of two simple moving averages", *Journal of the Operational Research Society*, vol. 50, no. 12, pp. 1199-1204, Dec. 1999.

[3] M. Aslam, M. Azam and C. H. Jun, "A new control chart for exponential distributed life using ewma". *Transactions of the Institute of Measurement & Control*, vol. 37, no. 2, pp. 205-210, Feb. 2015.

[4] J. D. Croston, "Forecasting and stock control for intermittent demands", *Operational Research Quarterly*, vol. 23, no. 3, pp. 289-303, Sep. 1972.

[5] A. Segerstedt, "Inventory control with variation in lead times, especially when demand is intermittent", *International Journal of Production Economics*, vol. 35, pp365-372, Jun. 1994.

[6] T. R. Willemain, C. N. Smart, J. H. Shockor and P. A. Desautels. "Forecasting intermittent demand in manufacturing: a comparative evaluation of croston's method", *International Journal of Forecasting*, vol. 10, pp. 529-538, Dec. 1994.

[7] K. Godfrey, "Simple linear regression in medical research", *New England Journal of Medicine*, vol. 313, no. 26, pp. 1629-1636, Jan. 1986.

[8] T. Isobe, E. D. Feigelson, M. G. Akritas, G. J. Babu, "Linear regression in astronomy", *Astrophysical Journal*, vol. 364, no. 1, pp. 104-113, Nov. 1990.

[9] K. Liu, X. Chen, L. Li, H. Chen, X. Ruan and W. Liu, "A consensus successive projections algorithm- multiple linear regression method for analyzing near infrared spectra", *Analytica Chimica Acta*, vol. 858, no. 9, pp. 16-23, Feb. 2015.

[10] S. Figura, D. M. Livingstone and R. Kipfer, "Forecasting groundwater temperature with linear regression models using historical data", *Groundwater*, vol. 53, no. 6, pp. 943-945, Nov. 2015.

[11] P. Ramos, N. Santos, and R. Rui, "Performance of state space and arima models for consumer retail sales forecasting", *Robotics and Computer-Integrated Manufacturing*, vol. 34, pp. 151-163, Aug. 2015.

[12] G. Li and Y. Wang, "Automatic arima modeling-based data aggregation scheme in wireless sensor networks", *Eurasip Journal on Wireless Communications and Networking*, vol. 2013, no. 85, pp. 1-13, Mar. 2013.

[13] T. S. Chang, C. Y. Ku and H. P. Fu, "Grey theory analysis of online population and online game industry revenue in Taiwan", *Technological Forecasting and Social Change*, vol. 80, no. 1, pp. 175-185, Jan. 2013.

[14] S. L. Ou, "Forecasting agricultural output with an improved grey forecasting model based on the genetic algorithm", *Computers and Electronics in Agriculture*, vol. 85, pp. 33-39, July. 2012.

[15] W. Zhou and J. M. He, "Generalized gm (1,1) model and its application in forecasting of fuel production", *Applied Mathematical Modelling*, vol. 37, no. 9, pp. 6234-6243, May. 2013.

[16] C. C. Yang and M. D. Shieh, "A support vector regression based prediction model of affective responses for product form design", *Computers & Industrial Engineering*, vol. 59, no. 4, pp. 682-689, Nov. 2010.

Research on the Selection of Business-to-Customer e-commerce Logistics Model Based on Analytic Hierarchy Process Method

Jun-feng BAI, Xiao-yan WEI* ,Jun-cun Yan
Changchun University of Technology, Changchun 130012,China,
(*e-mail:baijunfeng@ccut.edu.cn)

Abstract- **In recent years, Business-to-Customer (B2C) e-commerce has become one of the largest e-commerce transaction models. As an important link in the e-commerce transactions, logistics has an increasing impact on e-commerce. Analysis for enterprise logistics mode becomes particularly important. This paper analyzes the domestic e-commerce transaction scale changes in recent years by using Analytic Hierarchy Process (AHP) to study factors which affecting logistics, and ultimately proposes a new approach to help companies choose their own suitable logistics model .**

Keywords- **AHP, B2C, e-commerce, logistics model**

I. Introduction

Currently, the development of Business-to-Consumer (B2C) e-commerce faces three bottlenecks which are network security, online payment and logistics distribution[1]. With the rapid development of computer technology and the third-party payment platform, the former two problems have been improved. How to choose a proper logistics distribution model is still a big problem. Logistics distribution as "the last mile" of B2C commerce e-commerce is an important part of the user experience[2,3]. This paper develops a practical approach for the selection problem of B2C e-commerce Logistics Model.

II. Development of E-Commerce

With the development of IT Industry, e-commerce has been becoming the most popular business model and even to be almost omnipresent in people's life, and will continue to develop at a faster speed. According to a survey in 2014, the number of Internet users in main land China has reached 632 million. The e-commerce market has reached 12.3 trillion RMB. The per capita consumption is nearly about 20,000 RMB, 21.3% more than that of 2013. See Fig. 1 for e-commerce transaction size during 2011-2014.

B2C, Business-to-Consumer e-commerce model, one of the most important e-commerce sales models, has been rising in recent years. As shown by iResearch report in the third quarter of 2013, China's Internet market transaction estimated at 454.76 billion RMB, of which 36.6% belonged to B2C model. As Fig.2 shows, the transaction size of China's online shopping in the first quarter of 2012 to the third quarter of 2013.

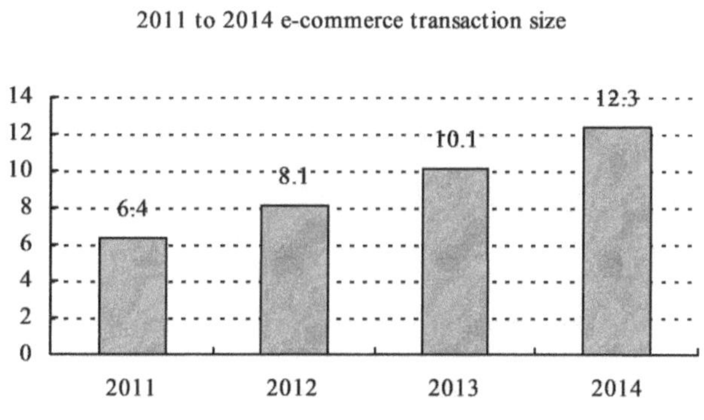

Fig.1 2011 to 2014 e-commerce transaction size

E. Qi et al. (eds.), *Proceedings of the 23rd International Conference on Industrial Engineering and Engineering Management 2016,* DOI 10.2991/978-94-6239-255-7_3

Fig.2 2012Q1 -2013Q2 China's online shopping market transaction size

III. Alternative logistics models of B2C

By a summarization and detailed analysis on the basis of above statistics, we can see that B2C market transaction scale is huge in recent years, and therefore results in a corresponding increase in the number of transactions. So how to choose a suitable logistics mode becomes an urgent and important issue with top priority. Such a logistics and distribution model can not only help cost saving in terms of business running and production for the enterprises, but also improve relationships & cooperation between enterprises and consumers[4]. Consumer spending will also be promoted.

Logistics and distribution model will directly affect the company's sales performance as an important hub between businesses and consumers. At present, most enterprises of B2C e-commerce model mainly adopt three types of logistics models as bellows.

A. Third-party Logistics Model

So called third-party logistics model is that the product supply-side sign contract with a third party Logistics Company who will professionally deliver materials to the demand-side. This mode mainly occurs in e-commerce transactions. Compared with traditional logistics, the third party logistics model is more faster and efficient and also cost saving. Whereas, the only side-effect is that over-reliance on this distribution model will weaken the contact and relations between businesses and customers.

B. Company's own Logistics Model

The company's own logistics model is that the supply-side builds own delivering system to accomplish its materials distribution. Currently, there are two kinds of e-commerce enterprises using this type of logistic model. One kind is the newly emerging relatively large professional e-commerce companies which provide logistics services for the demand side through a scientific logistics management systems to reduce product costs. The other is the traditional large-scale manufacturing factories which build their own e-commerce sales system, and their own logistics model for material distribution. By this way, the supply-side can directly contact with the demand-side so as to complete the transactions much more coordinated and simpler.

C. Logistics Alliance Model

Logistics Alliance Model is something like a strategic logistics cooperation among supply-side, demand-side and professional logistic company on the basis of mutual benefit. A formal cooperation agreement among the three sides will be signed to achieve a win-win situation.

IV. Logistics Hierarchy Model

Analytic Hierarchy Process (AHP) is a kind of method trying to more effectively solve a complex problem by analyzing the related various factors which can be divided into different levels through systematic and hierarchical analysis in order to find the key factors. Generally the problem to be resolved can be put into three levels, the first layer is the target layer (Goal) which corresponding to the logistics mode selection G; the second layer is criterion layer (Norm) that is the factor layer which corresponding to the middle part of Fig.3 (the third layer F1-F8[5,6]; the final layer is the scheme layer (Scheme) which is the feasible option to

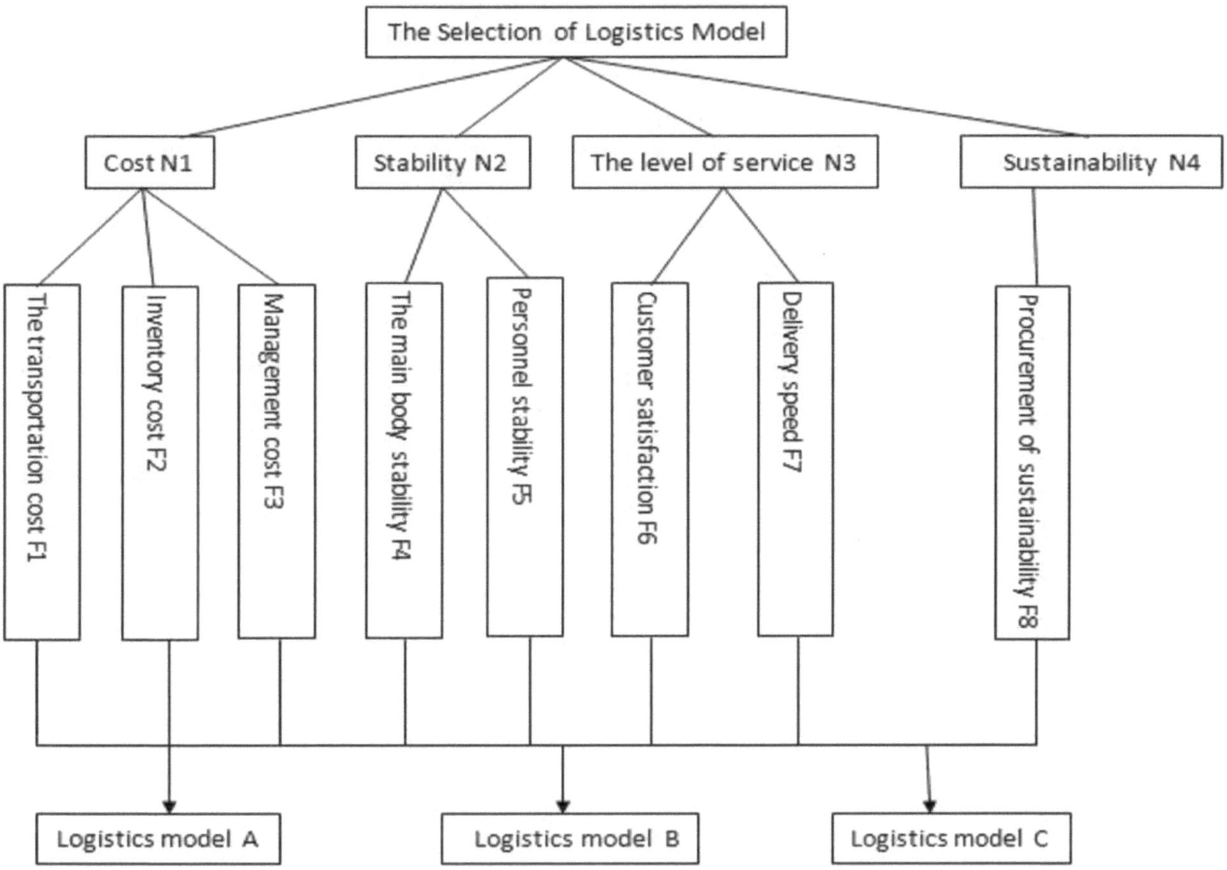

Fig.3 Logistics Hierarchy Model

solve this complex problem, corresponding to Fig.3 in the final layer.

After the systematically analysis so as to select the best logistics mode of B2C for the enterprises, we finally determine the following criteria: (1) Choose the lowest logistics costs as much as possible[7]; (2) Higher stability needed for the logistics[8]; (3) Logistics service qualification to meet the customer's requirements[9,10]; (4) Sustainability of the logistic services and long term good cooperation with customer[11].

Finally, we can get the hierarchical model as Fig.3 shows after decompose the factors of each criterion.

V. Selection of the optimal logistics model by AHP

Hierarchical model built in the previous section total has four criteria (N1-N4) and eight impact factors (F1-F8).The first step is to determine the weights of the four criteria as bellows:

TABLE I

CRITERIA LAYER MATRIX ELEMENT

Norm	N1	N2	N3	N4
N1	N11	N12	N13	N14
N2	N21	N22	N23	N24
N3	N31	N32	N33	N34
N4	N41	N42	N43	N44

In the Table I, Nij represent the degree of importance Ni relative to Nj, e.g. N12 = 3 indicates N1 is 3 times important than N2. To determine the weight of each index weight coefficient, we can use the eigenvector method in this article, see equation (1):

$$N\omega = \lambda_{\max}(t) \tag{1}$$

Where N is the judgment matrix .Substituting it back into the matrix after obtained the largest eigenvalue (λ_{max}) of the matrix N, then calculation the eigenvectors of the matrix N , eventually seeking the weight of each features vector in whole features vector, finally get the weights of all indicators in four criterions:

$$M_N = \omega_N = [\omega_1^N, \omega_2^N, \omega_3^N, \omega_4^N]^T \quad (2)$$

In the same manner, the corresponding weights of factor layer (F1-F8) in Fig.3 can be calculated:

$$M_F = \omega_F = [\omega_1^F, \omega_2^F, \omega_3^F, \omega_4^F, \omega_5^F, \omega_6^F, \omega_7^F, \omega_8^F]^T \quad (3)$$

We can use the feature vector method to find A, B, C three logistics model right corresponding weight of each factor after received the weigh of the criteria and factor. The following example is to calculate customer satisfaction with three corresponding logistics model weight:

TABLE II

MATRIX ELEMENTS OF CUSTOMER SATISFACTION TABLE

Logistics	L1	L2	L3
L1	L11	L12	L13
L2	L21	L22	L23
L3	L31	L32	L33

In the Table II, Nij represents that logistics mode Li is important than Lj for the degree of customer satisfaction photogenic.

With the above methods, we can obtain the largest eigenvalue and eigenvectors about different logistics pattern to F6 (the degree of customer satisfaction). In the same way, determine the largest eigenvalues and eigenvectors other modes of different logistics relative to each factor, then combine the eigenvectors of each factor into decision matrix M_L:

$$M_L = \begin{matrix} L1 \\ L2 \\ L3 \end{matrix} \begin{bmatrix} \omega_{11}^L & \omega_{12}^L & \omega_{13}^L & \omega_{14}^L & \omega_{15}^L & \omega_{16}^L & \omega_{17}^L & \omega_{18}^L \\ \omega_{21}^L & \omega_{22}^L & \omega_{23}^L & \omega_{24}^L & \omega_{25}^L & \omega_{26}^L & \omega_{27}^L & \omega_{28}^L \\ \omega_{31}^L & \omega_{32}^L & \omega_{33}^L & \omega_{34}^L & \omega_{35}^L & \omega_{36}^L & \omega_{37}^L & \omega_{38}^L \end{bmatrix} \quad (4)$$

In the formula (4), ω_{ij}^L represents the degree of importance that the impact factors j of the logistics mode i. Eventually, the calculation of the total score for each vector of logistics mode, calculated as follows:

$$D = M_L M_F = [L1 \quad L2 \quad L3]^T \quad (5)$$

The resulting maximum value corresponding to the logistics business model is the best alternative distribution model.

VI. CONCLUSION

With the increase of B2C business volume and sales revenue, the logistics mode of economic efficiency of enterprises become increasingly important. In this paper, logistics model of AHP method has been successfully used, and then presents a new algorithm for logistics models' selection. Organizations can select a suitable model referred to the method discussed in this paper in order to save production cost and provide better logistic services.

REFERENCES

[1] Wang Jing. The solution of B to C e-commerce logistics bottlenecks based on third party logistics [J]. CO-Oerativeconomy & Science, 2005, 02: 34-35.

[2] Haul Lee, Seungin Whang. Winning the last mile of E-Commerce [M].McGraw-Hill Companies, Inc.2002.106-121

[3] Tennessee State University. Research of TPL. Journal of Business, 2003, (4):34-35

[4] Beamon B.M. designing the green supply chain. Logistics Information Management, 1999, (2):33-35

[5] Li Bing, Wei Jia, Liao Ren-kui. Overview of e-commerce based logistics research [J]. Higher education and academic research, 2009.4

[6] Du Dan-qing , Yu Ruo-xi. Enterprise E-Commerce Logistics Mode Selection Method [J]. Jiangsu trader theory, 2010.11.

[7] Zhang Fang. E-commerce logistics mode selection in China [D]. Dongbei University of Finance and Economics, 2007.

[8] Wei Xiu-jian, Yan Jian-yuan, Wang Yan. E-commerce logistics [M]. Posts & Telecom Press, 2004

[9] Chang Jia-dong. Research on e-commerce logistics development countermeasure [J].Modern commerce industry, 2013(9)

[10] Liu Lian-hua. E-commerce environment on the development countermeasures of the three party logistics enterprises [J]. China market,2013 (30)

[11] Bai Xiu-yan. Research on the quick response of enterprise's logistics under the environment of e-commerce [J]. Technology management, 2010(22)

Content Framework and Evaluating Method of the Energy Efficiency Assessment Standard for Machining period

Qiu-lian. Wang[1,*], F. Liu[2], T. Hu[1], Y. H. Zeng[1]
[1] School of Economics and Management, Nanchang University, Nanchang, China
[2] State Key Laboratory of Mechanical Transmission, Chongqing University, Chongqing, China
(*wangqiulian@163.com)

Abstract - **Energy efficiency assessment is the precondition of energy optimization of machining period, and a set of standardized energy efficiency assessment system can help the optimization of machining period; therefore, the energy efficiency assessment standard for machining period is studied, and the content framework and evaluating method of energy efficiency assessment standard of machining period are proposed. The first step of the evaluating method is to establish the mathematic model of energy efficiency assessing standard, including the energy equation and power equilibrium equation of machining, which are the theoretical basis of the standard; the second step is to set up the energy efficiency assessing indexes including energy utilization ratio and specific energy consumption; the third step is to give the method for obtaining energy efficiency assessing indexes based on the combination of off-line basic data acquiring and on-line input power monitoring. The case study shows that the energy efficiency indexes can be gotten by the proposed evaluating method with enough precision, and only input power of the machining period should be monitored on the site. The results can support the energy efficiency assessment and optimization of machine tools and work-pieces.**

Keywords – **machining, energy efficiency assessment, content framework, evaluating method**

I. INTRODUCTION

Machining is a kind of operation using cutting machine to change the shape, size or performance of work-piece, including turning, drilling, boring, milling, planing, slotting, sawing, broaching, grinding, finishing, gear machining, thread machining and so on [1]. Energy Efficiency of Machining Period (EEMP) is the quantitative expression of effective use ratio of machining period [2]. Evaluating the Energy Efficiency of Machining Period (EEEMP) is to analyze and measure the energy efficiency of machining period.

Recently, many researchers and companies have paid attention on EEMP issue. For example, the Environmentally Benign Manufacturing (EBM) research group at the Massachusetts Institute of Technology (MIT) studies the energy consumption model and energy efficiency calculation model of machine tools; the Laboratory for Manufacturing And Sustainability (LMAS) at University of California, Berkeley focuses on the energy efficiency monitoring system of machining period; the Institute for control engineering of machine tools and manufacturing units at Stuttgart University researches the energy efficiency modeling and simulation of machining period. In addition, Rockwell Automation released the white paper named "the Manufacturing Execution System (MES) used to achieve the sustainability: Expand the scope of MES to achieve energy efficiency and sustainability" in 2009, and energy efficiency improvement of manufacturing process was recognized as a new aim of MES in this white paper [3].

In conclusion, the EEMP issue catches people's eyes. And we can find that the current researches emphasis on energy efficiency modeling [4-5], energy efficiency analysis and improvement [6-7], etc. Few study about energy efficiency assessment standard of machining period has been carried out.

Energy efficiency assessment is the precondition of energy efficiency improvement of machining period, and set up a standardized energy efficiency assessment system can support the energy efficiency improvement of machining period. Until now, there is not international or national standard about EEEMP, excepting some energy efficiency assessment standards for certain typical products. For example, the International Organization for Standardization (ISO) published the international standard ISO 14955-1:2014 " Machine tools -- Environmental evaluation of machine tools -- Part 1: Design methodology for energy-efficient machine tools" on 13th May, 2014 [8]. This standard would influence on the machine tool industry greatly as we can see. What is more, some organizations also implement energy efficiency assessment and have formed their own assessing system, such as the Industrial Assessment Centers (IACs) established by Energy Department of the USA. IACs aim to improve the energy efficiency and productivity of a plant site by in-depth assessment of its energy consumption and energy efficiency following a five step industrial assessment protocol [9].

Summarily, EEMP issue is hot and some energy efficiency standards about certain products have published, but the energy efficiency assessment standard for machining period is absent, especially the researches about the mathematical model of the energy efficiency assessment standard for machining period, the index selection and the index acquiring of the energy efficiency assessment standard for machining period. Therefore, the content framework and evaluation method of energy efficiency assessment standard for machining period are studied in this paper with the financial support of the National High-Tech. R&D Program of China (863 Program).

E. Qi et al. (eds.), *Proceedings of the 23rd International Conference on Industrial Engineering and Engineering Management 2016,* DOI 10.2991/978-94-6239-255-7_4

II. CONTENT FRAMEWORK OF ENERGY EFFICIENCY ASSESSMENT STANDARD FOR MACHINING PERIOD

The use of energy efficiency assessment standard is to give a standardized work instruction for evaluators. The energy efficiency assessment standard should include energy mathematical model, energy efficiency index and the acquiring method of energy efficiency index. Hence, the content framework of energy efficiency assessment standard for machining period is proposed as shown in Fig.1.

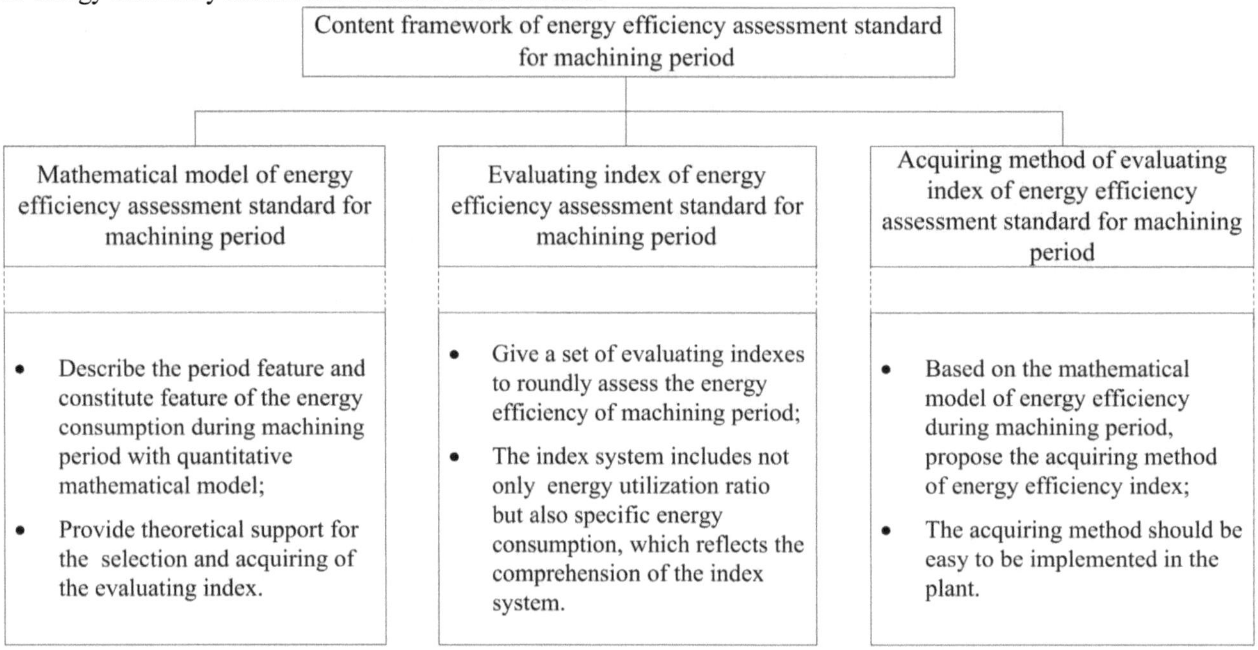

Fig.1 The content framework of energy efficiency assessment standard for machining period

In Fig.1, the mathematical model of energy efficiency assessment standard for machining period is the theoretical basis, which supports the energy efficiency analysis and quantization. With the theoretical basis, the application content of energy efficiency assessment standard can then be implemented, i.e. the selection of evaluating index and the acquiring of evaluating index.
The content framework just gives us a guideline to understand the energy efficiency assessment standard for machining period. And then the three contents in the content framework will be explained further in the following sections.

III MATHEMATICAL MODEL OF ENERGY EFFICIENCY ASSESSMENT STANDARD FOR MACHINING PERIOD

To establish the mathematical model of energy efficiency assessment standard for machining period, we should know the energy breakdown of the machining period at first, and then propose the energy equation and power equilibrium equation of machining period.

A. The drawing of energy consumption during machining period

View from the period feature of machining period, the machining period can be divided into four kinds of periods, which are starting period, unload period, cutting period and break period [10]. One kind of period may appear multiple times during a machining period. A drawing of the energy consumption of one machining period is shown in Fig.2.

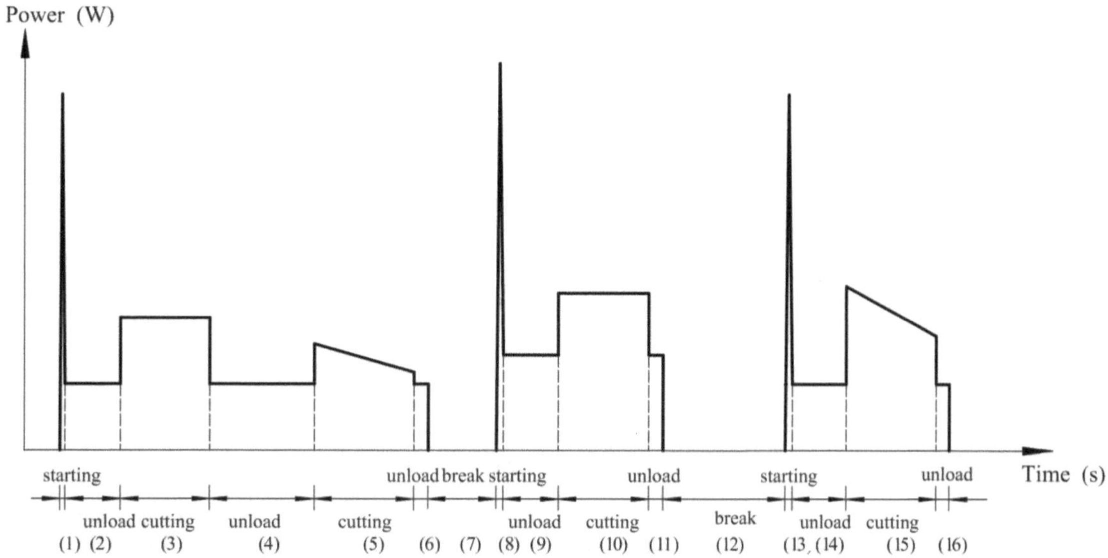

Fig.2 The drawing of energy consumption of a machining period [10]

B. Energy equation of machining period

Refer to section A, the energy equation of machining period can be expressed by (1).

$$E_t = E_s + E_u + E_C + E_b \quad (1)$$

Where,

E_t ——the total energy of the machining period;

E_s ——the total energy of all starting periods;

E_u ——the total energy of all unload periods;

E_C ——the total input energy of all cutting periods;

E_b ——the total energy of all break periods.

C. Power equilibrium model of machining period

(1) Power equilibrium diagram of machining period

The power equilibrium diagram during the machining period is shown as Fig.3. Particularly, the cutting power during the starting, unload and break periods should be zero.

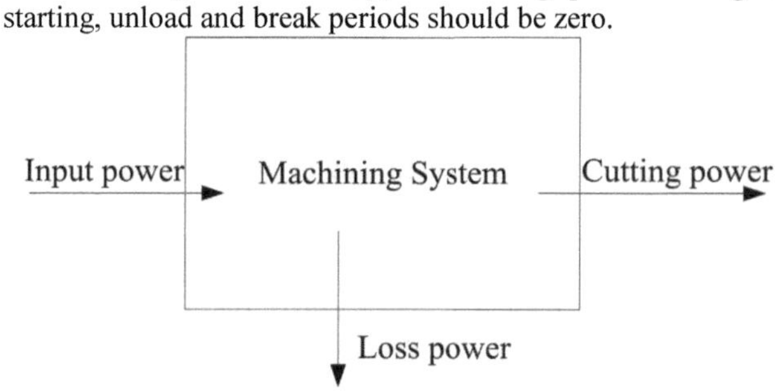

Fig.3 Power equilibrium diagram of machining period

(2) Power equilibrium equation of machining period

① Power equilibrium equation during the cutting period of machining period

The power equilibrium equation during the cutting period of machining period can be expressed as (2).

$$P_{in} = P_c + P_u + P_a \quad (2)$$

Where,

P_{in} ——input power of the machining system;

P_c ——cutting power;

P_u ——unload power of the machining system;

P_a ——additional load loss power of the machining system.

② Power equilibrium equation during the other periods of machining period

During the starting, unload and break periods, the cutting power is zero, and the input power is equal to the loss power.

Ⅳ THE EVALUATING INDEX OF THE ENERGY EFFICIENCY ASSESSMENT STANDARD FOR MACHINING PERIOD

EEMP is the quantitative expression of effective energy use ratio of machining period. Usually, energy efficiency is thought to be the ratio of effective energy to input energy, namely energy utilization ratio. In this paper, energy efficiency can be defined as the ratio of input energy to effective output as well, which is named specific energy consumption.

A. Energy utilization ratio

Energy utilization ratio is the ratio of effective energy to the total input energy during one machining period or a certain period of time. And it can be expressed by (3).

$$\mathrm{E}_e = \frac{E_c}{E_t} \quad (3)$$

Where,

E_e ——energy utilization ratio;

E_c ——the total cutting energy of all cutting periods.

B. Specific energy consumption

Specific energy consumption (SEC) is the ratio of total input energy to effective output during the machining period, which can be expressed as (4).

$$S_e = \frac{E_t}{E_o} \tag{4}$$

Where,

S_e ——specific energy consumption;

E_o ——effective output, including material removal volume, number of work-piece and so on.

V ACQUIRING THE EVALUATING INDEX OF THE ENERGY EFFICIENCY ASSESSMENT STANDARD FOR MACHINING PERIOD

A. Acquiring the energy utilization ratio

(1) Acquiring the total input energy

The total input energy can be monitored by the power meter installed at the energy input port of the machining system. We can monitor the input energy of the whole shift period or a machining period of a work-piece.

(2) Acquiring the cutting energy

It is easy to obtain the cutting power during the machining period in laboratory through measuring the cutting force and rotating speed of the main spindle. However, in the plant, it is not practical and un-allowed to install cutting force sensor at the cutting area of the machine tool usually. Therefore, the cutting power is difficult to obtain in real time. Based on our research before, one acquiring method of cutting energy based on off-line basic energy data and real-time input power of the machining system is proposed [4]. The cutting energy can be calculated by quadratic linear regression or simple linear regression with the input power from real-time monitoring, and unload power and additional load loss coefficients from off-line measuring. The calculation equations are shown as (5) and (6).

$$\begin{aligned} E_c &= \int_0^{t_T} P_c(t)\,dt = \sum_{i=1}^{Q_C} E_{ci} = \sum_{i=1}^{Q_C} \int_0^{t_{Ci}} P_c(t)\,dt \\ &= \sum_{i=1}^{Q_C} \int_0^{t_{Ci}} \frac{-a_{1i} + \sqrt{a_{1i}^2 + 4a_{2i}\left[P_{in}(t) - P_{Ui}\right]}}{2a_{2i}}\,dt \end{aligned} \tag{5}$$

$$\begin{aligned} E_c &= \int_0^{t_T} P_c(t)\,dt = \sum_{i=1}^{Q_C} E_{ci} \\ &= \sum_{i=1}^{Q_C} \int_0^{t_{Ci}} P_c(t)\,dt = \sum_{i=1}^{Q_C} \int_0^{t_{Ci}} \frac{P_{in}(t) - P_{ui}}{a_i}\,dt \end{aligned} \tag{6}$$

Where,

a —simple load loss coefficient;

a_1 —quadratic load loss coefficient 1;

a_2 — quadratic load loss coefficient 2;

i — serial number of period;

t_T — total time of the machining period;

Q_C —total number of cutting period;

t_{Ci} —time consumption during cutting period i.

(3) Calculating the energy utilization ratio

The energy utilization ratio of a cycle time and a work-piece can be calculated respectively. The former is the ratio of the total cutting energy to the total input energy during a cycle time, and the latter is the ratio of the cutting energy to the total input energy for machining a work-piece.

B. Acquiring the specific energy consumption

(1) Acquiring the total input energy

With the same method mentioned in Section A, the total input energy for calculating the specific energy consumption index can be got.

(2) Acquiring the effective output

The number of finished work-piece can be counted through observing or automatic counter. And the material removal volume can be calculated by the ratio of cut weight to material density, where the cut weight is gotten by weighting the raw material and the machined work-piece.

(3) Calculating the specific energy consumption

The specific energy consumption of a cycle time and a work-piece can be calculated respectively. The former is the ratio of total input energy to the finished work-piece number or the material removal volume during a cycle time. The latter is the ratio of the total input energy to the material removal volume for machining a work-piece.

VI CASE STUDY

To verify the energy efficiency standard introduced above, the energy efficiency of machining a plate-part is evaluated following the standard, and the detailed description of the experiment was given in [10].

A. Evaluating step

(1) Acquiring the cutting energy

As (5) and (6) shown, some basic data should be known before for calculating the cutting energy, including the unload power and load loss coefficients of the machine tool.

There are two ways to set up the unload power function of machine tools. Firstly, we can measure the corresponding unload power of the machine tool under its rating rotating speeds. Secondly, we can get the unload power function by regression analysis based on the statistical unload powers measured under typical rotating speeds of the machine tool. The machine tool used in this case study is grading control machine tool, therefore, the first way is adaptive. It is necessary to notice that we need to measure the unload power function of a machine tool just once, and the energy efficiency evaluation of machining by this machine tool in the future can always use the data.

To acquire the load loss coefficients, multiple set of cutting experiments with different cutting parameters under each grade of rotating speed of the machine tool need be implemented. By measuring the input power and cutting power of each cutting scheme, integrating with the unload power measured in last paragraph, the load loss coefficients can be calculated through simple linear regression or quadratic linear regression. In this case study, three cutting schemes are carried out under each rating rotating speed of the machine tool. And the machining period is separated into cylindrical turning, face turning, inner hole turning and second face turning in sequence. The detailed cutting schemes and measured powers can refer to [10]. As the same as the acquiring of unload powers, the measurement of load loss coefficients is implemented just once, and the energy efficiency evaluation of machining at the machine tool in the future can always use the data.

(2) Acquiring the effective output

The effective output in this case study is material removal volume. According to the cutting parameters of this machining period and the feature of the work-piece, we can calculate that the material removal volume of the machining period is 41069.63mm^3.

(3) Acquiring the total input energy

The input energy is measured by the power sensor HIOKI 3390 which is installed at the energy entrance of the machine tool, and the sampling interval of input power is 20ms.

Through real-time monitoring of the input power during the machining period on the site, the energy consumption drawing of the machining period can be obtained as shown in Fig.2.

B. Result analysis

As Fig.2 shown, there are 16 periods during the whole machining period, and 4 cutting periods distribute among the periods. The sequence number of the 4 cutting periods are 3, 5, 10 and 15. The related date for calculating the energy efficiency index of the machining period are shown in Table I.

TABLE I.
RELATED DATE FOR CALCULATING THE ENERGY EFFICIENCY INDEX OF THE PLATE-PART MACHINING PERIOD

indicator	Quadratic approximate cutting energy (J)	Material removal volume (mm^3)	Quadratic approximate energy utilization ratio	Specific energy consumption (J/mm^3)
value	1.19E+05	41069.63	25.38	11.42

As shown in Table I, the energy utilization ratio and specific energy consumption of the machining period using quadratic linear regression method are 25.38 % and 11.42 J/mm^3 respectively. To verify the results, a torque sensor TQ201 is installed at the cutting area of the machine tool to measure the actual energy utilization ratio. In details, to get the torque and actual rotating speed during the machining period by the torque sensor at first; and then to calculate the actual cutting power with the measured torque and rotating speed; to calculate the energy utilization ratio at last, which is 24.96%. Comparing the estimating result with the actual result, we can find that the error is 1.68%.

V. CONCLUSION

Energy efficiency assessment is the fundament of energy efficiency optimization for machining period. Set up a standardized energy efficiency assessment system is to provide support for energy efficiency optimization for machining period. As a result, the content framework and evaluating method of the energy efficiency assessment standard for machining period are studied in this paper. The evaluating method contains three main parts, which are the energy mathematic model for providing theoretical support, the evaluating index system including energy utilization ratio and specific energy consumption, and the acquiring method for energy efficiency index with monitoring the input power on site.

The case study shows that using the evaluating method proposed in this paper can get the energy efficiency index with small error. The standard can be applied into the energy efficiency evaluating, monitoring and optimization of the machine tool and work-piece, having a broad application prospect.

ACKNOWLEDGMENT

We gratefully acknowledge the financial supports of the National High-Tech. R&D Program of China (863 Program) (2014AA041506), Scientific and Technological Research Projects of Education Department of Jiangxi Province (No. 150088).

REFERENCES

[1] Department of Mechanical Industry of China. Mechanical engineering handbook. Beijing: China Machine Press, 1997, pp. 159.

[2] F. Liu, Q. L. Wang, and G. J. Liu. "Content architecture and future trends of energy efficiency research on machining systems". Chinese Journal of Mechanical Engineering, vol. 49, issue 19, pp. 87-94, 2013.

[3] Rockwell Automation. Extending the scope of mes to achieve energy efficiency and sustainability goals. Publication SUST-WP001A-EN-P, May 2009.

[4] F. Liu and S. Liu. "Period energy model of electro-mechanical main driver system during the service process of machine tools". Chinese Journal of Mechanical Engineering, vol. 48, issue 21, pp. 132-140, 2012.

[5] Q. L. Wang and F. Liu. "Research on mathematical model of multi-source energy flows for CNC machine tools". Chinese Journal of Mechanical Engineering, vol. 49, issue 7, pp. 5-12, 2013.

[6] T. G. Gutowski, J. B. Dahmus, and A. Thiriez. "Electrical Energy Requirements for Manufacturing Processes". 13th CIRP International Conference on Life Cycle Engineering, May 31st – June 2nd, 2006, Leuven, Belgium.

[7] A. Dietmair and A. Verl. "A generic energy consumption model for decision making and energy efficiency optimization in manufacturing". International Journal of Sustainable Engineering, vol. 2, issue 2, pp. 123–133, 2009.

[8] ISO/TC Machine tools Committee. ISO/DIS 14955-1. Environmental evaluation of machine tools — Part 1 : Design methodology for energy-efficient machine tools. Winterthur: ISO/TC39 , 2012.

[9] IAC. Industrial assessment centers. [2014-3-20]. www1.eere. energy.gov/industry/bestpractices/iacs.html.

[10] S. Liu, F. Liu, and Q. L. Wang. "Energy Efficiency Acquisition Method for Electromechanical Main Driver System during the Service Process of Machine Tools". Chinese Journal of Mechanical Engineering, vol. 48, issue 23, pp. 111-116, 2012.

Combining Problem and Lecture Based Learning for Production System Modeling and Simulation Course in Industrial Engineering Education

Jun-feng WANG*, Shi-qi LI, Yan FU, Shi-ping LIU, Hui JIANG
School of Mechanical Science and Engineering, Huazhong University of Science and Technology, Wuhan, China
(*wangjf@hust.edu.cn)

Abstract - **Production system modeling and simulation is an important technical course of industrial engineering education for undergraduate students. This paper presents an initial implementation of hybrid problem based learning (hPBL) method for the course, which combine problem-based learning (PBL) and lecture based learning (LBL) in teaching and learning process. The learning object and content of the course is introduced. An hPBL curriculum schedule is design for the course within a large classroom environment. The various open end problems of modeling and simulation are formulated for group collaborative learning. During learning procedure, the basic concepts and problem solving steps are introduced at the LBL classes, while the PBL classes focus on solving of selected problems. The hPBL greatly arouses the learning interest and improve the learning efficiency of students. The course is evaluated based on the students' survey. The examination score of hPBL is analyzed and compared with pure LBL classes.**

Keywords – **Industrial engineering education, lecture based learning, problem based learning, production system modeling and simulation**

I. INTRODUCTION

In the new millennium, innovation will be the very important motivation for China development. Besides the technical ability personnel, more and more person with creative thinking ability will be needed. The engineering education must enable students to meet the growing challenges and the increasingly more complex demands for the work. Professional problem-solving skills dealing with uncertainty and interpersonal communication ability are two important qualities for the new era students. However, today's engineering graduates lack these skills and have difficulty applying their fundamental knowledge to problems of practice. This requires that teaching, learning and assessment are conceived in such a way that they provide students with several opportunities to support the development of these competencies [1].

Problem-based learning (PBL) originated in medical education to enable a smoother transition of students into clinical education at hospitals, and prepare them better for professional practice [2]. PBL represents a major shift in educational paradigm to problem-based, process-oriented, discipline-integrated, and student-centered learning in collaborative small groups. The goals of PBL include fostering active learning, interpersonal and collaborative skills, open inquiry, real-life problem solving, critical thinking, intrinsic motivation, and the desire to learn for a lifetime. The traditional lecture-based learning (LBL) method is deductive, which begin with theories and progressing towards application of those theories. The teacher presents information without a discussion of why the mathematical models are being developed and what practical problems they will solve [3].

PBL is a philosophy that has to be adapted to the specific situation of a university and the nature of the discipline or subject field in which it is applied. Problem-based approaches to engineering courses may better prepare engineers for the types of work they will actually perform upon graduation and entry into professional practice. A number of studies have reported engineering implementations of PBL and some have identified challenges that were experienced by the engineering student.

The integration of the PBL method into the overall teaching methodology was implemented within the Department of Electrical Engineering [4]. The role of PBL facilitators and the characteristics of the PBL problems were posed in the courses. The usual way of learning technical knowledge about a microcontroller is by reading relevant handbooks and textbooks. Tse and Chan [5] proposed using the problem-based learning to convey such engineering knowledge. The case study in biomedical materials course [6] showed that students made significant improvements in their problem-solving skills, written communication and self-directed learning, which are defined as the desirable professional engineering skills for engineering students. Sanjeev et al. [7] presented the design and construction of a PBL-based course in materials science at the junior level in a mechanical & aerospace engineering department. They assessed the ability of a PBL course based on longer complex problems to enable students to learn both fundamental knowledge of the subject matter and also problem solving skills and contrast it with outcomes in a traditional lecture based course.

Barrows [2] admitted that varying degrees of PBL are often necessary, given the wide variety of contexts in which the method was attempted. One of the challenges the teachers face when implementing PBL is student resistance and discomfort when transitioning from the traditional curriculum to a PBL curriculum. Therefore a hybrid problem based learning (hPBL) maybe an eclectic choice. More importantly, the studies delivered through a combination of PBL and lecture format also showed a significant improvement of student problem-solving skills [8]. The hybrid model is favored as an attractive model in consideration of time, efforts and resources for its acceptance and implementation. Retention of a significant

This study was financial supported by Huazhong University of Science and Technology teaching reform project (15005).

E. Qi et al. (eds.), *Proceedings of the 23rd International Conference on Industrial Engineering and Engineering Management 2016*, DOI 10.2991/978-94-6239-255-7_5

portion of the didactic lectures in hPBL puts the teachers and students in a comfort-zone and thus faces less resistance to curriculum change [9]. It is believed that the hPBL method may provide a more challenging, motivating and enjoyable approach to education [10].

In China, the LBL approach still dominates as the preferred teaching approach from the primary schools to the universities, in which most of the focus is on content coverage. The students are habitual to traditional teaching and evaluation methods. In order to make the students progressively accommodate to the transition, the hPBL approach is a compromising appropriate choice. This paper is the preliminary attempt to adopt the hPBL to the course of industrial engineering education.

II. COURSE DESIGN

A. *Course object and content*

Industrial engineering is concerned with the development, improvement, and implementation of integrated systems of people, money, knowledge, information, equipment, energy, materials, analysis and synthesis, as well as the mathematical, physical and social sciences together with the principles and methods of engineering design to specify, predict, and evaluate the results to be obtained from such systems or processes [11]. Production system modeling and simulation (PSMS) is one of the main courses offered to the third year students of industrial engineering in the college of mechanical science and engineering, Huazhong University of Science and Technology (HUST).

The main objective of PSMS course is to provide an insight into how simulation modeling can aid in effective decision-making for discrete event system (such as production, manufacturing, logistics facilities, supply-chains, service et al). The students should understand the general concept and be provided with the primary problem solving ability for the practical different types of real life complex problems of business and industry after completing the course. Therefore, the course covers topics and issues such as the fundamental logic, structure, components and management of simulation modeling for discrete event system, the analytical techniques for interpreting input data and output results pertinent to simulation models based on probability theory, the modeling and simulation methods for the queue system and inventory system. It is also demonstrated how computer simulation can be used to successfully model, analyze and improve systems under study. Some simulation software (Arena, Flexsim and Quest) is used to build and execute the models.

There are 32 academic hours for this course. It was taught using traditional didactic teaching which included four hours of lectures in seven weeks. In the eighth week, four hours modeling and simulation exercise will be carried out for an example system by using chosen commercial discreet event system simulation software. In the classroom, additional lecture notes (PowerPoint slides) are used as the main media to teach the course content. Some home works are delivered after each class in order to force the students comprehending and reviewing the knowledge. The final exam including main memorial knowledge will be hold to give the score of every student. According to the past teaching experience, the students will have ideal scores for the course. But the problem-solving ability cannot be confirmed when students face with the new practical cases.

Problem-based learning in engineering is a natural fit since it espouses developing students' ability to solve ill-structure problems, increasing critical thinking skills, and broadening their communication skills [12]. Considering the education situation in China and its practical application for the future engineering career, the hPBL learning and teaching method is adopted for PSMS in order to motivate the problem solving ability of the industrial engineering undergraduate students of HUST.

B. *A hPBL method for PSMS*

Because of the limited academic hours for PSMS, the teaching process should be more effective for course content and practical exercise. Engineering students feel that less content is covered when inductive teaching is used compared to deductive, lecture-based approaches. Students very often are unable to solve a problem if they do not know the steps or sequence to follow [6]. The teacher has to carefully think about what content to teach with the problem based approach and what to cover with lecture in hPBL process.

Engineering has a hierarchical knowledge structure so that students need to have adequate basic knowledge to be successful at learning later concepts [13]. For the PSMS course, the basic concepts and the main steps of modeling and simulation are challenges to students without any preliminary knowledge about the course. Consequently, it is important to teach these knowledge by lecture before using problem-based learning, as the PBL method will not compensate for students' lack of proficiency in those basic concepts [14].

For our teaching reform, Fig. 1 shows the schedule of the PSMS course combined the LBL class (grey block) and PBL (white block) method. It can be seen that the content proportion of the teacher centered and the students centered are changed gradually with the proceeding of the course. In the first half of the schedule, LBL method is mainly used to teach the basic concepts and techniques of modeling and simulation, which will quickly set up the essential foundation for the course. In the second half of the schedule, PBL method will be dominated to give the students chance for exploring the solution of the simulation problems. At the beginning of the class, a set of modeling and simulation problems are presented to the students. The problems mimic the real modeling and simulation scenario and correspond to the job specification of the assigned role as industrial practitioners. It is more like a project, but will be solved paralleling to the course. Besides the project problems, some example questions specific to modeling and simulation aspects are also presented for every LBL class, which served as the training of the analysis ability for the project problems. It is expected that after solving the integrated project problems and several small independent problems through a series of LBL and PBL cycles, students will transform from "novice" modeling and simulation problem solvers to "experts" within the course duration. The executing process of the course realizes the modeling and simulation knowledge acquisition, understanding, and application based on the hPBL method.

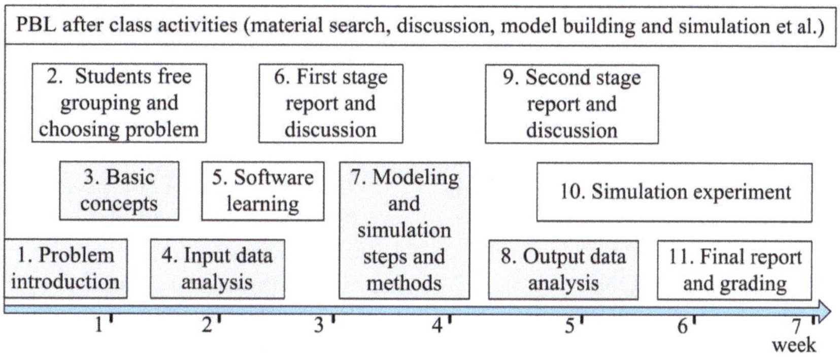

Fig. 1. The schedule of PSMS class in an hPBL method.

C. Problem design for PSMS

Before the hBPL classes, the students groups will be present with messy, unstructured problems, which will be solved within the context of the future profession. In general, there are three different types of problems, i.e. fictional, authentic and real. However, real engineering problems are hard to find, and if any, they cannot be directly used for academic perspectives [15]. Therefore, crafting the authentic problems is always a good alternative and is the most preferred by PBL practitioners. Effective PBL problems should consider students' prior knowledge, have suitable complexity, be in the context of relevant future professional practice, stimulate critical thinking and meta-cognitive skills, promote collaboration among group members, and cultivate self-directed and life-long learning [16].

For the PSMS course, we have design a problem set including twenty-six problems, which mainly from the manufacturing industry and service industry. There are job shop scheduling, assembly system, logistics and inventory system problems for manufacturing industry. The problems of service industry are classified as bank, supermarket, amusement park and medical care system. Each modeling and simulation problem covers the main learning content, i.e. input data analysis, system logic modeling, simulation modeling with specific software, output data analysis and system improvement. With the progress of the course, the student groups should solve the problem collaboratively by themselves under the supervising of the tutors.

Apart from the problem set used as the professional ability training, some after class homework are also design for the warm up activity of the problem and necessary exercise not included in the problem. The quantity of the homework is less than the pure LBL teaching method because the main task of the students is the solving of the project problem with group discussion and learning.

III. PILOT IMPLEMENTATION

Fifty-two third year undergraduate students enrolled in industrial engineering major participate in this study at the first semester of the 2015-2016 academic years. Participants included forty males and twelve females. The mean age of students is 20.48 (SD = 0.48). There will be 4 hours each week for the course, which is separated to two times with 2 hours. In the sixth week, there is a simulation experiment class for all groups in a large classroom in order to consult some practical errors during simulation. The students are allowed to choose their team mates by themselves. Every group will have 3 to 5 five members and chose one modeling and simulation problem. There are 13 groups in the class.

The group members should aim to collaboratively solve the problem step by step during the whole course hours with the schedule in Fig1. Even in the LBL class, the teacher also firstly presents the problems related with course contents, and then introduces the concepts and solving techniques for the corresponding problems. Some time will be reserved after the teaching hour and used to discussion the project question based on groups. Individual group discussions are not supervised since only one instructor is present in the classroom. Several classes later, an open stage presentation and discussion class with all groups participating will be followed. The teacher will moderate the discussion and evaluation the stage achievement of the project question. The initial reporting group is picked in a random manner so that all groups are attentive and ready to report. At this stage, the teacher may ask guiding questions if clarifications are needed or if the discussion needs to be redirected or refined. This may involve directing the right thinking and establishing stage application of the concepts and techniques.

At the end of the PBL module, each team presents their modeling and simulation solutions of selected project problems to other teams and the teacher in

PowerPoint. The simulation models with animation are also shown to give the vivid impress for the problems. The groups are awarded the project problem grades/scores based on their performance in the presentations by mutually evaluating among groups and the teacher. With the stage and the final report in open class, the communication (written and verbal) and presentation skill of the students are taken exercise.

The anonymous questionnaire is administered on the end of the course after the class. There are two sections in the survey. The first section includes four aspects rated according to the students' experience. In the second section, the students can provide his/her open suggestions and comments about the class.

IV. COURSE EVALUATION

The traditional lecture based learning course of PSMS was offered to the third year first-semester bachelor of industrial engineering degree program before the academic year 2015-2016. There are no questionnaire surveys for the LBL course before year 2015. For the course of year 2015, we make a survey. Of the 52 students in the class, 51 students responded to the survey (response rate of 98%) and 41 chose to provide comments (response rate of 80%).

A. *Course learning evaluation*

There are four aspects about the course learning with total 16 questions evaluated by students, the overall impress about the class, the teaching material, the hPBL class method and the group performance. On the questionnaire, students were asked to indicate the effective level with 4-point response choice. The higher the score the more likely the students considered the item to be more effective, satisfactory or consentient. Table I shows the 16 problems and the evaluation results.

From the Table I, overall, students have a more or very impression on hPBL course (item mean 3.25±0.66). The comments from the survey also illustrate the acceptance of the hPBL method to students. The lower score of overall experience is the hPBL effect on learning motivation (item mean 3.06±0.61). In the teaching materials items, the after class homework has the more score (item mean 3.67±0.55), which reveal that the good balance between homework and problem solving activities. The understanding of the course content is less satisfied (item mean 3.06±0.68). The reason lie in it is the first experience of hPBL course for the students. The learning style, material searching and problem solving activities mainly relay on themselves. They have more stress on content understanding than traditional LBL course. For the group performance, the evaluation on self behavior is also less satisfied (item mean 3.12±0.68), which indicate that most students believe they can perform better in the later hPBL course.

TABLE I
STUDENTS' EVALUATION OF THE COURSE

Survey items	Mean	SD
Overall experience		
1.General impression of HPBL	3.25	0.66
2.HPBL effect on learning object	3.29	0.64
3.HPBL effect on learning motivation	3.06	0.61
4.Feeling of HPBL versus LBL	3.55	0.67
Teaching materials		
5.The problems set for HPBL	3.33	0.84
6. Difficulty of your problem	3.24	0.93
7.The PPT content for LBL class	3.16	0.70
8.The homework for LBL class	3.67	0.55
The HPBL class		
9.Understanding of the content	3.06	0.68
10.Schedule of the HPBL class	3.71	0.50
11.environment of the HPBL class	3.45	0.54
12.interaction on the HPBL class	3.27	0.57
Group performance		
13.Participation of your member	3.41	0.70
14.Interaction among group members	3.41	0.70
15.Your behavior in group	3.12	0.68
16.Your solution of the problem	3.33	0.55

Figure 2 shows the effective level percentage of 16 questions. The satisfied level (very and more) of all items is 88.73%. For the difficulty of the problem (item 6), the student quantities of the foul levels are 26, 14, 8 and 3 respectively. The result indicates that the problem is difficulty for most students. Nevertheless, there are 49 students (96%) consider that they successfully give the solution of the chosen problem (item 16).

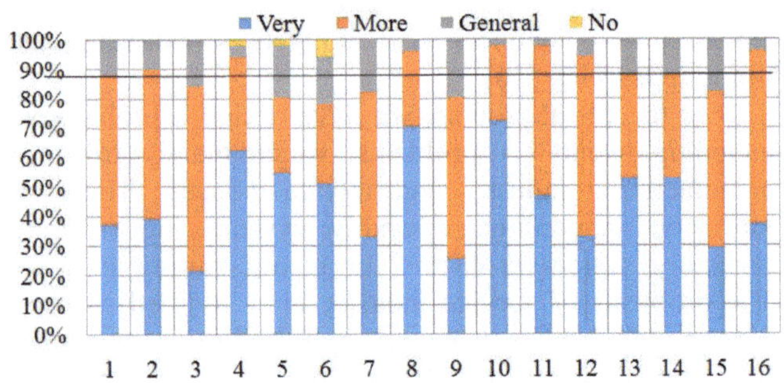

Fig. 2. The response percentage of the survey items.

B. *Test score comparison*

The hundred mark system is used to grade five common levels, i.e. excellent for score not less than 90, good for score between 80 and 89, medium for score between 70 and 79, passed for score between 60 and 69, and failed for score less than 60. The test scores in final course exam with similar examination questions for 2015 hPBL class and 2012 to 2014 LBL class are shown in Table II. The result of independent sample t test for hPBL class and every LBL class is all less than 0.05, which indicate that there is no significance difference by comparing the mean score of different classes.

Figure 3 shows the score grade distribution for the four years classed. It is interesting that the percentage of excellent and failed students in 2015 hPBL class is lower than 2012-2014 LBL class. But the percentage of good and medium students in 2015 hPBL class is higher than 2012-2014 LBL class. The overall pass rate of the test in 2015 hPBL and 2012-2014 LBL class is 84.31% and 92.33%. The statistics results will be further analyzed based on the hPBL practice in the following several years.

TABLE II
TEST SCORE EVALUATION FOR DIFFERENT CLASSES

Class type	Students	Mean	SD	*p* value
2015 HPBL class	51	75.33	12.88	
2014 LBL class	52	76.63	11.16	0.29
2013 LBL class	44	73.55	10.87	0.23
2012 LBL class	51	77.31	12.79	0.21

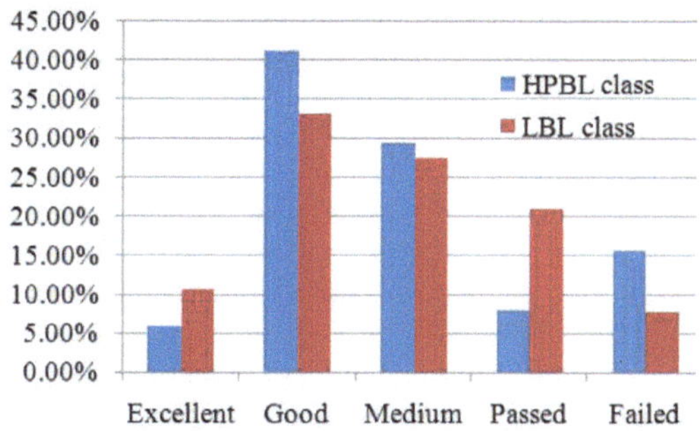

Fig. 3. The score grade percentage of the different classes.

V. CONCLUSION

Problem based learning and lecture based learning have their respective advantages in engineering education. For the technical foundation course in undergraduate industrial engineering students, production system modeling and simulation, a preliminary implementation of hybrid problem based learning is described in this paper. A problem set with various open end simulation questions is build for group collaborative learning. A hybrid curriculum is present which gradually increase the student centered class proportion. The survey indicates that the hybrid method is accepted by students. From the current implementation, the mean test score has no different between hybrid class and pure lecture based class. But the students' ratio of good and medium score is higher under the hybrid learning method. The ongoing work will be completed with several next academic years to further improve and explore the effect of the hybrid method.

REFERENCES

[1] S. Fernandes, M. A. Flores, R. M. Lima, (2012) "Students' views of assessment in project-led engineering education: findings from a case study in Portugal", *Assessment & Evaluation in Higher Education*, Vol.37, NO. 2, pp. 163-178, 2012.

[2] H. S. Barrows, "Problem-based learning in medicine and beyond: A brief overview", *New Directions for Teaching and Learning*, Vo. 68, pp. 3-12, 1996.

[3] M. Prince, R. Felder, "Inductive teaching and learning methods: Definitions, comparisons, and research bases", *Journal of Engineering Education*, Vol. 95, No. 2, pp. 123-138, 2006.

[4] S. M. Said, F. R. Mahamd Adikan, S. Mekhilef, N. Abd Rahim, "Implementation of the problem-based learning approach in the Department of Electrical Engineering", *European Journal of Engineering Education*, Vol. 30, NO.1, pp. 129-136, 2005.

[5] W. L. Tse, W. L. Chan, "Application of problem-based learning in an engineering course", *International Journal of Engineering Education*, Vol.19, NO.5, pp.747-753, 2003.

[6] N. W.James, M. J. Mohammadi-Aragh, "Case study: use of problem-based learning to develop students' technical and professional skills", *European Journal of Engineering Education*, DOI: 10.1080/03043797.2015.1040739, 2015.

[7] K. K. Sanjeev, A. W. Robert, H. J. David, A. Tawfik, H. Henry, "Effectiveness of Introducing Problem Based Learning in the Undergraduate Engineering Curriculum", *in Proceedings ASME 2012 International Mechanical Engineering Congress and Exposition*, Houston, Texas, USA, November 9–15, pp. 449-457.

[8] J. Lian, F. He, "Improved performance of students instructed in a hybrid PBL format", *Biochemistry and Molecular Biology Education*, Vol.41, No.1,pp.5-10, 2013.

[9] M. Carrió , P. Larramona , J.E. Baños, J. Pérez, "The effectiveness of the hybrid problem-based learning approach in the teaching of biology: a comparison with lecture-based learning", *Journal of Biological Education*, Vol. 45, No. 4, pp. 229-235, 2011.

[10] C. Y. Kwan, "Problem-based learning and teaching of medical pharmacology", *Naunyn-Schmiedeberg's archives of pharmacology*, Vol. 366, No. 1, pp. 10-17, 2002.

[11] G. Salvendy, "Handbook of industrial engineering: Technology and operations management", *Wiley-Interscience*, 2011.

[12] P. A. Johnson, "Problem-based cooperative learning in the engineering classroom", *Journal of Professional Issues in Engineering Education and Practice*, Vol. 125, No. 1, pp. 8-11, 1999.

[13] J.E. Mills, D.F. Treagust, "Engineering education-Is problem-based or project-based learning the answer", *Australasian Journal of Engineering Education*, Vol. 3, No. 2, pp. 2-16, 2003.

[14] A. Yadav, D. Subedi, M. A. Lundeberg, C. F. Bunting, "Problem-Based Learning: Influence on Students' Learning in an Electrical Engineering Course", *Journal of Engineering Education*, Vol. 100, No. 2, pp.253-280, 2011.

[15] M. Z.Jamaludin, K. M.Yusof, N. F.Harun, S. A. H. S. Hassan, "Crafting Engineering Problems for Problem-Based Learning Curriculum", *Procedia-Social and Behavioral Sciences*, Vol. 56, pp. 377-387, 2012.

[16] W. Hung, K. Mehl, J. B. Holen, "The relationships between problem design and learning process in problem-based learning environments: Two cases", *The Asia-Pacific Education Researcher*, Vol. 22, No. 4, pp. 635-645, 2013.

Schedule Risk Management of IT Outsourcing Project Using Negotiation Mechanism

Hua-ling Bi, Xiang Jia, Fu-qiang Lu*, Min Huang
(College of Information Science and Engineering, Northeastern University, Shenyang, China 110819)
(*e-mail fqlu@neuq.edu.cn)

Abstract - **Schedule risk control is a very important issue for the success of IT outsourcing project. There are two independent agents in IT outsourcing project, which is the client and the vendor, the two sides need to reach an agreement on the schedule and cost of the project. However, traditional schedule risk control methods haven't taken this characteristic of IT outsourcing into account. Given that, this paper analyze the characteristic of IT outsourcing with negotiation mechanism. We formulate the negotiation protocol that the two sides need to obey in the process of negotiation, design the decision making model and concession functions. Based on the simulation results, the effectiveness of the negotiation mechanism is verified. We also analyzed the influence of maximum negotiation rounds and concession factors on the negotiation results.**

Keyword - **IT outsourcing, Schedule risk management, Negotiation, Project management**

I. INTRODUCTION

In recent years, information technology plays an increasingly important role in business process and routine work of enterprises. The main advantage of IT outsourcing is reducing cost, focusing on core competition and get access to latest technology, which is also the motivation of enterprises to choose IT outsourcing[1-2]. Meanwhile, enterprises might encounter many potential risks when outsourcing their IT work, which include business uncertainty, loss of innovative capacity and cost overrun caused by project delay [3-5]. Among them, the success of IT outsourcing projects depend on whether we can control the schedule effectively, which gains widespread attention in recent years.

PERT/CPM, critical chain, event chain, Gantt chart and milestones method are the traditional schedule risk control methods[6-8]. However, IT outsourcing project is different from the traditional software project. Therefore, the schedule risk control methods mentioned above can't apply to the field of IT outsourcing directly, the main reasons are as follows. Firstly, schedule and cost of traditional software projects are determined by the enterprise itself, the project are implemented by the enterprise itself as well. But the schedule and cost of IT outsourcing project should meet the requirement of the two sides and at the same time, the project is implemented by vendor [9]. Secondly, traditional software project management tends to take only one factor into account, which is schedule or cost. Traditional method tends to satisfy the requirement on schedule but with a high cost, or satisfy the requirement on cost but with a high schedule. Therefore, if we give a comprehensive consideration of both schedule and cost, the enterprise's interest will be better served [10]. Finally, the effect of the schedule risk control method ensures the project implemented according to the plan. But if schedule and cost of the project is unreasonable, the project delay is inevitable even though we apply the schedule risk control method mentioned above. Therefore, determination of reasonable schedule and cost is also important.

The client and vendor in an IT outsourcing project have varied preference over schedule and cost. They all expect that the distribution scheme of schedule and cost are as favorable to them as possible [11-12]. Put differently, the client prefers that the value of schedule and cost is as large as possible, vendor prefers that the value of schedule and cost is as small as possible. However, the difference between the two sides' preference is non-confrontation. Generally, we can get a result that is beneficial to both sides. The two sides should take some measures in order to reach an agreement. The effective ways to solve this kind of non-confrontation disputes and conflicts are negotiation, mediation and arbitration [13-14]. Negotiation is an effective way that agents communicate and compromise to reach a mutually beneficial agreement [15]. Available literatures apply negotiation to the field such as electronic commerce and multi-agent systems mostly [16-19]. The model that widely used is bargain model [20]. Bargaining theory mainly research on the situation that two parties expect to reach an agreement but have contradiction in how to cooperate.

This paper describes the interaction process by negotiation, designs the negotiation protocol of both sides, and establishes the decision-making model correspondingly. The two parties' concession strategy was given, the model also give a comprehensive consideration of schedule and cost. Finally, we make experimental analysis under different concession strategy of both sides and maximum negotiation rounds.

II. BILATERAL NEGOTIATION MODEL

A. Basic assumptions

There is a situation in practice that the client and vendor have a long-term friendly relations and cooperation in IT outsourcing. As a result, client tends to designate a vendor to undertaken the IT outsourcing project. Generally speaking, the IT outsourcing project consists of some sub-projects, both sides should reach an agreement on schedule and cost by negotiation [21]. In this condition, we analyze the problem with an alternating offers protocol and gives out a bilateral multi-issue negotiation model. Before establishing the model, we need to make some basic

E. Qi et al. (eds.), *Proceedings of the 23rd International Conference on Industrial Engineering and Engineering Management 2016,* DOI 10.2991/978-94-6239-255-7_6

assumptions as follows:

1. The cost of negotiation is neglected.

2. The offer evaluation function is monotonous. The preference for both sides on the schedule and cost is fixed.

3. Before the start of negotiation, the two sides have complete information about each other, except the reservation utility and utility of each round.

4. Time is valuable to both sides. The two parties will set the maximum negotiation rounds for the negotiation; they all expect to reach an agreement in the limited rounds.

5. Both sides are self- interested, all of them expect to maximize their utility.

B. Negotiation issues and Standardization

The IT outsourcing project is divided into m sub-projects. Before negotiation, they often set bounds for schedule and cost of each sub-project. Both sides need to negotiate between the bounds [22-23]. The two sides exchange their schedule and cost of each sub-project, schedule and cost of the whole project consists of each sub-projects' schedule and cost. Client prefers that the value of schedule and cost as large as possible. Instead, vendor prefers that the value of schedule and cost as small as possible. Therefore, we can consider that the lower limits of schedule and cost depend on vendor's lower limits, the upper limits of schedule and cost depend on client's upper limits. The parameters are defined as follows:

s : Client

b : Vendor

$x^j_{\max}$: The maximum value of schedule d or cost c for sub-project j, $x^j_{\max} = (c^j_{\max}, d^j_{\max})$.

$x^j_{\min}$: The minimum value of schedule d or cost c for sub-project j, $x^j_{\min} = (c^j_{\min}, d^j_{\min})$.

$d^j_{\min}$ 、 $d^j_{\max}$: The minimum and maximum value of schedule for sub-project j respectively.

$c^j_{\min}$ 、 $c^j_{\max}$: The minimum and maximum value of cost for sub-project j respectively.

$x^{j,t}_s$: The client's offer for sub-project j in round t, $x^{j,t}_s = (c^{j,t}_s, d^{j,t}_s)$.

$x^{j,t}_b$: The vendor's offer for sub-project j in round t, $x^{j,t}_b = (c^{j,t}_b, d^{j,t}_b)$.

X^t_s : The client's offer for the whole project in round t, $X^t_s = (x^{1,t}_s, x^{2,t}_s, \cdots, x^{m,t}_s)$.

X^t_b : The vendor's offer for the whole project in round t, $X^t_b = (x^{1,t}_b, x^{2,t}_b, \cdots, x^{m,t}_b)$.

Client and vendor have varied weights on each sub-projects and on each sub-projects' schedule and cost. Different level of weight can be expressed by a number which is called weight. The larger the number is, the more important this item is. Meanwhile, the value of weights are constant which won't change with time. The parameters are defined as follows:

w^j_s 、 w^j_b : The weight of sub-project j for client and vendor respectively.

wd^j_s 、 wc^j_s : Client's weight on schedule and cost for sub-project j respectively.

wd^j_d 、 wc^j_d : Vendor's weight of schedule and cost for sub-project j respectively.

The dimensions of schedule and cost are different. In order to calculate and compare those issues, we need to make them in non-dimensional state which is called standardization. This paper applies transform formulas below to standardize those issues, making them in the range of [0, 1]. Put differently, for client, the smaller schedule and cost, the larger its interest. Therefore, its standardized formulas is shown as formula (1). For vendor, the greater schedule and cost, the greater its interest, its standardized formulas is shown as formula (2).

$$V^s(x^{j,t}_s) = \frac{x^j_{\max} - x^{j,t}_s}{x^j_{\max} - x^j_{\min}} \tag{1}$$

$$V^b(x^{j,t}_b) = \frac{x^{j,t}_b - x^j_{\min}}{x^j_{\max} - x^j_{\min}} \tag{2}$$

C. Negotiation framework

The whole negotiation framework includes three parts, which include negotiation issues, interaction mechanism and utility evaluation mechanism. Interaction mechanism describes the alternating offers process, this paper adopts the alternating offers model of Rubinstein to describe this process. The process is shown in fig.1.

Before negotiation, the two sides should discuss the maximum negotiation round $t_{\max}$ of the project, provide their negotiation range of each sub-project, and alternate their weights for each sub-project and weights for schedule and cost of each sub-project. Among those, their concession strategies and reservation utilities are private information. That is to say, they won't inform the other side of their concession strategies and reservation utilities. In this way, they can force the other side to provide a reasonable price. Neither side will benefit from a failure negotiation, therefore, both sides hope to reach an agreement in a limited rounds.

Figure 1 can be divided to the following steps:

Step 1: The rounds add 1 which is described as $t = t + 1$. If $t \le t_{\max}$, client put forward it's utility demand according to its concession strategy and go to Step 2. Otherwise, go to Step5.

Step 2: Vendor received client's utility demand, generating

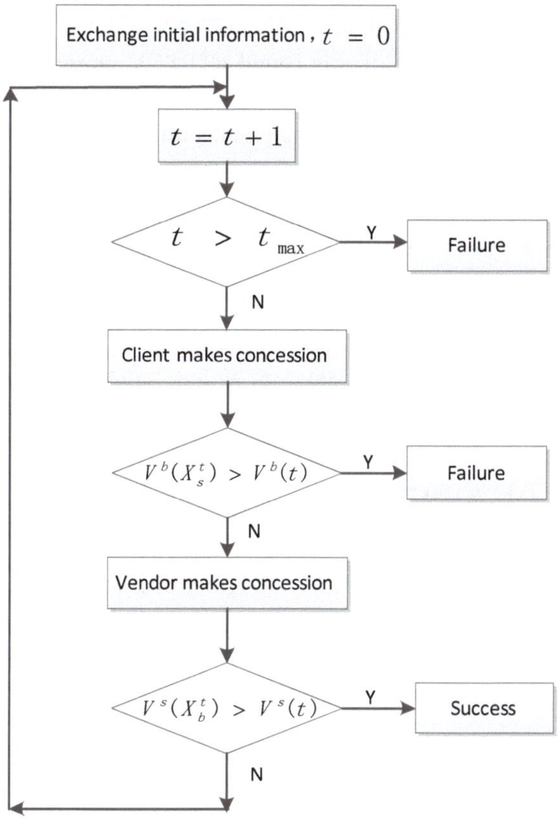

Fig.1. Negotiation process

their schedule and cost according to its decision-making model and judge whether the schedule and cost satisfy its minimum utility demand, which is defined as $V^b(X_t^s) \geq V^b(t)$.The negotiation succeed if satisfy the utility demand. Vendor inform the client of the schedule and cost, utility of this round is the final utility and the schedule and cost are the final schedule and cost. If dissatisfy, go to Step 3.

Step 3: Vendor put forward its utility of this round according to its concession strategy, and then go to Step 4.

Step 4: Client should make concession after receiving vendor's utility. Analogously, client generates its schedule and cost according to its decision-making model and judge whether the schedule and cost satisfy its minimum utility demand. The demand can be defined as $V^s(X_t^b) \geq V^s(t)$. The negotiation succeed if satisfy the utility demand, then terminate the negotiation. Client inform vendor of the schedule and cost, utility of this round is the final utility and schedule and cost are the final schedule and cost. Otherwise, go to Step 1.

Step 5: Terminating the negotiation and negotiation failure.

Concession strategy and decision-making strategy are the very important parts in the negotiation mechanism. Concession strategy is used to make concession and put forward new utility when negotiation didn't reach an agreement. In order to maximize their interests, they won't put forward their final minimum utility in the first round. However, the negotiation can't proceed infinitely. If the two sides didn't reaching an agreement when reaching the maximum negotiation rounds, then negotiation failed. The two sides need to do some concessions and reach an agreement because negotiation failure has negative impact on the interests of both sides. The model adopts bilateral concession strategy and the concession function is defined as (3) [24]. The two sides can select several concession strategies which is called multi-strategies concession based on their actual situation. This method can simulate agents' concession strategy better and can also avoid the other side speculating its concession strategy. So the agent can maintain its interest better. The parameters are explained as follows:

$$V^i(t) = V_i + (1 - V_i)\left[\frac{t_{max} - t}{t_{max}}\right]^{\beta} \tag{3}$$

i : Denoting client s or vendor b

β : Concession coefficient

V_i : Final reservation utility of Agent i

$V^i(t)$: Reservation utility of Agent i in round t

Concession strategy usually can be divided into three types, which is conservative strategy, linear strategy and aggressive strategy. We provide corresponding utility chart when $t_{max} = 10$ and $V_i = 0.4$. In figure 2, agent applies conservative strategy when $\beta < 1$. At the beginning, utility changes slowly. With negotiation proceed, the agent makes more and more concessions. Agent applies linear strategy when $\beta = 1$. Agent's utility changes with the process of negotiation with constant rate. Agent applies aggressive strategy when $\beta > 1$. At the beginning, agent's utility changes with negotiation fleetly. The utility reduce slowly with the proceeding of negotiation.

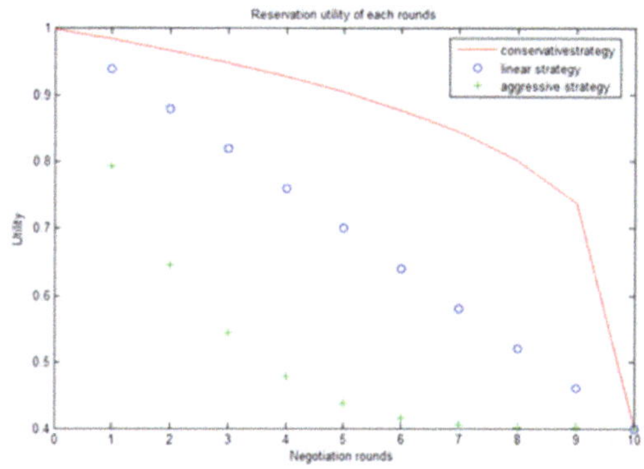

Fig.2. Concession function

Decision-making model is applied to generate new offers when the two sides didn't reach an agreement. Traditional negotiation applies face-to-face communication ways which need the two sides to assign representatives. They try to narrow down their demand gap of each issue and reach an agreement finally. However, when the two sides didn't reach an agreement and counter-advice should be put forward. The two sides tend to give offers according to their experience before. Therefore, the offers are subjective and are often not optimal. So, we model the decision-making process in order to generate the optimal offers. As mentioned above, there are two main factors to be satisfied for the agents. One is that those offers satisfy opposite side's utility demand, the other one is that the offers satisfy the constraint of schedule and cost for each sub-projects.

Vendor's decision-making model

$$\max \sum_{j=1}^{m} \omega_j^s V^s(x_s^{j,t}) \quad (4)$$

$$\text{s.t.} \sum_{j=1}^{m} \omega_j^b V^b(x_s^{j,t}) \geq V^b(t) \quad (5)$$

$$d_{\min}^j \leq d_s^{j,t} \leq d_{\max}^j,\ j = 1, \cdots, \mathrm{m} \quad (6)$$

$$c_{\min}^j \leq c_s^{j,t} \leq c_{\max}^j,\ j = 1, \cdots, \mathrm{m} \quad (7)$$

Client's decision-making model

$$\max \sum_{j=1}^{m} \omega_j^b V^b(x_b^{j,t}) \quad (8)$$

$$\text{s.t.} \sum_{j=1}^{m} \omega_j^s V^s(X_b^{j,t}) \geq V^s(t) \quad (9)$$

$$d_{\min}^j \leq d_b^{j,t} \leq d_{\max}^j,\ j = 1, \cdots, \mathrm{m} \quad (10)$$

$$c_{\min}^j \leq c_b^{j,t} \leq c_{\max}^j,\ j = 1, \cdots, \mathrm{m} \quad (11)$$

III. CASE ANALYSES

This paper designed a case to analyze the effectiveness of the model. We assume that the IT project can be divided into 3 sub-projects. Table 1 gives the range of schedule and cost for each sub-project. The information also includes the weights of each sub-project.

TABLE I
INITIAL INFORMATION OF IT OUTSOURCING PROJECT

p^j	c^j	d^j	w_b^j	w_b^j	wc_s^j , wd_s^j	wc_b^j , wd_b^j
1	[10,20]	[6,10]	0.3	0.1	(0.3,0.7)	(0.3,0.7)
2	[40,60]	[20,40]	0.5	0.4	(0.2,0.8)	(0.3,0.7)
3	[30,50]	[15,30]	0.2	0.5	(0.4,0.6)	(0.5,0.5)

A. Negotiation process and its analyses

We assume that the maximum negotiation rounds $t_{\max} = 10$ and they all apply the linear concession strategy $\beta = 1$. The final reservation utility of client and vendor is $V_s = 0.5$ and $V_b = 0.4$ respectively. The decision-making model of this paper is a linear programming problem, so Lingo is applied to solve this problem. The process and result of negotiation are shown in table 2. For vendor, $V^b(t)$ represents the minimum reservation utility of this round. $V^b(X_s^t)$ represents the maximum utility generated by the vendor's decision-making model that satisfying client's utility demand.

TABLE II.
NEGOTIATION PROCESS AND RESULTS

rounds	$(V^s(t), V^s(X_b^t))$	$(V^b(t), V^b(X_s^t))$
1	(0.95,0.18)	(0.94, 0.156)
2	(0.9,0.329)	(0.88,0.292)
3	(0.85,0.414)	(0.82,0.396)
4	(0.8,0.5)	(0.76, 0.5)
5	(0.75, 0.586)	(0.7,0.56)
6	(0.7, 0.67)	(0.64, 0.62)
7	(0.65, null)	

As we can see from the table II, the negotiation reach an agreement after 7 rounds of negotiation. At the same time, the result not only satisfy vendor' reservation utility demand which is $V_b(t) = 0.64$, but also satisfy client's reservation utility which is $V_s(t) = 0.67 > 0.65$.

B. Analyses of negotiation rounds

In general, smaller negotiation rounds results in larger utility in each round and larger randomness of negotiation results. We can reduce the change of concession utility of each round by adding negotiation rounds, then the randomness will reduce correspondingly. In order to verify this, Fig.3 depicts the variation of negotiation results according to the variation of negotiation rounds while the other parameters are fixed. This paper quantify the fairness of the negotiation, the fairness we defined in the paper is depend on the extent of gaining profit. The fairness under different negotiation rounds is shown as figure 3.

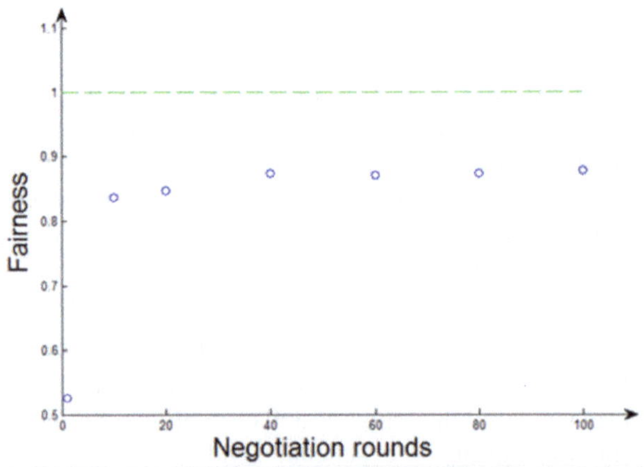

Fig.3. Analyses of negotiation rounds

We can see that the positive correlation between R and the negotiation rounds. This is a practical conclusion for IT outsourcing enterprise. If enterprises expect a fair result, then they should negotiate with more rounds. If enterprises have a preference to reach an agreement quickly and don't care about the fairness, then enterprises can negotiate with less rounds.

IV CONCLUSION

This paper proposed a negotiation framework for schedule risk control of IT outsourcing project. Decision-making model and concession strategies are designed. The decision-making model is a linear programming problem, so Lingo is applied to solve it. For the concession strategies, there are mainly three types, which include aggressive, linear and conservation concession strategy. The experimental results show that the framework can increase the utilities of both sides. This reflect the win-win characteristic of negotiation.

ACKNOWLEDGMENT

This work is supported by the National Science Foundation for Distinguished Young Scholars of China under Grant No. 71325002, No. 61225012; the National Natural Science Foundation of China under Grant No. 71401027, 71401027, 71071028; the Fundamental Research Funds for State Key Laboratory of Synthetical Automation for Process Industries under Grant No.2013ZCX11. Natural Science Foundation of Hebei Province under Grant No. G2016501086, Key Project of Science and Technology Research of Higher Education in Hebei Province under Grant No. ZD2016202.

REFERENCES

[1]. Mary C. Lacity, Shaji A. Khan, Leslie P. Willcocks. A review of the IT outsourcing literature: Insights for practice [J]. Journal of Strategic Information Systems, 2009, 18:130-146.

[2]. Chitrasen Samantra, Saurav Datta, Siba Shankar Mahapatra. Risk assessment in IT outsourcing using fuzzy decision-making approach: An Indian perspective[J]. Expert Systems with Applications, 2014, 41: 4010-4022.

[3]. Narasimhaiah Gorla, Toni M. Somers. The impact of IT outsourcing on information systems success [J]. Information & Management, 2014, 51(3): 320-335.

[4]. Earl, Michael J. The risks of outsourcing IT[J] . Sloan Management Review, 1996 ,37:26- 32.

[5]. Liangjuan Qin, Harris Wu, Nan Zhang, Xiang Li. Risk identification and conduction model for financial institution IT outsourcing in China, 2012, 13(4): 429-443.

[6]. Shanlin Yang, Lei Fu. Critical chain and evidence reasoning applied to multi-project resource schedule in automobile R&D process[J]. International Journal of Project Management. 2014, 32(1): 166-177.

[7]. Barbara Gadysz, Dariusz Skorupka, Dorota Kuchta, Artur Duchaczek. Project risk time management- A proposed model and a case study in the construction industry[J]. Procedia Computer Science. 2015,64: 24-31.

[8]. James M. Wilson. Gantt charts: A centenary appreciation[J]. European Journal of Operational Research, 2003,149(2): 430-437.

[9]. Xiaowei Zhu. Management the risks of outsourcing: Time, quality and correlated costs[J]. Transportation Research Part E: Logistics and Transportation Review, 2015.

[10]. Shahryar Monghasemi, Mohammad Reza Nikoo, Mohammad Ali Khaksar Fasaee, Jan Adamowski. A novel multi-criteria decision making model for optimizing time–cost–quality trade-off problems in construction projects[J]. Expert Systems with Applications, 2015, 42(6): 3089-3104.

[11]. Fatima S S, Wooldridge M, Jennings N R. Bargaining with incomplete information [J]. Annals of Mathematics and Artificial Intelligence, 2005, 44: 207-232.

[12]. Shaheen S. Fatima, Michael Wooldridge, Nicholas R. Jennings. An agenda-based framework for multi-issue negotiation[J]. Artificial Intelligence, 2004: 152(1): 1-45.

[13]. Huan-Jyh Shyur, Hsu-Shih Shih. Designing a multi-issues negotiation support system based on prospect theory[J]. Information Sciences, 2015, 322: 161-173.

[14]. Fabrice Lumineau, James E. Henderson. The influence of relational experience and contractual governance on the negotiation strategy in buyer–supplier disputes[J]. Journal of Operations Management, 2012, 30(5): 382-395.

[15]. Ilana Zohar. "The Art of Negotiation" Leadership Skills Required for Negotiation in Time of Crisis[J]. Procedia - Social and Behavioral Sciences, 2015: 209: 540-548.

[16]. Gong Wang, T.N. Wong, Chunxia Yu. A computational model for multi-agent E-commerce negotiations with adaptive negotiation behaviors[J]. Journal of Computational Science, 2013, 4(3): 135-143.

[17]. Norberto Eiji Nawa, Katsunori Shimohara, Osamu Katai, On fairness and learning agents in a bargaining model with uncertainty, Cognitive Systems Research, 2002,3(4): 555-578.

[18]. Mukun Cao, Xudong Luo, Xin (Robert) Luo, Xiaopei Dai. Automated negotiation for e-commerce decision making: A goal deliberated agent architecture for multi-strategy selection[J] Decision Support Systems, 2015, 73: 1-14.

[19]. Fatima S S, Wooldridge M, Jennings N R. An agenda-based framework for multi-issue negotiation [J]. Artificial Intelligence, 2004, 152:1-45.

[20]. Rubinstein A. Perfect equilibrium in a bargaining model[J]. Econometrica, 1982, 50(1): 91-109.

[21]. Sunil Adhau, M.L. Mittal, Abhinav Mittal. A multi-agent system for decentralized multi-project scheduling with resource transfers[J]. International Journal of Production Economics, Volume 146, Issue 2, December 2013, Pages 646-661.

[22]. Shaheen S. Multi-issue negotiation with deadlines[J]. Journal of Artificial Intelligence Research, 2006, 27(1): 381-417.

[23]. Sascha Alavi, Jan Wieseke, Jan H. Guba. Saving on Discounts through Accurate Sensing – Salespeople's Estimations of Customer Price Importance and Their Effects on Negotiation Success[J]. Journal of Retailing, 2015.

[24]. Gregory E. Kersten, Rustam Vahidov, Dmitry Gimon. Concession-making in multi-attribute auctions and multi-bilateral negotiations: Theory and experiments[J]. Electronic Commerce Research and Applications, 2013, 12(3): 166-180.

The Effect of Reciprocity on Revenue Sharing Contract Coordination and its Dynamics

Ru-jing HOU [1,*], Chu-bing ZHANG [2]

[1]School of Economic and Business, Tianjin University of Technology and Education, Tianjin, China

[2] Business School, Tianjin University of Finance and Economics, Tianjin, China

(*hrj_yy99@163.com)

Abstract - **To analyze the effect of reciprocity on revenue sharing contract's coordination and its dynamic effect on agents' decisions, this paper established the extended utility function including the preference of reciprocity based on the model. Besides, controlled experiments were designed to test the above results. Unlike research results of wholesale price contract, reciprocity caused the failure of coordination. Supplier's decision did not meet the condition for coordination. Therefore, the retailer's order quantity is lower than the benchmark under centralized decision, which means the efficiency of supply chain could not be promoted. During exchange process, participants not only purse their own interests, while the relationship is also part of their goal.**

Keywords - **Reciprocity, revenue sharing contract, social preferences, supply chain coordination**

I. INTRODUCTION

Under revenue-sharing contract, a retailer pays a supplier a wholesale price for each unit purchased, plus a percentage of the revenue the retailer generates. Such contracts have become more prevalent in the videocassette rental industry [1]. By setting a relatively low wholesale price and adjusting sharing revenue from retailer, overall efficiency of supply chain members could be promoted.

Traditional economics assumes that people are self-interested, rational agents. Besides, all people are exclusively motivated by their material self-interest. However, recent developments in behavioral economics suggest that actors may care about reciprocity, fairness, and status in addition to economic benefits [2-5].

Based on revenue sharing contract, we investigated the effects of reciprocity on decision behaviors as well as contract coordination. It focused on uncertain demand, which is consistency with real situation. Firstly the theoretical model about decision-making was built to obtain research hypothesis. Unlike previous articles, we design experiments to collect data for testing. Both the motivation and results of members' decision-making as well as the evolution of preferences were analyzed to conclude the founding, which would expand theoretical research and guide practical activity.

II THEORETICAL MODEL AND HYPOTHESIS

A. The Theoretical Model of Revenue Sharing Contract

There is a dyad supply chain with one supplier and one retailer here. The retailer faces a newsvendor problem with uncertain demand *D*(random integer from 0 to 100) and an exogenous price p in the market. He must decide an order quantity q from the supplier before selling season. With revenue sharing contract, the supplier charges the retailer w for per unit purchased. But if the retailer sell out products, the supplier shares r per unit. For the supplier, there is a production cost, c. for each unit. Π_s, Π_r and Π_T separately represents the profit of supplier, retailer and whole supply chain.

From the theoretical model, revenue sharing contract could coordinate the supply chain only if formula (1) was satisfied. Thus, retailer's order quantity q^* equals the order quantity q^0 under centralized decision-making: $q^* = q^0 = 75$.

$$r = (c - w)/c \times p \qquad (1)$$

B. Extended Model and Research Hypothesis

The extended model including status considerations was built based the model from Loch[6].

This research was supported by emergency management project of Chinese National Natural Science Foundation(71540030), School Project（KJ15-22）and School Project for Talent Scholars(KYQD14045)

E. Qi et al. (eds.), *Proceedings of the 23rd International Conference on Industrial Engineering and Engineering Management 2016,* DOI 10.2991/978-94-6239-255-7_7

The retailer's utility is written as follows:

$$U_r = \Pi_r + a_r \Pi_s \quad (2)$$

Similarly, the extended utility of supplier represents as follows:

$$U_s = \Pi_s + a_s \Pi_r \quad (3)$$

Here, a_r and a_s separately represent the parameters of reciprocity for retailer and supplier (a_r<1, a_s<1). The optimal strategy of retailer was as follows by optimizing its utility function (2):

$$q^* = 100 \times \frac{p - a_r c - (1-a_r)(w+r)}{p-(1-a_r)b} \quad (4)$$

As first mover, knowing the retailer's choice about q^*, the supplier's optimal strategy would deviate from the benchmark for coordination. Thus, we can get:

Hypothesis 1: The preference of reciprocity changed the results for supply chain coordination: a) The correlation between wholesale price and sharing revenue deviated from theoretical predictions. b) The order quantity decided by retailers is significantly lower than order quantity under centralized decision-making.

Social preferences are not forward looking but backward looking. It is the past transactions that affect the current price choice through the emotions. Our participants were matched in a partner manner, such that they interacted with the same player for 20 rounds. Their other regarding parameter could change over time based on past interactions. Because other-regarding parameter subsumes relationship, this suggests the following updating function over the interaction periods indexed by t, in its simplest linear form:

$$a_{s,t+1} = a_{s,t} + \theta_{s,1} q_{r,t} \quad (5)$$

Supplier's parameter about reciprocity increases if the other side behaves positively (higher order quantity), and decreases if order quantity is lower. Analogously, retailer's reciprocity parameter represents as follows:

$$a_{r,t+1} = a_{s,t} - \mu_{r,1} w_{s,t} - \mu_{r,2} r_{s,t} - \mu_{r,3} w_{s,t+1} - \mu_{r,4} r_{s,t+1} \quad (6)$$

Thus, we can get:

Hypothesis 2: a) Sharing revenue has negative relationship with order quantity in last period. b) There is positive correlation relationship between order quantity and wholesale price as well as sharing revenue in both last period and current period.

III. RESEARCH DESIGN

A. *Experimental Design*

To verify above hypothesis, two experimental conditions were designed under revenue sharing contract. This game represents decentralized decision making including one supplier and one retailer, and the latter faced "newsboy problem". In all treatments we set the supplier's production cost to be c = 3 and the retail price to be p = 12 to create a setting with potential high supply chain profits [7-8].

The followings occurred in order: subjects as supplier firstly made decisions about wholesale price w as well as sharing revenue r. The game would end with zero profit if retailers rejected. If they accepted supplier's decision, the system would ask them to decide order quantity q and finish the game. If rejected, both screens will display information including w, r, "reject", zero profit and current demand D (integer between 1 and 100 randomly generated by system). If accepted, the extra information would include q and both profits calculated by system.

We conducted the experiment at Selten Laboratory of Nankai University. All players were students recruited through the website. Totally, there are 32 participants including 20 males and 12 females. The experimenter controlled the whole process including introduction, testing and decision-making. In the end, the software calculated their total profit over all rounds and the value multiplied 0.01 was their final payoff. We paid them by cash right after the experiment. The average payoff is about ￥22.

B. *Manipulation of Reciprocity*

In the control condition, participants were randomly matched. No one knew whom they will play with.

The second experimental condition implies a salient relationship; before the start of the game, the two players briefly meet face to face and introduce themselves. This process initiates a relationship, which is reinforced by the following written paragraph handed to each participant before the game starts: "You have already met the person with whom you will play the game. Now the person is no longer a stranger to you. You can imagine that the other player is a good friend. You have a good relationship and like each other." This relationship perception is not associated with any economic benefits and the players make their subsequent decisions separately and without further communication, as in the control condition.

Under both condition, subjects were separate and anonymous. They interacted through computers without communication. The experiment was programmed and conducted with the software z-Tree[9]。

Ⅳ. ANALYSIS AND RESULTS

A. *The Results of Coordination*

We have 8 independent observations by taking averages of each individual's 20 decisions. The comparison between groups indicated that the two order quantity was significantly different（$p<0.05$）. Thus, revenue sharing contract failed to coordinate supply chain.

To verify supplier's behavior in H1, the relationship between wholesale price and sharing revenue could be modelled as:

$$r = \beta_0 + \beta_1 \times w + \beta_2 \times R + \beta_3 \times Rw \qquad (7)$$

R is dummy capturing reciprocity conditions (it is set to one in its condition and zero otherwise). *Rw* was new variable after *R* multiplied *w* and its result represents whether the strategy for supplier was significantly different. The regression model fits well ($R^2 = 0.725$), showing it can effectively explain the 72.5% variation of sharing revenue. The effect of *Rw* on buyback was significant ($p<0.05$), indicating suppliers' decision strategy had changed due to reciprocity under different conditions.

B. *Dynamic Process of Reciprocity*

(1) dynamic process of reciprocity for supplier

A test of Hypothesis 2 requires a regression shows that supplier's current wholesale price and sharing revenue are driven by last-period's order quantity. Ordered logit analysis was used here. Since sharing revenue are mostly between two and eight, we treat prices two and below together, and prices eight and above as one choice. The generalized ordered logit model for supplier was written as follows:

$$\text{logit}[P_r(r_{s,t} > j)] = \alpha^j + \beta_1^j w_{s,t} + +\beta_2^j \times q_{t-1} + +\beta_3^j R, j = 2, \ldots, 7, t = 2,3 \ldots, 20 \qquad (8)$$

R was the same as before. The parameter β_2^j captures the influence of last-period's order quantity on the probability sharing revenue is larger than *j*. β_1^j captures the relationship between sharing revenue and wholesale price. The generalized ordered logit model allows different coefficients for different price levels j[10-11]. The result was showed in TABLE I.

TABLE I

ESTIMATED COEFFICIENTS AND REGRESSION PARAMETERS FOR SUPPLIER

variable	r	variable	r	variable	r
j=2		j=3		j=4	
$w_{s,t}$	-2.3688***(0.5684)	$w_{s,t}$	-1.6875***(0.2614)	$w_{s,t}$	-2.5387***(0.3281)
q_{t-1}	-0.0342(0.0184)	q_{t-1}	-0.0327*(0.0137)	q_{t-1}	0.0016(0.0109)
R	-1.2804(1.4613)	R	.8793(0.6457)	R	0.7368(0.6011)
Constant	17.042***(4.1215)	Constant	8.9904***(1.3944)	Constant	9.5387***(1.1860)
j=5		j=6		j=7	
$w_{s,t}$	-2.2989***(0.2823)	$w_{s,t}$	-2.9837***(0.3357)	$w_{s,t}$	-2.5140***(0.4079)
q_{t-1}	-0.0022(0.0068)	q_{t-1}	-0.0215**(0.0073)	q_{t-1}	-0.0309*(0.0121)
R	-1.0275*(0.4112)	R	-1.2469**(0.4358)	R	-1.5047*(0.6661)
Constant	6.8945***(0.8895)	Constant	6.1441***(0.8630)	Constant	3.4833***(0.9456)
Number of obs	400				
Log likelihood	-396.3478				
LR chi2(30)	615.41				
Prob>chi2	0.0000				
Pseudo R2	0.4371				

***p<0.001;**p<0.01;*p<0.05.

Sharing revenue decreases with the increasing of wholesale price, which indicates that supplier was rational. The correlation between order quantity in last period and sharing revenue was significantly positive while j=3,6 and 7. Thus, the supplier would sacrifice its own earing if order quantity was relatively high in last period. In other word, this is a kind response to other's behavior, leading more profits for both. All those support supplier's behavior of reciprocity in hypothesis 2.

(2) dynamic process of reciprocity for retailer

Analogously, the generalized ordered logit model for retailer is given by:

$$\text{logit}[P_r(q_{r,t} > j)] = \alpha^j + \beta_1^j ratio_{s,t} + \beta_2^j ratio_{s,t-1} + \beta_3^j R,\ j = 10,20,\ldots,70,\ t=2,3\ldots20 \qquad (9)$$

R was the same as before. β_1^j and β_2^j capture the influence of supplier's decision in current and last period. he result was showed in TABLE II

As second mover, the behavior of retailer is more complex than supplier. Its preference of reciprocity obviously exists during the game because regression coefficient for R is significant while j equals to 10, 30, 40, 50, 60 and 70. It is concluded that individuals care about relationship in addition to economic interests. Furthermore, reciprocity is based on mutual motives other than expectations for future earnings.

The impact of price ratio on order quantity is negative when j = 20 &70, indicating that increasing level of wholesale price is greater than the reducing for sharing revenue. Therefore, the increasing of this ratio will bring the reduction for order. However, the impact is positive when j = 30, which showed the reducing for sharing revenue is greater than increasing level of wholesale price. The retailer judges whether the behavior of supplier is friendly and vice versa. Hypothesis 2 is verified.

TABLE II
ESTIMATED COEFFICIENTS AND REGRESSION PARAMETERS FOR RETAILER

variable	q	variable	q	variable	q
j=10		j=20		j=30	
$ratio_t$	-0.1031(1.1232)	$ratio_t$	0.8586(0.5901)	$ratio_t$	0.1472(0.2332)
$ratio_{t-1}$	-0.5719(0.9563)	$ratio_{t-1}$	$-1.3096^{*}(0.6247)$	$ratio_{t-1}$	$0.7240^{*}(0.3223)$
R	$-2.1810^{*}(1.0983)$	R	-.3859(0.4347)	R	$0.7629^{*}(0.3309)$
Constant	$4.9404^{***}(1.0913)$	Constant	$2.1307^{***}(0.4553)$	Constant	0.0550(0.3020)
j=40		j=50		j=60	
$ratio_t$	-0.0709(0.1722)	$ratio_t$	$-.5632^{*}(0.2447)$	$ratio_t$	-0.8259(0.4977)
$ratio_{t-1}$	0.2939(0.2420)	$ratio_{t-1}$	.1611(0.1553)	$ratio_{t-1}$	-0.2084(0.5633)
R	$0.9671^{**}(0.3052)$	R	$.9107^{*}(0.3552)$	R	$1.3022^{*}(0.6595)$
Constant	-.0451(0.2754)	Constant	$-.9849^{**}(0.3718)$	Constant	$-2.5114^{***}(0.7176)$
j=70					
$ratio_t$	-1.4063(1.8428)				
$ratio_{t-1}$	$-3.8150^{*}(1.8710)$				
R	$-7.4148^{**}(2.2423)$				
Constant	$5.4766^{*}(2.5146)$				
Number of obs	348				
Log likelihood	-579.37479				
LR chi2(35)	149.73				
Prob>chi2	0.0000				
Pseudo R2	0.1144				

$^{***}p<0.001;^{**}p<0.01;^{*}p<0.05$

V. CONCLUSION

This paper built extended revenue sharing contract model including the preference of status seeking. Empirical data was collected according to experimental design, the main conclusions are as follows:

First, revenue sharing contract did not reach theoretical coordination because of reciprocity preference. Data analysis showed that members' decision-making motivation has changed. The supplier's strategy did not satisfy coordination condition, thus as second mover, the retailer would reduce its order.

Second, the influence of reciprocity on subjects' decision-making is dynamic. They maximized their own interests, but also care about relationship others. Furthermore, maintaining relationship would be part of their goals. Because of the extended utility functions, individuals' equilibrium behavior has changed.

REFERENCES

[1] Dana Jr J D, Spier K E. Revenue sharing and vertical control in the video rental industry[J]. The Journal of Industrial Economics, 2001, 49(3): 223-245
[2] Rabin, M. Incorporating fairness into game theory and economics[J]. Amer. Econom. Rev., 1993, 83(5): 1281–1302
[3] Camerer, C. Behavioral economics: Reunifying psychology and economics[J]. Proc. Natl. Acad. Sci.USA, 1999,96(19):10575–10577
[4] Gary E Bolton , Axel Ockenfels. A theory of equity, reciprocity, and competition[J]. The American Economic Review, 2000, 90(1): 166-193
[5] Fiske, A. P.. Socio-moral emotions motivate action to sustain relationships[J]. Self Identity,2002,(1): 169–175
[6] Loch, C.H., Wu Y. Social prederence and supply chain performance: An experimental Study[J]. Management Sci.,2008,54(11):1835-1849
[7] Elena Katok, Diana Yan Wu. Contracting in Supply Chains: A Laboratory Investigation[J]. Management Science, 2009, 55(12): 1953–1968
[8] Schweitzer, M., G. P Cachon. Decision bias in the newsvendor problem: Experimental evidence[J]. Management Science, 2000, 46(3) :404–420
[9] Urs Fischbacher. z-Tree: Zurich toolbox for ready-made economic experiments[J]. Experimental Economics, 2007, 10(2):171-178
[10] Peterson, B., F. Harrell. Partial proportional odds models for ordinal response variables[J]. Appl. Statist. 1990, 39: 205–217
[11] Richard, W. Generalized ordered logit/partial proportional odds models for ordinal dependent variables[J]. Stata J. 2006, 6(1):58–82
[12] Fehr, E., S. Gachter, G. Kirchsteiger. Reciprocity as a contract enforcement device: Experimental evidence[J]. Econometrica ,1997.64(4) :883–860

Intelligent Rescheduling System for Steelmaking and Continuous Casting Production

Xin-fu Pang[1, 2, *], Liang Gao[1], Quan-ke Pan[1], Sheng-ping Yu[3]
[1]State Key Laboratory of Digital Manufacturing Equipment Technology, Huazhong University of Science and Technology, Wuhan, China
[2]School of Automation, Shenyang Institute of Engineering, Shenyang, China
[3]State Key Laboratory of Synthetical Automation for Process Industries, Northeastern University, Shenyang, China
(*e-mail: pangxinfu@163.com)

Abstract - **There are many types of disturbances that upset the plan in steelmaking and continuous casting (SCC) production, including processing time variation, temperature variation, quality variation, machine failures, which could cause the static schedule to become inefficient and even infeasible. It is necessary to adjust the schedule or generate a new executable schedule upon the occurrence of unanticipated disruptions and changes. Two rescheduling methods are presented: the partial rescheduling and complete rescheduling. The former refers to the right shift partial rescheduling based on continuous casting, the partial rescheduling based on case-based reasoning and man-computer interaction. The latter refers to the complete rescheduling considering changeable processing time, the complete rescheduling considering changeable processing time and production path. The mathematical models of two class complete rescheduling are formulated, and the solving methods are also given. The intelligent rescheduling system for SCC production with these methods is successfully applied to one large steel plant, and rapidly optimal rescheduling is achieved when disturbances occur.**

Keywords – **Complete rescheduling, intelligent rescheduling system, rescheduling strategy, steelmaking and continuous casting, the partial rescheduling,**

I. INTRODUCTION

The steelmaking and continuous casting process is the major bottleneck in iron and steel production [1], including steelmaking, refining and continuous casting, is a flexible manufacturing process with multiple parallel machines in each stage. Most modern iron and steel companies apply a computer-integrated manufacturing system (CIMS) to improve the productivity, reduce production costs and enable efficient material and energy utilization [2]. There are high expectations of molten steel time, temperature and quality ingredients, and the production capability is sufficiently exerted on the basis of continuous production. Rescheduling of SCC process is the process of updating an existing production schedule on the basis of practical production status when the parameter and schedule data change, which ensure continuity and stability of SCC schedule under satisfying production constraint.

Although there is an extensive research on SCC static scheduling [1, 3-10], there are limited studies on SCC rescheduling [11-14]. The utility and stability [11] that revises schedule or rescheduling against disturbance are presented. The knowledge model consisted of tasks, inference and domain [13] is presented to managing schedule disturbance in steel-making.

The rescheduling methods, including the right shift partial rescheduling method based on casting optimized, the partial repair rescheduling methods that combine CBR and man-computer interaction, complete rescheduling considering changeable processing time, complete rescheduling considering changeable processing time and production path, are presented to update (repair) the plan of scheduling after the disturbances of production time delays, temperature changes or quality ingredients viration, caster failure occurring in practical production process. The two rescheduling models of complete repair is formulated and the solving methods are given. Finally the rescheduling subsystem with the rescheduling methods is successfully applied to one large steel plant, and realizes rapidly optimal rescheduling.

II. DESCRIPTION OF SCC RESCHEDULING

The SCC process consists of three major stages: steelmaking, refining and continuous casting. In the steelmaking stage, some substances such as carbon and silicon in the molten iron are reduced to desirable levels after burning off in a converter (LD). The concurrent smelting in the same furnace is called a charge. After steelmaking, a charge is poured into a ladle and is transported to a refining furnace. In the refining stage, the molten steel is to be further refined to get desired grade or chemistry by the high-precision adjustments of the chemical composition of the steel. The refining stage includes three process steps that are Rushrstal Heraeus (RH) process for vacuum degassing, Ladle Furnace (LF) process for desulfurization, and Injection Refining-up Temperature (IR_UT) process for heating. Due to the technique constraints, some charges are refined from RH to LF, while others from RH to IR_UT or from LF to IR_UT. After refining, a charge is taken to a continuous caster, which can process multiple charges consecutively in the form of a cast (batch). All charges in a cast share the same crystallizer. There exists setup time between two consecutive casts on the same caster since the crystallizer needs to be changed. In particular, all charges in the same cast have the same chemical composition and undergo the same refining route. The refining routes of the charges in different casts may be different. Furthermore, according to the predefined processing sequence, all charges in a cast must be processed consecutively on a specified

E. Qi et al. (eds.), *Proceedings of the 23rd International Conference on Industrial Engineering and Engineering Management 2016,* DOI 10.2991/978-94-6239-255-7_8

machine in the last stage. For convenience, the charge is called job and the cast is called batch in our study.

The SCC production scheduling process is a production scheduling problem with multi-mode, multi-working-procedure, multi-device, multi-disturbance, the multi-objective, and has high require of molten steel time, temperature and quality ingredients. On the basis of continuous casting, scheduling task is to determine in what sequence, at what time and on which machine molten steel should be arranged to enter various production stages from steelmaking to continuous casting and is to achieve some one optimal evaluating function (minimum waiting time, maximum completion time: Makespan, et al). The produce process of steelmaking and continuous casting is shown Fig.1, and the refining stage is from 1 to 4 stages.

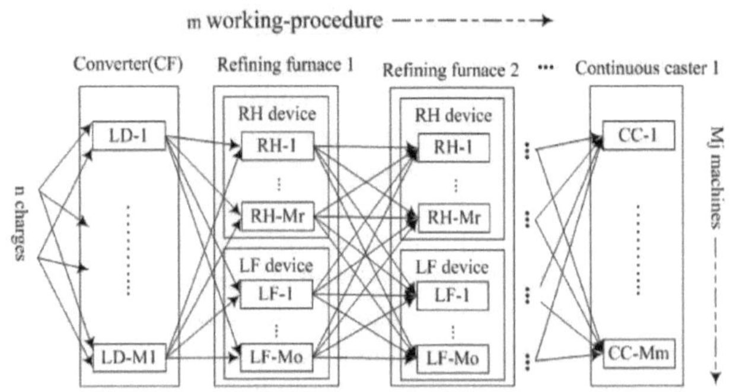

Fig. 1. The produce process of steelmaking and continuous casting

In SCC process there are many types of disturbances, and the production scheduling of SCC is a predictive-reactive scheduling problem. Practical production time severely departures from the predictive scheduling, adding the refining working procedure or prolonging the refining processing time because of lower temperature, changing steel grade because of quality ingredients not accord with criterion, which bring production time and production path changing, and make the plan of scheduling infeasible. The rescheduling is the process of updating an existing production scheduling plan in repose to production time, temperature changes or quality ingredients, and which ensures continuity and stability of SCC schedule.

III. THE RESCHEDULING METHODS

The optimal scheduling strategy framework is presented in paper [15], the rescheduling methods is a part of the scheduling strategy framework and is applied to the dynamic scheduling subsystem. The rescheduling problem is described by rescheduling environment, rescheduling strategy and rescheduling method [16]. Because in this paper the number of rescheduling charge is certain and the practical production time, temperature and quality are uncertain, then the rescheduling environment is a stochastic, static rescheduling problem[16-19]. Because the purpose of rescheduling is for updating existing production schedule in response to disruption or other changes, the rescheduling strategy is event-driven predictive-reactive schedule. Several category of partial rescheduling and complete rescheduling methods are engaged and are shown in Fig.2.

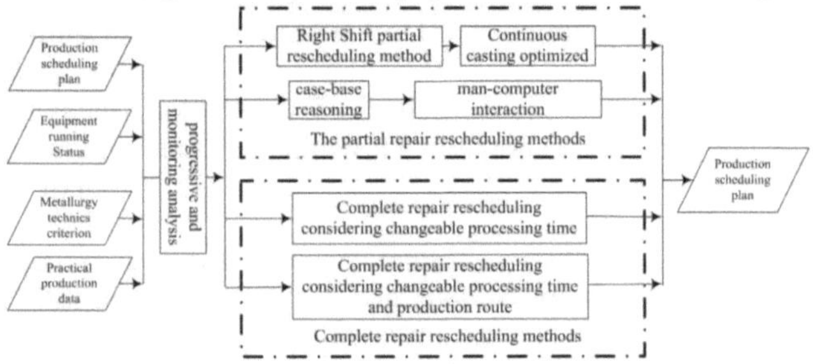

Fig.2 Rescheduling methods of intelligent scheduling system for SCC

A. Right Shift partial rescheduling method based on casting optimized

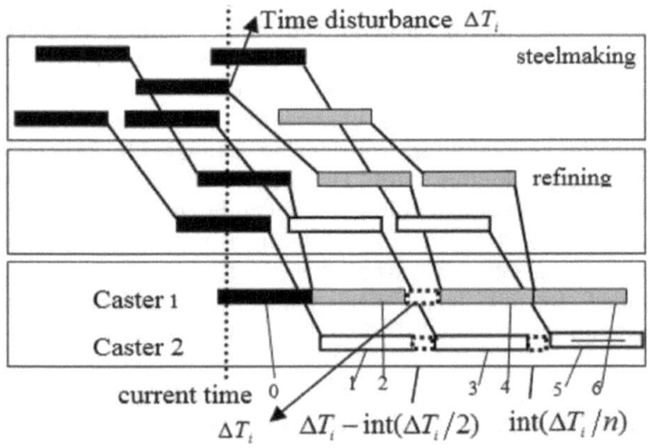

Fig.3 Right Shift partial rescheduling method based on casting optimized

In Fig.3. the time disturbance ΔT_i ($5 \le \Delta T_i \le 10$, if $\Delta T_i < 0$, then left shift) occurs in the ending time point of the smelting, and is within a certain scope (the definition of time warp scope depends on the specific problems in a plant, the time warp is 5-10 minutes in the paper). The scheduling system automatically shifts production units that charge 4 and later charges of caster 1, charge 5 and later charges of caster 2 to the right with ΔT_i on Gantt chart. The continuous casting optimized : the set $\{0,2,4,6\}$ denotes the charges on caster 1, foregoing charge of charge 4 is charge 2 that can not start casting, and the casting ending time is postponed with ΔT_i . The set $\{1,3,5\}$ denotes the producing charges on caster 2, the latest casting charge is charge 5, foregoing charge of charge 5 are charge 3 and charge1 that can not start casting, and the casting ending time of charge 3 is postponed with $\text{int}\left(\Delta T_i / n\right)$. If many charges are foregoing charges of charge 5 that can not start casting, the casing ending time of every charge is postponed with

$\text{int}\left(\left(\Delta T_i - \text{int}\left(\Delta T_i / n\right)\right) / (n-1)\right)$, and the continuous casting optimized is a iterative process. Because $n = 2$, the casting ending time of charge 1 is postponed with $\Delta T_i - \text{int}\left(\Delta T_i / 2\right)$.

B. The partial repair rescheduling methods that combine case-base reasoning and man-computer interaction

There are time disturb adjustment, temperature compensation adjustment, abnormal casting adjustment and steel grade changing adjustment. The adjustment strategy is reasoned by CBR and the adjustment methods is selected to repair predictive schedule plan by scheduler on man-computer interaction platform, which make the effect of disturbing schedule plan minimum after disturbance occurring.

The partial repair rescheduling methods that combine CBR and man-computer interaction [15, 20] are efficient for eliminating effect of disturbing schedule plan, and which prevents the computing time and stability that frequently use complete rescheduling and prevents frequent changing of predictive production path (device) and time. However, complete rescheduling methods are used to overall optimal scheduling problem, which is finitely used makes the effect of rescheduling optimal, thus there is used by combining the partial repair rescheduling methods and complete rescheduling methods.

C. Complete repair rescheduling methods

In SCC process, there are three instances that need engage complete rescheduling methods. (1) The large time disturbance induces large confliction of later charge. (2) The production path is changed for postpone refining time or append refining stage because lower temperature. (3) The production time and path is changed for steel grade disturbance because quality ingredient is failure. On three instances the partial repair rescheduling methods can not satisfy the require of optimal objective, the complete repair rescheduling methods are needed. There are two class of complete repair rescheduling methods: one is the complete rescheduling considering changeable processing time (CR-CT), and the other one is the complete rescheduling considering changeable processing time and production path (CR-CTP).

(1) Complete repair rescheduling considering changeable processing time

Considering changeable processing time and unchangeable production path, sequence on the device of the production schedule, CR-CT method is complete optimal rescheduling, and its optimal objects are which makes time of breaking off continuous casting and waiting time minimum. That minimizes time of breaking off continuous casting (BOCC) means that the charge/jobs of a CAST in a continuous caster must continuously cast, otherwise production cost is increased and continuous caster is scathed. That minimizes waiting time means that the long waiting time makes temperature of liquid (molten steel) descending and quality of liquid changing, and badly affects not only the production time of the charge but also the production time of other charges plan.

This problem is multi-constraint and multi-objective, and constraints and objective function are linear. Lexicographic Order is engaged to deal with precedence of multi-objective linear programming, the fist stage objective is minimum time of BOCC and the second stage objective is minimum waiting time. The variable x($x > 0$) is the beginning time of production unit, and current time denotes as zero. The constant c is the beginning time of producing unit and produced unit, which ensures the correct constraint relation. Because the variables of continuous casting stage compose of the first stage objective function, and are set as constant to compose of the second stage objective after solving the first stage objective, which ensures solution of the fist stage objective not destroyed.

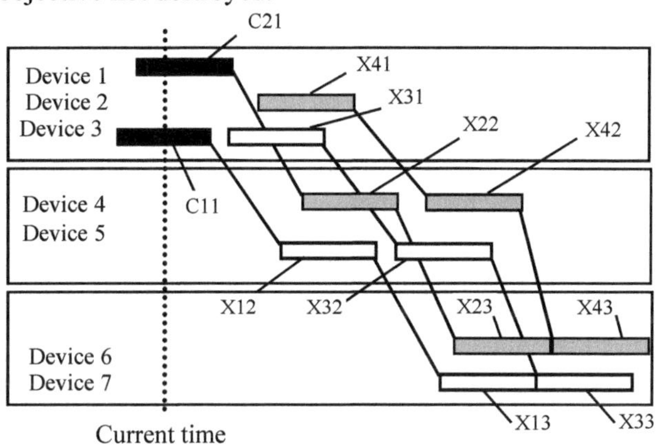

Fig.4 The definition of variable and constant

The variables and constants of the first stage optimal objective (minimum time of BOCC) is in Fig. 4, and the model can be defined as follows:

$$f_1(x) = \min\left\{\sum_{i=1}^{n} x_{i,m}\right\} \tag{1}$$

S.T.

$$x_{i,j'} - x_{i,j} \geq d\left(x_{i,j}\right) + t\left(x_{i,j}, x_{i,j'}\right), j' = j+1 \tag{2}$$

$$x_{i,j} = c_{i,j}, j \leq jc(i) < m \tag{3}$$

$$x_{i,j,k} - x_{SI(i,J,k),J,k} \geq d\left(x_{SI(i,J,k),J,k}\right), k = 1,.., K \tag{4}$$

$$x_{i,j,k} = \begin{cases} c_{i,j}, j \leq jc(i) \\ x_{i,j}, j > jc(i) \end{cases} \tag{5}$$

$$x_{i,j,k} \geq 0, 1 \leq i \leq n, 1 \leq j \leq m \tag{6}$$

Constraint (2) ensures that for contiguous operation of the same charge, only when the preceding operation is completed and the charge is transported to the next machine, the immediately next operation on the charge can be started. Constraint (3) represents that producing unit and produced unit of the charge i are set as constant. Constraint (4) ensures that for two contiguous charges processed on the same machine, only when the preceding charge is finished, the immediately next one can be started. Constraint (5) represents that the value of the charge i on the machine k relies on unit j, the value of producing unit or produced unit is variable $x_{i,j}$, and the value of production unit is constant. The n is the total number of charges. The m($3 \le m \le 6$) is the stage number of a charge. The i($1 \le i \le n$) is the number of charge. The j, j', J ($1 \le j, j', J \le m$, $j' = j+1$) is the number of the stage. The k is the number of machine, K is the total number. The $SI(i,J,k)$ is the immediate foregoing charge of charge i on machine k. The $jc(i)$ ($jc(i) < m$) is the stage number of producing unit or immediate produced unit on charge i. The $x_{i,j}$ is the variable of model, and denotes the beginning time of charge i on the j stage. The $c_{i,j}$ ($j \le jc(i)$) is the constant of model, and denotes the beginning time of producing unit and produced unit on charge i. The $d(x_{i,j})$ denotes the processing time of charge i on the machine of stage j. The $t(x_{i,j}, x_{i,j'})$ denotes transportation time from the machine of stage j to the machine of stage j'.

After the values of producing unit and produced unit are set as constant, each stage objective is a linear programming (LP) model of single objective, and is solved by standard LP. After the first stage objective is solved, $x_{i,m}$ is set as constant $c_{i,m}$ and is saved. The model of minimum waiting time can be defined as follows:

$$f_2(x) = \max\left\{\sum_{j=1}^{m-1}\sum_{i=1}^{n} x_{i,j}\right\} \tag{7}$$

S.T.

$$x_{i,j'} - x_{i,j} \ge d(x_{i,j}) + t(x_{i,j}, x_{i,j'}), j' = j+1 \tag{8}$$

$$x_{i,j} = \begin{cases} c_{i,j}, j \le jc(i) < m \\ c_{i,m}, j = m \end{cases} \tag{9}$$

$$x_{i,j,k} - x_{SI(i,J,k),J,k} \ge d\left(x_{SI(i,J,k),J,k}\right), k = 1,..,(K - K_m) \tag{10}$$

$$x_{i,j,k} = \begin{cases} c_{i,j}, j \le jc(i) \\ x_{i,j}, j > jc(i) \end{cases}, \; k = 1,..,(K - K_m) \tag{11}$$

$$x_{i,j,k} \ge 0, 1 \le i \le n, 1 \le j \le m-1 \tag{12}$$

Because the value of continuous casting stage is set as constant, the objective is maximizing $\sum x_{i,j}, (j \ne m)$ ($x_{i,j}$ moving to the right on a Gantt chart). Constraint (8) is same to constraint (2), but not including the continuous casting stage. Constraint (9) is same to constraint (3), and the value of continuous casting stage is set as constant $c_{i,m}$. Constraint (10) is same to constraint (4), not including machines of the continuous casting stage. Constraint (11) is same to constraint (5), not including machines of the continuous casting stage. The K_m is number of machine of the last stage (continuous casting stage).

After solving the model of the second stage, the starting times of production unit on other stage except for continuous casting stage are obtained and CR-CT problem is solved. The value of producing unit and produced unit is set as constant, which is an important technique and makes the model solved by standard LP considering stochastic time parameter.

(2) Complete repair rescheduling considering changeable processing time and production path

When starting time or finishing time is largely delayed, it can lead to unexpected changes so that the original scheduling plan becomes unrealizable, and the problem of BOCC may appear by CR-CT method. For solving the problem, CR-CTP method [21] is presented by considering stability and utility, which does not change assigned machines of processing charge. A decomposition strategy is proposed, and the original problem is split into subproblems. This is then followed by the establishment of repairing time model that repairs starting time of the processing charges by introducing production schedule parameter β. The establishment of a multi-objective nonlinear 0-1 integer programming model is used to solve the scheduling problem of not starting production charges, and a two-stage optimal scheduling method is proposed including the production path planning of charges and the production time scheduling.

When converter or refining furnace is breaked, the rescheduling method with breakdown of converter or refining furnace [22] is presented considering changeable processing time and production path.

IV. APPLICATION IN ENGINEERING

The steel factory in Shanghai, three BOFs can contain 260 tons, 6 refining equipments with three attributes (3 RH, 2 LF, 1 IR_UT), 3 caster equipments, The number of refining processes are ranging from a 0 to 4, there are variety of products, 1000 kinds steel. The computer systems have the level 2 computer system of process control and the level 3 computer system of region management. The actual scheduling system of SCC only realizes information management, and there are not optimal rescheduling according to disturbance category and degree.

The optimal rescheduling methods presented in the paper have been applied to a steel plant in Shanghai. The dynamic scheduling subsystem software [15], including progress and monitoring module, time category adjustment, temperature category adjustment, casting abnormal type adjustment, Changed Steel category adjustment, the overall situation rescheduling, is developed with VC, Delphi, Visual Prolog language and database technology. The complete repair rescheduling conditioned by changeable processing time is shown in

Figure 6 and Figure 7. The complete repair rescheduling conditioned by changeable processing time and production path is shown in Figure 8 and Figure 9. The computing time of solving rescheduling is within 25s (scheduling system obtains the optimal result for one day for the largest output 80 stoves within 30 seconds), and which may satisfy the real time requirement.

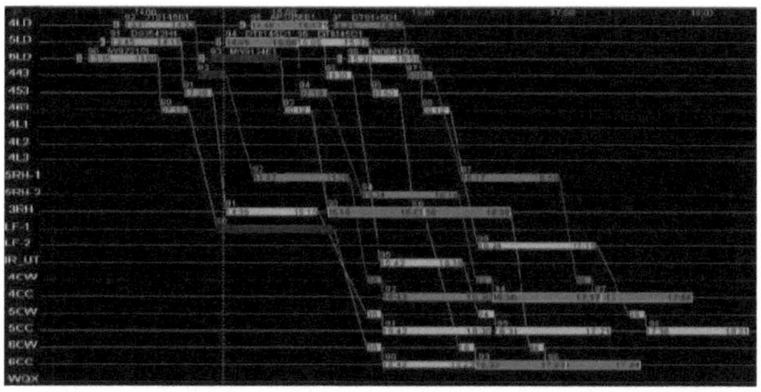

Fig. 6. Schedule plan before rescheduling considering changeable processing time

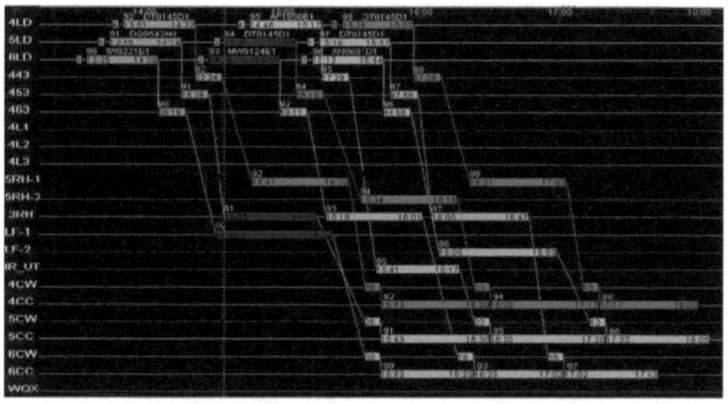

Fig.7. The result of rescheduling considering changeable processing time

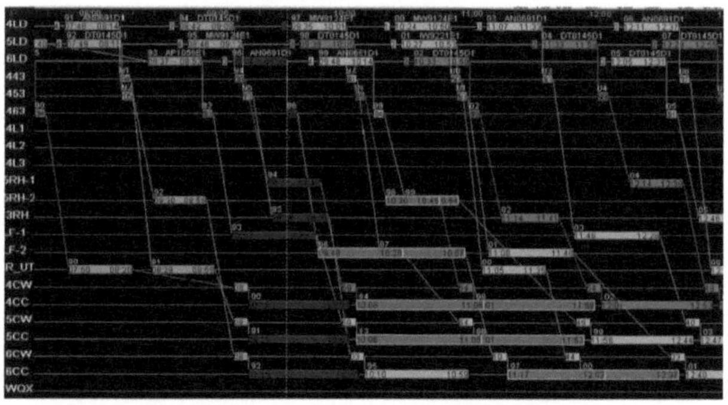

Fig. 8. Schedule plan before rescheduling considering changeable processing time and production path

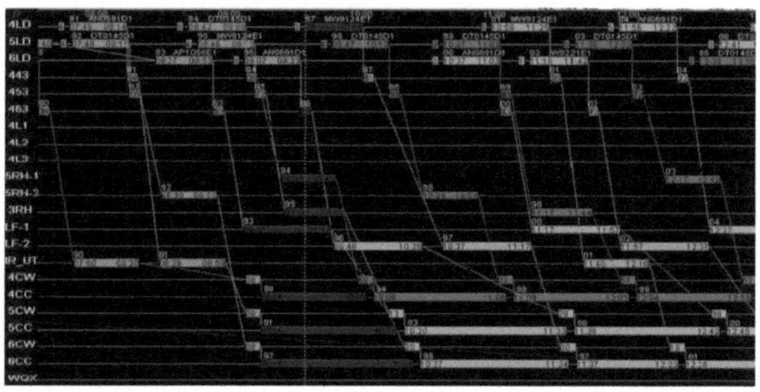

Fig. 9. The result of rescheduling considering changeable processing time and production path

V. CONCLUSION

Right Shift partial rescheduling method based on casting optimized, the partial repair rescheduling methods that combine CBR and man-computer interaction, CR-CT method and CR-CTP method are designed for disturbance adjustment. The dynamic scheduling system based rescheduling methods is applied to a steel factory. The computing time of rescheduling is with 25s, and the system is operating well online, has friendly interface, is suitable for dispatchers. Its design thought, rescheduling methods of solving problems can apply to scheduling problems of other steel plants. The system is easy to popularize.

ACKNOWLEDGMENT

This research is partially supported by National Science Foundation of China (51435009, 51575212, 61573249, 61573178), Postdoctoral Science Foundation of China (2015T80798, 2014M552040, 2014M561250), and Liaoning Province Education Project (L2015372).

REFERENCES

[1] VIKAS KUMAR, SHASHI KUMAR, M.K.TIWARI, "Aucton-based approach to resolve the scheduling problem in the steel making process", *International Journal of Production Research*, vol.44, no.8, pp.1503-1522, 2006.

[2] Hubert Missbauer, Wolfgang Hauber, Werner Stadler, "A scheduling system for the steelmaking-continuous casting process. A case study from the steel-making industry", *International Journal of Production Research*, vol.47, no.15, pp.4147-4172, 2009.

[3] Dario Pacciarelli, Marco Pranzo, "Prodution scheduling in a steelmaking-continuous casting plant", *Computers and Chemical Engineering*, vol.28, no.12, pp.2823-2835, 2004.

[4] A. Bellabdaoui, J. Teghem, "A mixed-integer linear programming model for the continuous casting planning", *International Journal of Production Economics*, vol.104, no.2, pp.260-270, 2006.

[5] Arezoo Atighehchian, Mehdi Bijari, Hamed Tarkesh, "A novel hybrid algorithm for scheduling steelmaking continuous casting production", *Computers & Operations Research*, vo.36, no.8, pp.2450-2461, 2009.

[6] Pan Quanke, "An effective co-evolutionary artificial bee colony algorithm for steelmaking-continuous casting scheduling", *European Journal of Operational Research*, vol.250, no.3, pp.702-714, 2016.

[7] TAN Yuanyuan, HANG Yinglei, LIU Shixin, "Two-stage mathematical programming approach for steelmaking process scheduling under variable electricity price", *International Journal of Iron and Steel Research*, 2013, vol.27, no.7, pp.1-8, 2013.

[8] Ye Yun, Li Jie, Li Zukui, Tang Qiuhua, Xao Xin, Christodoulos A. Floudas, "Robust optimization and stochastic programming approaches for medium-term production scheduling of a large-scale steelmaking

continuous casting process under demand uncertainty", *Computer and Chemical Engineering*, vol.66, no.10, pp.165-185, 2014.

[9] Mao Kun, Pan Quanke, Pang Xinfu, Chai Tianyou, "A novel Lagrangian relaxation approach for the hybrid flowshop scheduling problem in a steelmaking-continuous casting process", *European Journal of Operational Research*, vol.236, no.1, pp.51-60, 2014.

[10] Hao Jinghua, Liu Min, Jiang Shenglong, Wu Cheng, "A soft-decision based two-layered scheduling approach for uncertain steelmaking-continuous casting process", *European Journal of Operational Research*, vol244, no.3, pp.966-979, 2015.

[11] Peter Cowling, Marcus Johansson, "Using real time information for effective dynamic scheduling", *European Journal of Operational Research*, vol.139, no.2, pp.230-244, 2002.

[12] Peter Cowling, Djamila Ouelhadj, Sanja Petrovic, "A multi-agent architecture for dynamic scheduling of steel hot rolling", *Journal of Intelligent Manufacturing*, vol.14, no.5, pp.457-470, 2003.

[13] R.ROY, B.A.ADESOLA, S.THORNTON, "Development of a knowledge model for managing schedule disturbance in steel-making", *International Journal of Production Research*, vol.42, no.18, pp.3975–3994, 2004.

[14] Tang Lixin, Zhao Yue, Liu Jiyin, "An improved differential evolution algorithm for practical dynamic scheduling in steelmaking-continous casting production", *IEEE Transactions on Evolutionary Computation*, vol.18, no.2, pp.209-225, 2014.

[15] Pang Xinfu, Zheng Binlin, Chai Tianyou, "Steel-casting dynamic scheduling subsystem and its application", *Metallurgical industry automation*, vol.30, pp.8-12, 2006. (Chinese)

[16] VIEIRA G E, JEFFREY W H, LIN E, "Rescheduling manufacturing system: a framework of strategies, policies, and methods", *Journal of scheduling*, vol.6, no.1, pp.39-62, 2003.

[17] I. Sabuncuoglu, O.M. Kizilisik, "Reactive scheduling in a dynamic and stochastic FMS environment", *International Journal of Production Research*, vol.41, no.17, pp.4211-4231, 2003.

[18] Haldun Aytug, Mark A. Lawley, Kenneth McKay, "Executing production schedules in the face of uncertainties: a review and some future directions", *European Journal of Operational Research*, vol.161, no.1, pp.86-110, 2005.

[19] Andras Pfeiffer, Botond Kadar, Laszlo Monostori, "Stability-oriented evaluation of rescheduling strategies, by using simulation", *Computers in Industry*, vol.58, no.7, pp.630-643, 2007.

[20] Pang Xinfu, Zheng Binlin, Chai Tianyou, "Dynamic scheduling method and its application for steelmaking and continuous casting", *Journal of petrochemical universities*, vol.20, no.3, pp.60-63, 2007. (Chinese)

[21] Pang Xinfu, Yu Shengping, Luo Xiaochuan, Pan Quanke, Chai Tianyou, "Hybrid job shop rescheduling method and its application for steelmaking-casting", *Systems Engineering-Theory & Practice*, vol.32, no.4, pp.826-838, 2012. (Chinese)

[22] Pang Xinfu, Gao Liang, Pan Quanke, Yu Shengping, "Rescheduling method of steelmaking and continuous casting with breakdown of converter or refining furnace", *Conrol and Decision*, vol.30, no.11, pp.11921-1929, 2015. (Chinese)

Application of RCA to the Data Analysis in Medication Errors of the TPR System

Yu-chun LO [1,*], M. C. HSIEH [1], E. M. - y. WANG [1], Y. H. FANG [2],
Y. T. HU [2], W. C. KUNG [2], M. H. HUANG [2]
[1]Department of Industrial Engineering and Engineering Management, National Tsing Hua University, Hsinchu, Taiwan
[2] Hsinchu Mackay Memorial Hospital, Hsinchu, Taiwan
(*yuchun_46@hotmail.com)

Abstract - **The purpose of this study is to evaluate the performance of data analysis in Taiwan Patient-safety Reporting system (TPR). The TPR system is evaluated by comparing the results of the Root Cause Analysis (RCA) and the collecting data from TPR system. We used 18 medication errors and recruited 30 nurses to report those errors to the TPR system. With systematical approach, Human Factors Analysis and Classification System model (HFACS) were applied to analyze the results. The study showed that the sum of the causal factors and the number of causal factors in all layers in HFACS model from the TPR system are less than the results from the RCA. It implied that the TPR system cannot offer useful and effective information for causal analysis and would need some further analysis and improvement.**

Keywords - **HFACS model, RCA, TPR system**

I. INTRODUCTION

Recently, patient safety has become a crucial issue. In the United States, almost ten thousand people died from the preventable medical errors annually. The cost from those medical error would be estimated around 9 billion dollar every year [1]. Avoiding the medical error can not only improve the medical quality but also reduce the cost [2]. In order to prevent these errors, the need for reporting system is quite urgent [3]. Some medical errors reporting systems have been established such as MedMARx and The Medication Errors Reporting Program (MERP) in the United States, The National Reporting and Learning System (NRLS) in the UK [4]. In Taiwan, the Taiwan Patient-safety Reporting System (TPR) has been set up in 2006. The purpose of the system is to encourage all the medical institutions to report the medical errors honestly and voluntarily without any punishment. By doing so, all other institutions can learn from those sharing experience.

When it comes to error analysis, the Swiss Cheese Model would be mentioned frequently. The model explains how the failure happened in the system and changes the flaw concept from the "character flaw" to the "system flaw". The Swiss cheese is used as a metaphor. The defenses against the failure in the system are represented as slices of cheese. The holes on the cheese would be the weakness in the system. The loss happens only when the hazard passes through all the holes in every slice [5] [6]. However, the defenses and weakness were not defined clearly in the Swiss Cheese Model. To deal with this shortage, more than 300 cases of aviation safety were analyzed and Human Factors Analysis and Classification System (HFACS) model was established. The defenses in the system were defined and categorized systematically in HFACS model. The model includes 4 levels which are Unsafe Acts, Precondition for Unsafe Acts, Unsafe Supervision and Organizational Influence. It folds the idea of Swiss Cheese model into application and defines 19 causal categories within 4 levels [7].

One practical method to analyze failure in medical field is Root Cause Analysis (RCA). It is a concept which the ultimate purpose is to figure out the root cause of the failure [8] [9]. There are some specific ways to conduct RCA such as tabular timeline, fish bone, causal factor analysis, barrier analysis [10]. The idea of causal factor analysis would be used for evaluate the TPR system in our research. The definition of the causal factors is the event or the condition that produces or contributes to the occurrence of the accident. It includes three kinds of causal factors as following: direct causes, contributing causes, and root causes [11].

In our research, we would like to evaluate the utility of the data analysis of the TPR system. Although the system collected more and more data in recent years, the data are not useful for causal analysis. The possible causes which the TPR system concluded on the annual reports seem like description of the failure or the superficial causes of the adverse events [12] [13].

The medication errors would be used as materials because it is one kind of medical errors with higher complexity and relates to more different medical staffs [14] [15]. We developed specific RCA method according to our needs. The RCA we used applied the idea of HFACS model and causal factor analysis. We conducted the RCA to the medication errors and used the results of the RCA as the standard. On the other hand, the participants would be recruited to report those same medication errors on the TPR system. The causal analysis results would be compared between RCA and TPR system reporting. By using this design, we could achieve the purpose of the research.

II. METHODOLOGY

There are two approaches included in the methodology. One approach is RCA; another is TPR system reporting.

E. Qi et al. (eds.), *Proceedings of the 23rd International Conference on Industrial Engineering and Engineering Management 2016,* DOI 10.2991/978-94-6239-255-7_9

A. Participants

This study was cooperated with the regional hospital in Taiwan. All the participants are the medical personnel from the cooperated hospital. Thirty nurses (30 females) with more than one-year working experience were recruited. They conducted the TPR system reporting. Four experts were recruited in an interdisciplinary analysis team performing the RCA of 18 medication errors. The member of the expert-group includes 2 nurses, 1 pharmacist and 1 human error analyst. All the medical staffs in the expert-group would have at least 10-year working experience in the cooperated hospital.

B. Materials and apparatus

Eighteen medication errors were used for these two analyzing approaches in the study. The public information on the TPR system would be our reference source of those 18 medication errors. All materials would be modified in order to simulate the real condition in the cooperated hospital. Microsoft excel visual basic application were be used to collect the data in TPR system reporting.

C. Procedure

In RCA approach, the expert-group performed the data reconstruction and RCA. For every medication error, the researcher offered the draft of the causal factor charting so that all the experts could quick overview the medication error in a minute. The experts needed to reconstruct the content of the medication error by answering some standard questions. The meeting would be videoed and taped. After the meeting, the researcher would finish the data reconstruction based on the content of the group discussion. The main purpose of the data reconstruction is to make those medication errors seemed more realistic and could be used for RCA analyzing and TPR system reporting. The procedure for performing the RCA applied the concept of HFACS model. In order to conduct the RCA of 18 medication errors, the experts verified the consequence of the data reconstruction at the first stage and identified the causal factors (CFs) from the causal factor charting. The standard procedures of verifying CFs would be followed [11] (see Table Ⅰ). Those verified CFs may be seen as CFs belonged to the Unsafe Acts in the HFACS model. In order to figure out the root causes for the medication errors, those CFs in level 1 would be analyzed and tried to find out the deeper CFs level by level in HFACS model (see Figure 1).

In TPR system reporting approach, every nurse reported 9 medication errors which were selected in random. They reported all the information based on the content of the cases. The content of the TPR system would be seen in Table Ⅱ. Part B-Content of Event would be the main part to focus on. This part contains the main information about the medication error such as: *the happening stage*, *the possible causes*, or *the related drug*, etc.

TABLE I
STANDARD PROCEDURES OF VARIFYING CFs

Procedure	Questions
1	If this event had not occurred, would the accident have occurred?" If the answer is, "The accident would have occurred whether this event happened or not" then the event is not significant.
2	If the answer to the evaluation question is, "The accident would not have occurred without this event," then determine whether the event represented normal activities with the expected consequences.
3	If the event was intended and had the expected outcomes, then it is not significant. If the event deviated from what was intended or had unwanted consequences, then it is a significant event.

The objectives which we used for comparing the results of these 2 approaches are as following: (1) the sum of the causal factors (2) the number of the causal factors from every level in HFACS model. The logic of establishing these two objectives is that the more causal factors to identify, the higher probability to figure out the root cause. The results from the RCA approach were seen as a standard because it contained much discussion on every possible aspect for every case. In general, it spent around 40 mins to 1 hour for every case.

In RCA approach, we calculated the sum of causal factors and the number of the causal factors from every level in HFACS model; in TPR system reporting approach, we categorized the *possible causes* offered from the TPR system into the HFACS model. Based on this categorization, we could calculate the sum of causal factors and the number of the causal factors from each level in HFACS model.

In our hypothesis, the original TPR system cannot suggest enough causal factors correctly. Based on this, the sum of CFs and the number of CFs in every level obtained from the TPR system would be less than the results from the RCA.

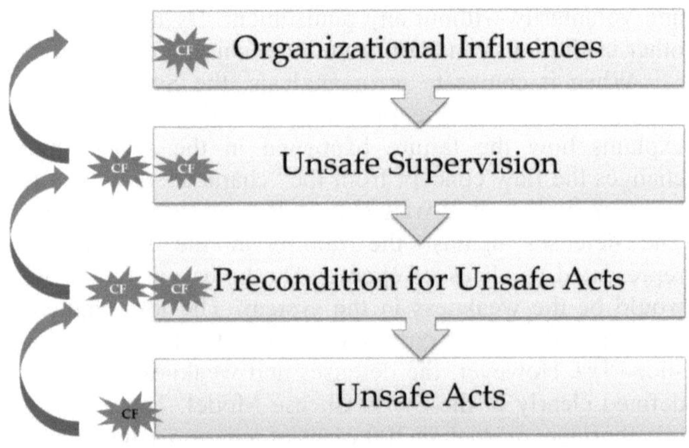

Fig. 1. An example of steps for conducting RCA

TABLE II
THE CONTENT OF THE TPR SYSTEM

Part	Subtitle
A	Basic Information (Location, Related People)
B	Content of Event
C	Immediate Actions
D	Precautions
E	Reoccurring Probability

III. RESULTS

The sum of CFs and the number of the CFs from every level in HFACS model by RCA would be calculated during the expert meeting.

In contrast, in order to calculate the number of CFs in TPR system reporting approach, we categorized all the *possible causes* offered by the TPR system into HFACS model. Because the participant was asked to report 9 cases in random, all the cases might obtain around 15 reporting records. The mean of the CFs for objective 1 and objective 2 would be counted.

To compare the utility for data analysis from the RCA and the TPR system reporting, a series of paired-sample *t* test would be used. Paired-sample *t* test is used because the difference and complexity is large from case to case. For objective 1, the sum of the CFs from the RCA is higher than the TPR system reporting; for objective 2, there was a similar result. The number of CFs from every level in RCA is more than which in TPR system reporting (see Table III and Figure 2).

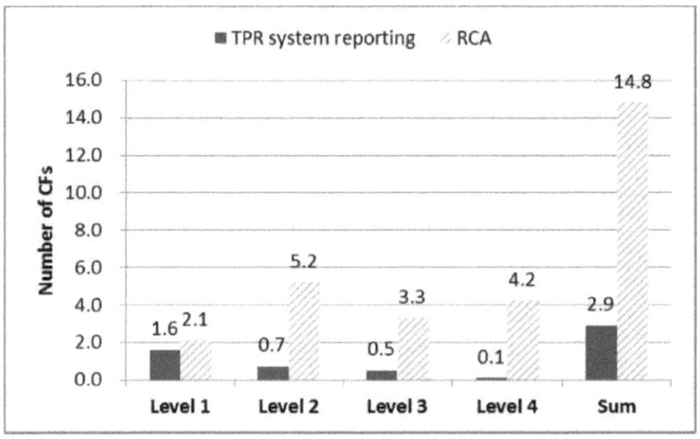

Fig. 2. The mean difference of the number of CFs between the RCA and TPR system reporting

TABLE III
RESULTS OF PAIRED-SAMPLE *t* TEST

Category	Paired Difference		*t*	*df*	*p*
	Mean	*Std*			
Level 1	-0.51	0.97	-2.24	17	0.039*
Level 2	-4.42	1.23	-15.23	17	0.000***
Level 3	-2.87	1.60	-7.61	17	0.000***
Level 4	-4.08	1.47	-11.82	17	0.000***
Sum	-11.89	4.10	-12.29	17	0.000***

$*p < .05$, $**p < .01$, $***p < .001$

IV. DISCUSSION

According to the results, the original TPR system cannot figure out enough causal factors as much as the RCA. The results suggested that the significant difference is small only in level 1, except that, the significant difference is quite large in all other levels and sum. This implies that the original TPR system obtains the causal factors which are more superficial and much closed to the active failure, which means that the number of causal factor extracted from the original TPR system is insufficient on the whole.

The difficulty that the TPR system hardly identifies the deep causal factors might result from the frustration for the most reporters. They might be struggled to identify the deeper causes without any considerations, especially for reporters working in primary care. Even the TPR system offered the deeper causal factors for the reporter to choose, it may still be tough for the reporters to answer the complete causal factors.

V. CONCLUSION

The results verified our hypothesis. The method which we used folding the idea of HFACS model and causal factors analysis in RCA. We combined with quantitative approach into the qualitative study and used objectives to evaluate the utility of the data analysis in TPR system. However, there are some limitations. First, because of the difficulties on obtaining the medication errors, the number of cases was only 18. Besides, it was only pay attention to the medication errors. It was insufficient to develop a more general conclusion. Moreover, the HFACS model was not suitable in all aspects when discussing about the medication errors. The HFACS model was defined based on aviation safety cases. It is well-used in aviation, but when using in medical field, there are some shortcomings such as the difficulty on the categorization of patients' factors.

In the future research, we can develop a better TPR system based on the conclusion in our study. An improved TPR system should identify sufficient causal factors completely and precisely. On the other hand, the

HFACS model can be modified in some levels so that it can be fitted well in the medical field.

Realizing about the TPR system can help people to know the nature of the medical error. Furthermore, based on the conclusion in our study, we can construct an improved TPR system which may be more valid on data analysis. This may help us to deal with the patient safety and the corresponding problems. Only when we figure out the root causes of all the medical errors as more as possible, the probability of recurrence could decrease. That is the step to improve the medical quality and offer a better and much safer environment for all patients and citizens.

ACKNOWLEDGMENT

This research was supported by National Tsing Hua University and Hsinchu Mackay Memorial Hospital. We thank the reviewers for their insights and all other people who had helped us during the course of our research.

REFERENCE

[1] L. T. Kohn, J. M. Corrigan, and M. S. Donaldson, "To Err is Human: Building a Safer Health System," in Washington DC: National Academy Press, 1999.

[2] P. Barach, and S. D. Small, "Reporting and preventing medical mishaps: lessons from non-medical near miss reporting systems," *BMJ*, vol. 320, pp.759-763, 2002.

[3] J. W. Beasley, K. H. Escoto and B. T. Karsh, "Design Elements for a Primary Care Medical Error Reporting System," *Wisconsin Medical Journal,* vol. 103, no. 1, pp. 56-59, 2004.

[4] C. S. Gau, C. L. Shih, H. H. Liao, H. C. Yang and C. Wung (converted by the authors), "Analysis of Medication Errors on the TPR system," *Formosan Journal of Medicine*, vol. *11*, no. 5, pp. 546-555, 2007.

[5] J. Reason, "Understanding adverse events: human factors," *Quality in Health Care,* vol. 4, pp. 80-89, 1995.

[6] J. Reason, "Human error models and management," *British Medical Journal,* vol. 320, pp. 768-770, 2000.

[7] S. A. Shappell and D. A. Wiegmann, "The Human Factors Analysis and Classification System–HFACS," Washington D.C.: Office of Aviation Medicine, 2000.

[8] J. J. Rooney and L. N. V. Heuvel, "Root Cause Analysis for Beginner," *Quality Progress,* vol. 37, no.7, pp. 45-53, 2004.

[9] J. Neily and G. Ogrinc, "Using Aggregate Root Cause Analysis to Improve Patient Safety," *The Joint Commission Journal on Quality and Patient Safety,* vol. 29, no. 8, pp. 434-439, 2003.

[10] C. L. Shih (Chief editor), "Root Cause Analysis: Manual for teacher version," New Taipei City, Taiwan: Ministry of Health and Welfare, 2004.

[11] Department of Energy, DOE Workbook: Conducting Accident Investigations, Washington D.C., 1999.

[12] P. C. Wang, C. Wung and C. L. Shih, "Analysis on Central Venous Catheter - Related Incidents in the Taiwan Patient - safety Reporting (TPR) System," *Fu-Jen Journal of Medicine,* vol. 12, no. 1, pp. 37-50, 2014.

[13] Ministry of Health and Welfare, "Taiwan Patient-safety Reporting System: Annual Reports 2014," in R.O.C. (Taiwan), 2014.

[14] R. E. Ferner and J. K. Aronson, "Clarification of Terminology in Medication Errors Definition and Classification," *Drug Safety,* vol. 29, no. 11, pp. 1011-1022, 2006.

[15] J. K. Aronson, "Medication errors: what they are, how they happen, and how to avoid them," *QJM,* vol. 102, no. 8, pp. 513-21, 2009.

An Investigation of Adverse Events in Medication Processes by HFACS and Conditional Probability

Min-chih Hsieh [1*], Y. C. LO[1], E. M.-y. WANG[1], Y. H. Fang[2], W. C. Kung[2], M. H. Huang[2], Y. T. Hu[2]
[1]Department of Industrial Engineering and Engineering Management, National Tsing Hua University, Hsinchu, Taiwan
[2]Mackay Memorial Hospital, Hsinchu, Taiwan
(*g9674019@cycu.org.tw)

Abstract – **The study analyzes the adverse events in medication process by HFACS and determines the frequency of AMEs using conditional probability. Four experts were recruited to form a committee for the study: one pharmacist, (working as manager of a pharmacy department), two nurses, and one human factors and ergonomics expert. 17 examples were selected from TPR and 52 causal factors were identified by the committee. The findings show that the traditional approach could underestimate the frequency of AMEs.**

Keywords – **Adverse events, Adverse medication events, Conditional probability, Human factors analysis and classification system (HFACS)**

I. INTRODUCTION

Adverse medication events (AMEs) are well-awarded in hospitals around the world [1-3], and it remain as threats to patient safety. AMEs occur under circumstances such prescribing, transcribing, dispensing, administering, and monitoring a drug. Ferner and Aronson [4] defined AMEs as failures in the treatment process that could potentially harm the patient. AMEs are one of the main causes for adverse events [5, 6]. Adverse medication events account for 21% of adverse events in Pennsylvania, the United States, hospitals [7], and 31% of adverse events in Taiwan [8]. In addition to personal harm, AMEs could impose a considerable financial burden to a hospital. For instance, AMEs contributed additional $2 billion dollars per year in healthcare expenses in the United States [9]. Thus, reducing AMEs not only increases patient safety but also provides a cost-saving opportunity.

Understanding and classifying AMEs allow researchers to identify broad categories of events and, further, to quantify them in order to predict and prevent them [4]. Several studies have analyzed and classified adverse events in medication process. Hynniman et al. [10] divided AMEs into three explicit categories: Errors of omission, errors of commission, and discrepancies. However, medication processes are complex, involving many departments in a hospital such as outpatient department, pharmacy, and nursing department. Thus, it is not easy to use three categories of error to represent the AMEs in a hospital.

Lisby et al. [1] investigated the frequency and type of AMEs in various stages of the medication process, such as ordering, transcription, dispensing, administering, and discharge. They analyzed a total of 1056 errors in 2467 incidences, which errors might occur, in 2003. The results showed that the most common types of error throughout the medication process were lack of drug form, unordered drug, omission of drug, and lack of identity control. Pham et al. [9] analyzed AMEs occurred in emergency departments, revealing that errors occurred most frequently in the administration phase and improper dose/quantity was the most common type. In addition to emergency departments, Kaushal et al., [11] surveyed 10778 medication orders of inpatients in pediatric departments to find potential AMEs. Six hundreds and sixteen AMEs have been identified from 10778 medication orders, and most AMEs occurred at the stage of ordering, which involved incorrect dosing, anti-infective drugs, and intravenous medications. Horsky et al., [12] focused the investigation on the AMEs related to computer-based physician order entry (CPOE) and characterized errors in several converging aspects of the drug ordering process. The research found that the AMEs related to drug ordering process were due to the confluence of several factors, including use of the clinical information system, the absence of automated safeguards that help prevent errors, and physicians' uncertainty when dealing with unusual ordering scenarios.

However, these error types were the outcomes of unsafe behaviors resulting from the mishaps in medication process, which could not represent the causes of the errors. Thus, in order to understand the context of AMEs, new approaches of analysis and classification necessary. The most important part of classifying AMEs is to explain the medication events rather than merely describing them. Hence, the results that produced from an explanatory manner could provide insight into the context of AMEs [4] and help researchers to improve patient safety.

Ferner and Aronson [4] utilized the errors identified category, proposed by Rasmussen [13], to explain the published examples of AMEs and classified errors into five categories, including knowledge-based errors, rule-based errors, slips, lapses, and technical errors. These categories can explain why the adverse events occurred and account for the shortage of personal behavior. However, according to the Swiss cheese theory [14], an accidence does not result from single latent human error. Thus, adverse events usually occur due to a series of unsafe conditions in relation to organizational issues, supervision issues and personal behavior. Therefore, to analyze and understand the causal relationship of unsafe behaviors among each AME is important.

E. Qi et al. (eds.), *Proceedings of the 23rd International Conference on Industrial Engineering and Engineering Management 2016,* DOI 10.2991/978-94-6239-255-7_10

Human factors analysis and classification system (HFACS) [15] has been used as the factors database of human errors for aviation maintenance tasks. It is derived from Reason's organizationally based model of human error [14, 16] and has provided an organizational framework for accident analysis [17]. It divides human error accidents into four categories based on the structure and levels of an organization, including organizational influences (level 4), unsafe supervision (level 3), preconditions for unsafe acts (level 2) and unsafe acts of operators (level 1). However, previous studies focused primarily on the unsafe acts of operators and overlooked the influence of organization and supervision in AMEs. Therefore, to understand the context of AMEs, HFACS were applied in this study.

In addition to classification of AMEs, frequency of AMEs is also important. It offers researchers information on probability of AMEs occurrence in healthcare system. However, previous studies recorded single type frequency of AMEs, but causality between different types of AMEs was left out. Kim et al., [18] applied HFACS structure to analyze 38 incidents of nuclear power plant and recorded the correlation between different levels of HFACS. Despite presenting the correlation between different levels of HFACS, Kim et al. failed to fully account for the causality of probability of occurrence. Therefore, this study set out to determine the frequency and causality of AMEs using conditional probability. It is expected that this research may enhance the depth and broadness of the methodology of AMEs analysis and improve the safety of medication process.

II. METHODOLOGY

A. Data collection and analysis

This is a retrospective study that focus on adverse events in medication process. The published AMEs examples from Taiwan Patient Safety Reporting System (TPR) were extracted for the current study. An approach, namely events and causal factors charting, was used to present the causality of each condition and event in each example.

Four experts were recruited to form a committee for the study, including one pharmacist who works as a manager of a pharmacy department, two nurses, and one human factors and ergonomics expert. The Pharmacist and the nurses have more than10 years of experience of working in hospitals in Taiwan. The human factors and ergonomics expert has more than 5 year research experience in National Tsing Hua University.

One AMEs example might contains two or more causal factors that lead to the occurrence of adverse events. Thus, the committee identified conditions and events that are likely to be the causal factors from the events and causal charting by going through the examples. The committee, then, analyzed the causal factors using HFACS structure from level 1 to level 4, step by step. For instance, one causal factor is "wrong dose" that is classified in level 1 of HFACS, then we were to identify the preconditions for "wrong dose" (level 2). The error paths, which were used to calculate the frequency of AMEs and conditional probability, based on HFACS structure were recorded in this study.

B. Conditional probability of AMEs

Conditional probability is a measure of the probability of an event given that another event has occurred. If the event of interest is A, and the event B is assumed to have occurred, the probability of A under the condition B is usually written as P(A|B). The formula of conditional probability is as follows.

Based on Swiss cheese theory and HFACS structure, the former errors, which belong to level 4 of HFACS structure, could lead to the present errors, which belong to level 3 of HFACS structure, and so on. Then, the accidents would happen in the near future. Thus, the first step of calculating the conditional probability is to estimate the probability of errors in level 3 given that errors in level 4 has occurred earlier. The second step is to calculate the probability of errors in level 2 given that errors in level 4 and 3 has occurred. Finally, in the same manner, the probability of errors in level 1 were calculated given that errors in level 4, 3 and 2 has occurred.

III. RESULTS

A. The individual probability of AMEs

This study analyzed the published examples which occurred during medication process. 17 examples were selected from TPR and 52 causal factors were identified by the committee. Out of these 52 causal factors of AMEs, 17.31% were categorized as decision errors and skill-based errors, 25% were categorized as perceptual errors, and 40.38% were categorized as violations in level 1 of HFACS. Based on the causes of casual factors in level 1, 28.85% were categorized as technological environment, 71.15% were categorized as adverse mental states and adverse physiological states, 78.85% were categorized as physical/mental limitations, 17.31% were categorized as crew resource management and 34.62% were categorized as personal readiness. Based on the causes of casual factors in level 2, 36.54% were categorized as inadequate supervision, 76.92% were categorized as planned inappropriate operations, 63.46% were categorized as failed to correct problem and 5.77% were categorized as supervisory violations. Based on the causes of casual factors in level 3, 75% were categorized as resource management, 67.31% were categorized as organizational

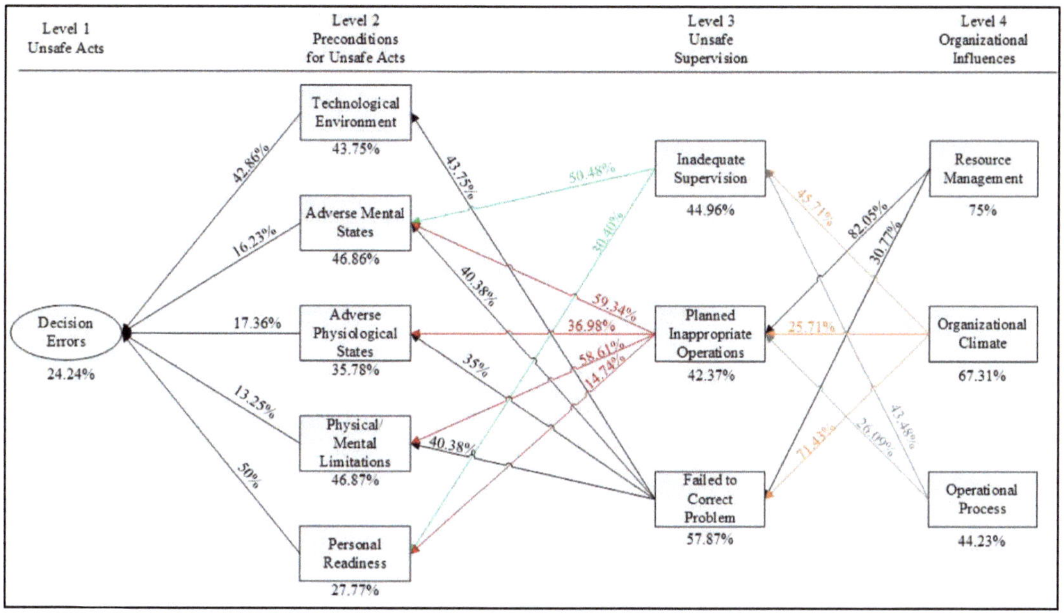

Fig. 1. The structure of conditional probability in decision errors.

climate and 44.23% were categorized as operational process.

B. The conditional probability of AMEs

The conditional probabilities were calculated using (1). In order to describe how the former events could influence the causal factors in level 1 of HFACS, the structure of conditional probability was divided into four main sections according to decision errors section, skill-based errors section, perceptual errors section and violations section. In each structure of conditional probability, two types of probability were calculated. One type locates near the different colors of solid line represent the probability of present errors under former errors (conditional probability). The other type locates under each error types represent the probability of the errors, which was averaged from different error types of conditional probabilities by geometric mean. The four conditional probability structures are demonstrated in Fig. 1, 2, 3 and 4.

IV. DISCUSSION

This study calculated two different types of probability in AMEs including individual probability, which is commonly reported in previous study [1, 9, 11], and conditional probability. Comparing the difference of individual probability and conditional probability, the results show that using individual probability without consider the causality between former and present errors

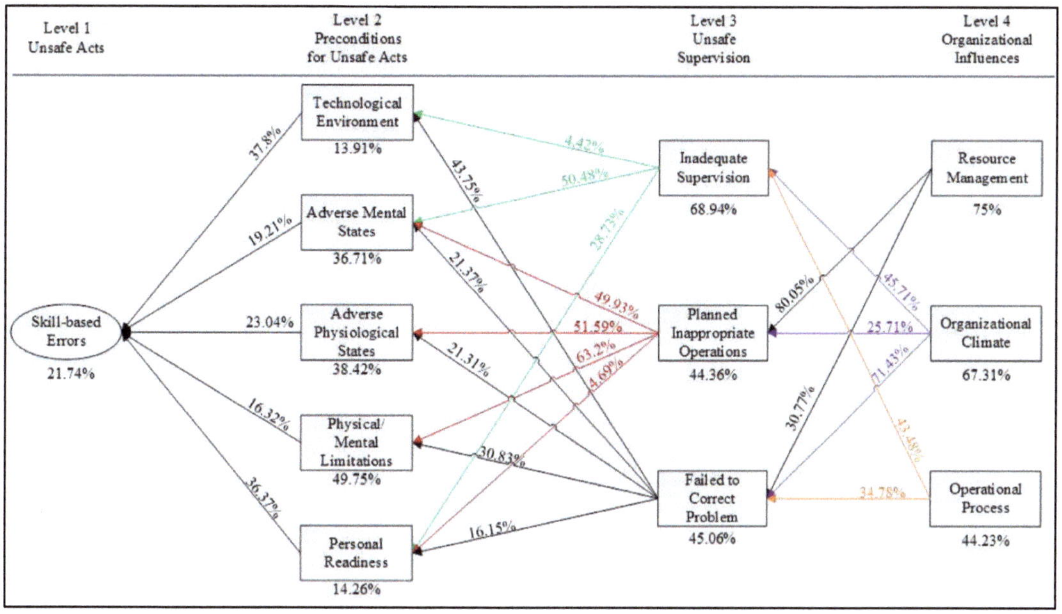

Fig. 2. The structure of conditional probability in skill-based errors.

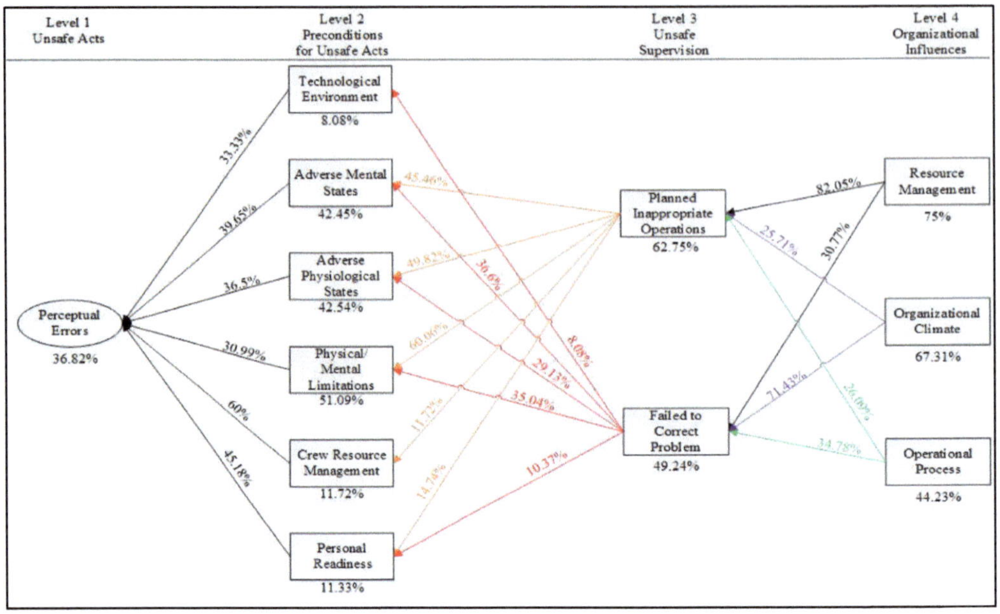

Fig. 3. The structure of conditional probability in perceptual errors.

could underestimate the occurrence rate of AMEs. For instance, the occurrence rate of decision errors without consider the former errors is 17.31%. However, the occurrence rate of decision errors is almost 25% when the former errors has been considered. Underestimating the occurrence rate might cause some serious consequences. Employees who work in hospital might not consider that the AMEs are an important issue or they might not think that some organizational policies (such as financial limitation and supervision issue) and unsafe acts (such as not executing 5 wrights well) could lead to AMEs and thereby harm to patients. Therefore, according to the results of this research, the occurrence rate of adverse events need to include potential threat in further analyzes and consider the influence of the threat, which can reflect the real states of adverse events and improve patient safety.

In accordance with the results of conditional probability, the highest probability of AMEs in level 1 of HFACS is violation error, which presented in Fig.4. Crew resource management (CRM) is the factors that mainly influences violations errors in level 2 of HFACS. Supervisory violations are the factors that mainly affect CRM in level 3 of HFACS. Further, these errors are from organizational climate that is a factors in level 4 of HFACS. This error path represents that poor organizational climate related to patient safety could lead the supervisory defense broken. Further, the workload is distributed unevenly. Consequently, some employees begin to violate the standard operation procedure (SOP) in order to get the work done.

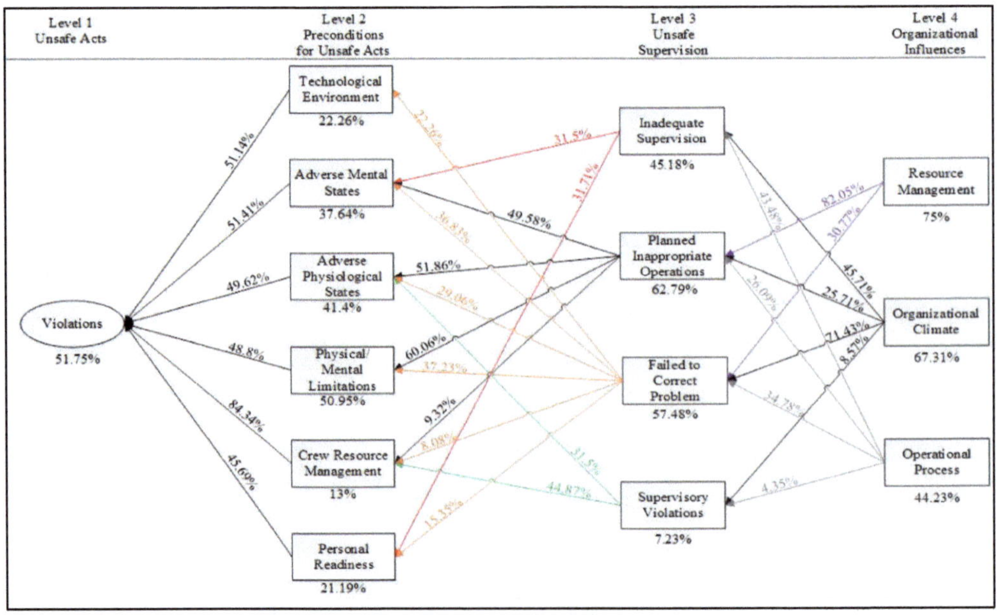

Fig. 4. The structure of conditional probability in violations.

In addition, the results also show that the most important problem in AMEs is that the amount of patients increase every year whilst the amount of medical employees do not accordingly. As the workload pushes medical employees to the physical and psychological limit, the AMEs occur easily. To prevent the situation from happening and increase patient safety, the related policies must be amended.

V. CONCLUSION

The purpose of this study is to analyze the adverse events in medication process in Taiwan and to reduce the frequency of AMEs. HFACS and conditional probability were used in this study. Findings show that traditional approaches underestimate the frequency of AMEs. Also, our findings suggest that the most common type of AMEs in level 1 of HFACS is violations and it is influenced largely by CRM in level 2, supervisory violations in level 3, and organizational climate in level 4 of HFACS.

This study complements the shortages of the adverse event analysis method, and provides the concept of conditional probability to determine the frequency of AMEs. Hence, this study suggests that HFACS should be used to analyze additional adverse events and conditional probability should be utilized to represent the frequency of adverse events in hospitals. Doing so, researchers could understand the situations of adverse events completely and increase the safety of patients. Furthermore, this method could be applied to improve safety in other areas such as aviation, nuclear power plant, military, and other industries in the future.

Two limitations exist in this study. First, there were only 17 published samples were analyzed, which could not fully reflect the real situations of AMEs in Taiwan. Second, HFACS might not explain all the events and conditions in AMEs. It needs to be modified to correspond the situations in hospitals in the future.

ACKNOWLEDGMENT

This paper would like to thank the Ministry of Science and Technology, Taiwan, for partially financially supporting this study under contract number MOST 104-2221-E-007-084-MY3.

REFERENCE

[1] M. Lisby, L. P. Nielsen, & J. Mainz, "Errors in the medication process: frequency, type, and potential clinical consequences," International Journal for Quality in Health Care, vol. 17, no. 1, pp. 15-22, 2005.

[2] Breeding, J., Welch, S., Whittam, S., Buscher, H., Burrows, F., Frost, C., ... & Wong, A. (2013). Medication Error Minimization Scheme (MEMS) in an adult tertiary intensive care unit (ICU) 2009–2011. *Australian Critical Care*, *26*(2), 58-75.

[3] Patanwala, A. E., Sanders, A. B., Thomas, M. C., Acquisto, N. M., Weant, K. A., Baker, S. N., ... & Erstad, B. L. (2012). A prospective, multicenter study of pharmacist activities resulting in medication error interception in the emergency department. *Annals of emergency medicine*, *59*(5), 369-373.

[4] Ferner, R. E., & Aronson, J. K. (2006). Clarification of terminology in medication errors. *Drug safety*, *29*(11), 1011-1022.

[5] Lisby, M., Nielsen, L. P., Brock, B., & Mainz, J. (2010). How are medication errors defined? A systematic literature review of definitions and characteristics. *International Journal for Quality in Health Care*, *22*(6), 507-518.

[6] Leape, L. L., Brennan, T. A., Laird, N., Lawthers, A. G., Localio, A. R., Barnes, B. A., ... & Hiatt, H. (1991). The nature of adverse events in hospitalized patients: results of the Harvard Medical Practice Study II. *New England journal of medicine*, *324*(6), 377-384.

[7] Pennsylvania Patient Safety Authority, "annual report for 2013", Pennsylvania Patient Safety Authority, 2014.

[8] Taiwan Patient-safety Reporting System, "annual report 2014", Joint commission of Taiwan, 2015.

[9] Pham, J. C., Story, J. L., Hicks, R. W., Shore, A. D., Morlock, L. L., Cheung, D. S., ... & Pronovost, P. J. (2011). National study on the frequency, types, causes, and consequences of voluntarily reported emergency department medication errors. *The Journal of emergency medicine*, *40*(5), 485-492.

[10] Hynniman, C. E., Conrad, W. F., & Parker, P. F. (1970). A comparison of medication errors under the University of Kentucky unit dose system and traditional drug distribution systems in four hospitals. *American journal of hospital pharmacy*, *27*(10), 802-14.

[11] Kaushal, R., Bates, D. W., Landrigan, C., McKenna, K. J., Clapp, M. D., Federico, F., & Goldmann, D. A. (2001). Medication errors and adverse drug events in pediatric inpatients. *Jama*, *285*(16), 2114-2120.

[12] Horsky, J., Kuperman, G. J., & Patel, V. L. (2005). Comprehensive analysis of a medication dosing error related to CPOE. *Journal of the American Medical Informatics Association*, *12*(4), 377-382.

[13] Rasmussen, J. (1982). Human errors. A taxonomy for describing human malfunction in industrial installations. *Journal of occupational accidents*, *4*(2), 311-333.

[14] Reason, J., 1990. Human Error. Cambridge University Press. Cambridge.

[15] Wiegmann, D.A., Shappell, S.A., 2001. Applying the Human Factors Analysis and Classification System to the analysis of commercial aviation accident data. In: Proceedings of 11th International Symposium on Aviation Psychology. Ohio State University, Columbus, OH.

[16] Reason, J., 1997. Managing the human and organizational response to accidents. Aldershot: Ashgate.

[17] Daramola, A.Y., 2014. An investigation of air accidents in Nigeria using the human factors analysis and classification system (HFACS) framework. J. Air Transp. Manag. 35, 39-54.

[18] Kim, S. K., Lee, Y. H., Jang, T. I., Oh, Y. J., & Shin, K. H. (2014). An investigation on unintended reactor trip events in terms of human error hazards of Korean nuclear power plants. *Annals of Nuclear Energy*, *65*, 223-231.

Adaptive Grasping and Adjusting Method of Virtual Hand Gestures with Semantics Information for Virtual Assembly

Wen-min ZHU, Xiu-min FAN*, Zhe-hao WU
Institute of Intelligent Manufacturing and Information Engineering,
Shanghai Jiao tong University, Shanghai, P.R.China
*xmfan@sjtu.edu.cn

Abstract –**To improve the grasping efficiency and posture fidelity of virtual hands in virtual assembly processes, a novel adaptive grasping and adjusting method is presented. The attributes and information of assembly parts and assembly gesture of virtual hands are described completely with semantic description models in advance. Basic gestures model base and gestures knowledge base are established to facilitate the generation of hand gestures. According to the spatial relationship between a part and a virtual hand and the semantic description model of the part, the transitions from initial gesture to basic gesture can be calculated. Grasping rules are defined and used to adaptively obtain the final gesture and grasping position of virtual hand. Based on the constraints dynamic-solving algorithm and reasoning of semantic information, the virtual hand is adjusted to the appropriate grasping posture automatically. The validity of this method is verified by the virtual assembly of a reducer.**

Keywords - **virtual assembly, virtual hand, semantic description model, adaptive grasping**

I. INTRODUCTION

In virtual assembly, a user completes operations of grasping and assembling of parts through a virtual hand. The fidelity of simulation process is very important, as it affects the interaction experience of a user directly [1-3]. In current virtual assembly applications, the operation efficiency of a virtual hand is not high enough [4-5], there still exists following problems:

(1) For different types of parts or the same type of parts but with different sizes, the grasping gesture and position cannot be determined quickly and efficiently.

(2) Grasping gestures cannot be changed automatically according to different assembly processes. The same part in different assembly processes may have different grasping gestures. If the user always interrupts the assembly operation from time to time to redefine the gesture when the gesture needs to be altered, it will cost a lot of time and increase the operation workload.

To solve the problems mentioned above, this paper studies the semantic description model of parts and virtual hand, then establishes basic gestures model base and gestures knowledge base. Matching and reasoning of user's intention is executed in real time to realize the adaptive grasping and adjusting method for virtual hands.

The remainder of this paper arranges as follows: Section 2 defines the research concepts and links between them. The semantic description model of parts and virtual hand are described in Section 3. Section 4 describes the basic gestures model base and gestures knowledge base. Section 5 introduces the adaptive grasping and adjusting method of virtual hand gestures. A case study is presented for further demonstration and justification of the method in Section 6. Conclusion and future research are outlined in Section 7.

II. OVERALL IDEA OF VIRTUAL ASSEMBLY BASED ON VIRTUAL HAND

A virtual hand is one of the important elements of virtual assembly and its main job is to simulate the process of grasping and assembling [6]. The realization of virtual assembly requires matching and reasoning in real time based on semantic description model of parts and virtual hand, basic gestures model base and gestures knowledge base. The process of assembly simulation includes different concepts and their interactions are shown as in Figure 1.

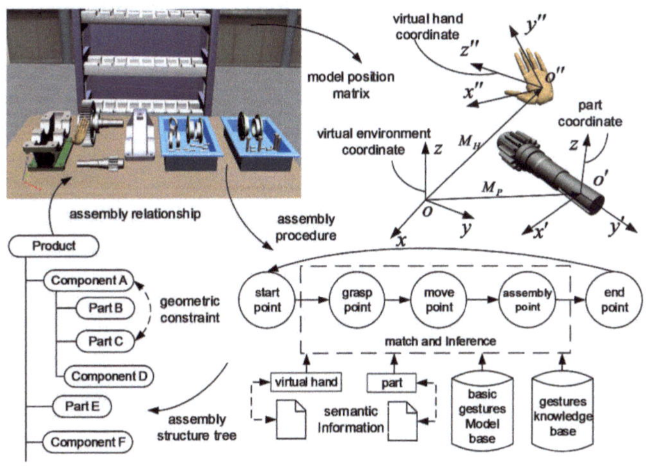

Fig. 1. The virtual assembly simulation based on virtual hand

III. THE SEMANTIC DESCRIPTION MODEL OF PARTS AND VIRTUAL HAND

A. Semantic description model of parts

The semantic description models of parts are established through extraction and reconstruction of the precise geometric information of parts. The structure of semantic description model is shown in Figure 2. The model includes basic information, physics information,

E. Qi et al. (eds.), *Proceedings of the 23rd International Conference on Industrial Engineering and Engineering Management 2016,* DOI 10.2991/978-94-6239-255-7_11

assembly status, and assembly hierarchies and constraints information.

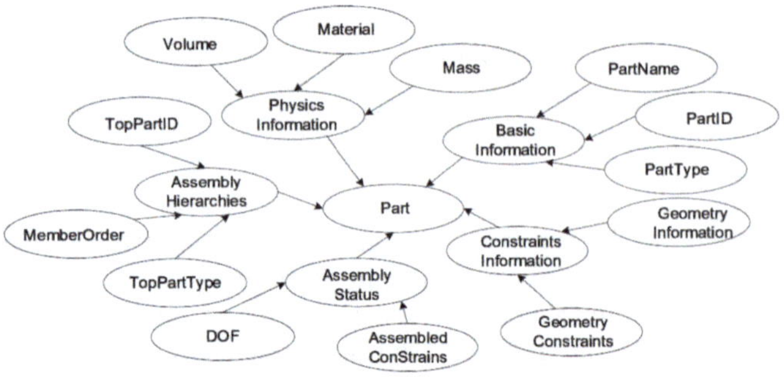

Fig. 2. The Semantic description model of parts

B. Semantic description model of virtual hand

According to the anatomic structure and motion characteristics of a hand, a 3D model of virtual hand is established with some simplified processes. The model includes one palm and five fingers; each finger has three joints as shown in Figure 3. The palm and every joint can be driven independently. The palm has three rotational degrees of freedom and three translational degrees of freedom; every joint has a rotational degree of freedom. The grasping gesture of a virtual hand is determined by its joints synthetically.

The structure of semantic description model of a virtual hand is shown in Figure 4. The model includes basic information, palm position, gesture information and grasped part.

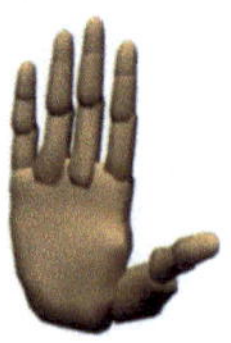

Fig. 3. The 3D model of virtual hand.

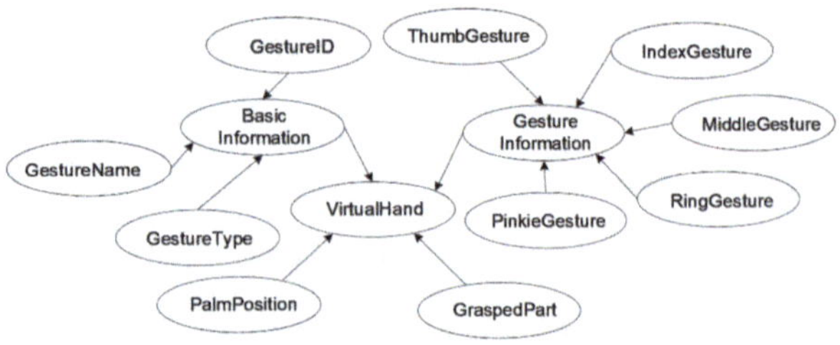

Fig. 4. The Semantic description model of virtual hand

IV. BASIC GESTURES MODEL BASE AND GESTURES KNOWLEDGE BASE

The determination of grasping gestures is a complex process [7-8]. According to the part features, mapping between gestures and features is established in reference [9] without regard to the situation that the same part may have different gestures. It is seemly that different parts may have different grasping gestures, but the same kind of parts may have different gestures also. In this paper, basic gestures model base $G=\{g_1,g_2,\cdots,g_n\}$ ($g_1,g_2,\cdots,g_n$ represents different gestures) for each type of parts is established. For example, there are 3 types of basic gesture model defined to grasp a bearing as shown in Figure 5.

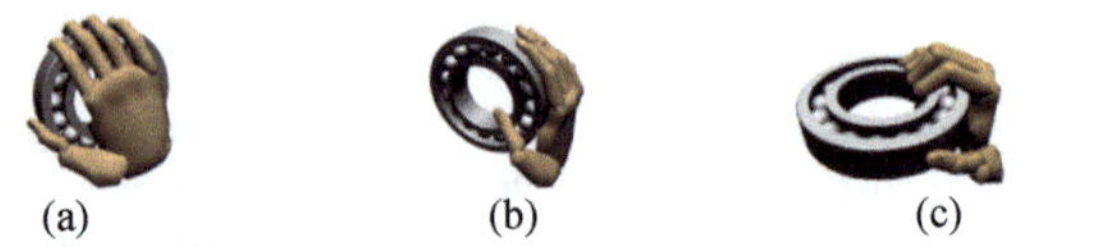

Fig. 5. Different gestures for a same part.

Gestures knowledge base is established based on semantic description model of parts and virtual hands. The building of gestures knowledge base includes two processes, namely initialization and dynamic expansion. The initialization process stores information of specific cases by designer. During the process of grasping a new part, gestures knowledge base is the basis of matching and reasoning for adaptive grasping and adjusting of virtual hand. After a user completes grasping and assembly operation, a new gesture case can be saved automatically to realize dynamic expansion of gestures knowledge base.

V. ADAPTIVE GRASPING AND ADJUSTING METHOD

A. Adaptive Grasping and Adjusting Method based on part features

In the process of virtual hand close to a part, the initial gesture will convert and match a basic gesture stored in the basic gestures model base when the bounding boxes of the part and the virtual hand intersect. However, which basic gesture is the most suitable one depends on the type of part and spatial relationship between the part and the virtual hand. Four steps are proposed to determine the most suitable basic gesture and the flow chart of the process is shown in Figure 6.

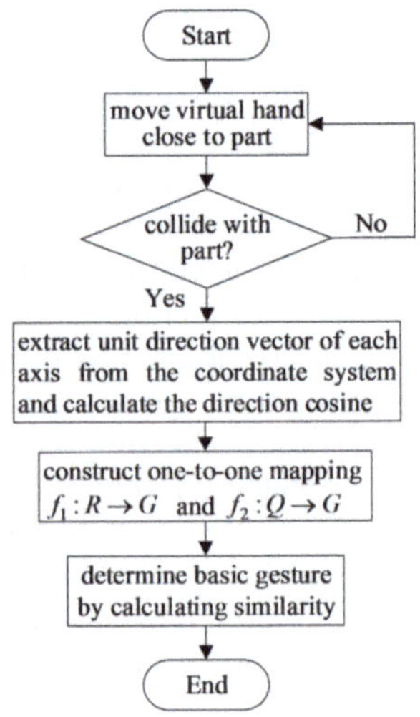

Fig. 6. Determine a basic gesture.

Step 1. Get the local coordinate system of part P_{xyz}, and virtual hand H_{xyz} when the bounding boxes of the part and the virtual hand have intersected. Extract unit vector of each axis direction from the coordinate system, which can be expressed as $(\vec{n}_{Px},\vec{n}_{Py},\vec{n}_{Pz})$ and $(\vec{n}_{Hx},\vec{n}_{Hy},\vec{n}_{Hz})$. Obtain position matrix M_P of the part and position matrix M_H of the virtual hand in the virtual environment.

Step 2. First, calculate the position of the unit vector of each axis direction, namely $(M_P\vec{n}_{Px},M_P\vec{n}_{Py},M_P\vec{n}_{Pz})$ and $(M_H\vec{n}_{Hx},M_H\vec{n}_{Hy},M_H\vec{n}_{Hz})$. Then, calculate values of direction cosine between the corresponding coordinate axes of the part and the virtual hand, as shown in equation (1). Final, define the judgment vector of gesture $\vec{j}(\cos\theta_x,\cos\theta_y,\cos\theta_z)$.

$$\cos\theta_m=\frac{M_P\vec{n}_{Pm}M_H\vec{n}_{Hm}}{|M_P\vec{n}_{Pm}||M_H\vec{n}_{Hm}|},\quad \text{m denotes x, y, z} \quad (1)$$

Step 3. Each grasping gesture g_i in the basic gestures model base G has a spatial relationship r_i with respect to a virtual hand. A set of relative position $R=(r_1,r_2,\cdots,r_n)$ can be constructed by spatial relationships, namely existing one-to-one mapping $f_1:R\to G$. $\vec{q}_i(\cos\theta_{xi},\cos\theta_{yi},\cos\theta_{zi})$ is calculated by every spatial relationship r_i according to the formula (1), then the set of judgment vector $Q=\{\vec{q}_1,\vec{q}_2,\cdots,\vec{q}_n\}$ can be constructed, so the one-to-one mapping $f_2:Q\to G$ can be obtained.

Step 4. Calculate which $\vec{q}_j(\cos\theta_{xj},\cos\theta_{yj},\cos\theta_{zj})$ in the set of judgment vector Q has the highest similarity with the judgment vector $\vec{j}(\cos\theta_x,\cos\theta_y,\cos\theta_z)$. Then basic gesture g_j can be determined through $\vec{q}_j$ according to the one-to-one mapping f_2.

When searching in the gestures knowledge base according to a basic gesture g_j and the semantic information of a part, there may have four results. First, there has a case which has the same semantic information of the part and adopts the basic gesture g_j. Second, there has a case which has the same part type and adopts basic gesture g_j, but the part dimension is different. Third, there has a case which has the same part type, but doesn't adopt basic gesture g_j. Fourth, there does not have a case which has the same part type. For the first result, the found case is then pushed to user directly. For the second result, the found case having the closest dimension with the part is pushed to user. For the third and fourth results, no applicable cases can be pushed to users.

Besides the first result, matching and reasoning is needed for the rest three results as so to obtain a final gesture. First of all, the virtual hand and part should satisfy spatial relationships, that is to say, the certain coordinate axes in the local coordinate system of the part and virtual hand should be aligned. Next, the distance between the virtual hand and part must be in a certain range, otherwise a distortion or even failure of grasping will appear. For the third and fourth results, the process of determining the final grasping gesture and position is as follows:

Step 1. Align the coordinate axes of the part and virtual hand according to the requirements of spatial relationships through matrix T, the position matrix of virtual hand becomes $M_{H1}=TM_H$.

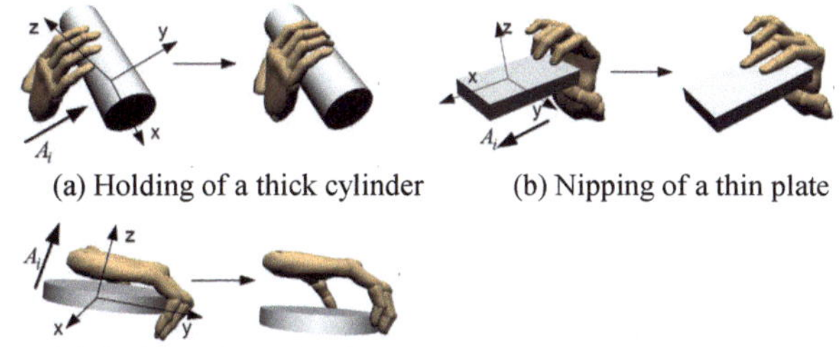

(a) Holding of a thick cylinder (b) Nipping of a thin plate

(c) Grabing of a small disc

Fig. 7. Adaptive adjustment principle

Step 2. Drive joints of fingers to execute grasping operation. Only when grasping process conforms to the grasping rules can virtual hand be thought to complete the grasping. Otherwise the user needs to adjust the position of virtual hand to judge again. Figure 7 shows the principle of adaptive adjustment of virtual hand for some typical distortion phenomena of grasping. The grasping rules and corresponding adjusting strategies are shown in Table 1. According to the adjusting strategies, the position of virtual hand is adjusted many times according to the working process as shown in Figure 8, and the adjustment amount is matrix $A_i(i=1,2,\cdots,n)$ each time. Among them, the drive-strategy of joints uses the average angle method [10]. Finally, the position matrix of virtual hand is $M_{H2}=A_1A_2\cdots A_nM_{H1}$.

Step 3. Calculate the relative position matrix of virtual hand and part $M_{ref}=M_PM_{H2}^{-1}$. Store the case in the gestures knowledge base.

For the second result, start from the step 2 after obtaining information pushed by gestures knowledge base.

B. Adaptive Grasping and Adjusting Method based on geometric constraint

Sometimes gestures are required to adjust in the process of assembly in order to meet the requirements of authenticity. According to different types of objects, there are four types of assembly: (1) a part assembles to another part, (2) a part assembles to sub-assembly, (3) a sub-assembly assembles to part, and (4) a sub-assembly assembles to a sub-assembly. The grasping and adjusting method of these four types are identical. The following steps are taken the first type of assembly as example to describe the method and the flow chart of the process is shown in Figure 9.

Step 1. Move virtual hand to make part close to the object which is assembled. Use the method in reference [11] to confirm the geometric constraints and complete the solution of the first geometric constraint. Then the right position matrix of part is M_P^c under the restriction of the first geometric constraint.

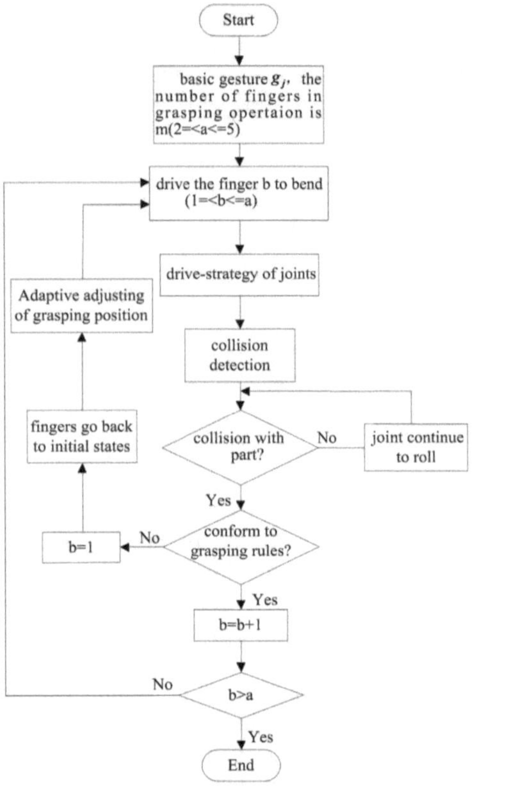

Fig. 8. Working process of grasping.

TABLE I
ADJUSTING RULES AND STRATEGY

part type	grasping rules	current problems	adjusting strategy
thick cylinder	three joints of each finger contact with the part	only the first joint of each finger contact with part	adjust virtual hand along the y axis of part
small disc	the first two joints of each finger can contact with the part	only the third joint of each finger contact with part	adjust virtual hand along the z axis of part
thin plate	two joints of each finger at least contact with the part	only the first joint of each finger contact with part	adjust virtual hand along the x axis of part

Step 2. Execute the process of constraint navigation [12-13]. The transformation matrix that virtual hand adjusts the position of part is M_{P_VH}, and calculate the component matrix of virtual hand under the geometric constraint, which is M_{P_VH}'. Then the actual position matrix of part $M_P^{c'} = M_P^c M_{P\_VH}'$ is obtained under the dual function of virtual hand and geometric constraint.

Step 3. The relative position matrix M_{ref} of virtual hand and part is constant during the constraint navigation, so the movement of virtual hand is restricted too. Traverse the transformation matrix $M_{H\_Control}$ which can be adopted by virtual hand in the restricted space. If virtual hand collides with other objects all the time, it indicates that the gesture is not suitable for the current assembly process, and need to be adjusted. Record the position matrix of virtual hand $M_H = M_{ref}^{-1} M_p^{c'}$ and execute the step 4. Otherwise, execute the step 5.

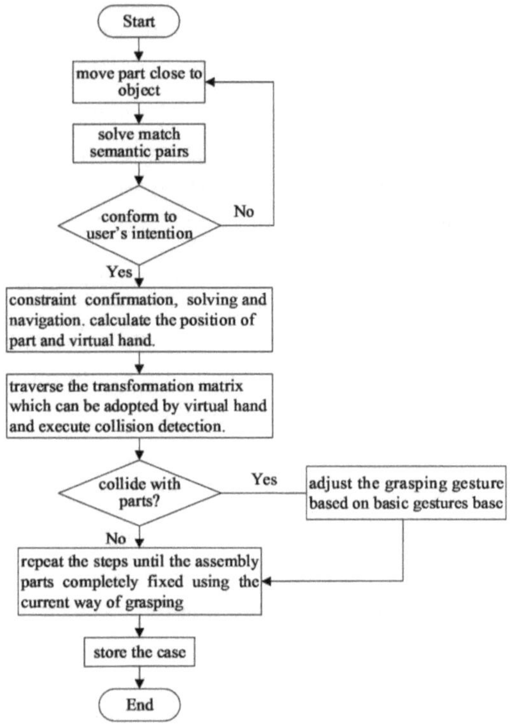

Fig. 9. Adjusting process based on geometric constraint.

Step 4. Search in the gestures knowledge base according to the basic gesture g_j, semantic information of part and geometric constraints. The case will be pushed to user only if the case is perfectly matched. Otherwise, select basic gesture $g_i (i = 1,2,\cdots n, i \neq j)$ and determine the final grasping gesture and position according to the method described in section *A*, then the position matrix M_H' is obtained. Traverse the transformation matrix $M_{H\_Control}'$ to execute collision detection. If virtual hand still collides with other objects, then use the other basic gestures to calculate, until there is no collision.

Step 5. Store the case in the gestures knowledge base.

Step 6. Repeat the steps until the assembled parts are fixed completely.

VI. CASE STUDY

Virtual Engineering Simulation Platform (VESP) is an interactive simulation system developed by Institute of Intelligent Manufacturing and Information Engineering, Shanghai Jiao tong University [14]. In VESP, virtual hand can be real-time driven by VR interactive devices based on semantic processes, which includes semantic recognition and confirmation, semantic solving and navigation. Combined with the VESP and the semantic definition module [15], adaptive grasping and adjusting method of virtual hand is developed based on the research works presented in this paper.

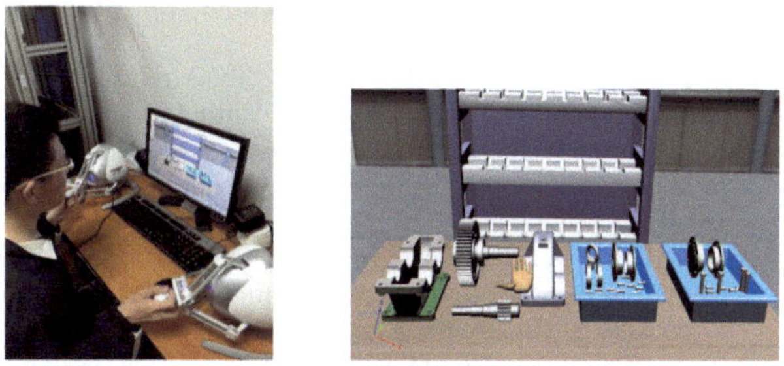

(a)Interactive with virtual hand (b) Environment of virtual assembly
Fig. 10. Virtual assembly demonstration of a reducer.

In this paper, we verify the proposed method through grasping parts in the reducer assembly workplace. Two Falcons (force feedback device) are utilized to drive virtual hand as shown in Figure 10(a). One is for controlling the movement of virtual hand, and the other for giving commands on grasping and assembling operation of parts. Figure 10(b) shows the environment of virtual assembly.

The semantic description model of part is defined in the semantic definition module as shown in Figure 11. The defined part can be used repeatedly, and has a good ability of reusing and sharing.

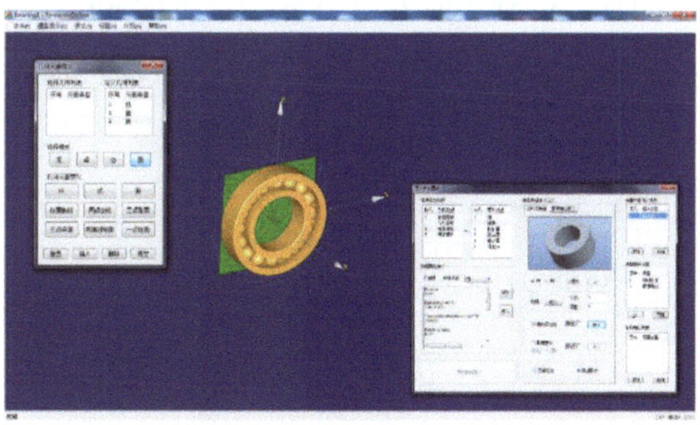

Fig. 11. Definition of semantic description model.

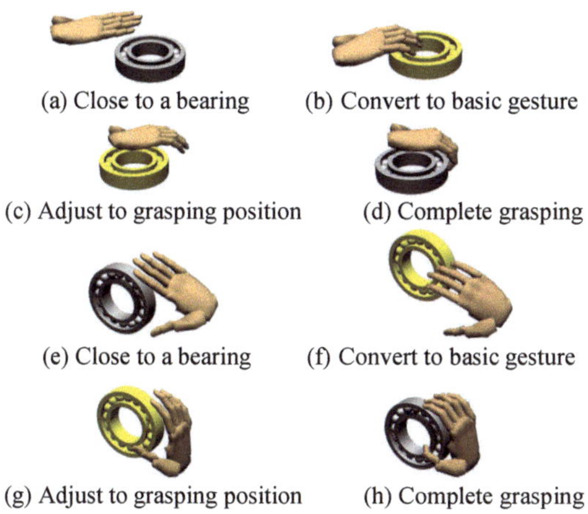

(a) Close to a bearing (b) Convert to basic gesture
(c) Adjust to grasping position (d) Complete grasping
(e) Close to a bearing (f) Convert to basic gesture
(g) Adjust to grasping position (h) Complete grasping
Fig. 12. Grasping processes of a bearing.

Figure 12 shows that the grasping processes of a bearing. The user adopts different grasping intentions to close to the bearing, and then grasps the bearing in different ways according to the spatial relationship between bearing and virtual hand.

Now, the user adopts the grasping gesture of Figure 12(d) to close the shaft. The recognition of objects and matching of geometric constraint semantic pairs are completed, then highlight the geometric elements of matched semantic pairs (Figure 13(a)). The solving of the first geometric constraint is carried out, and the position of part and virtual hand is adjusted according to the result of constraint solving. Virtual hand collides with the shaft in restricted space all the time (Figure 13(b)). It means that the grasping gesture is not suitable for the current assembly process, and then virtual hand will be adjusted automatically to suitable grasping gesture and position (Figure 13(c)). The bearing is assembled to the final position through constraint navigation and constraint solving (Figure 13(d) and Figure 13(e)), and then the user releases the bearing finally (Figure 13(f)).

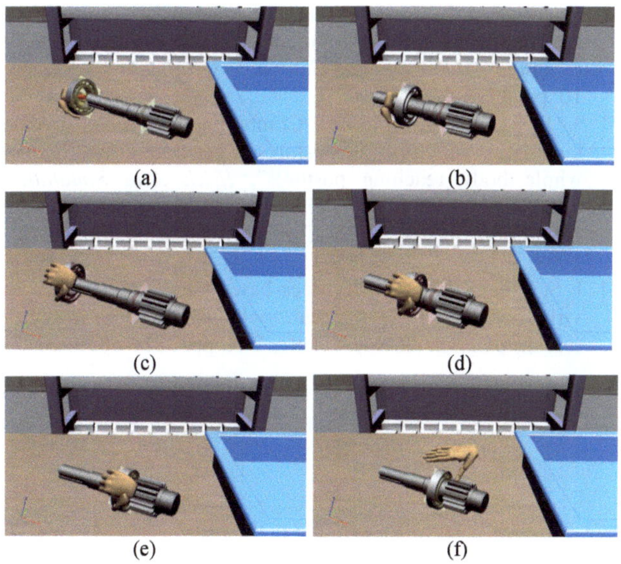

(a) (b) (c) (d) (e) (f)
Fig. 13. Assembling processes of a bearing.

VII. CONCLUSION AND FURTHER RESEARCH

In this paper, the grasping operation of simple parts which have common features is summarized through semantic description model of parts and virtual hands. Gestures knowledge base is established to store the cases of grasping and assembling. According to the assembly context and results of matching and reasoning, gestures knowledge base can push case of highest similarity to user when a new operation is carried out, which greatly improves the working efficiency.

Further research of basic gestures model base and semantic description model which has complex parts and features are ongoing. In addition, based on the physical information of a part defined with a semantic description model, the integration of force feedback needs to be solved in future.

ACKNOWLEDGMENT

The work is supported by the National Natural Science Foundation of China (No. 51475291). The authors are also grateful to editors and anonymous reviewers for helpful comments.

REFERENCES

[1] L. Figueiredo, R. D. Anjos, J. Lindoso, "Bare hand natural interaction with augmented objects", *IEEE International Symposium on Mixed & Augmented Reality*, Adelaide, SA, 2013.

[2] M. Prachyabrued, C. W. Borst, "Virtual grasp release method and evaluation", *Int. J. Human-Computer Studies*, vol. 70, no. 11, pp. 828-848, 2012.

[3] M. Prachyabrued, C. W. Borst, "Effects and optimization of visual-proprioceptive discrepancy reduction for virtual grasping", *IEEE Symposium on 3D User Interfaces*, Orlando, FL, pp. 11-14, 2013.

[4] S. Li, Y. Liu, N. D. Wang, "Research on multi-finger grasping in virtual environment" (in Chinese). *Journal of Computer-Aided Design&Computer Graphics*, vol.22, no. 10, pp. 1728-1733, 2010.

[5] S. Li, J. Liang, G. Liu, "Combining analytical inverse kinematics with example postures to generate virtual human whole body reaching postures", *IEEE Asia Simulation Conference-7th International Conference on System Simulation and Scientific Computing*, Beijing, pp. 45-52, 2008.

[6] J. S. Kim, J. M. Park, "Physics-based hand interaction with virtual objects" *IEEE International Conference on Robotics and Automation (ICRA)*, Seattle, WA, pp. 3814-3819, 2015.

[7] S. L. Li, J. H. Liang, B. Wu, "Survey of virtual character motion generation and control", (in Chinese), *Journal of System Simulation*, vol. 23, no. 29. pp. 1758-1771, 2011.

[8] M. Billinghurst, T. Piumsomboon, H. Bai, "Hands in Space: Gesture Interaction with Augmented-Reality Interfaces", *IEEE Computer Graphics & Applications*, vol. 34, no. 1, pp. 3814-3819, 2014.

[9] X. J. CAI, C. CHENG, A. Marwah, "Research on Behavior Simulation of Virtual Hand in Virtual Manufacturing Environment" (in Chinese), *Journal of Computer-Aided Design & Computer Graphics*, vol. 27, no.3, pp. 499-507, 2015.

[10] H, LIU, X. X. LI, "Design and Implementation of Virtual Human Grasp Motion in Jack" (in Chinese), *Computer Engineering*, vol. 33, no. 3, pp. 213-217, 2007.

[11] Z. X. Xia, H. M. Zhu, D. L. Wu, "Reasoning method based on semantics for virtual assembly operation", (in Chinese) *Computer Integrated Manufacturing Systems*, vol. 15, no. 8, pp. 1606-1613, 2009.

[12] R. D. Yang, X. M. Fan, D. L. Wu, "Virtual Assembly Technologies Based on Constraint and Dof Analysis", *Robotics and Computer-integrated Manufacturing*, vol. 23, no. 4, pp. 447-456, 2007.

[13] H. M. Zhu, D. L. Wu, X. M. Fan, "Assembly Semantics Modeling for Assembling Process Planning in Virtual Environment", *Assembly Automation*, vol.30 , no.3, pp.257-267, 2010.

[14] S. G. Qiu, X. M. Fan, D. L. Wu, "Virtual human modeling for interactive assembly and disassembly operation in virtual reality environment" ， *International Journal of Advanced Manufacturing Technology*， vol.69, no.9, pp. 2355-2372, 2013.

[15] L. Tian, X. M. Fan, K. Y. Liu, "Semantic Modeling and Reasoning Method for Dynamic Generation of Assembly Relationship", (in Chinese), *Journal of System Simulation*, vol.26 , no.9, pp.1901-1906, 2014.

Dynamic Medical Risk Assessment based on Bow-tie Approach and Bayesian Network

Zi-xian LIU[1], Mei LI[1,*]
[1]College of Management and Economics, Tianjin University, Tianjin 300072, China
(*lm314835237@163.com)

Abstract–**Healthcare is regarded as an industry with high risk that has an important influence on patient safety and medical quality. Traditional bow-tie method lacks the quantitative applications in medical areas and has limitations on dynamic medical risk assessment. Through mapping a bow-tie into a corresponding Bayesian network, a framework of quantitative analysis in healthcare using bow-tie approach is built in this paper and dynamic risk assessment is conducted by updating probability based on Bayes' theorem. The feasibility and effectiveness of the proposed method are verified by an example of aeroembolism in the process of hemodialysis.**

Keywords–**Medical risk assessment, bow-tie model, Bayesian network, dynamic risk assessment**

I. INTRODUCTION

The sophisticated medical technologies, diversity influencing factors and uncertain medical outcomes make healthcare a high-risk industry. Medical risks exist everywhere of the medical activities such as diagnosis and treatment, and have an important influence on patient safety and medical quality. Consequently, it is essential and crucial for healthcare to conduct risk assessment to avoid adverse events and reduce medical costs, further improve the medical quality.

Reliability analysis methods commonly used in manufacturing industry have been increasingly used to assess medical risk [1].Among them, tree models including event tree analysis (ETA) and fault tree analysis (FTA) are more popular in medical risk assessment for their flexibility and adaptability [2]. These models can display the interrelationships between failure events in an accident by Boolean logic directly and conduct the qualitative and quantitative analysis simultaneously. They have some applications in healthcare to determine the likelihood of risk in the radiation treatment with a fault tree [3], as well as to analyze the risk of patients in rehabilitation with a fault tree and event tree integrated model [4].

Based on these two tree methods, a bow-tie (BT) model is obtained and becomes widely used in industries such as aviation and petrochemistry. In healthcare, Wierenga et al. [5] and Kerckhoffs et al. [6] apply this method in the risk analysis of medication safety and patient safety in ICU. However, the applications of BTs in healthcare have only qualitative analysis which lacks the quantitative aspects that are more critical. Moreover, BTs have the limitations on dynamic risk assessment due to the static nature of the components, i.e. fault tree and event tree. Some research have taken dynamic factors into consideration by mapping the BT into a Bayesian network (BN) and used the integrated method in chemical industry and offshore drilling [7, 8]. In medical areas, this dynamic risk assessment method has not been used. Therefore, the purpose of this paper is to apply the quantitative BT in medical risk assessment and to incorporate dynamic factors into static BT to improve its ability by converting it to BN. A medical accident of aeroembolism in the process of hemodialysis is served as the case study. The remaining parts of this paper are organized as follows: section II gives a brief introduction to BT and BN, as well as the mapping rules from BT to BN; section III applies the proposed model to a specific example; section IV concludes this study.

II. RISK ASSESSMENT METHODS

A. *BOW-TIE APPROACH*

A bow-tie is a graphical tool that can illustrate a complete accident scenario, starting from accident causes and ending with its consequences. BT consists of a fault tree on the left side identifying the possible events causing the top event, and an event tree on the right side showing the possible consequences of the top event based on the failure or success of safety barriers [9]. Fig. 1 shows a generic BT model.

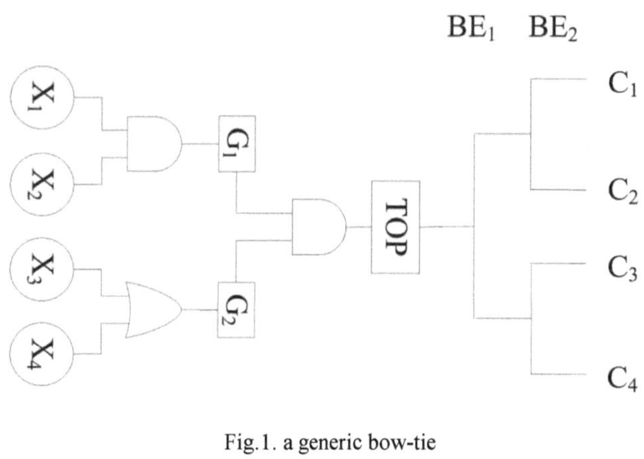

Fig.1. a generic bow-tie

B. *BAYESIAN NETWORK*

A BN is a directed graph including a set of nodes representing the system variables and a set of directed arcs representing the dependencies or the cause-effect relationships among the variables [10]. Each node has an

E. Qi et al. (eds.), *Proceedings of the 23rd International Conference on Industrial Engineering and Engineering Management 2016*, DOI 10.2991/978-94-6239-255-7_12

assigned Conditional Probability Table (CPT) which stands for conditional dependencies. Based on the d-separation and the chain rule, BN represents the joint probability distribution *P(U)* of variables U = {A_1, A_2,…, A_n}, included in the network as: *P(U)* = where $Pa(A_i)$ is the parent set of A_i[11].Additionally, BNs have a powerful inference mechanism based on Bayes' theorem that can update the prior probability of events according to the observed facts during the operation of a process to realize risk assessment dynamically. Fig. 2 shows a simple BN.

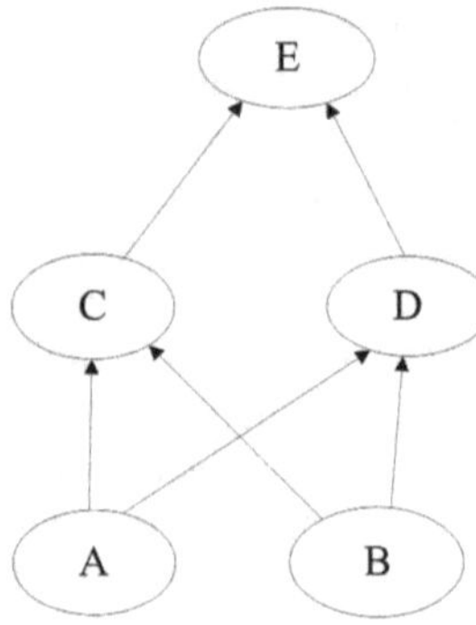

Fig.2. a simple BN

C. MAPPING RULES

Mapping a BT into a BN comprise two parts including mapping from a fault tree and an event tree.

Bobbio et al. [12] propose a mapping method suitable for fault trees. On one hand, the basic events, intermediate events and the top event in a fault tree are represented as root nodes, intermediate nodes, and the leaf node in the corresponding BN, respectively. The nodes of BN are connected in the same way as the corresponding events in the fault tree. On the other, the prior probability of root nodes are the same as basic events' and each intermediate mode has a CPT according to the logical relationships of events in the fault tree.

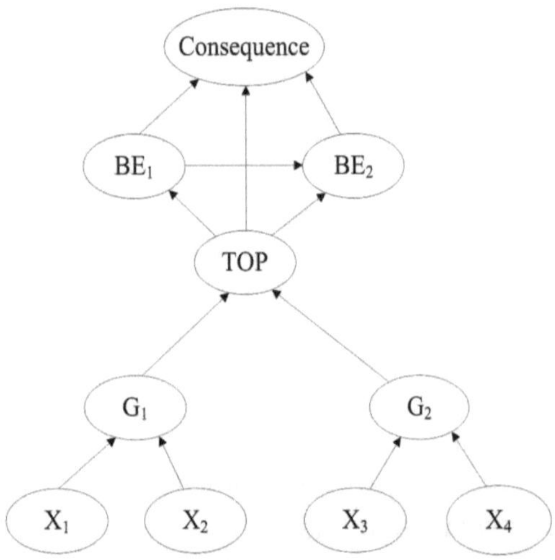

Fig.3. the corresponding BN for BT in Fig. 1

Mapping rules for event trees are put forward by Bearfield et al [13]. For one thing, each branching event (BE) in the event tree is converted to a safety node having two states, one for the occurrence and the other for the non-occurrence of the branching event. A consequence node is adding to the BN and has many states as the number of the event tree consequences. The consequence node is connected to all safety nodes. For another, every safety node has the prior probability same as the probabilities of branching events.

After the mapping of these two parts, they are connected to each other via the top event. The corresponding BN for BT in Fig. 1 is shown in Fig. 3 using the methods described before.

III. THE CASE STUDY

A. BOW-TIE ANALYSIS

An example of aeroembolism is illustrated to verify the feasible and effectiveness of the proposed method. Aeroembolism is a serious complication in hemodialysis that has a damaging effect on patient safety. Aeroembolism is occurred because air entering the body in the process of hemodialysis due to equipment failures or human errors. The event of air entering the body is selected as the top event. Then, based on relevant research, the BT for aeroembolism is established as is shown in Fig. 4. The physical meanings of the elements, as well as some of their prior probabilities, are shown in Table I – IV respectively.

TABLE I PHYSICAL MEANINGS OF EVENTS IN THE FAULT TREE

Event	Description
TOP	Air entering the body
G_1	Devices failure
G_2	Human error
G_3	Pipelines failure
G_4	Hemodialysis machine failure
G_5	Air detector failure

TABLE II BASIC EVENTS AND THEIR PRIOR PROBABILITIES

Event	Description	Probability(10^{-3})
X_1	Hemodialysis pipelines failure	1.12
X_2	Fistula needle failure	2.64
X_3	Dialyzer failure	1.43
X_4	Air alarm failure	1.79
X_5	Air-removal equipment failure	3.33
X_6	Detector pollution	3.65
X_7	Lack of experience	2.50
X_8	Wrong connections between pipes	2.23
X_9	Fail to observe the air entering	1.05
X_{10}	Error in fluid infusion	2.75

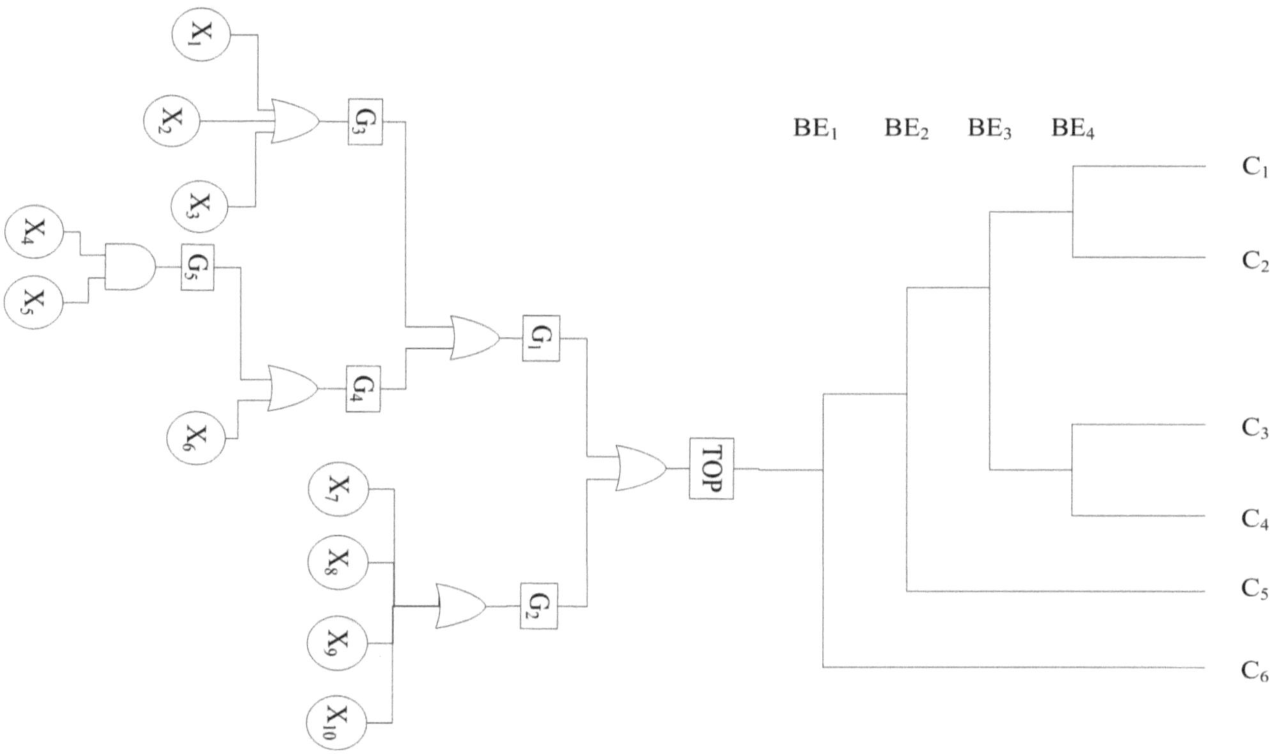

Fig.4. BT for aeroembolism

TABLE III BRANCHING EVENTS AND THEIR PRIOR PROBABILITIES

BE_i	Description	Probability(10^{-3})
BE_1	Fail to cut off the air source	5.62
BE_2	The air entering into ventriculus dexter	3.31
BE_3	Fail to remove the air from the body	3.17
BE_4	Complications	4.75

TABLE IV PHYSICAL MEANINGS OF CONSEQUENCES

Consequence	Description
C_1	Safety
C_2	Complications
C_3	Cardio-pulmonary failure
C_4	Serious complications
C_5	Aeroembolism, death
C_6	Aeroembolism, death

The probability of the top event can be obtained using minimal cutset method. Based on the probability of the top event and the probabilities of BEs listed in Table III, the probabilities of consequences is computed as is shown in Table V.

TABLE V PRIOR PROBABILITIES OF THE TOP EVENT AND CONSEQUENCES

Symbol	Prior probability
TOP	1.72×10^{-2}
C_1	1.69×10^{-2}
C_2	8.07×10^{-5}
C_3	5.38×10^{-5}
C_4	2.57×10^{-7}
C_5	5.66×10^{-5}
C_6	9.67×10^{-5}

B. *BAYESIAN NETWORK ANALYSIS*

Fig. 5 shows the corresponding BN mapping from the BT in Fig. 4 using the methods described in section II. According to the BN model, the prior probabilities can also be obtained and are the same as the outcomes obtained by the BT (Table V).However, the main task of mapping BT to BN is not the probability computation but the probability updating which is the key to dynamic risk assessment by the inference ability of BN based on Bayes' theorem, a task which BT is incapable of doing.

This paper considers probability updating of two types of elements: basic events and branching events. Both types are based on Bayes formula:

(1)

where *P(U)* is the prior probability of a basic event or a branching event, is the likelihood function and *P(E)* is the probability of new observations.

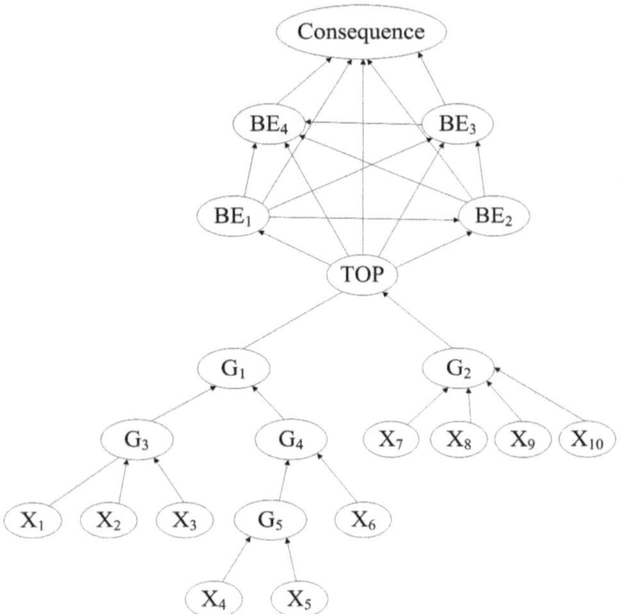

Fig.5. BN for aeroembolism

For basic events, probability updating is to update the prior probability of a basic event in fault tree given one or more events have occurred. The most widely used posterior probability is , which stands for probability of event X_i known the top event has occurred. This type of posterior probability helps to diagnose fault. When an accident has occurred, events with high posterior probability are the most likely causes that need attention. Table VI shows the posterior probabilities of basic events. As can be seen from Table VI, X_6, X_{10} and X_2 are the possible causes for the top event's occurrence. Conversely, X_4 and X_5 have the least probabilities probably because of "AND" logical relationship between them.

TABLE VI POSTERIOR PROBABILITIES OF BASIC EVENTS

Event	Probability	Event	Probability
X_1	0.0651	X_6	0.2122
X_2	0.1535	X_7	0.1453
X_3	0.0831	X_8	0.1297
X_4	3.43×10^{-4}	X_9	0.0610
X_5	3.43×10^{-4}	X_{10}	0.1599

As for branching events, probability updating refers to adjusting the probability of a branching event according to occurrence number of consequences over a period of time, that is . Table VII shows the number of each consequence during 5 months except C_1 which is not an accident. These data is used to construct the likelihood function, which is the ratio between number of adverse consequences caused by a certain branching event and the total number of adverse consequences [14, 15]. Different from probability updating of basic events, *P (E)* of equation (1) is the normalizing factor.

TABLE VII NUMBER OF CONSEQUENCES DURING 5 MONTHS

C_i	Month1	Month2	Month3	Month4	Month5
C_2	1	2	1	-	1
C_3	-	-	3	2	1
C_4	1	-	2	-	2
C_5	-	-	1	1	2
C_6	-	1	-	2	-

TABLE VIII POSTERIOR PROBABILITIES OF BRANCHING EVENTS

Month	BE_1	BE_2	BE_3	BE_4
1	5.62×10^{-3}	3.31×10^{-3}	3.17×10^{-3}	1
2	1.41×10^{-3}	3.31×10^{-3}	7.94×10^{-4}	1.87×10^{-2}
3	5.14×10^{-4}	3.02×10^{-4}	3.17×10^{-3}	6.64×10^{-3}
4	1.21×10^{-3}	4.43×10^{-4}	2.82×10^{-3}	3.33×10^{-3}
5	8.47×10^{-4}	6.99×10^{-4}	2.91×10^{-3}	3.66×10^{-3}

For example, the posterior probability of BE_1 at the end of month 2 can be computed as follows:

Likelihood function = 1/5 = 0.2.

Normalizing factor = $0.2 \times 5.62 \times 10^{-3} + (1 - 0.2) \times (1 - 5.62 \times 10^{-3}) = 0.7955$.

Then, = $(0.2 \times 5.62 \times 10^{-3})/0.7955 = 1.41 \times 10^{-3}$.

Based on equation (1) and Table VII, posterior probabilities of BEs are calculated and listed in Table VIII.

IV. CONCLUSIONS

The present paper takes dynamic factors into consideration to conduct dynamic risk assessment of aeroembolism in the process of hemodialysis. Bow-tie is used as the fundamental model of the accident scenario for aeroembolism and the corresponding Bayesian network is obtained for further analysis. The likelihood of the top event and consequences can be obtained using both BT and BN. Through mapping from BT into BN, the ability of risk assessment is enhanced by Bayesian inference that can update prior probability and get posterior probability. Thus, the limitations of static BT can be solved.

REFERENCE

[1] Lyons, M., Adams, S., Woloshynowych, M., and Vincent, C.. "Human reliability analysis in healthcare: a review of techniques." *International Journal of Risk & Safety in Medicine,* vol.16, pp.223-237, 2004.

[2]Z X Liu, X Ni, Y L Liu, Q L Song, and Y K Wang. "Gastric esophageal surgery risk analysis with a fault tree and Markov integrated model." *Reliability Engineering & System Safety*, vol. 96, pp.1591-1600, 2011.

[3]E Ekaette, R C. Lee, D L. Cooke, S Iftody, and P Craighead. "Probabilistic fault tree analysis of a radiation treatment system." *Risk analysis*, vol. 27, pp. 1395-1410, 2007.

[4]J Wreathall, and C Nemeth. "Assessing risk: the role of

probabilistic risk assessment (PRA) in patient safety improvement."*Quality & Safety in Health Care,* vol.13, pp. 206-212, 2004.

[5]P C Wierenga, L L Huen, S E Rooij, N S Klazinga, H J Guchelaar, and S M Smorenburg. "Application of the bow-tie model in medication safety risk analysis." *Drug Safety An International Journal of Medical Toxicology & Drug Experience*, vol. 32, pp. 663-673, 2009.

[6] M C Kerckhoffs, A F Sluijs, J M Binnekade, and D A Dongelmans. "Improving patient safety in the ICU by prospective identification of missing safety barriers using the bow-tie prospective risk analysis model." *Journal of Patient Safety*, vol. 9, pp. 154-9, 2013.

[7]N Khakzad, F Khan, and P Amyotte. "Dynamic safety analysis of process systems by mapping bow-tie into Bayesian network." *Process Safety & Environmental Protection*, vol. 91, pp. 46-53, 2013.

[8]N Khakzad, F Khan, and P Amyotte. "Quantitative risk analysis of offshore drilling operations: A Bayesian approach." *Safety Science*, vol. 57, pp. 108-117, 2013.

[9]N Khakzad, F Khan, and P Amyotte. "Dynamic risk analysis using bow-tie approach." *Reliability Engineering & System Safety*, vol. 104, pp. 36-44, 2012.

[10]P Weber, G M Oliva, C Simon, and B Iung. "Overview on Bayesian networks applications for dependability, risk analysis and maintenance areas." *Engineering Applications of Artificial Intelligence*, vol.25, pp. 671-682, 2012.

[11]Jensen, and V. Finn. "Bayesian networks and decision graphs, --2 ed." 2007.

[12]A Bobbio, L Portinale, M Minichino, and E Ciancamerla. "Improving the analysis of dependable systems by mapping fault trees into Bayesian networks." *Reliability Engineering & System Safety*, vol. 71, pp. 249-260, 2001.

[13]G Bearfield, and W Marsh. "Generalising event trees using Bayesian networks with a case study of train derailment." *Lecture Notes in Computer Science*, vol.3688, pp. 52-66, 2005.

[14]A Meel, and W D Seider. "Plant-specific dynamic failure assessment using Bayesian theory." *Chemical Engineering Science*, vol. 61, pp. 7036-7056, 2006.

[15]M Abimbola, F Khan, and N Khakzad. "Dynamic safety risk analysis of offshore drilling." *Journal of Loss Prevention in the Process Industries*, vol. 30, pp. 74-85, 2014.

Rational Thinking of the Marketing Strategy Based on Customer Equity

Jian-jun Li *
School of Business Administration, Shanghai University of International Business and Economics, Shanghai, China
(*276007663@qq.com)

Abstract - **The target of marketing is to maximize the customer equity. Seen from this angle, this article interprets the content of marketing strategy based on the analysis of the customer equity. We analyze the content of marketing strategy at last, rationally analyze inner connection in content of marketing strategy.**

Keywords - **marketing goal, customer equity, marketing strategy**

I. INTRODUCTION

The enterprises without strategic will lost its direction, the enterprise with no marketing strategy will also lose itself in the vast market, and the marketing strategy is a prerequisite as well as a guarantee to the success of the enterprise. It is very essential for the enterprise to set its general target, get well know of the relationship between the elements and their interaction. Therefore, this article, based on the perspective of customer equity, explores the inner relationship between the elements of marketing strategy and its centering on the ultimate aim of enterprises marketing strategy.

II.MARKETING GOAL: THE MAXIMIZATION OF CUSTOMER EQUITY

The theoretical exploration about marketing target has been on for a long time, and made a lot of valuable achievements. Some scholars believe that the goal of marketing is realizing the exchange, and some others believe that the marketing target is to reduce transaction costs. There are also scholars insisting that the marketing target is to provide customers with more value. We think the objective of enterprise marketing is to maximize customer equity.

Ronald•Coase thinks that market transactions cost and thus in order to effectively reduce transaction costs and enhance the efficiency of the trade, enterprise which is the main body of the market-enterprise comes into being. In this sense, the enterprise is the product of reducing the market transaction cost. However, as the main body of market, enterprise is a rational economic organization; its existing must be based on a solid economic foundation. Therefore, maximized profit is the pursuit of every enterprise.

The corporate profits come from the customers. A enterprise should have more loyal customers so that its profits can not only be guaranteed but also be the winner in the fierce market competition. Therefore, the enterprise marketing activities should center on providing customers with more transfer value and making customers recognize the value of their enterprise assignment, thus the more loyal customers will be attracted only to provide customers with more value the transfer; enterprise customer lifetime value will increase. So, the enterprise marketing target is to maximize customer equity.

III. THE NECESSITY OF CENTERING ON THE MAXIMIZATION OF CUSTOMER EQUITY

With the rapid development of economy, science and technology as well as the increasingly fierce market competition, the goals of enterprise marketing should be pointed directly at the source of its profits. The customers are not only the source of enterprise profit, but also the guarantee of long-term competitive advantage of enterprise. Enterprises provide customers with the most maximized transfer value. Making the maximization of its customers' equity is the necessity in marketing practice. Therefore, the aim of maximizing the customer's equity is not only the theory logical deduction but also the product of marketing practice. This paper will explain the necessity of maximized-customer-equity-centered marketing target from several perspectives.

A. The arrival of Hyper-competition

The acceleration of the globalization of economy and the rapid development of science and technology short the life cycle of technology. The process of information exchange has also undergone a fundamental change. The production method is experiencing constant innovation. All of these demonstrate the the arrival of (Hyper-competition).

Aveni (1994) put forward the concept of the competitive environment in "Hyper-competition: Manage the strategic Maneuvering", which had caught broad attention at that time. He thought that enterprise survival space has become more and more unrest, and that no organization can establish a lasting competitive advantage, and that each kind of competitive advantage could be quickly be eroded. Therefore, competitors must make innovations and do instead of just insisting on their competitive advantage to master the market. They should also broke their own advantages, meanwhile at the same time also destroy the competitive advantage. The success of a company depends on whether it can quickly change itself from an advantage to the other.

B.The reacquaints of the sources of competitive advantage

E. Qi et al. (eds.), *Proceedings of the 23rd International Conference on Industrial Engineering and Engineering Management 2016,* DOI 10.2991/978-94-6239-255-7_13

The sources of enterprise competitive advantage is also a hot topic which has been the focus of enterprise researches, As for the origin of competitive advantages, people have different points of view. Some people think that the competitive advantage comes from the enterprise internal management; some scholars argue that the brand is the origin of competitive advantages, there are also people think the enterprise culture is the core competitiveness of the enterprise and it is the enterprise competitive advantage. However, in the competitive environment, other competitive advantage are very easy to change But customers, the solid foundation for the enterprise survival and the development, are the most precious resource and the most valuable equity for the enterprise, Leyland, Ewing, Berithon (1998) pointed out in "Turning competitive advantage into customer equity" that we should re-awake of the origin of competitive advantages, and that the customer equity is the ultimate source of enterprise competitive advantage.

C. The demand of rational resources-allocation

Enterprise's resources are limited, thus the question of how to allocate the limited resources reasonably and produce the maximum benefit has called great attentions. As the main body of market, enterprise always follow the rational creed and search the optimum standards and norms in decision-making.. the precise master of the enterprise marketing goal is the premise of the maximization of customers equity and the reasonable configuration of the enterprise resources, the aim of the maximization of customer equity makes the enterprise resource flow rationally, allocate effectively, and pave way for the enterprise marketing practice activity Therefore, it is the inner demand for enterprise to seek the maximization of customer equity as the marketing target.

D. Real return to customer-orientation

In fact, market marketing theory had pay attention to customers from a very early time. John Chambers, the chief executive of Cisco Systems', said that put the customer in the center of your culture. In the early 1960s, Peter•Drucker also thought that the priority of a company is to "create more customer". Obviously, the customer - oriented theory has carried great weight all the time.

However, because of the underdevelopment of economy and the backwardness of technology in the old economy era, the marketing theory and practice had been replaced by production orientation, the product orientation, and the sales promotion orientation, and the importance of customer had been ignored by enterprise managers while. the development of social economy and the arrival of the era of competition have impacted the enterprise management and marketing theory and practice More and more people have realized the importance of customers "customers are the key to the success of business " , "the enterprises which own more valuable customer are the winners of the market" and "customer equity are the source of enterprise competitive advantage", "the development of society has made the customer be the core of the enterprise management and marketing " ; Really back to the customer- orientation is the necessity of the rational regression of marketing strategy Therefore the aim of the maximization of customer equity is the request of returning to customer orientation.

IV. THE CONTENT OF MARKETING STRATEGY

Marketing strategy is a integral and long-term cogitation a enterprise made based on the market marketing environment and dynamic change trend, their own marketing conditions and the resources they have, etc, Speaking of marketing strategy, people tend to compare it with company strategy. However, in the enterprise management course, marketing strategy really plays a pivotal role. it must be admitted that, marketing strategy is a functional strategy in a enterprise strategy, and it can never replace the enterprise strategy As an organic system, marketing strategy which is full of rich connotation, mainly consist of five subsystems: STP strategy, market development strategy market competition strategy, brand strategy and customers strategic.

A.STP strategy

STP strategy is one of the more important contents in the marketing activities. In the early 1950s,the American market experts Wendell Smith put forward the concept of Segmentation. Professor Philip systematical put forward STP strategy: in his masterpiece" the marketing management"(the ninth edition) ——S-Segmentation, T-Targeting, P-Positioning).STP strategy plays a very important role in the whole marketing strategy system, and all of the marketing strategy are built on the basis of STP strategy namely segment, target and positioning. The three part of market segmentation, target market and market positioning is organic connection and gradually sequential to each other. Market segmentation provides the basis, for determination of the target market, and market positioning is done in the target markets. What's more, it is just the differentiation positioning made in the target market can make the company product and services be occupied a unique position. In the mind of the target customers.

B. market development strategy

Market development is a plan planning and the corresponding market behavior. Which takes place on the basic of the existing market and product (service), and the analysis of the enterprise's circumstances and as to the future development of the enterprise prospect Market development strategy is a very important part in enterprise marketing strategy. Domestic marketing scholars have made positive exploration about the market development strategy. Such as Chi YingQing (2000) who proposed the market development strategy of roll-ability and concentric.

Market development strategy with the most authority belongs to I.H.Ansoff. In 1957, professor Ansoff, the ancestor of strategic management, published a paper about diversification business strategy in the famous "Harvard business review according to his own research results and put forward the "concept. Of " product market matching.He thought the essence of enterprise management strategy are the reasonable combination of. Four factors namely the existing products, the future products, the existing market and future market. Accordingly, Ansoff. put forward four basic strategy in 1965 in his the company strategy according to the different combination of product and market as following the market penetration strategy-existing products and the existing market combination, product development strategy-the existing market and the future of the product portfolio, market development strategy-and the future market existing product combination, diversification strategy--the future and the future of the combination products market. These four kind of strategies is the famous "Ansoff matrix", which became the main component of Ansoff's theoretical system.

C. the market competition strategy

In the competitive environment, market competition intensity is self-evident. Michael E.Porter made detailed analysis on the general framework of competitors based to the analysis of the industrial structure, Potter points out the five Competitive forces--entering threat, replacing the threat, customer price negotiation skills, supplier price negotiation ability and existing competitor- has reflected the fact that industry competition goes far beyond the scope of the existing participants. The customers, suppliers, vendors, substitutes and potential entrants are the industry's "competitors", and they will more or less show their importance in accordance with the specific conditions. Therefore, potter described the competition strategy as: the ability of taking offensive or defensive action, establishing a flexible position in the industry, successfully dealing with five competitive forces, thus won extra investment income. Potter put forward that there are three basic strategies for industry to win the competition, on the foundation of the analysis about the five competition force: strategy of Overall cost leadership, Differentiation strategy and Focus strategy.

D. brand strategy

There are two representative schools of thought about brand strategy in the marketing strategy: the school of product combination and the school of brand equity. The school of product combination research on the contribution of brand to the product from the perspective of product differentiation or the improvement of product competitive advantage. The school of brand equity studies the style and way to invest a brand from in the perspective of creating customer value. This paper has discussed the brand strategy in marketing from the perspective of the brand equity. Brand equity is linked with brand, brand name and its mark which can increase or decrease the value of its product or service and (or) a series of brand equity and liabilities which is related with the customer value. The equity and liabilities on which the brand equity are based must be associated with brand name and/or the brand logo. Brand strategy is a part of the enterprise marketing strategy. It is a series of plans and activities which centers on the brand creation, maintenance, operation and management and the constant improvement of the total brand equity with the rational configuration of enterprise resources, The essence of the brand strategy is to constantly enhance the brand equity worth. Great achievements have been made in the research about the brand equity both at home and abroad, among which David•Aaker is a model. Aaker held the opinion that brand equity mainly consist of brand awareness, brand association, brand loyalty, the quality of the cognitive and other proprietary rights (such as trademark, etc)

E. The customer strategy

Since the marketing concept put forward in 1950s,the customer has been regarded as the center of the marketing activities. Different from the product-orientation and manufacturing and sales-orientation, we turn to the philosophy of customer-orientation and feeling and response" (Philip p.m., 2003). Corporate profits comes from enterprise customers instead of its product. Enterprise customers replacing share market share became the measure to its competitive advantages And with the increased competition of customer share, more and more enterprises already put it in the position which equal to or even higher than traditional tangible or intangible equity namely customer equity rather than treat it as social public equity (Jan-Xin Liu , Xue-Yang Chen , 2005).

Blattberg and Deighton (1996) thought that customer equity refers the sum of all customer net present value (NPV) in their lifetime, and that is the most important value of the enterprise equity. Rust (2001) even thought that customer equity influence enterprise development strategy greatly. Companies should enhance customer equity, through the ascension of value equity, brand equity and relational equity and make use of the customer lifetime value to shape the future of the company strategy. Blattberg (2001) put forward three strategies about customer equity management: customers obtain strategy; customers sustain strategy, and additional sales strategy. He also thinks companies should promote customer equity through the following three ways: customers obtain management, customer sustain management, additional sales management.

To sum up, the customer strategy is the home of enterprise marketing strategy. The represent of customer strategy marked real returnees of the customer orientation.

V.THE RELATIONSHIP BETWEEN THE CONTENTS OF THE MARKETING STRATEGIES

Taking the maximization of customer equity as the marketing target is of inherent inevitability. Enterprise's marketing strategy is an organic system consisting of STP strategy, market competition strategy, brand strategy and customer's strategy. The four contents of marketing strategy are not independent. They interdependence, mutual association, mutual influence, and promote the overall development of the enterprise marketing strategy and he benign operation of enterprise manages jointly. The relationship between the content of marketing strategy can be represented as the following figure1:

From the content of development strategy and t competitive strategies we can see that both the two kinds of strategy have a very strong execution tendency, and both the two kind of strategy service for the customers, the strategy. The effective implementation of market development strategy and t the market competition strategy is guarantee of the achievement of the ultimate enterprise marketing target of the maximization of customer equity. Therefore, in figure 4, we take those two strategies as two wings of the marketing strategic rocket figure.

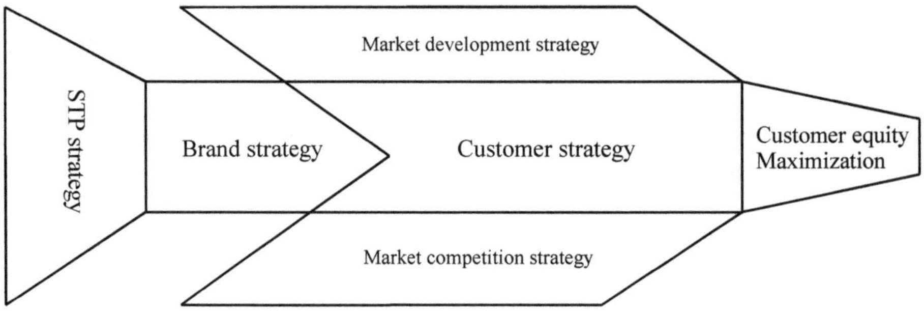

Figure 1: rocket graph of the relationship between elements of marketing strategy
(Source: the author finishing)

A. Marketing goal: the maximization of customer equity

From the above figure, we can see that the whole marketing strategy always struggle for enterprise's marketing goal,the planning and implementation of enterprise marketing strategy are should focus on maximizing the customer equity. The standard of enterprise marketing success is no longer the size of the market share but the size of the customer share. The customer is not only the stable source of enterprise profit, but also the source of enterprise core competitive advantages. Therefore, as one of the enterprises functional strategies, enterprises marketing strategy should be aimed at the maximization of customer equity.

B. STP strategy: the foundation of the marketing strategy

STP strategy, which takes the market segment, target market and market positioning as the core content, is the starting point of the marketing strategy and it plays a fundamental role in the whole marketing strategy. Other strategies can only work in the basis of STP strategy otherwise, it will go wrong.

C. Marketing development strategy and market competition strategy are two wings of marketing strategy

According to Ansoff's point, market development strategy mainly contains four kind of market development combination namely the market penetration, product development, market development and diversification while according to potter's point of view, the main competition strategy consists of cost leadership strategy, the differentiation strategy and target strategic focus.

D. brand strategy is the assistor of marketing strategy

In the modern society the meaning of the brand has completely gone beyond the mark connotation of a product or service. Brand not only includes the rich culture of a product or service, also represents the status of modern people. The brand is a every important connection between the products and services and the customers. On the one hand, it is the embodiment of the product or service marks and their abundant connotations; on the other hand, it is a sign of social status of the customer. A brand can only have internalized practical meaning. when it gain customers' recognition. Therefore, the brand strategy is a engine and a assistor for the marketing strategy which drive the constant development of customer strategy, starting from the STP strategy and operating under the influence of development strategy and the competition strategy. It is just the brand strategy that makes the whole marketing strategy system always so energized and dynamic.

E. customer strategies is the destination of marketing strategy

A enterprise can only gain competitive advantage in the market when it has stable loyal customer groups.. Customer-orientation strategy is the foothold and home of marketing strategy for the enterprise. No matter what it is: the STP strategy, brand strategy or market development strategy or market competition strategy, it is only practical when it comes to customers. If the enterprise marketing strategy cannot offer customers high value, enterprise customers total value life cannot be guaranteed, thus the constant appreciation of the enterprise customers equity also won't be achieved therefore the whole marketing strategy also will lose its significance and become worthless.

To sum up, the aim of the marketing strategy of a enterprise is to maximize the customer equity, and its subsystem strategy are organic component, They on one hand different from one another, on the other hand mutual influenced, jointly promote the development of the whole enterprise marketing strategy.

REFERENCE

[1] Robert.C.Blattberg,John.Deighton, *"Managing marketing by the customer equitytest"*, Harvard Business Review, vol.28,no.7,pp.136-144,1996.

[2] Rust, T.Roland, Valarie A. Zeithaml, and Katherine N. Lemon, *"Driving Customer Equity: How Customer Lifetime Value Is Reshaping Corporate Strategy"*,New York: The Free Press,2000,pp.37-38.

[3] Robert.C.Blattberg, Gary.Getz, Jacquelyn.S.Thomas, *"Customer Equity: Building and Managing Relationships as Valuable Assets"*,Boston:Harvard Business School Press,2001.

[4] Roland,T.Rust, KatherineN.Lemon, Das Narayandas, *"Customer Equity Management"*,Peking University Press,2006.

[5] I.H.Ansoff, *"Corporate Strategy: An Analytic Approach to Business for Growth and Expansion"* MaGraw-Hill. New York.1965.

[6] Michael•Porter, *"Competitive strategy"*,Trans. Xiao-yue Chen, Beijing: Huaxia Press, 2005. Pp.33-34.

[7] Philip•Kotler, *"Marketing management"* (Eleventh Edition) ,Trans.Qing-hao Mei, Shanghai: Shanghai people's Press, 2003.pp.65,340.

[8] David•Aaker, *"Managing Brand Equity"*, Trans.Wei-hua Xi, Beijing: Mechanical Industry Press, 2006.

[9] Doyle,P., "Marketing Management and Strategy", Trans.Yu-min Zhu,Ai-lin Yang,Yuan-huai Wang, Posts Telecom Press,2007.

[10] Boyett,•J•H, *"Classical marketing"*,Trans.Yang Yue, Beijing: Mechanical Industry Press, 2004.pp.41-42.

An Empirical Study on Manufacturing Strategy Formulation in Turbulent Environment: Dealing with Changes and Uncertainties

Jian WANG [1], *Zhi-xin CHEN[1], Pei-pei WANG[1]
[1]School of Management, Shanghai University, Shanghai, China
(jwang@t.shu.edu.cn, *catherinechen2014@foxmail.com)

Abstract - **In the turbulent environment, how to deal with changes and uncertainties has become the core competence of firms. Many manufacturing systems have been proposed with the capability to deal with changes and uncertainties. Facing with such many kinds of systems, it is difficult to set up the linkage between business strategy and manufacturing systems, which is the mission of manufacturing strategy. Some common approach shared by these systems are concluded, and the linkage is built based on the common approach. An empirical study was executed in Chinese manufacturing industry with a questionnaire linking with uncertainty. The common approaches and change sources were empirically tested with principal component analysis (PCA), and the relationships between the sources of change and the common approach were examined with regression correlation. The results found some significant relationships which can give suggestion on the strategic formulation of manufacturing strategy in the turbulent environment.**

Keywords - **Changes and uncertainties, manufacturing system, manufacturing strategy, regression analysis**

I. INTRODUCTION

The role of manufacturing strategy is to set up linkage between business strategy and manufacturing system. However, there are so many changes and uncertainties in the manufacturing systems in today's competitive turbulent environment. The sources of changes and uncertainties are different and dozens of manufacturing systems can be found with the capabilities to deal with the changes and uncertainties. The firms without the capabilities need pay for the serious consequences caused by mishandling the changes and uncertainties. Different systems can suggest the same capability with different names or support it with different tools. Therefore, these capabilities can be grouped into several common approach. The relationships between the changes and uncertainties and common approach are supposed to be factored into the formulation of manufacturing strategy and it is necessary to be identified. An empirical study was executed in Chinese manufacturing industry with a questionnaire. The relationships between the common approach and different changes and uncertainties needs to be further examined. The regression correlation was used in the process of examination.

The remainder of this paper is organized as follows. In the next section, a literature review on the research topic was summarized. Section 3 described the survey content and analysis steps. In section 4 the process of the analysis was discussed. The conclusion was summarized in the last section.

II. LITERATURE REVIEW

Manufacturing was firstly elevated to a strategic level by pointing that there was a critically and sensitively connection between manufacturing decisions and firm strategy [1]. It is implemented in a sequence of actions that begins with strategy formulation and ends at business performance [2]. Its mission is to found the linkage with business strategy [3] and manufacturing systems [4]. However, in today's turbulent environment, customers become more critical and expect better customized products and service than in the past with the growing competitive market [5]. Therefore, the linkages between strategy and systems have to become closer to get more adjustment [6] and it is important to pay more attention to changes and uncertainties in the manufacturing systems [7]. The probabilistic nature of these changes and uncertainties may not always be known [8] so that it is difficult for managers to profile the specific changes. However, it is necessary to learn the general concepts before decision-making. And changes can be concluded by the frequency of its occurrence. Several typical sources of changes and uncertainties are listed in TABLEI based on the literature.

Manufacturing flexibility associated significantly with agile production [14] is a basic responsive capability of manufacturing system in the dynamic, non-linear and chaotic environments and help firms to achieve good performance [15-17]. In order to work well in the turbulent environment, there are many manufacturing systems with flexible capabilities designed to deal with changes and uncertainties [18]. For example, a reconfigurable manufacturing system is designed to adjust production capability and functionalities within a part family in response to sudden changes in the market or in regulatory requirements [19]. Agile manufacturing system focus on speed and flexibility [20] while lean manufacturing system focus on efficiency and cost [21]. However, it is difficult to choose the specific system or specific approach during the formulation of manufacturing strategy.

E. Qi et al. (eds.), *Proceedings of the 23rd International Conference on Industrial Engineering and Engineering Management 2016,* DOI 10.2991/978-94-6239-255-7_14

TABLE I
THE SOURCES OF UNCERTAINTIES

Varieties	Specific concepts
1.Supply instabilities	1.1 Material mishandling[9]
	1.2 Fluctuations in demand[9]
2. Product variety uncertainties	2.1 Market acceptance of kinds of products[10]
	2.2 Length of product life cycle[10]
	2.3 Product mix[9]
	2.4 Ever-shorter product life cycle[11]
	2.5 Increasing product variety in the same time[11]
	2.6 Frequent change in product type[12]
3.Labor force uncertainties	3.1 Labor absenteeism and turnover[9]
4.Technology upgrade uncertainties	4.1 Machine downtime[10]
	4.2 Order priority, rush orders arrival and order cancellations[13]
	4.3 Equipment breakdown[9]
	4.4 Queuing delays, rejects and reworks[9]
	4.5 The steady advance of new technologies[11]

This problem can be studied in two steps. The first step is to classify manufacturing systems with capabilities into several "best practices", which is the way to research on manufacturing strategy [22] and can be utilized widely in many conditions such as the best performing firms [23] or some lower performing firms to improve their performing and competitive but some of them are constrained considering the specific contexts [24]. For example, it was concluded from several manufacturing systems that decentralization of company functions, exploitation of creativity, experience and competence of the employees can help firms to respond to increasing complexity [25]. Some other researchers concluded the manufacturing systems with the capabilities as: (1) enhancing the reactivity and proactivity of traditionally structure system by sophisticated new control systems; (2) developing adaptive systems; (3) developing adaptive systems, which are able to learn from past history [26-27]. The second step is to relate different "best practices" to different uncertainties to choose the specific approach for the specific firm and add it in the formulation of manufacturing strategy.

Some researchers completed the first step. Over 80 manufacturing systems with capabilities were listed by text mining software such as adaptive production, anthropocentric production and cellular manufacturing [18]. Some other researchers concluded 46 manufacturing system with capabilities and classified them into six common approach [29] which can be described as "best practices". Six common approach were chose to be discussed in this paper, the details are as following: (1) rapid adjustment of manufacturing system (C1). It represents a kind of system that responses quickly to changes and uncertainty by utilizing modularization in the turbulent environment; (2) expanding authorization in the manufacturing system and enhancing autonomy (C2). It represents a kind of system that encourages employees participating in the management positively and controlling the manufacturing process more; (3) improving integration of information system (C3). It represents a kind of system which has integrated the information system and has more control on the information collection system; (4) improving utilization of production system (C4). It represents a kind of system that forms functional units based on group technology and responses to different orders through those units; (5) function diversification and cooperation (C5). It represents a kind of system which is formed by the high-quality resources all over the country and the collaboration of different parts is valued in this system; (6) people-centric production system (C6). It represents a kind of system which strengthens the responsibility of employees by the knowledge and cooperation of them.

In light of the above, one hypothesis could be proposed as follows:

Hypothesis: These six common approach could be all effective in dealing with all sources of changes and uncertainties.

III. METHODOLOGY

The empirical study was conducted based on a questionnaire. All items in the questionnaire were designed by a five-point Liker scale. The survey responses can be classified into a first and second group, in accordance with the time received. We started the first group in October 2005 and completed it in March 2006 and started the second one in January 2016 and completed it in May 2016. The first survey was conducted through the part time students in several cities, such as Beijing, Shanghai, Wuhan, and Xian. These students were recommended by the universities for the manufacturing background and longtime manufacturing industrial experience. The filled firms in the second survey were selected randomly from the firm list provided by one web station named as "firms' yellow book".

Totally, 157 questionnaires were collected and 28 invalid questionnaires were identified for different reasons, for example, nineteen filled questionnaires absent firms' message and it is impossible to confirm the filled contents. We remained 129 completed questionnaires for the continuous analysis. The gross sales range from less than 20 million to over 3.3 billion Yuan. The size of firms ranges from less than 100 employees to over 1000 employees. Either big firms or small firms are more than medium scale ones. And PASW Statistics software was used in the analysis.

There are three steps to analyze the relationships. Details are as follows:

(1) The PCA was selected to extract uncertainty and common approach factors. These factors are linear combinations of the original variables and contain most of the information [30].

(2) The Pearson correlation was used to examine the significant differences between two response groups and the reliability of the relationships between the common approaches and the uncertainties. It is a measure of the linear dependency between two random variables [31].

(3) Regression analysis is a statistical technique that is very useful to explore the relationships between two or more variables. And multiple regression model is a regression model that involves situations with more than one regressor variable [32]. Therefore, it was used to confirm relationships between each uncertainty factor and common approach factors.

IV. DATA ANALYSIS

A. PCA

Eleven observation variables of uncertainty are selected and three uncertainty sources are identified according to the PCA results and the definition in literature review. The reasons why we deleted other observation variables were as follows: (1) observation variables which have over 30% load and under 15% gab load on two or more factors; (2) under the premise of over three observation variables, we deleted the observation variables to ensure the reliability of Cronbach's coefficient alpha. The reliability and validity of results were proved to be effective based on the eligible coefficient, e.g., the Cronbach's coefficient alpha, the KMO coefficient alpha.

Based on the literature review and the results of PCA, the three uncertainty sources can be described as: (1) supply instability (U1). It reflects uncertainties that firms met during the supply process including employees' absence, the unqualified rate of raw material; (2) product variety uncertainties (U2). It reflects the present product manufacturing condition and uncertainties in varieties every year; (3) technology upgrade uncertainties (U3). It reflects refresh frequency of upgrade.

B. Pearson correlation analysis

The results of the Pearson correlation found that there were almost no significant differences between two response groups and proved the consistency of survey response and suggest that the data from these two surveys could be used together.

The correlation relationship matrix of uncertainty factors and common approach factors is shown in TABLE II. Based on the table, it is easy to know different relationships existing in different combinations of factors, e.g., the correlation relationship between U2 and C3 and the independence relationship between U2 and C2.

TABLE II
PEARSON CORRELATION COFFICIENTS BETWEEN COMMON APPROACH FACTORS AND UNCERTAINTY FACTOR

	C1	C2	C3	C4	C5
U1	- 0.086	**-0.189**	**-0.169**	**-0.148**	-0.249#
U2	**0.185**	-0.06	0.265#	0.291#	0.256#
U3	0.258#	0.19	0.213#	0.137	0.264#

Note: Bold letters indicate significance level $p<0.05$; #represents significance level $p<0.01$; others have significance level $p>0.05$.

C. Regression analysis

Six common approaches were chose according to the literature [28]. The relationships between the three uncertainty factors and six common approach factors were analyzed with multiple regression analysis and the regression coefficients are listed in the TABLE III. The independent positive variables which have qualified significant level are listed with expected regression coefficients. Every equation has its own coefficients of determination (R^2) and F-statistics which are listed in the bottom of form. No D-W values fall into range of [0, 1.48] and [2.52, 4], which should be rejected in the Durbin-Watson with 1%. The results found that U2 can be influenced by C3, C4and C5 while U3 can be influenced by C1, C2 and C3 and C4. U1 and C6 can't be influenced by any factors so that not listed in the table.

TABLE III
RGRESSION COEFFICIENTS BETWEEN COMMON APPROACH FACTORS AND UNCERTAINTY FACTORS

Independent variables (Uncertainty)	Dependent variables(Common approach)				
	C1	C2	C3	C4	C5
U2			0.244	0.29	0.229
U3	0.247	0.244	0.191		0.260
R^2	0.102	0.092	0.144	0.127	0.207
F-statistics	4.721	4.199	7.004	6.063	10.846
Significance(F)	0.004	0.007	0	0.001	0
D-W	1.807	1.547	1.886	1.720	1.838

Note: Data in rows 2nd-3rd represents standardized regression coefficients with the significance level $p<0.01$.

V. CONCLUSION

It is important to consider the changes and uncertainties in the formulation of manufacturing strategy today. The method is to conclude the common approach of several manufacturing systems and the relationships between the uncertainties and common approach were researched. And the results can be applied in the formulation of manufacturing strategy.

The results of regression indicated that there are certain significant relationships between uncertainties and common approach, which means the six common approach are not effective in dealing with all sources of changes and uncertainties. The hypothesis was proved to be false. However, several specific findings are worthy to be highlighted: (1)Improving integration of information system, improving utilization of production system and function diversification and cooperation can help to deal with product variety uncertainty. (2)Rapid adjustment of manufacturing system, expanding authorization in the manufacturing system, enhancing autonomy and improving integration of information system and improving utilization of production system can help to deal with technology upgrade uncertainty. For the future work, researchers may seek to further quantify the relationships, especially model the relationships between other changes and uncertainties and common approaches.

ACKNOWLEDGE

This research was supported by NSFC(71572104).

REFERENCE

[1] W. Skinner, "Manufacturing – missing link in corporate strategy," *Harvard Business Review*, vol. 47, pp. 136-145, 1969.

[2] M. John, "Setting manufacturing strategy for a company's international manufacturing network," *International Journal of Production Research*, vol. 47, pp. 6179-6203, 2009.

[3] K. A. Chatha, I. Butt, "Themes of study in manufacturing strategy literature," *International Journal of Operations & Production Management*, vol. 35, pp. 604-698, 2015.

[4] J. Wang, D. B. Cao, "Relationships between two approaches for planning manufacturing strategy: A strategic approach and a paradigmatic approach," *International Journal of Production Economic*, vol. 115, pp. 349-361, 2008.

[5] M. Ulf, W. T. Ulrich, "Supply chain flexibility, uncertainty and firm performance: An empirical analysis of German manufacturing firms," *International Journal of Production Economics*, vol. 130, pp. 43-53, 2011.

[6] J. Wang, D. B. Cao, "Relationships between two approaches for planning manufacturing strategy:A strategic approach and a paradigmatic approach," *International Journal of Production Economics*, vol. 115, pp. 349-361, 2008.

[7] M. P. Joshi, P. Kathuria, S. J. Porth, "Alignment of strategic priorities and performance: an integration of operations and strategic management perspectives,"

Journal of Operations Management, vol. 21, pp. 353-369, 2003.
[8] A. K. Sethi, S. P. Sethi, "Flexibility in Manufacturing: A Survey," *The International Journal of Flexible Manufacturing Systems*, vol. 2, pp. 289-328, 2008.
[9] S. Rajeev, "Interplay between uncertainty and flexibility across the value-chain: Towards a transformation model of manufacturing flexibility," *Journal of Operations Management*, vol. 24, pp. 476-493, 2006.
[10] D. Gerwin, "Manufacturing flexibility: A strategic perspective. *Management Science*," *Management Science*, vol. 39, pp. 395-410, 1993.
[11] R. Sven, "Factory design and process optimization with flexibility,"*International Journal of Production Research*,vol. 50, pp. 6060-6071, 2012.
[12] N. He, D. Z. Zhang, Q. Li, "Agent-based hierarchical production planning and scheduling in make-to-order manufacturing system," *Production Economics,* vol.149, pp. 117-130, 2014.
[13] A. K. Jain, H. A. Elmaraghy, "Production scheduling/ rescheduling in flexible manufacturing," *International Journal of Production Research*, vol. 35, pp. 281-309, 2010.
[14] C. Camisón, A. V. López, "An examination of the relationship between manufacturing flexibility and firm performance," *International Journal of Operations & Production Management*, vol. 30, pp. 853-878,2010.
[15] D. M. Upton, "Flexibility as process mobility: The management of plant capabilities for quick response manufacturing," *Journal of Operations Management,* vol. 12, pp. 205-224, 1995.
[16] C. R. Duguay, S. Landry, F. Pasin, "From mass production to flexible/agile production," *International Journal of Operations & Production Management,* vol. 17, pp. 1183-1195, 1997.
[17] P. Leitão, "A holonic disturbance management architecture for flexible manufacturing systems," *International Journal of Production Research*, vol. 49, pp. 1269-1284, 2011.
[18] J. Wang, Q. Z. Wang, H. A. Matsukawa, "Configuration Study on Manufacturing Systems in Turbulent Environment Based on Text Mining," INFORMATION-An International Interdisciplinary Journal, vol. 16, pp. 4627-4638, 2013.
[19] Y. Koren, U. Heisel, F. Jovane, T. Morivaki, G. Pritschow, G. Ulsoy, V. H. Brussel, "Reconfigurable manufacturing system," *CIRP Annual*, vol. 48, pp. 527-540, 1999.
[20] A. Gunasekaran, K. H. Lai, T. C. E. Cheng, "Responsive Supply chain: a competitive strategy in a networked economy," *Omega: The International Journal of Management Science,* vol. 36, pp. 549 - 564, 2008.
[21] D. Vazquez-Bustelo, L. Avella, E. Fernandez, "Agility drivers, enablers and outcomes," *International Journal of Operations and Production Management*, vol. 27, pp. 1303 - 1332, 2007.
[22] C. A. Voss, "Alternative paradigms for manufacturing," *International Journal of Operations & Production Management*, vol. 25, pp. 1211-1222, 2005.
[23] T. L. Bjørge, A. Nuran, B. Harry, F. Jan, "Best manufacturing practices,What do the best performing companies do," *International Journal of Operations & Production Management*, vol. 25, pp. 131-155, 2005.
[24] C. A. Voss, "Paradigms of manufacturing strategy re-visited," *International Journal of Operations & Production Management*, vol. 2, pp. 1223-1227, 2005.
[25] H. P. Wiendahl, P. Scholtissek, "Management and control of complexity in manufacturing," *CIRP Annual*, vol. 43, pp. 533 - 540, 1994.
[26] L. Monostoril, E. Szelke, B. Kadar, "Management of changes and disturbances in manufacturing systems," *IFAC Manufacturing Systems: Modeling Management and Control*, Vienna, Austria, pp. 27-38, 1997.
[27] L. Monostoril, B. Kadar, J. Hornyak, "Approaches to managing changes and uncertainties in manufacturing," *CIRP Annual*, vol. 47, pp. 365-369, 1998.
[28] J. Wang, N. Makayi, D. B. Cao, G. Q. Shao, P. Qu, H. X. Liu, "A contingency Model of Manufacturing Strategy Based on Advanced Manufacturing Paradigms: an Empirical Study," *Industrial Engineering Journal*, vol. 17, pp. 55-59, 2014.
[29] M. Hubert, P. J. Rousseeuw, K.V, Branden, "ROBPCA: A New Approach to Robust Principal Component Analysis," *Technometrics*, vol. 47, pp. 64-79, 2010.
[30] A. Ly, M. Marsman, E. Wagenmakers, "Analytic Posteriors for Pearson' s Correlation Coefficient," *Statistics*, 2015.
[31] D. Montgomery, G. Runger, "Applied Statistics and Probability for Engineers," *Wiley & Sons* , vol. 37, pp. 878-878, 2007.

The Impacts of Trust Relationship on Knowledge Absorptive Capacity: An Empirical Study from Service Outsourcing Enterprises

Zhi-xiong XIAO*
School of Management, Anhui University, Hefei, China
(*e-mail: zhixiongxiao@126.com)

Abstract - **Knowledge absorptive capacity is essential for the development of enterprises in the era of knowledge economy. Trust relationship is the key influencing factor of knowledge absorptive capacity in the service outsourcing enterprises. Building a trust relationship between service outsourcing enterprises must meet four conditions: understanding, profitable, predictable and faithworthy. Above all, this paper proposes two dimensions of trust relationship: interpersonal communication and organizational commitment. Then this paper discusses the impact of the interpersonal communication and organizational commitment on the knowledge absorption capacity. Finally this paper does an empirical research as example of service outsourcing enterprises by using the structural equation model. The results of empirical study show that interpersonal communication, organizational commitment and knowledge absorptive capacity is positive correlation. The conclusion of study has an important role to improve the knowledge absorptive capacity of enterprises.**

Keywords - **Interpersonal communication, knowledge absorptive capacity, organizational commitment, service outsourcing enterprises, trust relationship**

I. INTRODUCTION

Since 90's of the twentieth century, with the great change of enterprise organization and the rapid development of modern technology, managers are aware of the necessity of developing and managing organizations with trust as the core elements. The trust relationship is introduced into the research of knowledge management. The trust relationship is regarded as a kind of important impact factor when scholars study knowledge absorption, knowledge transfer and knowledge sharing.

Agent enterprise and commissioned enterprise in service outsourcing is competition- cooperation relationship. On the one hand, commissioned enterprise need the cooperation with agent enterprise and provides the information and knowledge outsourcing enterprise needs for agent enterprises. On the other hand, agent enterprise and commissioned enterprise is competition relationship because the agent enterprise may implement the vertical integration strategy or diversified development strategy to become commissioned enterprise's competitor if they absorb enough knowledge or technology. Therefore, the commissioned enterprise always sets some limits in the process of imparting knowledge to agent enterprise. For example, when the two sides signed the principal-agent contract, they have a clear agreement with rights and obligations, set clear provisions on the ownership of intellectual property, prohibit the agent enterprise to disclose trade secrets or used for commercial purposes. But a lot of problems in the cooperation cannot be completely foreseen when enterprises sign a contract. The task of outsourcing will be difficult to be completed if the two sides in the cooperation have not necessary trust. The relational contract based on trust is increasingly concerned by the two sides in the cooperation because of the characteristics of service outsourcing.

II. LITERATURE REVIEW

The following is the point of view of some scholars. Inkpen (1998) pointed out in studying the knowledge transfer between the strategic alliances that the mutual trust between the members of alliances can reduce knowledge protection measures of the enterprise and promote mutual learning [1]. Von Hippel (1988) pointed out that trust relationship between the two sides is the prerequisite for cooperation, especially when the system arrangement cannot completely eliminate opportunism behavior, trust relationship is particularly important [2]. Anderson and Narus (1990) showed that the degree of trust and the effect of knowledge transfer of cooperative enterprises is a positive relation [3]. Jones and George (1998) studied the knowledge transfer in the organization. The research showed that the trust relationship can promote internal members to freely exchange the information and knowledge, improve their willingness of sharing and transferring knowledge, make them pursue a common goal [4]. Li, Liu, Li and W (2008) pointed out that the trust in outsourcing cooperation is an effective mechanism to reduce the environmental uncertainty, reduces the risk of knowledge transfer, helps to realize the internalization of goal and culture, and promotes each other's knowledge sharing [5]. Li, Zhang and Liu (2009) pointed out that trust is a necessary precondition for the success of outsourcing in the study of the influencing factors of knowledge absorptive capacity of outsourcing. The two sides of outsourcing are more willing to exchange and cooperation in the trust mechanism. It is conducive for the agent enterprise to absorbing the tacit knowledge [6]. Li-An Ho, Tsung-Hsien

E. Qi et al. (eds.), *Proceedings of the 23rd International Conference on Industrial Engineering and Engineering Management 2016,* DOI 10.2991/978-94-6239-255-7_15

Kuo, Binshan Lin (2012) studied the influence of social identification and trust on organizational online knowledge sharing. The results show that: social identification and trust at workplace have a mediating effect on online knowledge sharing within organizations; trust at workplace is the stronger mediator than social identification; social identification has a positive and significant impact on trust at workplace; and there is significant correlation between knowledge management system quality and social identification as well as trust [7]. Anna Sankowska (2013) presents the role of trust as a mechanism facilitating transfer and creation of knowledge within the company. Results indicate that knowledge creation partially mediates the relationship between trust and innovativeness, and knowledge transfer partially mediates the relationship between trust and knowledge creation [8]. Faizuniah Panqil, Joon Moi Chan (2014) studied the mediating effect of knowledge sharing on the relationship between trust and virtual team effectiveness. The results indicated that knowledge sharing and all the three types of trust are significantly related to virtual team effectiveness. However, only personality-based trust and institutional-based trust are significantly related to knowledge sharing, but knowledge sharing only partially mediates the relationship between these two types of trust and team effectiveness [9]. Werner Rutten, Joyce Blaas-Franken, Harry Martin (2016) analyzed the impact of low trust on knowledge sharing. The study found that the differences in the level of knowledge sharing in high versus low trust situations are significant. The effect is larger for affect-based trust and for implicit knowledge [10].

III. THEORY AND HYPOTHESIS

Levin (2008) thought that trust is a kind of expectation and dependence of individuals to interact with other individuals or groups, and a state of mind or attitude of individuals who is willing to bear the risk for them based on the reliability judgment of other person or group [11]. In the study, trust is usually regarded as a multi dimension concept. McAllister (1995) divided trust into cognition-based trust and affect-based trust [12]. Bart Nooteboom (1996) divided trust into non self-interested trust and intentional trust [13]. Lewicki and Bunker (1996) divided trust into instrumental model of trust and relational model of trust [14]. Vathsala Wickramasinqhe, Ruvini Widyaratne (2012) analyzed the effects of interpersonal trust, team leader support, rewards, and knowledge sharing mechanisms on knowledge sharing in project teams. The study found that Interpersonal trust and rewards have significant positive effects on knowledge sharing, "work-group communications" and "Personal interactions" had significant positive effects on knowledge sharing [15].

The trust relationship is built on the continual interpersonal communication. We know each other's ability and character through the interpersonal relationships and produce emotional connection in the long-term exchanges. So the establishment of trust relationship cannot do without interpersonal communication. However, the establishment of trust relationship must have a credible commitment when the organization activities involve the interests. Therefore, interpersonal communication and organizational commitment are regarded as the two important dimensions of trust relationship in this study.

The trust relationship has effect on knowledge acquisition, knowledge digestion, knowledge transformation, knowledge utilization in the process of knowledge absorption. Good trust relationship helps to form and improve the knowledge absorptive capacity of service outsourcing enterprise.

A. Interpersonal communication and knowledge absorptive capacity of service outsourcing enterprises

The social embeddedness theory thinks trust is always embedded in the structure of interpersonal relationship. If the mutual trust increases, the opportunism and the prisoner's dilemma is fewer, the strict control and the protection of contract can be reduced in transactions. Therefore, the establishment of the trust relationship based on interpersonal communication will reduce the intellectual protection of nominee enterprise. Especially when the the guarantee of formal contract is limited, the establishment of interpersonal trust relationship will makes a lot of economic activities carry out smoothly. In a harmonious business relationship, the reduction of knowledge transfer limition and the increasement of high frequency and deep communication will improve the efficiency of the acquisition and absorption of tacit knowledge. So this study puts forward the hypothesis H1: there is a positive correlation between interpersonal communication and knowledge absorptive capacity.

B. Organizational commitment and knowledge absorptive capacity of service outsourcing enterprises

Building trust relationship cannot do without organizational commitment. Organizational commitment includes formal contract that has a legally binding (e.g. contract) and relational contract that has no legal binding (e.g. oral agreement). Whether the formal contract and relational contract can guarantee smooth cooperation depends on the reliability of commitment. Cooperation between the two sides is actually a game relationship. If the commitment is not credible, the side feels unfair in cooperation may break the contract and lead to the break-up. The credible organizational commitment and the hundred-percent execution organizational commitment is the basic guarantee of trust relationship. The credibility of organizational commitment is higher, the trust relationship is stronger, the knowledge absorption of service outsourcing enterprise is more conducive. So this study puts forward the hypothesis H2: there is a positive correlation between organizational commitment and knowledge absorptive capacity.

IV. STUDY DESIGN

The design of the questionnaire in this study refers to a number of research results, especially the results of empirical testing. In this study, there are two level-one variables, those are trust relationship and knowledge absorptive capacity. There are six level-two variables. The level-one variable trust relationship consists of two level-two variables: interpersonal communication and organizational commitment. The level-one variable knowledge absorptive capacity contains four level-two variables: knowledge acquisition capacity, knowledge digestion capacity, knowledge transformation capacity, knowledge utilization capacity. Each of the level-two variables set the number of observed variables based on the above theoretical analysis. All is 18 observed variables.

This study selects as much as possible some of the service outsourcing enterprises with obvious characteristics in making the selection of sample because these outsourcing enterprises have a higher demand for knowledge and the knowledge absorptive capacity is more important for the development of enterprises. This study selected 16 service outsourcing enterprises. 220 questionnaires were issued by field interviews, e-mail, letters, and finally 192 questionnaires recovered. Recovery rate is 87%. The valid questionnaires are 168 and account for 88% of the questionnaire recovered.

V. DATA ANALYSIS AND RESULTS

A. The reliability analysis of the data

Reliability is used to test the correctness and accuracy of the measurement. This study uses the Cronbach alpha values as reliability criteria. Generally Cronbach alpha value greater than 0.7 indicates that the scale has a relatively high reliability.

1) *The reliability analysis of trust relationship*: The CITC index of the six questions of trust relationship is greater than 0.3 that is the minimum standard. The maximum value is 0.829 and the smallest value is 0.679. Total reliability coefficient of all questions of interpersonal communication is 0.825, total reliability coefficient of all questions of organizational commitment is 0.825. Total reliability coefficient of all questions of trust relationship is 0.917. All reliability coefficient is greater than 0.7. It can be seen from TABLE I. As shows that the survey data is reliable and meets the requirements of research.

TABLE I
CITC AND ALPHA COEFFICIENT OF THE MEASUREMENT ITEMS OF TRUST RELATIONSHIP

Measurement Items	Corrected Item-Total Correlation	Cronbach's Alpha if Item Deleted	Alpha Coefficient	Total Alpha Coefficient
TR11	.756	.904	.825	.917
TR12	.767	.905		
TR13	.791	.898		
TR21	.829	.893	.825	
TR22	.817	.898		
TR23	.679	.913		

2) *The reliability analysis of knowledge absorptive capacity*: The CITC index of the twelve questions of knowledge absorptive capacity is greater than 0.3 that is the minimum standard. The maximum value is 0.871 and the smallest value is 0.521. Total reliability coefficient of all questions of knowledge acquisition capacity is 0.825, total reliability coefficient of all questions of knowledge digestion capacity is 0.732, total reliability coefficient of all questions of knowledge transformation capacity is 0.713, total reliability coefficient of all questions of knowledge utilization capacity is 0.857. Total reliability coefficient of all questions of knowledge absorptive capacity is 0.939. All reliability coefficient is greater than 0.7. It can be seen from TABLE II. As shows that the survey data is reliable and meets the requirements of study.

TABLE II
CITC AND ALPHA COEFFICIENT OF THE MEASUREMENT ITEMS OF KNOWLEDGE ABSORPTIVE ABILITY

Measurement Items	Corrected Item-Total Correlation	Cronbach's Alpha if Item Deleted	Alpha Coefficient	Total Alpha Coefficient
KA11	.871	.928	.825	.939
KA12	.836	.929		
KA13	.595	.938		
KA21	.549	.939	.732	
KA22	.672	.935		
KA23	.827	.930		
KA31	.628	.937	.713	
KA32	.773	.932		
KA33	.521	.940		
KA41	.746	.933	.857	
KA42	.820	.930		
KA43	.822	.930		

B. The validity analysis of data

In order to verify whether the variables designed suit the structural equation model, this study will further analyze the scale by convergent validity and discriminant validity. The convergent validity tests the measure results of the significant variables on the latent variables. Discriminant validity tests the difference between significant variables, the difference between the latent variables.

1) *The validity analysis of trust relationship*: The Goodness of fit statistics of confirmatory factor analysis of trust relationship show $\chi 2/df$ equals 4.4, RMSEA equals 0.13. It can be seen from TABLE III. As shows

that the survey data is valid and meets the requirement s of research.

TABLE III
THE CONFIRMATORY FACTOR ANALYSIS OF TRUST RELATIONSHIP

Latent Variable	Measurement Items	Standardized Factor Load	Standard Error	t Value
interpersonal communication	TR11	0.79	0.06	12.18
	TR12	0.78	0.07	11.95
	TR13	0.80	0.06	12.39
organizational commitment	TR21	0.86	0.06	13.83
	TR22	0.82	0.06	12.95
	TR23	0.70	0.07	10.25
Goodness of fit statistics:				
χ^2=35.43	χ^2/df=4.4	RMSEA=0.13	GFI=0.94	
AGFI=0.85	NFI=0.97	NNFI=0.95	CFI=0.97	

2) *The validity analysis of knowledge absorptive capacity*: The Goodness of fit statistics of confirmatory factor analysis of knowledge absorptive capacity show χ2/df equals4.595, RMSEA equals 0.16. It can be seen from TABLE IV. As shows that the survey data is valid and basically meets the requirements of study.

TABLE IV
THE CONFIRMATORY FACTOR ANALYSIS OF KNOWLEDGE ABSORPTIVE ABILITY

Latent Variable	Measurement Items	Standardized Factor Load	Standard Error	t Value
knowledge acquisition capacity	KA11	0.90	0.06	14.85
	KA12	0.87	0.06	14.10
	KA13	0.62	0.07	8.82
knowledge digestion capacity	KA21	0.59	0.07	8.17
	KA22	0.69	0.07	9.80
	KA23	0.86	0.06	13.31
knowledge transformation capacity	KA31	0.69	0.07	9.82
	KA32	0.84	0.07	12.64
	KA33	0.52	0.08	6.93
knowledge utilization capacity	KA41	0.75	0.07	11.24
	KA42	0.84	0.06	13.23
	KA43	0.86	0.06	13.79
Goodness of fit statistics:				
χ^2=220.56	χ^2/df=4.595	RMSEA=0.16	GFI=0.80	
AGFI=0.67	NFI=0.94	NNFI=0.93	CFI=0.95	

C. SEM model and hypothesis testing

1) *The checking results of model*: The results of model testing of trust relationship influencing on knowledge absorptive capacity are shown in Fig.1.

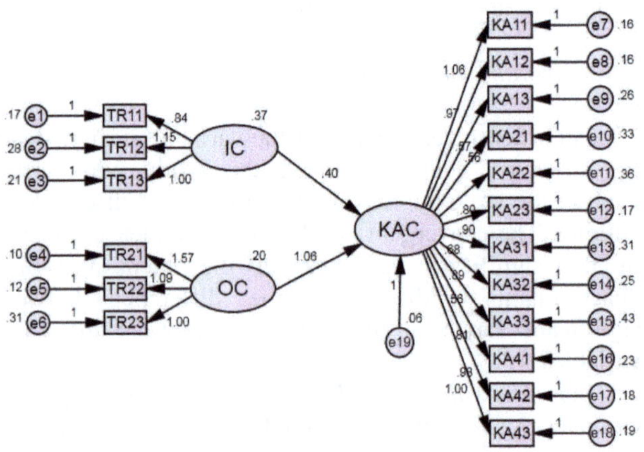

Fig.1. The impact of trust relationship on knowledge absorptive capacity

2) *The results of hypothesis testing*: The results of hypothesis testing show that all assumptions are true.

VI. CONCLUSION

From the results of structure equation model, the positive correlation between interpersonal communication, organizational commitment and knowledge absorptive capacity are basically confirmed, but the effect of interpersonal relationship was not significant. There are several possible reasons. Firstly, from the test result of the relationship between interpersonal communication and knowledge absorptive capacity, Chinese regards communication as a way to build relationships, but the communication on the level of knowledge not only is a communicative skill, but also needs more knowledge and skills as a strong support. So the interpersonal communication we usually think has obvious effect to obtain the agency business, but is not significant for the improvement of knowledge absorptive capacity. Secondly, from the test result of the relationship between organizational commitment and knowledge absorptive capacity, enterprises are more willing to believe the legally enforceable contract .On one hand, it shows the achievements of legal system construction in China, but on the other hand, it also shows that enterprise's credibility and integrity should be strengthened.

Building a trust relationship between service outsourcing enterprises must meet four conditions: understanding, profitable, predictable and faithworthy. Firstly, building a trust relationship must strengthen the interpersonal communication. Interpersonal communication is the process of transmitting information of both sides in cooperation, is a kind of efforts and coordination of both sides to achieve their goals, is a means to maintain the relationship between the two sides. Noorderhaven (1996) thought the inner core of people is in a state of division, including two kinds of trust and opportunism [16]. This division makes the trustworthy person not necessarily to abide by the commitments in any case. Interpersonal communication can create a transaction situation. Although the core of people can't completely be changed, the transaction situation can cause the two properties in a certain extent. So the interpersonal communication can change the degree of subjective trust. Secondly, building a trust relationship is inseparable from the economic benefit or other benefits. The trust relationship between enterprises always occurs in economic exchange activities between enterprises. The direct target of enterprises engaged in these activities is to complete the deal or task, the ultimate goal is to reduce the cost to gain a competitive advantage, or obtain the technology, knowledge and management experience for the development of enterprise, or extend the market. In short, if unprofitable, the establishment of trust will lose the foundation. Thirdly, Building a trust relationship requires a party of cooperation must give the other party predictable interest returns. The procurability of interest depends on the probability, the time limit and reliability of gaining interest. In the fierce market competition environment, no enterprise is willing to cooperate for a long and unavailable interest. Moreover, building a trust relationship requires a party of cooperation must give the other party faithworthy commitments.

ACKNOWLEDGMENT

This research was supported by the National Social Science Fund of China (NO. 12CTQ033) and the Humanities and Social Sciences Research Project of Anhui (No. SK2016A0067).

REFERENCES

[1] A.C. Inkpen, "Learning, knowledge acquisition, strategic alliances," European Management, Vol.16, pp.223-229, 1998.
[2] E.Von Hippel, "The sources of innovation," New York: Oxford University Press, 1998.
[3] J.C. Anderson and J.A. Narus, "A model of distributor firm and manufacturing firm working partnerships," Journal of Marketing, Vol.54, pp.42-58, 1990.
[4] G.R. Jones and J.M. George, "The experience and evolution of trust: Implications for cooperation and teamwork," Academy of Management Review, Vol.23, pp.531-546, 1998.
[5] Y. Li, Y. Liu, M.F. Li, and H.B. W, "Transformational offshore outsourcing: Empirical evidence from alliances in China," Journal of Operations Management, Vol.26, pp.257-274, 2008.
[6] X.Y. Li, X.W. Zhang and H. Liu, "Analysis of influencing factors of knowledge absorptive capacity of outsourcing," Progress and Countermeasures in Science and Technology, Vol.26, No.5, pp.110-113, 2009.
[7] Li-An Ho, Tsung-Hsien Kuo, Binshan Lin, "How social identification and trust influence organizational online knowledge sharing," Internet Research, Vol. 22, no.1, pp.4-28, 2012.
[8] Anna Sankowska, ""Relationships between organizational trust, knowledge transfer, knowledge creation, and firm's innovativeness," The Learning Organization, Vol. 20, no.1, pp.85-100, 2013.
[9] Faizuniah Panqil, Joon Moi Chan, "The mediating effect of knowledge sharing on the relationship between trust and virtual team effectiveness," Journal of Knowledge Management, Vol. 18, no. 1, pp.92-106, 2014.
[10] Werner Rutten, Joyce Blaas-Franken, Harry Martin, "The impact of (low) trust on knowledge sharing," Journal of Knowledge Management, Vol. 20, no. 2, pp.199-214, 2016.
[11] D.Z. Levin, "Trust in international encyclopedia of organization studies," Newbury Park CA: Sage Publications Inc, 2008.
[12] D.J. McAllister, "Affect and cognition-based trust as foundation for interpersonal cooperations," Academy of Management, Vol.38, pp.24-59, 1995.
[13] Bart Nooteboom, "Trust, opportunism and governance: A process and control model", Organization Studies, VOL.17 NO.6, pp. 985-1010, 1996.
[14] R. J. Lewicki and B.B. Bunker, "Developing and maintaining trust in work relationships", R.M. Kramer and T.R.Tyler (Eds), Trust in Organizations: Frontiers of Theory and Research. Thousand Oaks: Sage Publications, pp.114-139, 1996.
[15] Vathsala Wickramasinqhe, Ruvini Widyaratne, "Effects of interpersonal trust, team leader support, rewards, and knowledge sharing mechanisms on knowledge sharing in project teams," VINE, Vol. 42, no. 2, pp.214 - 236, 2012.
[16] N.G. Noorderhaven, "Opportunism and trust in transaction cost economics," Norwell, MA: Kluwer Academic Publishers, pp.105-128, 1996.

Determination of price and warranty length under claim rate of customer

Zi-xian LIU*, Danning LIU
College of Management and Economics, Tianjin University, Tianjin 300072, China
(*liuzixian@tju.edu.cn)

Abstract–**In this paper, a decision model is developed for producers to determine the optimal price and warranty length to maximize profit based on the pre-determined life cycle and the rate of customer claim. The pro-rata warranty (PRW) policy is considered under which producers refund a fraction of sale price if an item fails in the warranty period. The objective function includes both demand and cost function. The variation tendency of producers' profit will be derived for two different life distributed products. Some conclusions are drawn based on the sensitivity analysis.**

Keywords–**pro-rata warranty, price, warranty length, claim rate**

I. INTRODUCTION

Price and quality of product are the two major competitive tools of producer [1]. Lowering price will stimulate consumption, while price too low can't cover the cost of product with high quality. The quality of product will influence the user experience and the sales volume in a long time. To obtain maximum profit in the long run, a producer has to optimize the product price and quality [2].

Since most of customers are lack of professional knowledge and time, they can't complete a comprehensive evaluation of a product quality. Therefore, customers usually consider price as an indicator of product quality [3]. In addition, the warranty policy is also a standard of product quality and reliability [4-5].A better warranty policy will improve the sales rate and also bring additional costs to the producer [6-7]. The product warranty cost is closely related to its reliability and warranty policy[8-9].Optimal designing the warranty policy is an important way to reduce the warranty cost and obtain maximum profit. Many researchers proposed different models to optimize the repair times, repair interval and warranty length. Wang Z [10] integrated warranty, repair and reliability and considered two different maintenance modes which are the perfect maintenance and the minimal maintenance to reach a higher reliability with a lower cost under a certain warranty length. Chang W and Lin J H [11] optimized the number and degree of preventative maintenance to maximize the producer expected total profit. In addition to maximize the producer profit, some researchers also worked on the maximum of customer utility. Mahoudi A and Shavandi H[12] analyzed price, warranty length and service capacity under a fuzzy environment to maximize the profit function and minimize customer waiting time in queue of sales services. Pulido H and Aguirre-Torres [13] quantified the relationship between warranty length and sales volume and the relationship between warranty cost and customer satisfaction levels based on the utility function to optimize the warranty length and maximize the expected utility.

To study the influence of lifetime distribution on the optimal solution, we will deal with the problem of determining the optimal price and warranty length for products with two different lifetime distributions. The two different lifetime distributions are exponential distribution (ED) and normal distribution (ND). The failure rate of product with exponential lifetime distribution is highest in the initial stage of using and then the failure rate will gradually reduce. The failure rate of product with normal lifetime distribution is low in the initial stage of using and the failure rate will increase at first and then decrease. We consider the pro-rata warranty which producers refund part of sale price if an item fails during the warranty period. And according to the warranty period we also adjust the rate of customer claim. There are many reasons for customers not performing maintenance when the product failed before the end of warranty length. For example, customer will purchase products from the competitor of primary producer after the product is failed because of the uncomfortable experience during the usage [14]. So in this paper, the rate of customer claim changed over time will be considered when optimizing price and warranty length.

The rest of the paper is organized as follows: in Section 2, the notations used in the formulation of the model and the descriptions of the model to maximize profit are explained. The economic sensitivity analysis will be presented to evaluate the effect of the model parameters on the optimal solution of the decision model in Section 3. Finally, some conclusions will be drawn based on the analysis.

E. Qi et al. (eds.), *Proceedings of the 23rd International Conference on Industrial Engineering and Engineering Management 2016,* DOI 10.2991/978-94-6239-255-7_16

II. MODEL DESCRIPTION

To determine the optimal price and warranty length, we developed a price-warranty model in the present paper. This study uses the following notation:

p	the price of the product
w	the warranty length of the product
J	the present worth of profit over the planning horizon
T	the planning horizon
r	the discount rate
k_1	a constant of amplitude factor
k_2	a constant of time displacement
a	the price elasticity
b	the displaced warranty length elasticity
c_0	the fixed cost
c_1	the variable cost
β	the learning rate
c_m	the maintenance cost of producer

By modifying the existing model and considering free renewal warranty, Wu CC and Lin P C (2006) proposed a price-warranty decision model, which include the discount rate and the planning horizon [15]. For this dynamic optimization problem, the objective function is

$$max\,J = \int_0^T e^{-rt}[p(t) - c(t)] \times d(t)dt \tag{1}$$

$$c(t) = c(w(t), Q(t)) \tag{2}$$

$$d(t) = d(p(t), w(t)) \tag{3}$$

$$s.t. \quad \frac{dQ(t)}{dt} = d(p(t), w(t), Q(t)) \tag{4}$$

where $p(t)$ is the unit price at time t, $c(t)$ is the unit cost at time t, $d(t)$ is the sale rate at time t, $w(t)$ is the warranty length at time t, $Q(t)$ is the cumulative sales volume at time t.

The function of sale rate originates from the demand function in Clickman and Berger's model [16], i.e.

$$d(t) = d\big(p(t), w(t)\big) = k_1 p^{-a}(w + k_2)^b \tag{5}$$

Wu CC and Lin P C proposed the model under the free renewal warranty, so the cost function in the model is

$$c(t) = c\big(w(t), Q(t)\big) = c_0(1 + Q^s M(w)) \tag{6}$$

$s = log\beta/log2$, and $M(w)$ is the excepted number of renewals of the product with warranty length w.

In the present paper, because we consider the pro-rata warranty (PRW) policy, we have to modify the cost function in Wu CC and Lin P C's model.

According to Blischke W R and Karim M R (2011)[7], the excepted maintenance cost of producer under PRW is

$$E\big(c_m(w)\big) = h(w)[F(w) - \tfrac{\mu_w}{w}]p \tag{7}$$

$$\mu_w = \int_0^w tf(t)dt \tag{8}$$

$$F(w) = \int_0^w f(t)dt \tag{9}$$

Where μ_w is the excepted value of product fail time during the warranty length w, $f(t)$ is the lifetime distribution of product. We will consider two different lifetime distributions which are the exponential distribution (ED) and the normal distribution (ND). And $h(w)$ is the average of the rate of customer claim of the product with warranty period w. When the warranty period is small, the produce fails in the warranty period will be a big loss for customers and most customers will make claim. But with the increase of the warranty period, some customers will give up claim because the guarantee was lose or the customer has already purchased products from the competitor of primary producer after the product was failed. So in this paper, we hypothesize the trends of $h(w)$ is as Fig. 1. As can be seen in Fig. 1, with the increase of the warranty length w, the number of uncertainties is increasing and $h(w)$ will decrease. After the warranty length w is bigger than a certain number, $h(w)$ will be at a stable level. So we hypothesize that $h(w)$ is

$$h(w) = \begin{cases} 1 & w \le w_1 \\ 1 - (w - w_1) \times 0.05 & w_1 < w \le w_2 \\ 1 - (w_2 - w_1) \times 0.05 & w_2 \le w \end{cases} \tag{10}$$

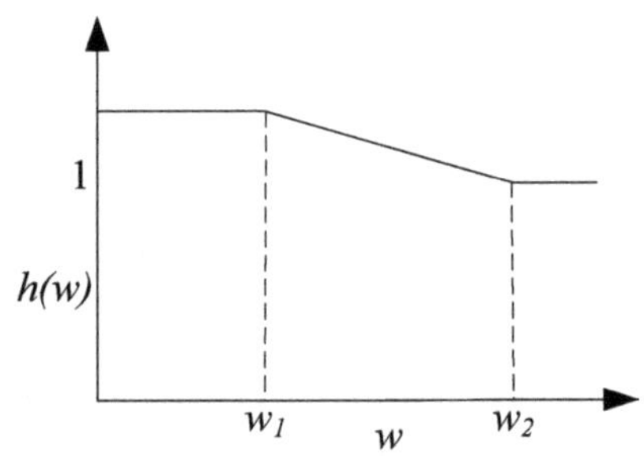

Fig.1 The average of the rate of customer claim

After obtain the excepted maintenance cost of producer, the unit cost at time t can be expressed by

$$c(t) = \frac{c_0}{Q} + c_1 + E[c_m(w)] \tag{11}$$

We are now concerned with the problem in which the present value of profit, in terms of two decision variables (i.e. price and warranty length) and two state variables (i.e. cumulative sale volume and the rate of customer claim), is to be maximized. The decision model with a discount rate r and a planning horizon T of the product can formulated as

$$max\,J = \int_0^T e^{-rt}[p(t) - c(t)] \times d(t)dt \tag{12}$$

$$s.t. \quad \frac{dQ(t)}{dt} = d(p(t), w(t), Q(t)) \tag{13}$$

The optimization problem in Eqs. (12) and (13) may be solved by applying the maximum principle. To apply the maximum principle, we first obtain the present value of Hamiltonian, and then apply the Heissian matrix (HM) for optimization. We also proved the HM is a negative definite matrix.

III. THE SENSITIVITY ANALYSIS

In this section, a sensitivity analysis is applied to show how the model parameters affect the optimal price and warranty length. To obtain the analysis result, we have to make certain assumptions.

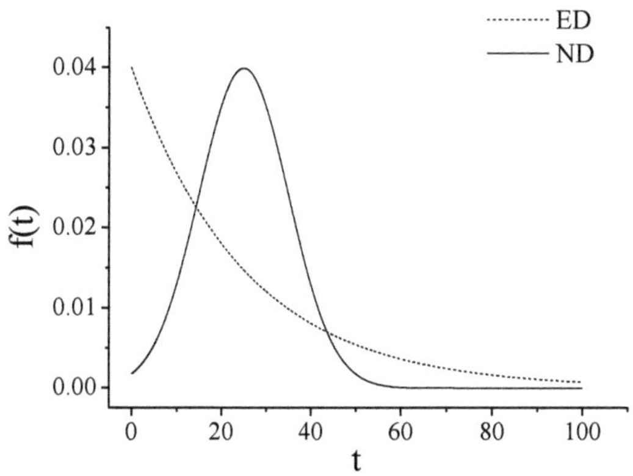

Fig. 2 The two different lifetime distribution

At first, the two different lifetime distributions ED and ND are displayed in Fig. 2 and they have the same mean value.

And we assume, as assumed by Glickman and Berger [16], that $k_1=10^3$ and $k_2=2$. The model parameters under the study include: the price elasticity (a); the warranty length elasticity (b); the fixed cost (c_0); the variable cost (c_1); the discount rate (r); and the average rate of the customer claim ($h(w)$), where a is from 1.5 to 2.5, b is from 0.1 to 0.9, c_0 is 50, c_1 is 0.4, r is from 0 to 0.1 and $h(w)$ is from 0.9 to 1. The results from sensitivity analysis are shown in TABLE I.

TABLE I
SENSITIVITY ANALYSIS UNDER PRW

r	(a, b)	ED	ND
		$(w, p, \text{Max } J)$	$(w, p, \text{Max } J)$
	(1.5, 0.1)	(1.4, 1.23, 6530)	(11.8, 1.24, 7501.9)
	(1.5, 0.5)	(30.0, 1.92, 16907)	(23.6, 1.37, 25120)
	(1.5, 0.9)	(57.8, 2.66, 73095)	(31.5, 1.55, 97943)
	(2.0, 0.1)	(0.5, 0.81, 6649)	(10.3, 0.82, 7608)
0.1	(2.0, 0.5)	(13.1, 1.02, 14780)	(20.7, 0.88, 24359)
	(2.0, 0.9)	(32.1, 1.32, 55087)	(27.7, 0.97, 90040)
	(2.5, 0.1)	(0.06, 0.67, 7811)	(9.1, 0.68, 8866.9)
	(2.5, 0.5)	(8.7, 0.79, 15695)	(18.8, 0.72, 27409)
	(2.5, 0.9)	(30.0, 1.07, 51122)	(24.5, 0.77, 97420)
	(1.5, 0.1)	(1.4, 1.23, 65906)	(11.8, 1.24, 75632)
0	(1.5, 0.5)	(30.0, 1.92, 169780)	(23.6, 1.37, 251980)
	(1.5, 0.9)	(57.8, 2.66, 732180)	(31.5, 1.55, 980900)

Comparing the optimal result of products with different life time distributions in TABLE I, the most different between the optimal result is the optimal warranty length. Fig. 3 is the comparison of the optimal warranty length for two lifetime distributions ED and ND by varying b.

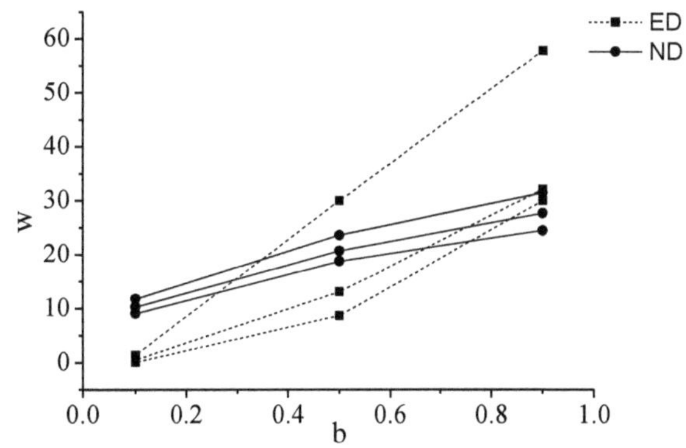

Fig.3 Comparison of the optimal warranty length for ED and ND by varying b

In Fig. 3, as to the failure rate of product with an exponential lifetime distribution is highest in the initial stage of using and then the failure rate will gradually reduce, to obtain a lower warranty cost the optimal warranty length will be short when b is small and the optimal warranty length will grow fast with the increasing of b. In addition, as to the failure rate of product with a normal lifetime distribution is low in the initial stage of using the optimal warranty length will still be long when b is small. But the following grow of warranty length is very slow.

The effect of $h(w)$ on the optimal solution can be seen in TABLE II. From TABLE II, we may conclude that: as a=1.5, b=0.1, the optimal price, warranty length and profit are same in the two conditions. Because in this case the optimal warranty length is 1.4 which is between 0 and w_1. As can be seen in Fig.3 the optimal warranty length is growing with b. When b=0.5 the optimal warranty length will be bigger than w_1 and the changeable $h(w)$ will be smaller than 1, since a smaller $h(w)$ corresponds to a lower warranty cost this condition is good for producer to get more profit.

TABLE II
EFFECT OF THE CUSTOMER CLAIM ON OPTIMAL SOLUTION (ED)

(a, b)	Changeable $h(w)$	$h(w)$=1
	$(w, p, \text{Max } J)$	$(w, p, \text{Max } J)$
(1.5, 0.1)	(1.4, 1.23, 6530)	(1.4, 1.23, 6530.3)
(1.5, 0.5)	(30, 1.92, 16907)	(16.8, 1.65, 16329)
(1.5, 0.9)	(57.8, 2.66, 73095)	(37.6, 2.32, 61979)

According to the sensitivity analysis, the effect of the model parameters on the optimal solution can be summarized in TABLE III.

TABLE III
EFFECT OF MODEL PARAMETERS OPTIMAL SOLUTION*

Parameter increased	w	p	Max J
a	(-)	(-)	(-)( )(+)
b	(+)	(+)	(+)
r	( )	( )	(-)
$h(w)$	(-)	(-)	(-)

*(+) indicates increasing, (-) indicates declining, and ( ) indicates unchanged

The effect of the model parameters on the optimal solution may be summarized in TABLE III. From TABLE III, the following conclusions can be drawn:

1. An increase in price elasticity leads to a declining price and a declining warranty length; however, the profit can be declining, increasing or unchanged. This result indicates that as the price elasticity of the market increase, demand is relatively sensitive to price, and the producer should reduce the price to attract more customers. And to make the lower price bigger than cost, the producers have to decline the warranty length. So when the decreased warranty cost is bigger than the decreased price, the present worth of profit over the planning horizon may increase. But when the decreased warranty cost is smaller than the decreased price, the present worth of profit over the planning horizon may decrease.
2. An increase in warranty length elasticity leads to an increasing price, an increasing warranty length and an increasing profit. This result indicates that as the warranty length elasticity of the market increase, demand is relatively sensitive to warranty length, and a longer warranty length will attract more customers. As can be seen in Fig.1, *h(w)* decrease as warranty length increasing. So with the increase of warranty length the warranty cost will increase at a lower rate. As a result, the increase of price will be bigger than the increase of cost and the cumulative sales volume will also increases, so does the profit.
3. The discount rate has no effect on the price and warranty length, and it can just influence the value of the present worth of profit over the planning horizon.
4. An increase in *h(w)* will leads to a declining price and a declining warranty length. Because an increase in *h(w)* will leads to an increasing in the warranty cost, the producer has to decline the warranty length to make balance. In the same time, because of the decline of warranty length, the producers need to decline the price to attract customer.

IV. CONCLUSION

In this paper, we modify the existing model for producers to determine the optimal price and warranty length for production with different lifetime distributions to maximize profit based on the pre-determined life cycle. The pro-rata warranty (PRW) policy and changeable rate of customer claim are also considered.

A low rate of customer claim is good to reduce the warranty cost for producer. But a decreasing rate of customer claim can reflect the dissatisfaction of customer to some extent. So producers should hypothesize a reasonable rate of customer claim when they optimize the price and warranty period. And producers should also find the reason of the low rate of customers claim and then satisfy customers through increasing warranty length or decreasing price.

When the lifetime distribution of productions is different, even they have the same mean value, the optimal strategies are different. Since the second derivative of cost with respect to warranty length (i.e., c_{ww}) will affect the warranty length's reflex of the warranty length elasticity, we can see the influence of lifetime distribution on the optimal solution through the analysis of the second derivative of cost with respect to warranty length (i.e., c_{ww}). When the lifetime distribution is exponential distribution, the c_{ww} will always be smaller than 0, so with the increase of warranty length the warranty cost will increase at a lower rate which will lead to a higher rate of the warranty length's increasing with the increase of warranty length elasticity. And when the lifetime distribution is normal distribution the c_{ww} will be larger than 0 at the first stage and then it will be smaller than 0, so with the increase of warranty length elasticity the warranty length will increase at a lower rate.

REFERENCE

[1] Wei J, Zhao J, Li Y. Price and warranty period decisions for complementary products with horizontal firms' cooperation/noncooperation strategies [J]. Journal of Cleaner Production, 2014, 105:86 - 102.
[2] Kapur P K, Sachdeva N. Managing Warranty Length & Price in the Presence of Customer Dissatisfaction [J]. Amity Global Business Review, 2015, 10: 64-74.
[3] Faridimehr S, Sta N, Faridimehr S. Optimal strategies for price, warranty length, and production rate of a new product with learning production cost [J]. Scientia Iranica, 2013, 20(6):2247-2258.
[4]Akdeniz M B, Calantone, R J, Voorhees, C M. Signaling Quality: an Examination of the Effects of Marketing- and Nonmarketing-Controlled Signals on Perceptions of Automotive Brand Quality [J]. Journal of Product Innovation Management, 2014, 31(4): 728-743.
[5]Stock A, Balachander S. The Making of a "Hot Product": a Signaling Explanation of Marketers' Scarcity Strategy [J]. Management Science, 2005, 51(8): 1181-1192.
[6]Darghouth M N, Chelbi A, Ait-kadi D. A Profit Assessment Model for Equipment Inspection and Replacement under Renewing Free Replacement Warranty Policy [J]. International Journal of Production Economics, 2012, 135(2): 899-906.
[7]Kim B, Park S. Optimal Pricing, EOL (end of life) Warranty, and Spare Parts Manufacturing Strategy Amid Product Transition [J]. European Journal of Operational Research, 2008, 188(3): 723-745.
[8] Anisur R, Gopinath C. Review of Long-term Warranty Policies [J]. Asia-Pacific Journal of Operational Research, 2006, 23(4): 453-472.

[9] Huang H Z, Liu, Z J, Murthy D N P. Optimal Reliability, Warranty and Price for New Products [J]. IIE Transactions, 2007, 39(8): 819-827
[10] Wang Z, Huang H Z, Du X. Optimal design accounting for reliability, maintenance, and warranty [J]. Journal of Mechanical Design, 2010, 132(1): 011007.
[11] Chang W L, Lin J H. Optimal maintenance policy and length of extended warranty within the life cycle of products [J]. Computers & Mathematics with Applications, 2012, 63(1):144-150.
[12] Mahmoudi A, Shavandi H. Analyzing price, warranty length, and service capacity under a fuzzy environment: Genetic algorithm and fuzzy system [J]. Scientia Iranica, 2013, 12(3):975-982.
[13]Pulido H G, Aguirre-Torres, J. A. Christen. A Bayesian approach for the determination of warranty length [J]. Journal of Quality Technology, 2006, 38(2):180-189.
[14] Blischke W R, Karim M R, Murthy D N P. Warranty Data Collection and Analysis [J]. Springer, 2011.
[15] Wu C C, Lin P C, Chou C Y. Determination of price and warranty length for a normal lifetime distributed product [J]. International Journal of Production Economics, 2006, 102(1):95-107.
[16] Glickman T S, Berger P D. Optimal price and protection period decisions for a product under warranty [J]. Management Science, 1976, 22(12): 1381-1390.

Research on Module Construction based on Product Lifecycle Management

Ming-kun Li[1, 2*], Jin-qi Shi[1, 2], Xiao-juan Cao[2], Zhong Wang[2]
[1] Science and Technology on Underwater Information and Control Laboratory
[2] Xi'an Precision Machinery Research Institute, Xi'an, China
([*]limingkun0675@163.com)

Abstract - **Facing the serious situation, in order to cope with the future challenge, scholars put forward many new development concepts, management modes and advanced manufacturing technologies. The common point is the rapid response to market demand has become the core of competition in the market. This paper summarized the method of modules construction, analyzed of the modules construction strategy based on product lifecycle management (PLM), and explained with example of table lamp.**

Keywords - **Product lifecycle management (PLM), module construction, strategy**

I. INTRODUCTION

Along with the formation of the global market, the rapid development of communication technology, biology and materials, fundamental changes have taken place in the external situation of product manufacturing. The producers are eager to develop the product in a low-cost and eco-friendly way. The optimal allocation of manufacturing resources has become a hot area of research for several years. And strategies based on modularity are the best way to deal with that change [1]. The modularity can divide the complex system into several sub-systems, which simplifies the difficulty of product analysis. PLM is of great significance as it can improve the development of new products and reduce manufacturing costs by controlling the products through their lifecycle [2]. Therefore, the research of module construction based on product lifecycle management (PLM) is of great importance.

II. PRODUCT LIFECYCLE MANAGEMENT

A. The concept of PLM

There are various concepts of the PLM. For example:

The company CIMdata thinks PLM integrate people, processes, and information together effectively by implementing a set of business solution. PLM plays an important role in the whole enterprise, from concept to scrap of products in lifecycle. Supporting the collaboration of definition information related to product in research, development, management, distribution and service.

Advanced Market Research (AMR) considers PLM as an auxiliary strategy of technology by integrating single point across the business process and different user groups.

Electronic Data Systems (EDS) believes PLM is a product as the core business strategy. It applies a series of business solutions to support the generation, management, distribution and use of product definition information in a coordinated way.

B. The characteristics of PLM

Despite the different emphases on concept of PLM, some typical characteristics can be extracted:

- PLM is an enterprise informatization strategy, which needs to plan PLM system from the strategic level, including the system, tools and implementation method, etc.;
- PLM ranges across all stages of the enterprise from product concept to product eliminate;
- The object of PLM is the product information, which not only includes the defined data, but also describes the process of product in design, manufacture, use and service;
- PLM is responsible for the acquisition, processing, transfer and storage of information produced by various applications such as computer aided design (CAD), computer aided manufacturing (CAM) and customer relationship management (CRM) [3];
- The purpose of PLM is to achieve the collaborative process of definition, manufacturing and management in product lifecycle through information technology.

C. The demands of the product module construction based on PLM

The core of PLM is a product data model, the data model integrates all kinds of data involving in the product lifecycle in a unified management platform, such as the data of product development, design, manufacture, sale, use, service and recycling [4]. The employees, partners and end users of enterprise can work together efficiently through this platform.

Combined with PLM, divide the product into several functional modules by utilizing the modular technology. Through this way, the product lifecycle management is converted to the lifecycle management function modules. Not only save the cost and time, but also improve the quality and reliability. Therefore, considering the factor of PLM in the initial module construction phase is of great significance for the maneuverability of product in circulation.

E. Qi et al. (eds.), *Proceedings of the 23rd International Conference on Industrial Engineering and Engineering Management 2016,* DOI 10.2991/978-94-6239-255-7_17

III. THE MODULE CONSTRUCTION METHOD

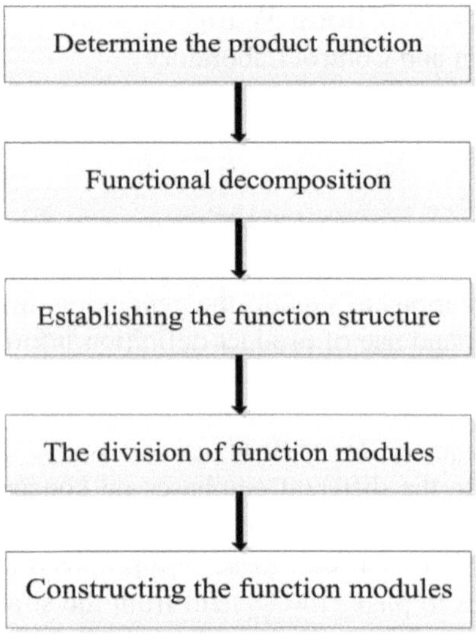

Fig.1. The method of module construction

A. Determine the product function

Function is an abstract description of the relationship between input and output in mechanical products. Through the analysis of needs of market and user, get the product design requirements and determine the design task, then getting into the conceptual design stage of the product.

B. Functional decomposition

It's often difficult to find the corresponding answer immediately after the total function of a system is determined, so it's necessary to decompose the difficult total function into the relatively simple sub-function or functional units, which is the process of functional decomposition [5]. The functional unit is the basic unit of the function.

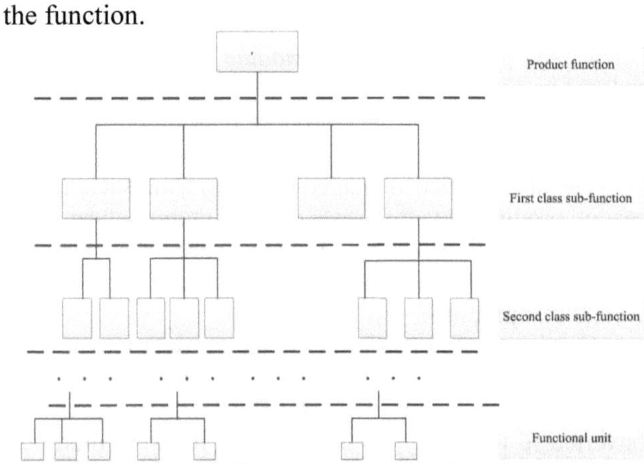

Fig.2. The schematic diagram of system functional decomposition

C. Establishing the functional structure

Function structure describes the relationship between the sub-functions of system. Or said differently, the function structure is the organic combination of sub-functions in the system. Function structure chart is used to describe the relationship commonly, which makes the system from abstract to concrete. It's important to make it clear that the link of sub-functions and functional units to meet total system function by the establishing the functional structure chart. These relatively simple sub-functions and functional units can be more easily to find the corresponding real structure.

D. The division of function modules

From the perspective of functional design, comprehensive interpreting the various kinds of product functions by carrying out the functional analysis on product, resolving and merging the functions according to unified principles. Then divide the product system into a series of function modules.

E. Constructing the function modules

Design the structure and interface according to the specific function modules. Considering the relevance of function, geometry, physics and information, treat the modules with complements and modifications.

IV. THE MODULES CONSTRUCTION STRATEGY UNDER PLM

A. Constructing the function modules

For the reliability and maintainability of product, the producer should choose the optimized module design. To ensure at the end of the technology and economic life, the product can restore original performance by using the advanced technologies such as surface engineering and module upgrade [6]. The remanufacturing process includes recycling, dismantling, sorting, cleaning, repair, upgrade, assembly and testing of waste product. The product lifecycle can be extended by determining which function module needs to be remanufactured in design stage [7]. To ensure the remanufacturing performances, put forward the indexes and requirements of remanufacturing in module design [8]. The module design for remanufacturing should meet many requirements. For example:

1) Transportability.

The transportation cost of waste product accounts for a large proportion of the overall cost in remanufacturing. The transportability needs to be considered in product module design.

2) Easy to clean.

The parts easy to clean can significantly improve the economy of remanufacturing. Currently, the cleaning methods include ultrasonic cleaning, solvent cleaning, etc. In the design stage, the surface parts easy to clean should be adopted, such as the flat surface, the appropriate surface materials and coatings, etc.

B. Product module design for recycling

To get the precious metal and valuable parts from a retired product, the producer should regard the recycling performance as one important goal in product design [9-10]. Recyclable design requirements for product modularity are as follows:

1) The modular structure should be easy to adjust and maintain.

The maintenance of product structure should be accomplished with the adjusting device as far as possible. According to the principle of task decomposition, put the vulnerable part on the layout or area that can be adjusted easily.

2) The compatible material should be adopted.

When choosing the material for a part of product, there may be a variety of types meeting the functional requirements. If choosing compatible materials, even if the connected parts composed of different materials difficult to disassemble, they can be recycling together.

3) Affix signs and labels.

It should be easy to classify and identify the reusable components material. According to the structure, connection size and material, the designer should make the corresponding marks. Therefore, the state of reusable parts can be identified easily with the function marks.

C. The module improvement design in advance

Due to the quickly technology updating, some expected developments of new technology and new component cannot be used in product for the reason that still not mature in product design phase. Some enterprises have no ability to use the new technology for insufficient funds. In order to make the product easy to occupy the market share by saving cost, they just develop the product of lower performance by taking some mature technologies and components temporarily.

However, in order to improve product performance continuously, the designer should consider the modular improvement plan in the design stage. When new technologies and components are mature, or enough funds have been raised, the existing product can be upgraded with the modular improvement. With the necessary technical interface due to the early arrangement, the cost and difficulties technical improvement will be reduced, moreover, the situations like the scrap of the whole system will not appear.

V. THE APPLICATION EXAMPLE

Take the desk lamp for example. Currently, there are various kinds of desk lamp, the types of light bulb include energy-saving lamps, incandescent lamps and led bulbs; and the control modes include switch control, touch control, brightness adjustable, and even voice control. The high speed development of science and technology make the desk lamp easy to out of date. However, consumers need a practical, energy saving and reliable desk lamp.

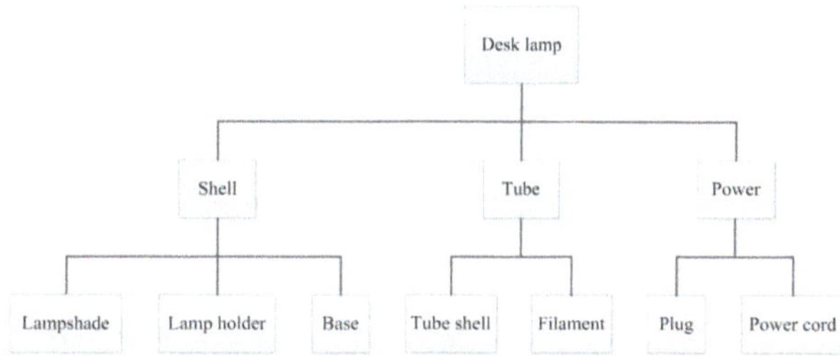

Fig.3. Function decomposition diagram of desk lamp

The function decomposition diagram of desk lamp is shown as Fig.3. To optimize the parts of desk lamp function module construction process by the strategies proposed in this paper, the results are as follows:

1) The module improvement design in advance.

The design of the lamp shell should be flexible, and consider the future development trend. For example, the remote control or voice control is the development trend of desk lamp. In the initial design of desk lamp, designer should plan the plug-in interface module in advance. When the remote control or acoustic technology is mature, users can achieve the function through the external access to corresponding function module, rather than to buy a new desk lamp.

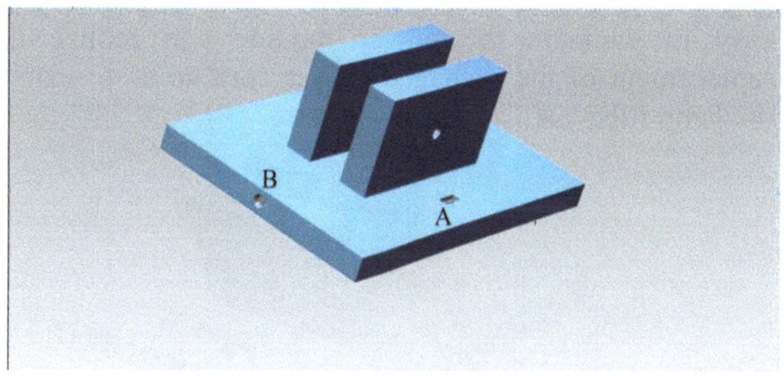

Fig.4. The plug-in interface in base module

As shown in Fig. 4, design two standard interfaces on the base module in advance. Interface A stands for the USB control interface, interface B stands for the standard interface of power module.

As shown in Fig.5, when it is ripe for voice control technology to upgrade the desk lamp, the corresponding module can be accessed to interface A, then voice control can be achieved in the existing product.

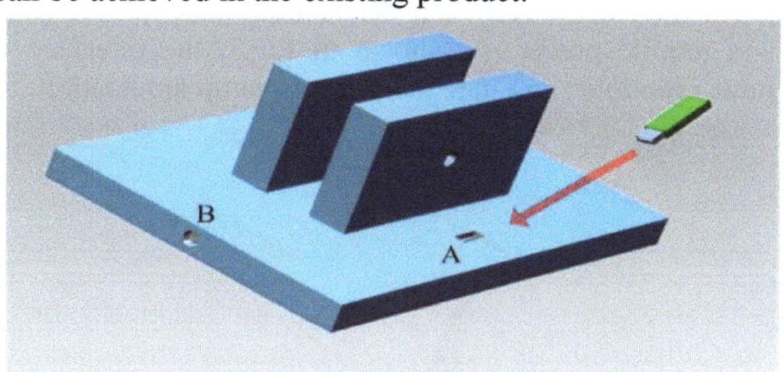

Fig.5. The upgrade module access to base module

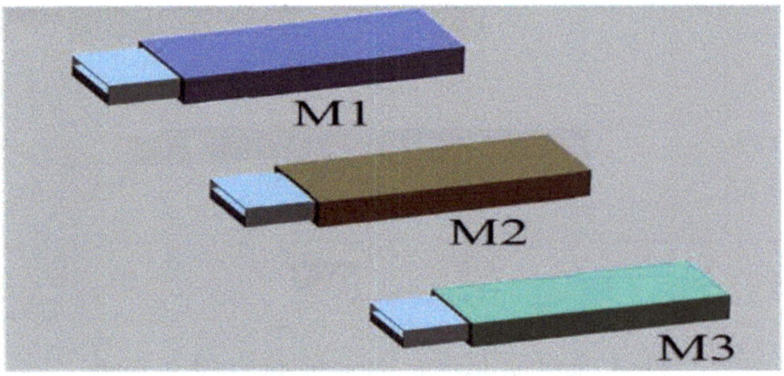

Fig.6. The upgrade modules

The upgrade modules access to Interface A are as shown in Fig.6, for example, M1 stands for remote control module, the M2 for light control module, M3 for temperature display module, etc. For any module of subsequent improvement in the product development, the producer can realize the function update of existing products by simply follow the existing design standards.

2) The module design for remanufacturing.

The modules of tube and power are more easily damaged. Therefore, the designer should fully consider the compatibility of the two modules and the desk lamp system, so as to achieve the replacement of new function module to damaged module. As shown in Fig.7, for the standard power standard interface and the power cord easy to damage, the power module can realize the replacement of the damaged power cord so as to extend the lamp lifecycle.

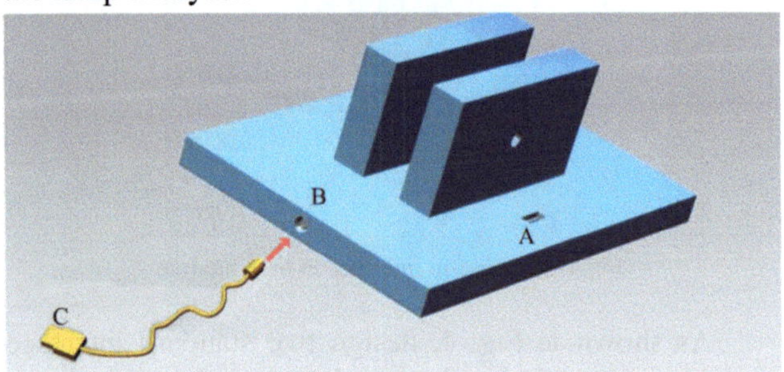

Fig.7. The power cord access to base module

Besides, with the function diversity development, the desk lamp not only limited to be used in the home, people still face a desk lamp use requirements in travel or absence of other power. Therefore, the power cord with stand interface C can connect to household power plug, and mobile power. As shown in Fig.7, the interface of power module can range the scope of lamp application.

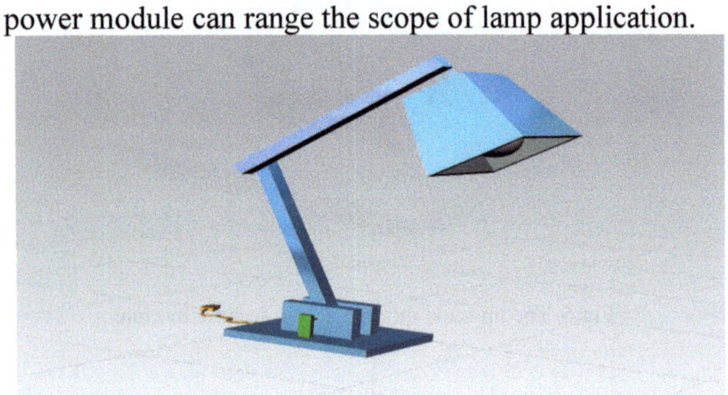

Fig.8. The integration of the upgrade module and power cord

Fig.8 shows the integration of the upgrade module and power module access to the desk lamp system. To optimize the process of access by considering the characteristics of beauty and economy in the process of product module construction, this will be more conducive to product promotion in the market.

VI. CONCLUSION

The consideration of PLM in module construction can realize the effective utilization of resources; improve the reliability of products, and extend the service life of products, moreover, to help the enterprises occupy the market share in time. However, the strong correlation of module construction and product lifecycle management (PLM) requires further research on the methods of effective integration and efficient data management.

REFERENCES

[1] C. Y. Baldwin and K. B. Clark, "Managing in an Age of Modularity," *Harvard Business Review*, vol. 75, pp. 84-93, May 1997.

[2] Li, J., Tao, F., Cheng, Y., and Zhao, L., "Big Data in product lifecycle management," *International Journal of Advanced Manufacturing Technology*, vol.81, no. 1, pp.667-684. 2015.

[3] Penaranda, N., Mejia, R., Romero, D., and Molina, A., "Implementation of product lifecycle management tools using enterprise integration engineering and action-research," *International Journal of Computer Integrated Manufacturing*, vol.23, no.10, pp.853-875, Oct. 2010.

[4] Chen, Jiena, L. Deng, and Q. Wu. "Research of Module Construction Based on Integration Theory," *Multimedia Technology (ICMT), 2010 International Conference on* IEEE, 2010,pp.1-4.

[5] Baldwin, Carliss Y., and K. B. Clark. "Design Rules, Volume 1: The Power of Modularity." *Mit Press Books* 1(2000).

[6] Worren, Nicolay, K. Moore, and P. Cardona., "Modularity, Strategic Flexibility and Firm Performance: A Study of the Home Appliance Industry," *Strategic Management Journal*, vol. 23, pp. 1123-1140, December 2002.

[7] Jun, H. B., Shin, J. H., Kiritsis, D., and Xirouchakis, P., "System architecture for closed-loop PLM," *International Journal of Computer Integrated Manufacturing* vol. 20, pp. 684-698, 2007.

[8] Schuh, G., Rozenfeld, H., Assmus, D., and Zancul, E., "Process oriented framework to support PLM implementation," *Computers in Industry* vol.59, pp.210-218, 2008.

[9] Lau, Antonio K.W., Yam, Richard C.M. and Tang, Esther P.Y., "Supply chain integration and product modularity: An empirical study of product performance for selected Hong Kong manufacturing industries," *International Journal of Operations and Production Management*, vol. 22, pp. 20-56, January 2010.

[10] Chiang, Tzu An, and A. J. C. Trappey. "Development of value chain collaborative model for product lifecycle management and its LCD industry adoption," *International Journal of Production Economics*, vol. 109, pp. 90-104, 2007.

Logistics Distribution Path Planning Based On Genetic Algorithm

Zhen-hua GAO , Wei-dong CHEN*
School of Management Science and Engineering, Anhui University of Technology, Anhui, China
(*1571803048@qq.com)

Abstract - **The logistics distribution path planning is of great significance for improving logistics distribution efficiency and saving the cost of distribution. Traditional logistics distribution path planning is usually a model that a single supplier serves several demands, which regards the minimum total distance distribution as the optimization goal and aims to pursue the optimal distribution route. But the logistics distribution model in this paper is that several suppliers serve several demands, then transforms into the classical TSP to solve optimization problems and mathematical model is established. Based on the mathematical model, the improved genetic algorithm is supported to use integer permutation encoding and adopt the optimal individual reserve strategy and roulette method on the individual selection. Finally, experiments have been carried out to calculate in this way, according to the calculation results, it shows that using the genetic algorithm to solve optimize logistics distribution route can obtain effectively the optimal solution or approximate optimal solution about this problem.**

Keywords - **Logistics, Genetic algorithm and Optimized**

I. INTRODUCTION

With the rapid development of economy, logistics industry presented a boom, in order to improve its competitiveness in the industry, logistics enterprises are constantly seeking reasonable distribution route to reduce the logistics enterprises costs that happens in the process of distribution, but one of the most effective ways to make the total distribution distance shortest is to reduce logistics cost. Optimization of logistics distribution path is a classic NP problem, so far, there are a lot of papers referred to it, but to sum up, some common algorithms for solving this kind of problems are genetic algorithm, simulated annealing algorithm, tabu searching algorithm, ant colony algorithm, etc. These algorithms have obtained certain achievements in the field of solving logistics distribution route optimization, and genetic algorithm is one of common algorithms to solve the problem of logistics distribution route optimization.

Genetic Algorithm (GA) is presented by J.Holland inspired by the evolution in 1975. It is the basic on the theory of natural selection and genetic, Genetic Algorithm is a highly parallel, random and adaptive optimization algorithm that combined survival of the fittest rules in biological evolution process with information transformation mechanism randomly in the same group of chromosomes, in this way, the problem of solving process can be expressed as "chromosome" survival of the fittest process, after selection, crossover and mutation operation, eventually, the best solution can converge to the individuals that adapt to the "environment" best, and then get the optimal solution of the problem or satisfied solution [1] .

II. THE MATHEMATICAL MODEL

Logistics distribution route optimization problem can be described as: several known distribution centers (outlets) serves for several demand points (demand), every distribution center location coordinate and distribution capability is known, and every demand point location coordinate and demand quantity must be known, asking for arrangement of distribution path reasonably is to make every demand point be satisfied and total distance shortest, what's more, the following conditions should be satisfied:

(1) every distribution center has certain distribution capability, the total demand of demand points that be served by a certain distribution center can't be more than its distribution ability;

(2) every demand point is served only by a distribution center;

(3) the distance between a distribution center and a demand point be calculated in a straight line distance;

E. Qi et al. (eds.), *Proceedings of the 23rd International Conference on Industrial Engineering and Engineering Management 2016*, DOI 10.2991/978-94-6239-255-7_18

(4) the demand of a demand point should not be more than capacity of a distribution center.

The following definitions are used in the mathematical formulations:

M	The number of distribution centers
A_i	The capacity of distribution center i for $i = 1, 2 \ldots M$
R	The collection of distribution centers
(a_i, b_i)	The location coordinates of distribution center i in collection of distribution centers
N	The number of demand points
E_i	The collection of demand points that are served by distribution center i for $i = 1, 2 \ldots M$
E	The collection of distribution centers
C_i	The number of demand points that are served by distribution center i for $i = 1, 2 \ldots M$
(x_k, y_k)	The location coordinates of element k in the collection E_i for $k = 1, 2, \ldots C_i$
$d_{i,k}$	the distance between distribution center i and demand point k in a straight line distance

If the shortest total distribution distance is taken as the objective function, the mathematical model of logistics distribution route optimization problem is establish, as follows:

$$\min Z = \sum_{i=1}^{M} \left[\sum_{k=1}^{C_i} d_{i,k} \right] \tag{1}$$

S.t

$$\forall C_i \le A_i \tag{2}$$

$$\sum_{i=1}^{m} C_i = N \tag{3}$$

$$R = \{(a_i, b_i) \mid i = 1, 2 \ldots M\} \tag{4}$$

$$E_i = \{(x_k, y_k) \mid k \in [1, C_i] \text{且} k \in Z^*\} \tag{5}$$

$$d_{i,k} = \sqrt{(x_i - x_k)^2 + (y_i - y_k)^2} \tag{6}$$

$$E_{i1} \cap E_{i2} = \phi, \qquad i1 \ne i2 \tag{7}$$

$$\bigcup_{i=1}^{N} E_i = E \tag{8}$$

The above models, the formula (1) is the objective function, which makes the distance between every distribution center and every demand point come to be shortest when the demand of demand points can be served; the formula (2) ensures that distribution capacity of a distribution center is more than or equal to the demand of demand points that are served by this distribution center; the formula (3) shows that the number of demand points that are served by all distribution centers are equal to the number of the whole demand points; the formula (4) shows all elements in the collection R of distribution center; the formula (5) shows the location coordinates in the collection of demand points that are served by distribution center i ; the formula (6) shows the distance between distribution center i and demand point k in a straight line distance; the formula (7) ensures that every demand point is served only by a single distribution center; the formula (8) shows the sum of demand points that are served by all distribution centers are equal to the total demand points [2-4] .

III. LOGISTICS DISTRIBUTION ROUTE OPTIMIZATION PROBLEM OF GENETIC ALGORITHM

A. The basic idea of genetic algorithm

Genetic algorithm is to simulate the process of organisms genetic and evolution in the natural environment and form a kind of adaptive probability searching algorithm that is global optimization. In the process of solving problem, firstly, the solution of the problem to solve is translated into be "chromosome" , meanwhile, before performing genetic algorithm (GA), a group of "chromosome" is generated randomly, which is the hypothesis. Then, these assumption solutions are put into the fitness function, and the optimum solutions are selected according to the principle of survival of the fittest, after selection, crossover and mutation, a new generation of "chromosome" group that is more adapt to the environment is produced. In this way, after the generation evolving, finally, it will converge to one of the most adaptive to the environment "chromosome", in other words, which is the optimal solution of the problem [5-6] . The basic framework of genetic algorithm is shown in fig. 1 below:

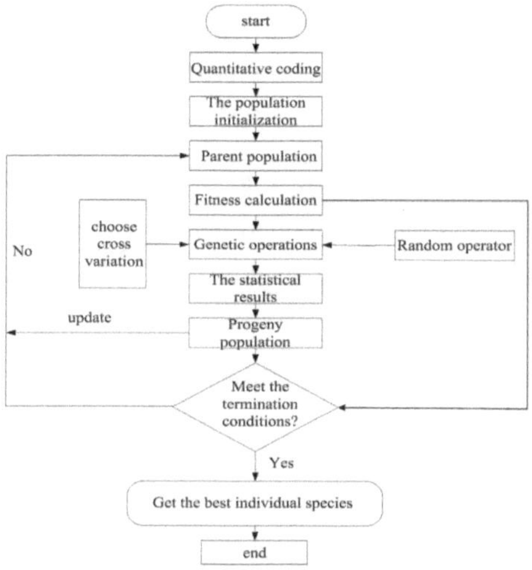

Fig. 1. The basic framework of genetic algorithm

B. The genetic algorithm structure of the optimization algorithm to optimize logistics distribution path

According to the characteristics of the optimization of logistics distribution path, this paper is to construct genetic algorithm for solving the problem on the basis of solving TSP problem.

(1) To determine coding method. This paper presents a direct permutation encoding method, due to there have N demand points currently , it is expressed in this method to product directly L arrangements between $1 \sim N$ natural numbers that do not repeat, which shall constitute an individual. According to the constraint conditions of logistics distribution route optimization, which is that distribution capacity of a distribution center is more than or equal to the demand of demand points that are served by this distribution center, and every demand point is served only by a single distribution center, every element of individual (demand point) should be delimited the collection of demand points. For example, there are two distribution centers to serve five demand points, assuming that the first distribution center distribution can serve two demand points, the second distribution center distribution can serve three demand points, assuming a demand point arrangement of individual is 23415, so which shows demand point for 2, 3 are served by the first distribution center and demand point for 4, 1, 5 are served by the second distribution center, with using this method to encode, the distribution route scheme can be gotten , under satisfying all constraint conditions of the problem, the optimal solution can be obtained.

(2) To determine the initial population. An arrangement of $1 \sim N$ natural numbers that don't be repeated are generated randomly, which is a formation of an individual. assuming population size is L , then L individuals like this are produced randomly, in other words, the initial population is formed.

(3) The fitness evaluation. For an individual distribution route scheme, which corresponds to determine their advantages or disadvantages. One way is to see whether they meet the constraints of distribution; the other way is to calculate the objective function value (that is, the total distance between every distribution center and every demand point). According to the encoding methods that depends on the characteristics of the distribution route optimization problem, which can imply to satisfy all constraint conditions that is every demand point can be served and be served only by a single distribution center, then every individual objective function value Z_j is calculated, finally, the formula of the fitness of a individual be shown $F_j = 1/z_j$.

(4) Selection operation. L individuals fitness of every generation are arranged from big to small, the first individual performance is optimal, copy it and put directly it into the first place of the next generation side by side. Other $(L-1)$ individuals in the next generation are produced by wheel selection method according to the fitness of L individuals in the former generation of population. Specifically, it is first to calculate directly the sum of all individual fitness $(\sum F_j)$ in the group, and then calculate the fitness of each individual proportion, which becomes its probability to be selected. This choice method not only ensures the best individual can exist to the next generation, but also can guarantee larger fitness individuals with greater probability access to the next generation.

(5) The crossover operation. Adopting a partial mapping and the parent of crossover operation is determined, and parent samples are to be grouped by two groups, according to the crossover probability p_c to match restructuring crossly. Each group repeats the following process: random integers r_1 and r_2 during two range are produced, after determining the two positions, to cross the middle data of the two position; if there are the same individual demand point number repeated, numbers that don't repeat are kept, numbers that have conflicts adopt part of the mapping method to eliminate conflicts, that is to use the middle section of the corresponding relation to map.

(6) Mutation. Due to adopting the way to keep the best sample in the selection mechanism, in order to maintain the diversification of individuals within population, the way to mutate is to pick randomly two points on the encoded string in this paper, according to the mutation probability p_m, to make mutation operation and exchange its position, it will produce two random integers r_1 and r_2 during two range, after determining this two positions, to exchange its position and the mutated chromosome coding is obtained [7-9].

IV. THE EXPERIMENT AND CALCULATION

According to the genetic algorithm (GA), a MATLAB language program was wrote, and the following example was calculated by the experiment.

Example: three distribution centers serve for nine demand points for the delivery. And distribution ability of these three distribution centers can respectively meet the demand of 2, 3, 4 demand points, the position coordinates of three distribution centers and nine demand points are shown in table 1 below.

TABLE I

POSITION COORDINATES

Category	Number	Location Coordinates
Distribution Center	1	(5, 6)
	2	(11, 14)
	3	(15, 6)
Demand Point	1	(2, 7)
	2	(4, 17)
	3	(7, 13)
	4	(8, 3)
	5	(11, 16)
	6	(10, 8)
	7	(17, 17)
	8	(16, 10)
	9	(18, 3)

In the experiment, the following parameter values are adopted: $L = 50$, $p_c = 0.65$, $p_m = 0.005$, $T = 100$. By MATLAB 2010a simulation software coding, the final simulation interface as follows: Fig 2 shows position coordinates of distribution centers and demand points; Fig 3 shows distribution path; Fig 4 shows the simulation results.

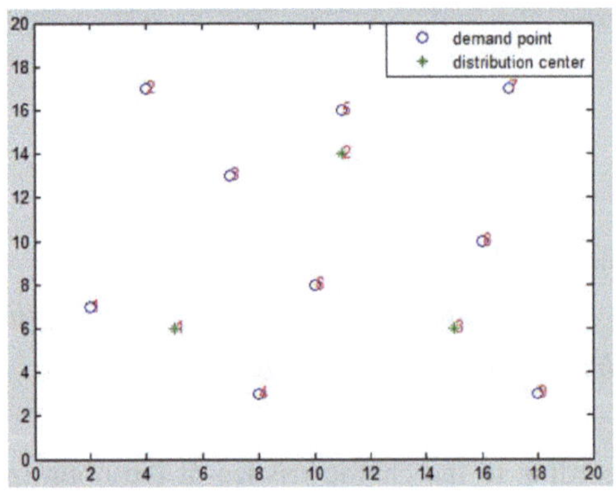

Fig. 2. Position coordinates of distribution centers and demand points

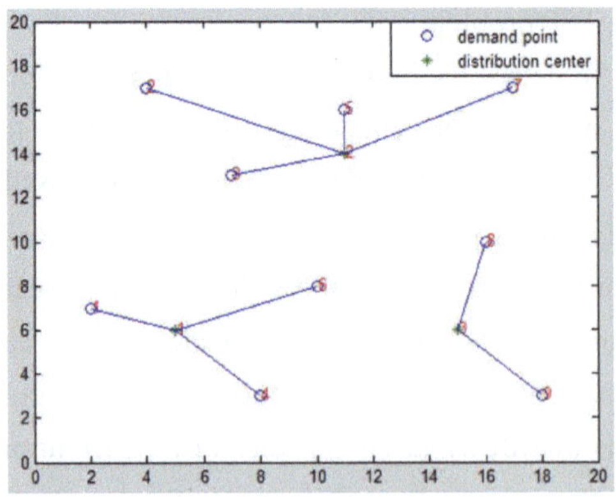

Fig. 3. Distribution path

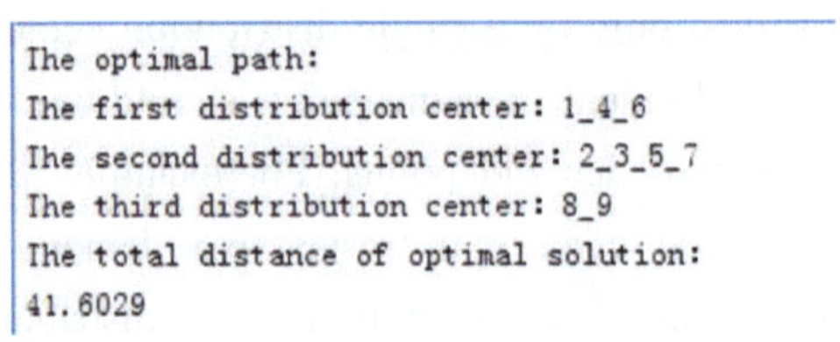

Fig. 4. The simulation results

From the simulation results to draw the following conclusion, summary into a tabular form are shown in table II below:

TABLE II

DISTRIBUTION RESULTS

Distribution Center	Demand Point	Total Distance Of Distribution
(5,6)	1,4,6	
(11,14)	2,3,5,7	41.6029
(15,6)	8,9	

V. CONCLUSION

Experiments shows that the mathematical model of optimization of logistics distribution path is established according to the description of the problem, and then on the basis of the TSP problem to make genetic algorithm optimization, which is a heuristic searching method with good performance. It can find the optimal path of logistics and the optimal solution of distribution quickly and efficiently.

But in this paper, there are some disadvantages in solving process that are based on the optimization of logistics distribution path, such as distribution capability of the distribution center is simple, and the construction of distribution network does not take into account the dynamic. Finally, we hope these problems could be solved in future research.

REFERENCE

[1] S. Zhu, W. Dong, W. Liu. Logistics distribution route optimization based on genetic ant colony algorithm[J]. Journal of Chemical & Pharmaceutical Research, 2014:6(6):2264-226 7.

[2] Jin Z, Teng F. Research of Genetic Algorithm in the Medical Logistics Distribution Routing Optimization[C]. International Conference on Intelligent Computation Technology & Automation. IEEE Computer Society, 2009:452-455.

[3] Yi Z. Solving Vehicle Routing Problem Based on Improved Genetic Algorithm[C]. International Symposium on Information Science & Engineering. IEEE Computer Society, 2011:590-594.

[4] Yan-cong Zhou, Xiao-chen Sun, Wei-xiang Yu. Logistics distribution route optimization based on improved genetic algorithm research[J]. Computer engineering and science, 2012,34 (10):118-122. (Chinese)

[5] Xiao B, Min W, Liu Y, et al. Improved Genetic Algorithm Research for Route Optimization of Logistic Distribution[C]. International Conference on Computational and Information Sciences. IEEE Computer Society, 2010:518-521.

[6] Wang Y. Distribution route optimization of logistics enterprise based on genetic algorithm[C]. World Automation Congress. IEEE, 2012:1-4.

[7] Pedro Larrañaga, Concha Bielza. Genetic Algorithm (GA)[M]. Dictionary of Bioinformatics and Computational Biology. 2004.

[8] Zhou Ming, Shu-dong Sun. Genetic Algorithm Principle and Application [M]. National Defense Industry Press, 1999.6. (Chinese)

[9] Cleve Moler. Experiments with MATLAB[M]. Beijing University of Aeronautics and Astronautics Press. 2013. (Chinese)

Research on the Mechanism of Risk Cascading Propagation across Interdependent R&D Networks

Yan-lu ZHANG*, Nai-ding YANG
Department of Management Science and Engineering, Northwestern Polytechnical University, Xi'an, China
(zhangyanlu@nwpu.edu.cn)

Abstract - **Considering that risks propagate not only within single R&D network, but also spread among interdependent R&D networks, we explore the risk cascading propagation mechanism across interdependent R&D networks. Firstly, we proposed the generation algorithm of individual R&D network based on R&D network has the characteristics of scale-free and small-world networks. Then we identified three types of interdependences (i.e., input interdependence, shared interdependence and co-located interdependence) between interdependent R&D networks. After that, we introduced cascading failures theory to model the risk propagation within single R&D network by defining the risk load and risk capacity on each firm, and building the rule of local triggering among risks. Finally, we propose the four-stage model to describe the process of risk cascading propagation across interdependent R&D networks.**

Keywords - **Interdependent R&D networks, risk cascading propagation, cascading failures, complex network**

I. INTRODUCTION

Facing the increasingly dynamic market, complex R&D technology and limited resources, building a R&D network to develop a complex new product has become common in those industries characterized by rapid technological innovation [1-2]. As a new R&D mode, R&D network enables firms to efficiently acquire the knowledge or resources they need [3-5]. However, owing to the complicated partnerships among firms, some firms join not only one R&D network but also other R&D network, or some firms in one R&D network supply technology support for dependent firms in other R&D network. The interdependences then cause interdependent R&D networks to be more vulnerable to risks. For instance, some existing risks on a firm will trigger not only the potential risks on some firms in the same R&D network but also the potential risks on dependent firms in other R&D network, and vice versa. This recursive process is called risk cascading propagation [6], which could cause firms to malfunction when the existing risks exceed their capacities and then might lead the interdependent R&D networks to collapse. Therefore, we will introduce cascading failures theory to explore the risk cascading propagation mechanism across interdependent R&D networks.

Over the past few years, some researchers have studied the risk propagation in many networks, such as supply chain network, computer network, financial network, etc [7-10]. However, these studies mainly focused on a single isolated network. In fact, many complex networks in real world are interdependent, and the failure of nodes in one network could spread within this network and then lead to the failure of dependent nodes in other network. Therefore, some researchers have started to study the cascading failures of interdependent networks. A groundbreaking work of this field is that Buldyrev et al. [11] proposed a theoretical framework to capture the phenomenon of cascading failures in interdependent networks under random attack, and found that the failure of a small fraction of nodes can result in complete fragmentation of all the networks, while a significant number of nodes must be randomly removed before degradation of a single network. Parshani et al. [12] studied the cascading failures of partially interdependent networks under random attack, and found that reducing interdependence could improve the robustness of networks against random attack. Brummitt et al. [13] further explored the cascading failures of interdependent power grids, and found that adding interdependence between two isolated networks is beneficial for improving the robustness of networks. Tan et al. [14] investigated the effect of three kinds of coupling link patterns, i.e., assortative link (AL), disassortative link (DL), random link (RL), on the vulnerability of interconnected data-packet transport networks to cascading failures. Wang et al. [15] further investigated these three link patterns on the interdependent networks to cascading failures, and found that each link pattern has different effects on the vulnerability of different interdependent networks. On this basis, Wang et al. [16] investigated the influence of coupled strength for these three link patterns (i.e., AL, DL, RL) on the vulnerability of interdependent networks to cascading failures under tow attacks. Chen et al. [17] then explored the effect of different coupling probability for these three link patterns on the vulnerability of interdependent scale-free networks under targeted attack.

However, the current studies mentioned above only focus on the cascading failures of interdependent networks based on the static percolation theory, while not considering the cascading failures induced by overload on nodes in interdependent networks. In fact, overload of node is a key factor in triggering cascading failures for many real-world networks, such as power grid, the

This work is supported by National Natural Science Foundation of China (No. 71501158; 71471146) and "the Fundamental Research Funds for the Central Universities" (No. 3102016RW005).

E. Qi et al. (eds.), *Proceedings of the 23rd International Conference on Industrial Engineering and Engineering Management 2016*, DOI 10.2991/978-94-6239-255-7_19

Internet, traffic network, etc. For R&D network, each firm often has to take the load of existing risks, and also has its own capacity of accommodating maximum risk load. When the risk load exceeds its capacity, this firm will fail and the existing risks will trigger other potential risks on its neighboring firms, which might then cause these neighboring participants to fail because of risk overload. Such cascading failures induced by risk overload may spread within a R&D network and then propagate to other R&D network through the interdependences. Therefore, we will consider the dynamics of risk triggering between firms to build a more realistic risk cascading propagation model of interdependent R&D networks

II. GENERATION OF INTERDEPENDENT R&D NETWORKS

A. Generation Algorithm of Individual R&D Network

So far, many researchers have investigated extensively the topology of R&D network[18-21]. There is a general consensus that R&D network has two main characteristics of scale-free network, namely, growth and preferential attachment. The reason why R&D network has the characteristic of growth is that some firms in R&D network always seek for collaboration with other firms out of R&D network due to their business development and expansion. Meanwhile, some firms out of R&D network actively seek to collaborate with some member firms in order to acquire the knowledge spillover of R&D network. The reason why R&D network has the characteristic of preferential attachment is that the firms that try to join the R&D network prefer to collaborate with those member firms that have already had more partners, which usually leads to an interesting phenomenon that those firms that have more partners and resources are more likely to become the targets with which other firms expect to collaborate. Consequently, a very few firms become the hubs that have a great number of partners in R&D network, whereas most of firms become the peripheral nodes that have a few partners. Besides, many researchers proposed that R&D network has also the characteristic of small-world network [19][21][22-23]. The main reason is that most of firms tend to select those firms that are geographically close to them as their partners given the communication cost and efficiency. Therefore, we propose the generation algorithm of individual R&D network, which is given as follows:

(1) At first, we assume that individual R&D network starts with m_0 fully connected but randomly distributed nodes (i.e., firms). So their coordinates, i.e., $<x_1, y_1>$, $<x_2, y_2>$, …, $<x_{m_0}, y_{m_0}>$ are m_0 pairs of independent and random numbers in the interval [0, 1].

(2) After each time step, we assume that a new firm i with a randomly distributed coordinate $<x_i, y_i>$ ($x_i \in [0,1], y_i \in [0,1]$) will be connected to m existing firms in R&D network based on the preferential attachment and geographical proximity. Therefore, the probability Π_{ij} that a new firm i connects to an existing firm j in R&D network is given as follows:

$$\Pi_{ij} \sim d_j^{\lambda} / D_{ij}^{\mu} \tag{1}$$

Where d_j is the degree of node j, i.e., the number of nodes directly linking to node j; D_{ij} is the Euclidean Distance between node i and j, i.e., $D_{ij} = \sqrt{(x_i - x_j)^2 + (y_i - y_j)^2}$; λ ($\lambda \geq 0$) and μ ($\mu \geq 0$) are the tunable parameters that govern the degrees of preferential attachment and geographical proximity, respectively. Obviously, a larger value of λ will favor connecting to firms with higher degrees, whereas increasing value of μ will discourage longer geographical distance.

(3) After N-m_0 time steps, R&D network will be generated with N member firms and average degree $\langle d \rangle = \frac{1}{N}\sum_i d_i \approx 2m$.

B. Building The Interdependences between R&D Networks

So far, many researchers have proposed several frameworks or methods to identify different types of interdependences among interdependent networks [24-27]. Rinaldi et al. [26] defined four types of interdependences: physical interdependence (An output from a system is an input to another system), cyber interdependence (The state of a system is dependent on information transmitted through the other information infrastructure), geographical interdependence (Two or more geographically proximate systems are affected simultaneously by a local event, e.g., an explosion, fire or flood), logical interdependence (Two systems are logically interdependent if the state of each depends on the state of the other via a mechanism that is not a physical, cyber, or geographical connection. For example, the power crisis from California's electricity network in late 2000 affected the financial infrastructures market via the logical interdependency). Then Lee et al. [27] proposed five types of interdependences between infrastructure systems: input interdependence (A system requires one or more services from another system to provide some other service), mutual interdependence (An example of mutual dependence occurs when an output of system A is an input to system B, and an output of system B is an input to system A), shared interdependence (Some components of the systems used in providing the services are shared), exclusive or interdependence (Only one of two or more services can be provided by a system. This dependence can occur within a single system or among two or more systems), co-located interdependence (Components of two or more systems are situated within a prescribed geographical region).

In interdependent R&D networks, we propose that there exist three types of interdependences, which are input interdependence, shared interdependence, co-located interdependence. Input interdependence means that some firms in one R&D network supply R&D technologies or other R&D resources for some firms in other R&D network. Shared interdependence means that some firms join not only one R&D network but also other R&D network. For example, AVIC (Aviation Industry Corporation of China) Chengdu Aircraft Company not only develops the components of nose landing gear door in Airbus 380 R&D network, but also joins the Boeing 787 R&D network to develop the rudder. Co-located interdependence means that although some firms belong to different R&D networks, they are located in a geographical region and easily influenced by a local event, e.g., local market fluctuation, financial crisis, etc.

For simplicity, we assume that interdependent R&D networks are only comprised of two networks, which are network A denoted by $G^A(V^A, E^A)$ and network B denoted by $G^B(V^B, E^B)$. Based on the three interdependences between interdependent R&D networks, we propose the interdependence matrix $I_{A\rightleftarrows B}=[I_{ij}]_{N^A\times N^B}$ (N^A is the number of all nodes in G^A, N^B is the number of all nodes in G^B) to describe the interdependences between interdependent R&D networks, which is expressed as follows:

$$I_{A\rightleftarrows B}: I_{ij}=\begin{cases} 1 & \text{if node } j\,(j\in V_I^B) \text{ requires services from node } i\,(i\in V^A) \\ -1 & \text{if node } i\,(j\in V_I^A) \text{ requires services from node } j\,(i\in V^B) \\ 2 & \text{if node } i\,(i\in V_S^A) \text{ and node } j\,(j\in V_S^B) \text{ are the same node} \\ 3 & \text{if node } i\,(i\in V_C^A) \text{ and node } j\,(j\in V_C^B) \text{ are located in a same region} \\ 0 & \text{otherwise} \end{cases} \quad (2)$$

Where V_I^A is the set of nodes in G^A that require services from nodes in G^B, and V_I^B is the set of nodes in G^B that require services from nodes in G^A; V_S^A and V_S^B are the same set of shared nodes that belong to two R&D networks simultaneously; V_C^A is the set of nodes in G^A that are located in a geographical region with the corresponding nodes (the set of which is defined as V_C^B) in G^B.

III. MODELING THE RISK PROPAGATION WITHIN SINGLE R&D NETWORK

Existing studies about risk propagation only described a chain reaction phenomenon of nodes' state (normal or failed) induced by risks, while not considering that risk consequences can be increasingly accumulated until they exceed the anti-risk capacity of nodes. Therefore, we consider risk in R&D network as the substance that can be quantified. Based on this, we define the consequence of risks that occurred on a firm as the risk load on this firm, and define the maximum ability to resist risk of as the risk capacity of this firm. Once the accumulated risk load of some firm exceeds its risk capacity, this firm will fail to function because of overload. Subsequently, the risks that occurred on this failed firm will trigger potential risks on neighboring partners, which might cause some partners to malfunction because of overload. Next, we will model the risk propagation within one R&D network by defining the risk load and risk capacity on each firm, and building the rule of local triggering among risks.

Firstly, we propose that each member firm in R&D network faces seven kinds of potential risks, which are market risk R_1 (market risk derives from demand fluctuation and market competition due to the interaction of R&D network and market), economic risk R_2 (economic risk is caused by change of economic cycle, fluctuation of interest rate, exchange rate and stock market), policy risk R_3 (policy risk is the probability of firms' loss caused by political change and intervention of government), ethical risk R_4 (ethical risk means some firms will supply false information, miss important information deliberately for getting high profits due to incomplete information), time risk R_5 (time risk is also called schedule risk, which is the probability that some tasks or projects are not completed on expected schedule), capital risk R_6 (capital risk is the probability of running out of fund or breakage of capital chain) and collaboration risk R_7 (collaboration risk is the probability that some strategies cannot be carried out effectively due to the cultural differences of firms) [28]. Considering that different risks have different occurrence probabilities and different consequences, we define the risk load $L_i(t)$ on firm i at time t with total consequences of the risks that have occurred on firm i, which is expressed as follows:

$$L_i(t)=\sum_{j=1}^{7} w_j\cdot\eta_j^i(t) \quad (3)$$

Where w_j is the consequence of risk R_j; $\eta_j^i(t)$ indicates whether risk R_j on firm i occurred or not at time t, $\eta_j^i(t)=1$ if risk R_j on firm i occurred, $\eta_j^i(t)=0$ if not occurred.

As we know, different member firms play different roles in making R&D network function well, so the risk capacities of different firms are different. We assume that the risk capacity C_i of firm i is related to its node degree, its partners' node degrees, and the average consequences of all potential risks [29], which is given as follows:

$$C_i=\beta\frac{(d_i\sum_{j\in\Gamma_i} d_j)^\alpha}{\max\{(d_k\sum_{l\in\Gamma_k} d_l)^\alpha \mid k=1,2,\cdots,N\}}\overline{V} \quad (4)$$

Where d_i is the node degree of firm i; Γ_i is the set of partners of firm i; $\overline{V}=\sum_{i=1}^{7} w_i p_i$ is the average consequence of all risks where p_i is the occurrence probability of risk R_i; α ($\alpha\geq 0$) is the tunable parameter that governs the uniform distribution degree of all firms' risk capacities. Apparently, the distribution of all firms' risk capacities become increasingly uniform when the value of α tends to 0. β ($\beta\geq 1$) is called the tolerance parameter against risk cascading propagation. Clearly, when the value of β is large enough, all firms in R&D

network are capable of accommodating the whole consequences of all risks. At time t, if the risk load $L_i(t)$ on the firm i exceeds its risk capacity C_i, i.e., $L_i(t)>C_i$, this firm i will fail to function because of risk overload. So we propose $\tau_i(t)$ to indicate whether firm i failed or not at the end of time t, which is expressed as follows:

$$\tau_i(t)=\begin{cases}0 & L_i(t)\le C_i\\ 1 & L_i(t)>C_i\end{cases} \tag{5}$$

When firm i failed to function at time t, the risks that occurred on this firm will trigger those potential risks on partners and then might cause some partners to fail because of risk overload. We assume that only the risks that occurred on the failed firms could trigger those potential risks on their partners. Besides, the risks that occurred on failed firms can not certainly trigger the occurrence of potential risks, but rather exists a certain trigger probability. Therefore, we propose the risk triggering probability $H=P(\text{occurrence of } R_j\mid \text{occurrence of } R_i)$ to indicate the conditional occurrence probability of any risk R_j given the occurrence of any risk R_i. Therefore, $L_j(t+1)$ on firm j at time t+1 is expressed as follows:

$$L_j(t+1)=L_j(t)+\sum_{R_k\in\Psi_j(t)}\left[w_k\cdot OR\{\varphi_{lk}^{ij}(t+1)\mid i\in\Gamma'_j(t), R_l\in\Phi_i(t)\}\right] \tag{6}$$

Where $\Psi_j(t)$ is the set of potential risks on node j by the end of time t; w_k is the consequence of risk R_k; $\Gamma'_j(t)$ is the set of failed partners of node j that have not been removed from network by the end of time t; $\Phi_i(t)$ is the set of risks that have occurred on failed partner i; $\varphi_{lk}^{ij}(t+1)$ indicates whether the potential risk R_k on firm j is triggered or not by the risk R_l that has occurred on failed partner i at time t+1, $\varphi_{lk}^{ij}(t+1)=1$ if the potential risk R_k on firm j is triggered by the risk R_l, $\varphi_{lk}^{ij}(t+1)=0$ if not triggered; OR is the "OR" logical operator.

At the end of time t+1, the firm j will continue to function normally if $L_j(t+1)\le C_j$, or fail to function if $L_j(t+1)>C_j$. If $L_j(t+1)>C_j$, the risks that occurred on failed firm j will further trigger the potential risks on its partners. This process continues recursively, which finally results in the risk propagation within single R&D network.

IV. MODELING THE RISK CASCADING PROPAGATION ACROSS INTERDEPENDENT R&D NETWORKS

Apart from the risk propagation within single R&D network, the risks that occurred on failed firms also might trigger potential risks on the coupled firms of other R&D network. Therefore, the next step is to describe the risk cascading propagation across interdependent R&D networks.

For simplicity, we assume that when one firm in one R&D network fail to function because of an initial attack or risk overload, the interdependent firm in other R&D network will fail to function. Then the risks that occurred on this coupled failed firm will trigger the potential risks on its partners, some of which will fail to work because of risk overload. Subsequently, the failure of some firms in this R&D network will in turn cause the coupled firms of other R&D network to fail through interdependences between interdependent R&D networks. Therefore, we give an example to describe the process of risk cascading propagation across interdependent R&D networks, which is illustrated in Fig. 1.

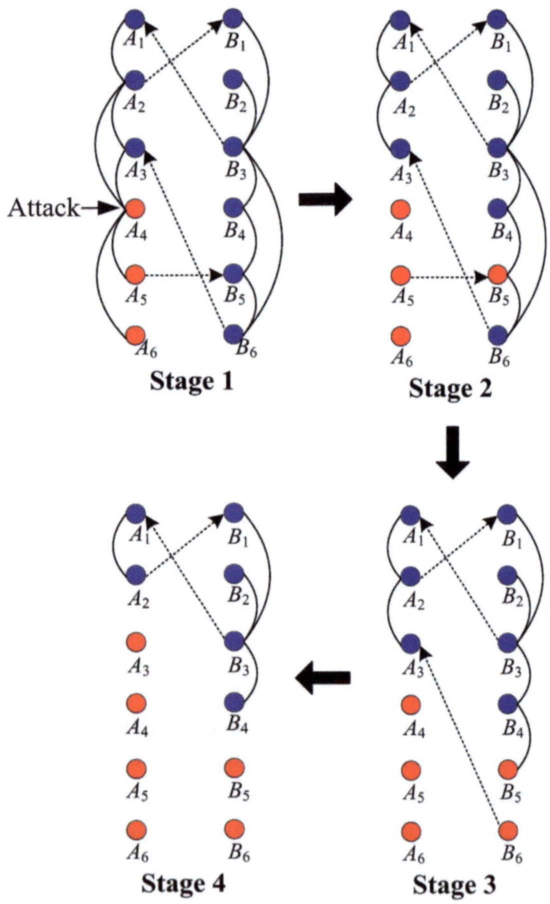

Fig. 1. An illustration of risk cascading propagation across interdependent R&D networks.

In Fig. 1, R&D network A include member firms A_1, A_2, A_3, A_4, A_5, A_6, and R&D network B include member firms B_1, B_2, B_3, B_4, B_5, B_6. The partnerships between firms within network A and B are shown by solid lines (e.g., A_2A_3 and B_1B_3), and interdependences between network A and B are shown by dash lines(e.g., A_2B_1 and B_3A_1). The blue nodes represent the normal firms that can function well, and the red nodes represent the failed firms that cannot function normally due to an attack. Then the process of risk cascading propagation across interdependent R&D networks can be illustrated by the following four stages:

Stage 1: Firm A_4 in network A is failed due to an attack and the risks that occurred on firm A_4 can trigger the potential risks on firms A_2, A_3, A_5, A_6. Here, we assume that firms A_5, A_6 are failed because their risk loads exceed their limited risk capacities, while firms A_2, A_3 are still normal because of the limited risk loads.

Stage 2: Due to the interdependence A_5B_5 from firm A_5 to firm B_5, firm B_5 will immediately fail to function.

Stage 3: the risks that occurred on firm B_5 in network B will trigger the potential risks on firms B_4 and B_6. Here, we assume that firm B_4 remains normal, but firm B_6 is failed because of the risk overload.

Stage 4: Due to the interdependence B_6A_3 from firm B_6 to firm A_3, the firm A_3 will immediately fail to function. We assume that no further risk triggering and node failure occurs subsequently.

In the end, the process of risk cascading propagation across interdependent R&D networks finally causes firms A_3, A_4, A_5, A_6, B_5, B_6 to fail leaving the giant interdependent component composed of normal firms A_1, A_2, B_1, B_2, B_3. Therefore, we use the giant component to measure the ultimate effect of risk cascading propagation across interdependent R&D networks.

Finally, the risk cascading propagation might cause either a few firms to fail or most firms in interdependent R&D networks to malfunction. Therefore, we measure the ultimate consequence of risk cascading propagation to the interdependent R&D networks with the indicator I^*, which is expressed as follows:

$$I^* = \lim_{t\to\infty} I(t) = \lim_{t\to\infty} \sum_{i\in V_A \cup V_B} \frac{\tau_i(t)}{N_A + N_B} \tag{7}$$

Where $I(t)$ is the proportion of the failed firms $\sum_{i\in V_A \cup V_B} \tau_i(t)$ at time t against all the firms $N_A + N_B$ of interdependent R&D networks. In this sense, the indicator I^* denotes the ultimate proportion when risk cascading propagation comes to an end. Clearly, if $I^* \to 1$, then most firms fail to function, and the interdependent R&D networks are completely broken down. But if $I^* \to 0$, then quite a few firms fail, and the interdependent R&D networks still function very well.

V. CONCLUSION

In the age of global economic integration and networking, forming a R&D network has become an inevitable trend where different firms can obtain their competitive edges through R&D collaboration. However, some firms join not only one R&D network but also other R&D network, or some firms in one R&D network supply technology support for other firms in other R&D network. Therefore, some R&D networks become increasingly interdependent, which makes the risks propagate not only within single R&D network, but also spread among interdependent R&D networks. Due to the interdependences between R&D networks, the risk cascading propagation might cause interdependent R&D networks to collapse. Therefore, we explored the risk cascading propagation mechanism across interdependent R&D networks. Firstly, we proposed the generation algorithm of individual R&D network considering that R&D network has the characteristics of scale-free network and small-world network. Then we identified three types of interdependences between interdependent R&D networks, which are input interdependence, shared interdependence and co-located interdependence. After that, we proposed the interdependence matrix $I_{A\rightleftarrows B}$ to describe the identified interdependences. Next, we introduced cascading failures theory to model the risk propagation within single R&D network by defining the risk load and risk capacity on each firm, and building the rule of local triggering among risks. Finally, we give four stages to describe the process of risk cascading propagation across interdependent R&D networks. Our research work can provide a theoretical basis on better understanding the process of risk cascading propagation in the real R&D networks

REFERENCES

[1] J. Hagedoorn. Inter-firm R&D partnerships: an overview of major trends and patterns since 1960[J]. Research Policy, 2002, 31(4):477-492.

[2] W. W. Powell, J. Owen-Smith. Network dynamics and field evolution: the growth of inter-organizational collaboration in the life sciences[J]. American Journal of Sociology, 2005, 110(4):901-975.

[3] F. Tödtling. Innovation networks, collective learning, and industrial policy in regions of Europe[J]. European Planning Studies, 1999, 7(6):693-697.

[4] G. Ahuja. Collaboration networks, structural holes, and innovation: a longitudinal study[J]. Administrative Science Quarterly, 2000, 45(3):425-455.

[5] M. D. König, S. Battiston, M. Napoletano, F. Schweitzer. The efficiency and stability of R&D networks[J]. Games & Economic Behavior, 2012, 75(2):694-713.

[6] Y. Zhang, N. Yang. Risk cascading propagation modeling and simulation in R&D network[J]. Systems Engineering Theory & Practice, 2014, 34(3): 723-731.

[7] K. Yang, Z. Zhang. The research on mechanism of supply chain network risk based on complex network theory[J]. Journal of Systems Science and Mathematical Sciences, 2013, 33(10): 1224-1232.

[8] Z. Li, G. Xu, X. Ban, Y. Zhang, Z. Hu. Complex information system security risk propagation research based on cellular automata[J]. Acta Physica Sinica, 2013, 62(20): 200203.

[9] S. Markose, S. Giansante, A. R. Shaghaghi. 'Too interconnected to fail' financial network of US CDS market: Topological fragility and systemic risk[J]. Journal of Economic Behavior & Organization, 2012, 83(3): 627-646.

[10] Y. Ma, X. Zhuang, L. Li. Susceptible-infected-removed (SIR) model of crisis spreading in the corrected network of listed companies and their main stock-holders[J]. Journal of Management Sciences in China, 2013, 16(7): 80-94.

[11] S. V. Buldyrev, R. Parshani, G. Paul, H. E. Stanley, S. Havlin. Catastrophic cascade of failures in interdependent networks[J]. Nature, 2010, 464(7291): 1025-1028.

[12] R. Parshani, S. V. Buldyrev, S. Havlin. Interdependent networks: reducing the coupling strength leads to a change from a first to second order percolation transition[J]. Physical Review Letters, 2010, 105(4):113-113.
[13] C. D. Brummitt, E. A. Leicht. Suppressing cascades of load in interdependent networks[J]. Proceedings of the National Academy of Sciences of the United States of America, 2012, 109(12): E680.
[14] F. Tan, Y. Xia, W. Zhang, X. Jin. Cascading failures of loads in interconnected networks under intentional attack[J]. EPL, 2013, 102(2): 28009-28014(6).
[15] J. Wang, C. Jiang, J. Qian. Robustness of interdependent networks with different link patterns against cascading failures[J]. Physica A: Statistical Mechanics & Its Applications, 2014, 393:535-541.
[16] J. Wang, Y. Li, Q. Zheng. Cascading load model in interdependent networks with coupled strength[J]. Physica A: Statistical Mechanics & Its Applications, 2015, 430:242-253.
[17] Z. Chen, W. Du, X. Cao, X. Zhou. Cascading failure of interdependent networks with different coupling preference under targeted attack[J]. Chaos Solitons & Fractals, 2015, 80:7-12.
[18] T. Roediger-Schluga, M. J. Barber. R&D collaboration networks in the European framework programmes: Data processing, network construction and selected results[J]. International Journal of Foresight and Innovation Policy, 2008, 4(3-4): 321-347.
[19] N. Hanaki, R. Nakajima, Y. Ogura. The dynamics of R&D network in the IT industry[J]. Research policy, 2010, 39(3): 386-399.
[20] M. Cloodt, J. Hagedoorn, N. Roijakkers. Inter-firm R&D networks in the global software industry: an overview of major trends and patterns[J]. Business History, 2010, 52(1): 120-149.
[21] Y. Zhang, N. Yang, X. Guo. Research on the model and simulation for the self-organization evolution of R&D network[J]. Journal of Management Science, 2012, 25(3): 10-20.
[22] M. E. J. Newman. The structure and function of complex networks[J]. SIAM Review, 2003, 45(2): 167-256.
[23] C. Autant-Bernard, J. Mairesse, N. Massard. Spatial knowledge diffusion through collaborative networks[J]. Papers in Regional Science, 2007, 86(3): 341-350.
[24] T. McDaniels, S. Chang, K. Peterson, J. Mikawoz， D. Reed. Empirical framework for characterizing infrastructure failure interdependencies[J]. Journal of Infrastructure Systems, 2007, 13(3): 175-184.
[25] J. Johansson, H. Hassel. An approach for modelling interdependent infrastructures in the context of vulnerability analysis[J]. Reliability Engineering & System Safety, 2010, 95(12):1335-1344.
[26] S. M. Rinaldi, J. P. Peerenboom, T. K. Kelly. Identifying, understanding, and analyzing critical infrastructure interdependencies[J]. IEEE Control Systems Magazine, 2001, 21(6): 11-25.
[27] E. E. Lee, J. E. Mitchell, W. Wallace. Restoration of services in interdependent infrastructure systems: A network flows approach[J]. IEEE Transactions on Systems, Man, and Cybernetics, 2007, 37(6): 1303-1317.
[28] Y. Zhang, N. Yang. Development of a mitigation strategy against the cascading propagation of risk in R&D network[J]. Safety Science, 2014, Safety Science, 2014, 68: 161-168.
[29] Y. Zhang, N. Yang. Research on robustness of R&D network under cascading propagation of risk with gray attack information[J]. Reliability Engineering & System Safety, 2013, 117: 1-8.

Research on Capabilities-based Optimization Matching of Vessel Maintenance Support Equipment

Le Wang [1], Xin-hao YUAN [2*], Ming-ran Deng[1]
[1] School of Management, Wuhan University of Technology, Wuhan 430070, China
[2] Department of Management, Naval University of Engineering, Wuhan 430033, China

(*18118630301@163.com)

Abstract - **This paper proposes an effective method for matching optimization of vessel maintenance support equipment, and applies the capabilities-based method in breaking down vessel maintenance support equipment and maintenance support mission, and building a comprehensive, reasonable and effective optimization matching model. In the end, MATLAB is utilized to calculate the results of optimization matching and fulfill the objective of optimization matching. This model is simple, easy to operate and highly extensible. The index system and weight can be adjusted based on actual condition to provide a new approach to the matching of maintenance support equipment.**

Keywords - **vessel maintenance support; capabilities-based; optimization matching**

I. INTRODUCTION

Maintenance support plays a crucial role in retaining and restoring the combat capability of equipment, so maintenance support system is an important part of the whole equipment system and a significant force to realizing the maintenance support function. Along with the continual progress of modern science and technology, the complexity and technology content of weapon equipment are constantly increased, while equipment maintenance support has become an important segment of combat capability of troops and holds the same important position as the combat performance of equipment. During peaceful period, weapon equipment in service or storage must be also maintained to keep operational readiness, extend service life and storage life, and satisfy the needs of troops for trainings, duties and operations.

Many studies have focused on the effectiveness of maintenance support, but rarely explored how to match the maintenance support equipment with vessels in the beginning. Based on the operational activity decomposition, Guo [1] analyzed the quantity demand of equipment system, and broken down the complex and macroscopic combat mission into relatively simpler combat activities with single functions. However, it the mission is too complex, it is very difficult to calculate in this way. Dong [2] paid attention to rate of equipment damage and analyzed the dynamic regularity of using maintenance support equipment during wartime to provide a scientific and reasonable solution for determining the quantity of maintenance support equipment. However, it is applicable to a narrow range, and can be only used to determine the quantity, not type, etc. Zhou [3] studied the structural design of the wartime equipment maintenance support system based on the utilization rate and number of the maintenance support units, determined the reasonable levels of support, performed the scientific and reasonable planning and optimization design inside the wartime maintenance support system. The research was conducted on the basis of allocated limited resources.

This paper studies the optimization matching of maintenance support equipment in the following procedure: break down the mission and maintenance support equipment respectively based on capabilities at first; build an optimization matching model based on the capabilities of mission and maintenance support equipment after breakdown; and eventually apply the MATLAB programming in simulation calculation.

II. CAPABILITIES-BASED BREAKDOWN

The "capabilities-based" idea [4] [5] [6] is borrowed to study the optimization matching of maintenance support equipment.

Capability demand analysis [7] [8] is the core of the analysis on the "capabilities-based" optimization matching of vessel maintenance support equipment. Maintenance support mission is supported by capabilities, while capabilities are revealed in the mission. Mission is a specific behavior to realize the maintenance support for units, while vessel maintenance support equipment are the carriers of capabilities and provide the physical basis for realization of maintenance capabilities. Hence, the capabilities-based optimization matching of vessel maintenance support equipment can greatly realize the military demand in maintenance equipment, and the conceptual model in the demand analysis is presented in Fig. 1.

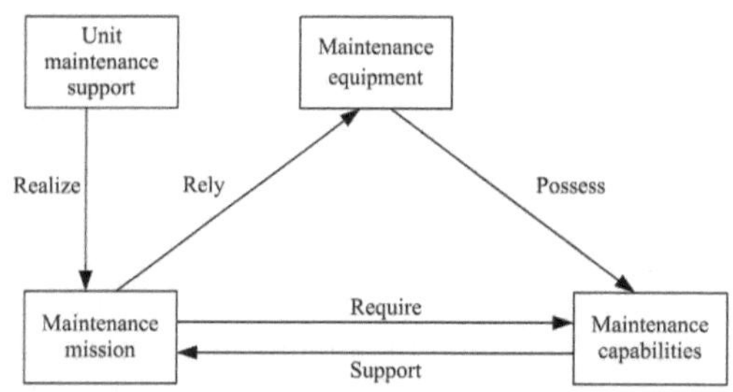

Fig. 1 Conceptual Model for Analysis on Capabilities-based Optimization Matching of Vessel Maintenance Support Equipment

The function breakdown of maintenance support equipment is a process of acquiring the capability demand. Nevertheless, we need to gain the capabilities of each maintenance support equipment. After obtaining the capability acquisition model for single function, we

E. Qi et al. (eds.), *Proceedings of the 23rd International Conference on Industrial Engineering and Engineering Management 2016*, DOI 10.2991/978-94-6239-255-7_20

should aggregate the capability demand for multiple functions, so as to obtain the capability index list of maintenance support equipment.

The capability index system matches with the hierarchical model, and there is the vertical or longitudinal relationship between indexes in the upper level and lower level. Hence, capability aggregation normally employs the "vertical aggregation" or "longitudinal aggregation" model for the analysis on the capability assessment, capability design and optimization and other problems. However, this paper studies the capability aggregation of functions, which is not a process of aggregating the index in the lower level to the index in the upper level, but aggregating the capabilities in the same level between different functions, so it is a "transverse capability aggregation" instead of "longitudinal capability aggregation". Different functions may have identical capabilities, so transverse capability aggregation model is utilized to integrate the capabilities in the same level between different functions.

III. OPTIMIZATION MATCHING MODEL

A. Capabilities-based Matching Process

The capabilities-based optimization matching of vessel maintenance support equipment keeps the fundamental idea as follows: make the capabilities needed to cope with the extensive challenges and changing environment in the modern time based on the vessel equipment for maintenance support and under some economic restraints, and put forward the optimization matching scheme of maintenance support equipment for a unit. The process is as follows: first, extensively identify the maintenance support missions that may be caused in a unit and should be completed by itself, and understand the demand for capabilities; second, breakdown the maintenance support mission, and analyze the capabilities, in other words, break down the functions of vessel equipment and analyze the required capabilities; third, optimize the matching of maintenance support equipment based on the capability demand. The capabilities-based optimization matching of vessel maintenance support equipment should not only take into account the maintenance support equipment needed in a mission, but also consider the maintenance support equipment that should be matched in a unit, so as to cope with the possible maintenance support missions in the unit. It follows the logical idea of "possible missions in a unit—necessary capabilities—capability demand scheme—optimization matching of equipment". Intrinsically, it is a process of summing up the abstract maintenance support missions of a unit, converting them into the specific demand for capabilities of maintenance support equipment, taking the capabilities as the bridge to convert the demand for maintenance support equipment through certain connections.

B. Assumptions and Explanations

Generally, a maintenance support mission can be handled when all the required conditions are achieved, including there are enough maintenance support equipment resources, maintenance support equipment resources are gathered in the region of maintenance mission, and the mission is ready for execution. Hence, maintenance support mission is handled with maintenance support equipment resources under the following conditions and constraints:

1) It does not take time or consume resources to deploy the maintenance support equipment to the place of mission.

2) Maintenance support equipment are all intact and function normally;

3) The number of maintenance support equipment operators satisfies the needs of equipment, and their competency and quality does not affect the use of equipment, and the consumption of resources by personnel is overlooked;

4) The maintenance is not affected by environment;

5) It is assumed that maintenance support equipment can provide 100% successful maintenance support;

6) No missions are executed simultaneously.

The optimization matching of maintenance support equipment is a combinatorial optimization problem, while the set covering problem is a typical problem in combinatorial optimization. Hence, it is feasible to solve the problem through the set covering problem.

We establish the matrix t_{ij} for the mission t and the generation frequency matrix f_i of each mission, and establish the maintenance support equipment matrix P_{kj}, time consumption matrix T_{kj} and resource consumption matrix C_{kj} for each type of maintenance support equipment.

The mathematical model for the optimization of maintenance support equipment is as follows:

$$\min\left\{\sum_{k=1}^{20}\sum_{i=1}^{10}\sum_{j=1}^{7} t_{ij} f_i x_k \left(0.8T_{kj} + 0.2C_{kj}\right) + NM\right\}$$

$$s.t.$$

$$t_{ij} \le p_{kj};$$

$$p_{kj} > 0, t_{ij} > 0;$$

$$x_k \in \{0,1\}$$

x_k is used to determine whether p_k maintenance support equipment is selected to complete the mission. If it is selected, the value is 1. If not, the value is 0.

IV. APPLICATIONOF GENETIC ALGORITHM INOPTIMIZATION MATCHING

Based on its non-convex, discontinuous and other features, the above problem cannot be effectively solved with the regular optimization approach. Genetic algorithm is a global optimization search algorithm, so it has some unique advantages to solve the discontinuous and non-differentiable optimization problems with non-convex feasible domain [9][10][11]. Based on the above model and the basic elements of algorithm, the procedure is determined as follows:

1) Type of coding

When problems are multidimensional and require continuous optimization and very high accuracy of solution, it is often inconvenient to employ the binary coding. To use the variables of problem for coding directly, we employ the real coding. Real coding is suitable for the problem of complex mechanical optimization design with continuous variables. Meanwhile, it avoids the coding and decoding process of standard genetic algorithm as well as the excessive code in binary coding of discrete variable. Hence, genetic algorithm with real coding is more applicable to numerical optimization than genetic algorithm with binary coding.

2) Generation of initial population

Normally, population size affects the final results of genetic optimization and the execution efficiency of genetic algorithm. When the population size is too small, genetic algorithm cannot achieve good performance of optimization. However, large population size may reduce the opportunities for genetic algorithm to obtain local optimal solution. Hence, we select multiple feasible solutions as the initial population. For each mission, the requirements of completion may be satisfied, but the final consumption will be affected by the number of types since each type of maintenance support equipment has different performance and consumption, so further optimization can be performed.

Based on its non-convex, discontinuous and other features, the above problem cannot be effectively solved with the regular optimization approach. Genetic algorithm is a global optimization search algorithm, so it has some unique advantages to solve the discontinuous and non-differentiable optimization problems with non-convex feasible domain. Based on the above model and the basic elements of algorithm, the procedure is determined as follows:

1) Selection of code and generation of initial population

Assuming that the mission set is $t=\{t_1,t_2,\cdots,t_{10}\}$; and the maintenance support equipment set is $p=\{p_1,p_2,\cdots,p_{20}\}$, so the chromosome is $x_1x_2\cdots x_{20}$.

To keep the randomly generated initial population within the range of constraint, feasibility analysis can be added when initial population is generated in computer. In other words, it analyzes whether the capabilities of the selected maintenance support equipment can complete all missions.

2) Selection of crossover and variation rules

For real coding, crossover operation employs the similar PMX method with the crossover probability p=0.08, which can effectively inherit some genes from father. Moreover, this algorithm tends to give more good genes to population, so as to achieve the purpose of priority seeking. Compared with other crossover methods, this method can achieve the variation effect to some extent when two father individuals are the same, which helps maintain the diversity of population to some extent. The variation strategy of exchanging the gene value between 2 points and the variation probability p_m =0.05 are employed, i.e. multiple random exchanges. Certain variation probability is used to determine whether variation is performed or not. After each variation, feasibility analysis is carried out.

3) Selection of termination criterion

It is impossible to know the optimal solution in solving the optimal allocation problem of multiple resources among missions, so the maximum iteration times is given as the termination criterion.

Genetic algorithm can search the region next to optimal solution very quickly, but it takes a long time to attain the optimal solution, and whether convergence can be performed or not relates to the selection of initial population and crossover and variation rules. When the accurate optimal solution is not required in a mission, genetic algorithm can certainly obtain the satisfying optimal solution at a very high speed. To present the effectiveness of the method and provide the calculation example for the matching of maintenance support mission, the MATLAB coding calculation is carried out to optimize and select the maintenance support equipment P1, P2, P3, P5, P9, P11, P12, P14, P15 and P17 when the iteration times is 1,000 times. The total comprehensive effectiveness is W=406.60. The matching results are shown in TABLE IV.

If completing the missions based on the experience, there must be many maintenance support equipment for option. For instance, maintenance support equipment p2, p3, p5, p6, p9 and p11 can be selected. The number of types of maintenance support equipment decreases, but the comprehensive effectiveness is improved considerably. The comprehensive effectiveness is more than doubled, w'=824.4. Clearly, the experience-based method has low efficiency and wastes resources. Therefore, the proposed matching method in this paper achieves the purpose of optimization, and is feasible to some extent. The specific experience-based configuration of maintenance support equipment is shown in TABLE V.

V. CONCLUSION

The optimization matching of vessel maintenance support equipment is a very complex problem. This paper breaks down the maintenance support equipment and maintenance support mission respectively based on capabilities, and takes a unit as the whole for optimization matching. When there are many types of vessel maintenance support equipment, the method is practical to some extent. However, the algorithm has lower efficiency, so it should be further optimized. To sum up, the study provides a new solution for vessel units to match maintenance support equipment, and helps troops make better use of maintenance support equipment and achieve high efficiency without waste of resources.

TABLE I
OPTIMIZATION MATCHING SCHEMEOFMAINTENANCE SUPPORTEQUIPMENT

ID	Mission	Maintenance Testing Equipment							ID	Mission	Maintenance Testing Equipment						
		r1	r2	r3	r4	r5	r6	r7			r1	r2	r3	r4	r5	r6	r7
1	*T1*	P14							*12*	*T5*	P14						
2	*T1*				P3				*13*	*T5*					P2		
3	*T1*					P2			*14*	*T6*	P12						
4	*T2*			P11					*15*	*T6*		P15					
5	*T2*						P11		*16*	*T7*			P17				
6	*T3*	P14							*17*	*T7*					P2		
7	*T3*		P15						*18*	*T8*		P15					
8	*T3*				P1				*19*	*T9*							P9
9	*T4*		P15						*20*	*T9*	P14						
10	*T4*			P11					*21*	*T10*				P1			
11	*T4*						P11		*22*	*T10*							P5

TABLE II
EXPERIENCE-BASEDMATCHING SCHEMEOFMAINTENANCE SUPPORTEQUIPMENT

ID	Mission	Maintenance Testing Equipment							ID	Mission	Maintenance Testing Equipment						
		r1	r2	r3	r4	r5	r6	r7			r1	r2	r3	r4	r5	r6	r7
1	*T1*	P2							*12*	*T5*	P2						
2	*T1*				P3				*13*	*T5*					P2		
3	*T1*					P2			*14*	*T6*	P2						
4	*T2*			P3					*15*	*T6*		P6					
5	*T2*						P11		*16*	*T7*			P3				
6	*T3*	P2							*17*	*T7*					P2		
7	*T3*		P6						*18*	*T8*		P6					
8	*T3*				P6				*19*	*T9*							P9
9	*T4*		P6						*20*	*T9*	P2						
10	*T4*			P3					*21*	*T10*				P6			
11	*T4*						P11		*22*	*T10*							P5

REFERENCES

[1] Q.S Guo, K. Wang, and Y.P. Fan. "Amount demand analysis method based on operational activity decomposition for equipment system". *Journal of Academy of Armored Force Engineering,* 2014.

[2] X.B. Dong, Y. Ma, and L.Y. Chen. "Research on the solution to question of maintenance support equipment number scale". *Journal of the Academy of Equipment Command & Technology*, 2011.

[3] F.J. Zhou, X.D. Zhu, H.Y. Gu, and L. Chen."The structural design of the wartime equipment maintenance and support system based on the utilization rate and number of the maintenance and support units". *National Defense Science & Technology*, 2014.

[4] J.B. Planeaux." Beyond The Task Force Conops: The Path to a Capabilities-based Modernization Framework for the Air Force". Lt Col, USAF, 2003.

[5] Q. Zhang, Y.D. Yan, and W. Chen. "Research on the problems of requirement analysis of maintenance equipment system based on capability". *Journal of Academy of Armored Force Engineering*, 2011.

[6] Y.H. Cao, S.X. Feng and X.F. Xu. "Research on conception representation of operational mission break". *Computer Simulation*, 2007.

[7] Y. Shu and Y.J. Tan. "Research on description of weapon equipment system structure based on the demand for capabilities". *Military Operations Research and Systems Engineering*, vol. 23, no. 3, pp. 51-55, 2014.

[8] J.F. Wang, J.C. Guo and Y.F. Zhou. "Research on demand analysis method for integrated combat capability". *Military Operations Research and Systems Engineering*, vol. 23, no. 2, pp. 30-34, 2009.

[9] Y.J. Dou. "Research on Capability Requirement Index Plan Generation of Weapon System-of-systems Meta-activity Decomposition-Oriented". M.A. dissertation, National University of Defense Technology, Changsha, China, 2011.

[10] W.X. Lin and Z. Wu. "Water resources allocation model based on the adaptive & vector evaluated genetic algorithm and its application". *South-to-North Water Transfers and Water Science & Technology*, vol. 6, no. 3, pp. 69-71, 2008.

[11] L.A. Ozdamar, "genetic algorithm approach to a general category project scheduling problem". *IEEE Trana Syst Man Cyber*, vol. 29, no. 1, pp. 44-59, 1999.

Analysis on Warship Technology Supportability Equipment System-of-Systems Based on System Dynamics

Le Wang [1], Chun-hui Yang[2*], Ming-ran Deng[1]

[1] School of Management, Wuhan University of Technology, Wuhan 430070, China

[2] Department of Management, Naval University of Engineering, Wuhan 430033, China

(*ych_ld@tom.com)

Abstract - **The configuration of Warship Technology Supportability Equipment(WTSE) is complex systems engineering, so it should comprehensively takes into account the impacts of warship system, organization, process, task, equipment and resource, etc., and be optimized based on overall input and demand to improve the capabilities considerably. First, this paper analyzes the structure of WTSE and builds a multi-view model for WTSE system-of-systems. Subsequently, a model based on system dynamics is constructed after analyzing the causal relations of index with input and task demand for warship technology supportability capabilities, while a growth curve is employed to elaborate the development and evolution process of capabilities. By means of simulation, this paper eventually analyzes the strategy for optimization configuration of WTSE system-of-systems.**

Keywords - **warship technology supportability; WTSE system-of-systems; system dynamics**

I. INTRODUCTION

Warship technology supportability is a generic term for the supporting measures and corresponding activities implemented to maintain and restore the good technical state of warship equipment. Warship technology supportability equipment refers to all kinds of equipment and tools used to carry out the warship technology supportability activities.

Warship technology supportability involves the activities of all kinds, e.g. maintenance support, apparatus support and corresponding training support and command support. However, warship equipment involve intensive high technology of wide range and professional field, and technology supportability is highly comprehensive and integrated. Also, warship equipment face a large variety of different tasks, multiple models of warship, and complex classification, while Warship technology supportability equipment (WTSE) have complex structure, take up a large amount of technology supportability costs and directly affect the realization and improvement speed of warship technology supportability capabilities. As a result, WTSE must be manufactured to satisfy the changes of tasks and warship equipment and developed according to the WTSE system-of-systems.

Along with the elapse of time, WTSE system-of-systems keeps welcoming new equipment and abandoning obsoleteequipment constantly, which is a dynamic and varying process. Correspondingly, warship equipment, technology supportability personnel, processes and technical documents, and warship technology supportability tasks, etc. are also changing and adjusting constantly. Thus, the relationship between the construction of warship technology supportability system-of-systems and the change of warship technology supportability capabilities must be analyzed, and the impacts of different construction schemes on warship technology supportability capabilities should be studied, in order to maximize the effectiveness of WTSE system-of-systems and facilitate the improvement of warship technology supportability capabilities.

First of all, this paper analyzes the structure of WTSE system-of-systems. Then, warship technology supportability capabilities are taken into account to analyze their interactive and feedback mechanism with the structure of WTSE system-of-systems, and the theories and methods of system dynamics are introduced into quantitative analysis. This paper aims to build a model for WTSE system-of-systems based on system dynamics, perform the simulation analysis on the effect that different cost allocation schemes can improve the warship technology supportability capabilities, and put forward the specific strategy for the construction and development of WTSE system-of-systems.

II. CONCEPT OF WTSE SYSTEM-OF- SYSTEMS

The basic elements of System-of-systems include environment, task, object, structure and process [1,2]. WTSE system-of-systems fulfills the warship technology supportability tasks jointly with and through cooperation with the object, organization, force, technology, information and other elements of warship technology supportability. Hence, WTSE system-of-systems consists of six elements, i.e. warship equipment, warship technology supportability organization, warship technology supportability task, warship technology supportability process, warship technology supportability resource and WTSE. From the approach of multi-view [3], the WTSE system-of-systems can be decomposed into a 6-view model. The views are interconnected closely. Some elements of technology supportability equipment system-of-systems may appear in different views in different forms.

The WTSE system-of-systems is constructed to further analyze the relationship between WTSE and other supportability elements for coordinating their development on one hand, and plan the construction and development of WTSE from the top level for improving the spending effectiveness of warship technology supportability costs and the construction level of warship technology supportability capabilities on the other hand.

E. Qi et al. (eds.), *Proceedings of the 23rd International Conference on Industrial Engineering and Engineering Management 2016*, DOI 10.2991/978-94-6239-255-7_21

A model for WTSE system-of-systems based on capability demand is built to describe the connections between system structure and capabilities and between demand and system structure, and reflect the degree of matching of WTSE system-of-systems with tasks and elements. In addition, the internal and external environments of WTSE system-of-systems are changing constantly, e.g. task adjustment, maintenance system reform, military and civil integration supportability, supportability process innovation, WTSE upgrading and reconstruction, etc. Hence, the system-of-systems also varies in structure, state, function and action, etc. along with the elapse of time. In other words, WTSE system-of-systems is evolving dynamically.

III. DYNAMIC EVOLUTION MODEL

The key problem in the dynamic evolution of WTSE system-of-systems is how to make full use of limited resources, better satisfy the demand of warship technology supportability tasks and improve the warship technology supportability capabilities.

Based on whether they directly serve warship equipment, WTSE can be classified into two categories. One category is the equipment directly providing the warship technology supportability activities for warship system, so as to form the model-oriented WTSE system-of-systems. The other WTSE belong to the second category, including the equipment needed in such activities as warship technology supportability command communication, warship technology supportability apparatus storage, warship technology supportability personnel training, etc. so as to form the basic WTSE system-of-systems.

Based on the relationship among elements inside warship technology supportability system, we can obtain the causal relationship diagram of the system as presented in Fig. 1. The model takes into account only the influence on the changes of warship technology supportability capabilities when warship equipment and WTSE are mobilized.

Based on the above analysis on causal relations, the vensim software can be applied to construct the corresponding stock flow graph [4], as presented in Fig. 1. Due to the changes of warship equipment and warship technology supportability missions, the increase of cost should be adjusted to guarantee warship technology supportability capabilities under the existing cost strategy. Meanwhile, the cost optimization allocation strategy can be adopted to improve the warship technology supportability capabilities as much as possible within the limited cost. For instance, warship technology supportability personnel capabilities can be improved first to change the influence of WTSE on capability level, while the input of WTSE in the first category is further increased on this basis.

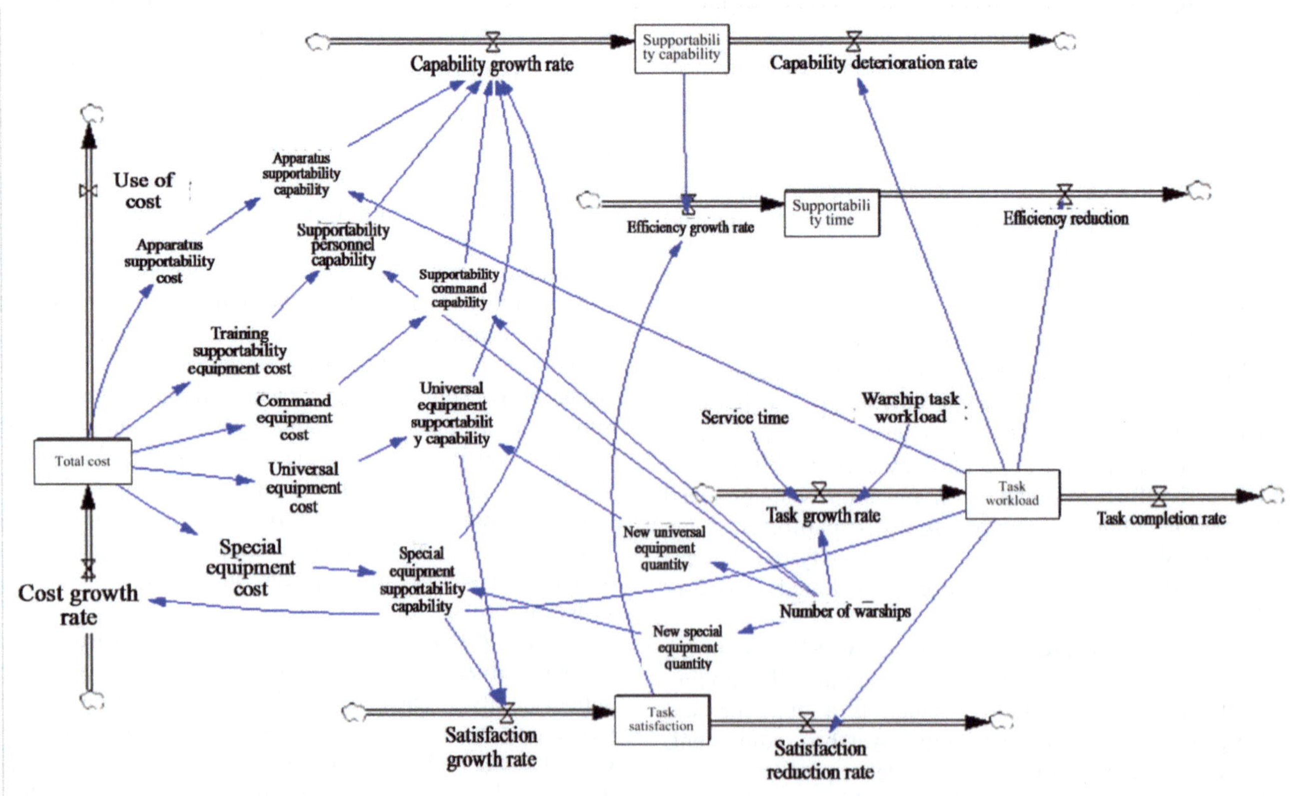

Fig.1. A Mode for Warship Technology Supportability System-of-Systems Based on System Dynamics

IV. SIMULATION ANALYSISONTHE MODEL

While satisfying the increasing number of warships and warship technology supportability tasks, WTSE system-of-systems is constructed to improve the warship technology supportability capabilities, so as to adapt to the external environment and satisfy the needs of tasks.

The capability levels of warship technology supportability are evaluated subjectively. The changes of capability level form a growth curve. We divide the life cycle of changing capabilities into four stages, i.e. infant, development, maturity and application [5][6], which correspond with the development states of capabilities in different periods, as presented in Fig. 2.

To help warship technology supportability personnel determine the current capability level of WTSE, we divide it into four stages.

As for some WTSE, the configuration may not always experience the above four stages and 10 levels. For instance, the comprehensive integration of WTSE for electronic equipment is high, so it can achieve high level after configuration, but the corresponding cost is also very high.

Along with the development of warship technology supportability and the increasing requirements of tasks, WTSE do not change, but their capability level decreases gradually.

The sensitivity of different WTSE to cost at each stage can be estimated by analyzing the conditions in the past years.

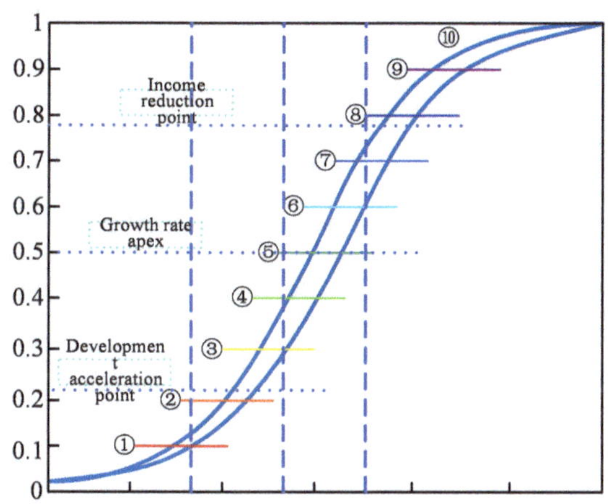

Fig.2. S Curve Knowledge Graph of WTS Capabilities

The model for growth curve is as follows:

$$y = \frac{y^*}{1 + a \times e^{-b(t-\tau)}}$$

The parameter b is the shape parameter of S curve, while τ is the displacement parameter of S curve. As the curve indicates capabilities, $y^* =$.

We define warship apparatus supportability capability, warship training supportability capability and warship command supportability capability as the basic capabilities. If the basic capabilities are lower than the universal/special equipment capabilities, it is necessary to increase the proportion of cost, so as to achieve the expected effect. If the basic capabilities are higher than the universal/special equipment capabilities, no adjustment should be made. For different growth rates, the growth curves have different time spans, as shown in Fig. 3. In addition, the cost of each content varies within a certain range to guarantee the change of growth curve. When growth rate is lower than a specific level, the capability value decreases constantly.

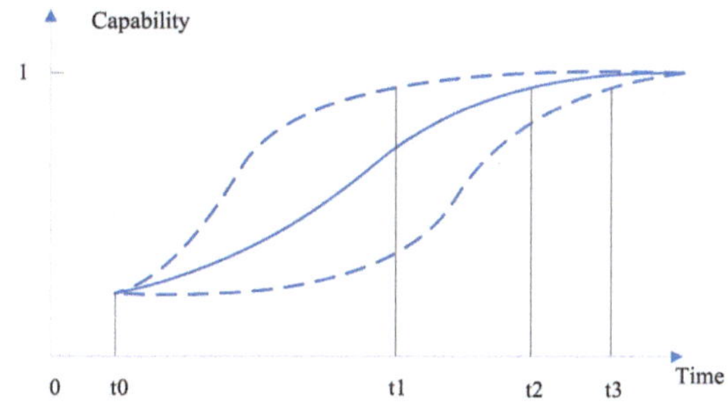

Fig. 3 Relationship between Capability Level Growth of Warship Technology Supportability and Cost Growth Rate

This paper employs the system dynamics simulation software vensim to perform the simulation operation and policy assessment for the constructed model. The simulation of model will focus on what input allocation strategy can improve the overall supportability capability level to the maximum.

The capability levels are analyzed to reveal that:

1) The level of warship technology supportability personnel capability is relatively low, so it will grow after input is made. The input is very high, and the growth has time delay.

2) The level of warship technology supportability command capability is relatively low, so it will grow fast after input is made. The input is very high, and the growth basically has no time delay.

3) The level of apparatus supportability capability is relatively high, so it will grow slowly after high input is made, but the growth has no time delay. It has limited influence on the overall capability.

4) Universal supportability equipment: A portion is used to maintain the use of current universal equipment and the completion of tasks, and the remaining portion is spent in purchasing new universal equipment. It is necessary to determine the proportion of universal equipment and special equipment and the proportion of affected capabilities. A large quantity of universal equipment should be purchased, so it affects the improvement of capabilities.

5) Special supportability equipment: A portion is used to maintain the use of current special equipment and the completion of tasks, and the remaining portion is spent in purchasing new special equipment. There is rigid demand for special equipment, so priority must be given to satisfying a portion of demand.

Hence, if the supportability cost is allocated first to satisfy the training supportability cost and improve its growth rate, capability level will grow fast at the late stage, and the growth rate will exceed the capability level under the existing proportion. Considering that the cost is normally planned for a period of five years, the training supportability cost is satisfied in the first 2-3 years of every 5-year plan, and the cost is allocated to the configuration of universal and special supportability equipment in the last two years, which can also achieve good effect.

The simulation results are presented in Fig. 3 and Fig. 4. In Fig. 3, the curve B represents the change of capability under the current strategy; the curve C means that if priority is given to guaranteeing the training supportability cost, the growth rate of capability level at the initial stage is very low, but increases fast at the late stage. Since the training supportability cost achieves some effect, training supportability capability is improved, which exerts very positive effect on the overall capability. In Fig. 4, the curve D means that, if the five-year plan is employed, priority is given to guaranteeing the training supportability cost in the first three years of every five-year plan, and the growth rate of capability level shows a trend of fluctuation, but the overall effect is still better than the existing allocation strategy.

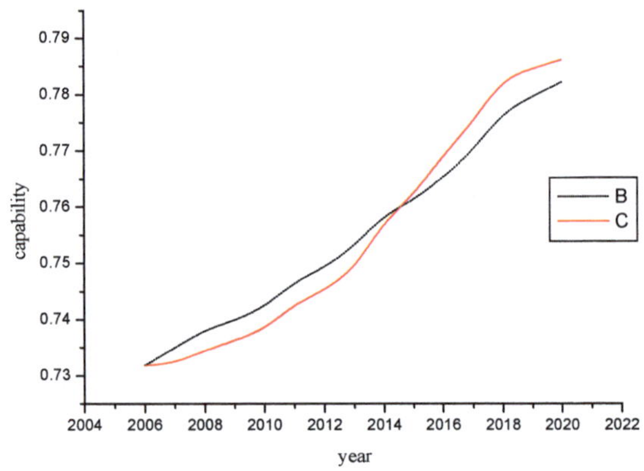

Fig. 3 Comparison of Simulation Results(a)

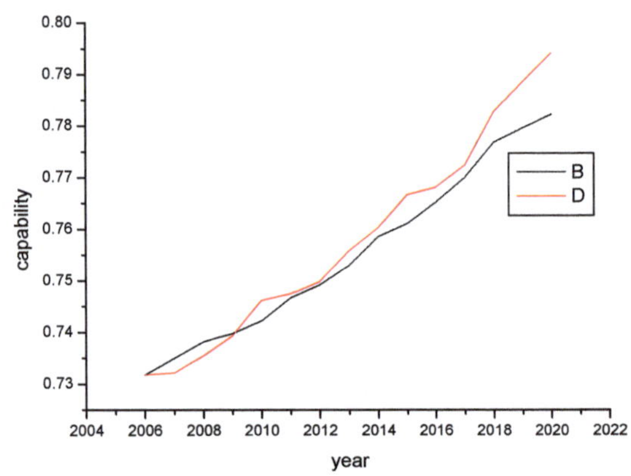

Fig. 4 Comparison of Simulation Results(b)

V. CONCLUSION

WTSE system-of-systems is an important part of warship technology supportability capabilities. Along with the increasing number of supportability tasks and the increasing requirements for supportability, the optimization construction of WTSE system-of-systems becomes more and more important. WTSE system-of-systems contains the model-oriented universal and special WTSE and the basic system formed by apparatus supportability equipment, training supportability equipment and command supportability equipment, which affect and promote each other. This paper further analyzes the logical relations of different types of WTSE with warship technology supportability tasks and capabilities, develops a model based on system dynamics and performs the simulation analysis on the strategy of cost optimization allocation.

REFERENCES

[1] B. Cheng, Y.J. Lu and Y. Zhou. "Research progress on weapon system-of-systems optimization methods". *Systems Engineering and Electronics*, vol. 34, no. 1, pp. 85-90, 2012.

[2] L. Liu. "Research on Evolution Modeling for Weapon Equipment System Development and Exploration of Analysis Methods". Ph.D. dissertation, National University of Defense Technology, Changsha, China, 2011.

[3] P.H. Deng, Y.M. Bi and O.Y. Zhang et al. "Multi-level and multi-view description framework of military requirements". *Systems Engineering and Electronics*, vol. 32, no. 11, pp. 2380-2384, 2010.

[4] H.X. Yang, J.D. Li and H. Zhang et al. "Research on the governance of urban traffic jam based on system dynamics". *Systems Engineering and Electronics*, vol. 34, no. 8, pp. 2135-2143, 2014.

[5] C.S. Ren. "Research on Assessment Methods for Weapon Equipment System Technology Maturity". M.A. dissertation, National University of Defense Technology, Changsha, China, 2010.

[6] L.X. Yang. "Research on Technology Maturity Multi-dimensional Assessment Model". M.A. dissertation, National University of Defense Technology, Changsha, China, 2011.

Simulation Model of Rail Transshipment System in Hydropower Development Project

Yi-xun Fang [1], Zhen-yuan Liu [1], Yan Kong[1]

[1] School of Automation, Huazhong University of Science and Technology, Wuhan, China

Abstract – **According to the operation process of rail transshipment systems, a discrete event-based simulation model is developed. With different wagon grouping rules and a set of wagon para-unloading schemes, the simulation model can record the penalty cost of wagons due to the beyond of the maximum tardiness time of wagons in the rail transshipment system. This model can be used for evaluating the rationality of the operation management of the rail transshipment system in hydropower development project, and providing decision support to managers in the real-time scheduling of the system.**

Keywords - **Hydropower development project, rail transshipment system, system modeling and simulation**

I. INTRODUCTION

The management of objectives such as the cost, progress and quality is restricted by the material supply. In order to ensure the material supply of the hydropower development project, the rail transshipment system should be established near the construction sites, used for the transshipment and storage of various kinds of engineering materials. Modeling and simulation are widely used for decision support in different business areas and management tasks[1].

Luis has built an overall system dynamics model of manufacturing enterprise and connected to it are a number of discrete event simulations for selected operational and tactical functions[2]. Abbas has built a simulation model of supply chain consisting of the three levels of raw-material supplier, manufacturer and end customers[3]. Salvatore has presented a simulation-based study of a coordinated, decentralized linear supply chain system[4]. Usha has used actual industrial data and simulation to help managerial decision-making on the number of collaborating partners, the level of investments and the involvement in supply chain processes[5]. Collaborative practices are proved to be very helpful, especially when the number of sources increases through building simulation models of Economic Order Quantity[6]. C. Vidalakis has adopted a discrete event simulation approach to assess the impact of demand fluctuations on two crucial logistics performance measures, lead time and cost efficiency[7]. Jung presented two different types of simulations, namely human-to-human and computer-to-human interactive simulation[8]. Vidalakis followed a field study approach to develop a conceptual logistics model facilitating experimentation using simulation modelling[9]. Ammar has used a process mapping methodology to model processes of precast concrete operations and to provide the framework and input required for simulation[10]. Yasser has presented a construction supply chain simulation toolkit that is capable of modeling different supply chain problems[11]. Navin used Data Envelopment Analysis approach to understand the optimal relative efficiency among different values of average fill rate performance obtained through simulation[12]. He has dealt with the modeling and optimization problem of multi-echelon container supply chain network[13]. Nihar used hybrid simulation based optimization frameworks to predict the optimal operation under the different scenarios[14].

The bottlenecks, such as the locomotive scheduling problem, existed in the daily operation of the rail transshipment system have been studied through building the discrete event-based simulation model.

II. OPERATION PROCESS OF RAIL TRANSSHIPMENT SYSTEM

A. *Operation Process*

According to the operation process of the rail transshipment system, the system can be divided into 3 logical areas: the grouping area, unloading area and storage area. The operation process of the rail transshipment system generally includes 9 steps: wagon grouping, allocating traction locomotive, allocating para-unloading locomotive, wagon para-position, wagon unloading, empty wagon grouping, material in-stock, material out-stock, material loading.

The quantity of locomotives is not much normally, but the locomotives are used during the whole operation process. So how to use and allocate locomotives reasonably is a key point in wagon logistics scheduling management. The dispatching mode of locomotives adopted by the rail transshipment system is special. In this mode, locomotives are divided into the traction locomotive and the para-unloading locomotive due to the different operation scopes and responsibilities of locomotives. The traction locomotive is responsible for the empty wagon grouping, the traction of wagons and empty wagons between the rail transshipment system and the railway station. The para-unloading locomotive is responsible for the traction of para-positon and unloading.

The following assumptions are proposed in the establishment of the simulation model of the rail transshipment system.

1) The operations of locomotives and unloading equipment are have no abnormalities;

E. Qi et al. (eds.), *Proceedings of the 23rd International Conference on Industrial Engineering and Engineering Management 2016*, DOI 10.2991/978-94-6239-255-7_22

2) The unloading equipment are enough and there is no resource constraint;
3) The manpower resources are enough.

B. Wagon Grouping and Sorting within the Group

Wagons can be grouped when they arrive at the railway station. If the quantity of wagons in a group achieves the limit of the locomotive capacity, the grouping is completed. If the quantity of wagons in a group doesn't achieve the limit of the locomotive capacity, but the waiting time of the wagon group achieves the limit, the grouping is also completed. When the grouping is completed, the wagon group can be sorted according to the wagon sorting model. Then a traction locomotive can be allocated to the wagon group. If there is no available traction locomotive in the rail transshipment system, the wagon group should wait until there is an available one. The railway station and rail transshipment system are connected by a special railway line. It only allows one wagon group to occupy itself. When a wagon group is completed and allocated a traction locomotive, it can be pulled into the rail transshipment system. The time of the wagon group shuttling from the railway station to the rail transshipment system obeys the normal distribution.

When a wagon group is completed, the para-unloading scheme should be established. In order to improve the efficiency of the para-unloading as much as possible, the optimization of wagon sorting within the group should be taken into consideration. According to the material records monthly, the set of wagon para-unloading schemes can be summed up. Each para-unloading scheme has a corresponding para-unlading time.

Kong has put forward a wagon sorting within the group model which can determine the quantity of unloading stages and corresponding quantities of each material at each unloading stage[15]. The para-unloading schemes of each wagon group and the corresponding para-unloading time can be obtained through this model.

The objective of this model is to minimize the whole para-unloading time of the wagon group. And it has 2 constraint conditions. The first one is for any kind of material g, the corresponding wagon quantities of the para-unloading scheme selected at each stage adding together is equal to the wagon quantity of material g in this group. The second one is for any stage p, it can choose only one para-unloading scheme k.

C. Para-unloading

When the wagon group is pulled into the rail transshipment system by the traction locomotive, the locomotive should be switched to the para-unloading locomotive. Then the wagon group should be pulled to corresponding storage tanks for the first time para-unloading. When it is completed, the wagon group should be pulled for the second time para-unloading until the whole para-unloading is completed for this wagon group. The para-unlading time of wagons obey the normal distribution.

D. Empty Wagon Grouping

Because each wagon group is unloaded in batches, each batch of empty wagons can be grouped when this batch completes para-unloading. When the empty wagon grouping is completed, the empty wagon group can be pulled to the railway station by the traction locomotive.

III. SIMULATION MODEL OF RAIL TRANSSHIPMENT SYSTEM

A. Inputs of Simulation

The corresponding symbol definitions and descriptions appeared in the simulation process are as shown in Table I.

TABLE I
INPUTS OF SIMULATION MODEL

Symbol	Unit	Description
w	-	The serial number of wagon
g	-	The serial number of material
m		The serial number of locomotive
j		The serial number of storage tank
k		The serial number of contractor
TR_w	-	The time of arriving at the railway station
CM_{wg}	-	The type of material loaded in the wagon
CN_{wg}	-	The weight of material loaded in the wagon
EC	*wagon*	The capacity of the locomotive
TF	*min*	The time of shuttling from railway station to rail transshipment system
TG	*min*	The grouping time of the wagon group
CT_{jg}	-	The type of material loaded in the storage tank
C_j	*t*	The volume of storage tank
S_j	*t*	The volume of storage tank at present
TC_k	*min*	The transport time from rail transshipment system to contractor
D_{kg}	*t*	The quantity of material demanded by contractor
t_{β_0}	*h*	The maximum tardiness time of wagon in the rail transshipment system
β_1	*h*	The rate of fees postponing from 1 to 10 hours
β_2	*yuan/h*	The rate of fees postponing from 11 to 20 hours
β_3	*yuan/h*	The rate of fees postponing from 21 to 30 hours
β_4	*yuan/h*	The rate of fees postponing over 30 hours

B. Outputs of Simulation

Under different simulation strategies, wagons go through arriving at railway station, wagon grouping, para-unloading, empty wagon grouping and empty wagon returning. The materials loaded in wagons go through in-stock, storage and out-stock. A set of output data will be obtained through the simulation model. The expressions are as follows.

$$Y = \sum_{w=1}^{W} y_w \tag{1}$$

$$y_w = \begin{cases} 0 & 0 \le t_w < t_{\beta_0} \\ \beta_1\left(t_w - t_{\beta_0}\right) & t_{\beta_0} \le t_w < t_{\beta_0} + 10 \\ 10\beta_1 + \beta_2\left[t_w - \left(t_{\beta_0} + 10\right)\right] & t_{\beta_0} + 10 \le t_w < t_{\beta_0} + 20 \\ 10(\beta_1 + \beta_2) + \beta_3\left[t_w - \left(t_{\beta_0} + 20\right)\right] & t_{\beta_0} + 20 \le t_w < t_{\beta_0} + 30 \\ 10(\beta_1 + \beta_2 + \beta_3) + \beta_4\left[t_w - \left(t_{\beta_0} + 30\right)\right] & t_w \ge t_{\beta_0} + 30 \end{cases} \tag{2}$$

$$t_w = TRE_w - TR_w \tag{3}$$

C. *Simulation Model*

In the actual operation, the decision maker needs to know the specific state of each locomotive, wagon and storage tank at each discrete time point to carry out reasonable transshipment scheduling. The discrete event simulation method can be used for simulating the actual situation of the rail transshipment system to achieve this purpose. Aiming at several common rules of the wagon grouping in the simulation, reasonable scheduling schemes and the corresponding values concerned by the decision maker can be obtained under different rules. Because there are only one traction locomotive and one para-unloading locomotive in this case, the wagon group can be pulled into the rail transshipment system when the traction locomotive is available. So does the para-unloading locomotive. The process of the simulation model is as shown in Fig. 1.

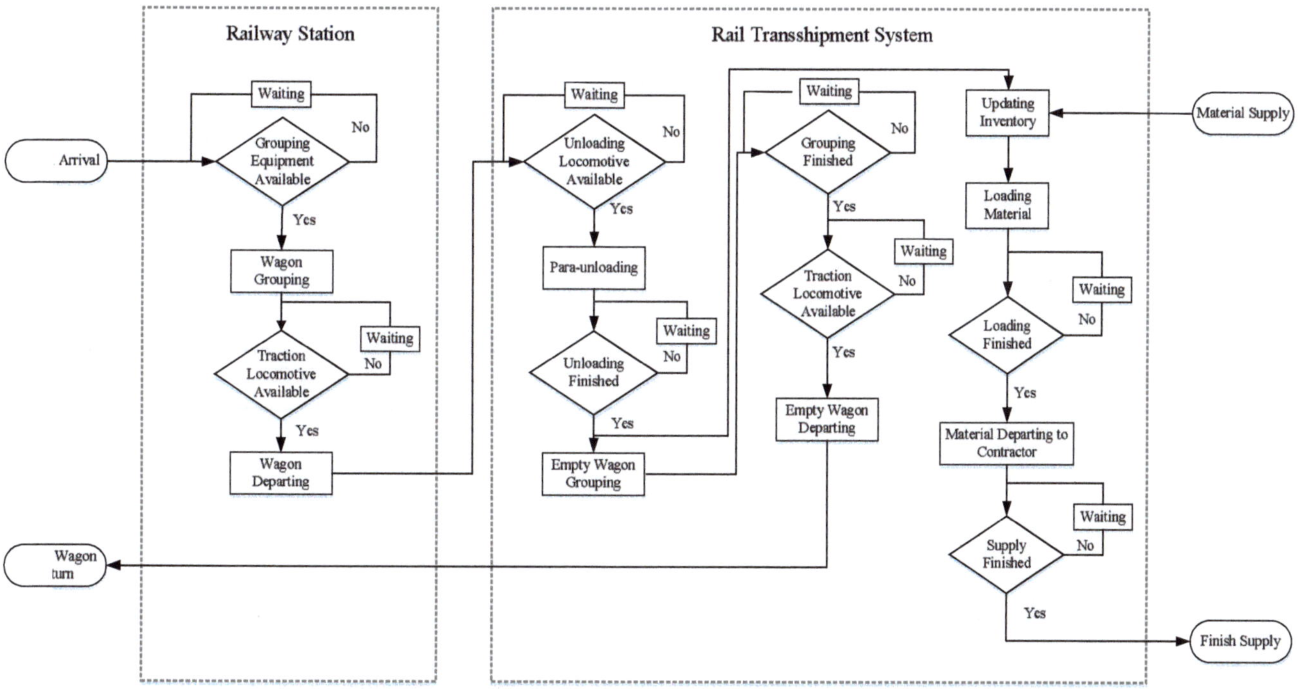

Fig. 1. The simulation process of rail transshipment system

In actual cases, rules of wagon grouping generally contain: the first wagon arrived being grouped first, grouping according to the type of material, grouping according to the distribution of storage tanks, the emergency material being grouped first, grouping to get the minimum tardiness time of wagon in the rail transshipment system.

The distribution of storage tanks and unloading special line is as shown in Fig. 2.

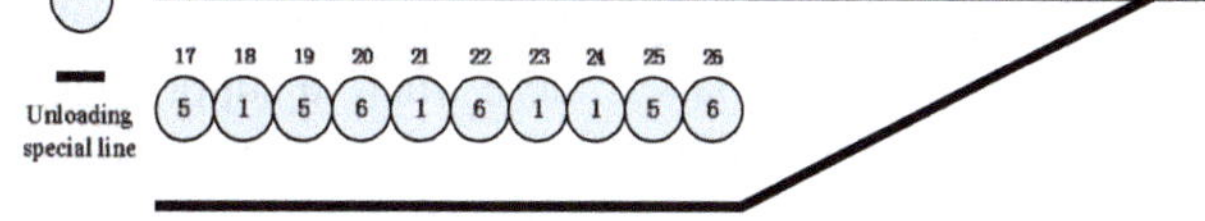

Fig. 2. The distribution of storage tanks and unloading special line

From the figure above, the distribution of storage tanks has certain regularity. Material 1 distributes on two unloading special lines averagely. Materials 2, 3 and 4 distribute only on the unloading special line at the top, the distributions of materials 5 and 6 are the other way round.

So in the simulation, two grouping rules are selected as follows:

1) GR1: Wagons are grouped immediately when they arrive at the railway station until the quantity of wagons in the group achieves the capacity limit of the traction locomotive;
2) GR2: Getting the distribution of storage tanks during the supply period, wagons are grouped according to the different unloading special lines.

For the grouping rule GR2, there is an example. If materials arrived within a period are 2, 3, 4, 5 and 6, materials 2, 3 and 4 can be grouped into the same group, materials 5 and 6 can be grouped into another group. Material 1 is grouped randomly.

IV. SIMULATION EXPERIMENT AND RESULT ANALYSIS

According to the research scope, two supply periods of the rail transshipment system of Yalong River Hydropower Development Company are selected as simulation cases. In these cases, there are 5 suppliers, 6 construction materials, 16 contractors and 31 storage tanks.

Supply period 1 is from 00:00 of December 20, 2011 to 00:00 of December 28, 2011. Supply period 2 is from 00:00 of June 1, 2011 to 00:00 of June 9, 2011. These two supply periods are the trough and the peak of the entire supply respectively. Enter the arrival of materials and supply plan of materials, as well as the initial inventories of storage tanks and input data as shown in Table II.

TABLE II
INPUT DATA OF SIMULATION MODEL

Symbol	Numerical Value
TF	$N\sim(30,5)$
TG	$N\sim(30,5)$
EC	30
t_{β_0}	120
β_1	4.4
β_2	8.8
β_3	13.2
β_4	17.6

The simulation times of two supply periods are 20. The simulation experiment results are as shown in Table III.

TABLE III
SIMULATION EXPERIMENT RESULTS

Supply Period		Supply Period 1		Supply Period 2	
		GR 1	GR 2	GR1	GR2
Tardiness Time	AVERAGE	3184.651	4445.03	4482.466	3185.527
	STDEV	10.32393	84.0254	11.6679	10.3887
	MAX	3204.338	4636.738	4501.296	3204.338
	MIN	3158.337	4234.106	4459.805	3158.337
Penalty Cost	AVERAGE	17217.89	25550.21	22421.64	17222.29
	STDEV	57.32924	620.0456	78.01176	58.6566
	MAX	17329.04	27116.18	22580.23	17329.04
	MIN	17088.38	24205.67	22254.45	17088.38

The actual tardiness time of these two supply periods are 4102.1 hours and 6097 hours. The penalty costs of these two supply periods are 19684.6 yuan and 31991.2 yuan. Compared with the simulation experiment results, we can get conclusions as follows.

1) When using the grouping rule GR1, the tardiness time and penalty costs all have been reduced in different degree, especially in supply period 2. Through the analysis, we can get the cause of this result: the arrival of materials is more when in supply period 2, the para-unloading efficiency is improved greatly by the reasonable choice of the para-unloading scheme.
2) When using the grouping rule GR2, the tardiness time and penalty costs present the different results in different supply periods. The tardiness time and penalty cost of supply period 1 increase greatly, but they are further reduced when in supply period 2. Through the analysis, we can get the cause of this result: the arrival of materials is less when in supply period 1. Grouping according to different unloading special lines will get more wagon groups. These wagon groups basically don't achieve the limit of locomotive capacity, leading to efficiency reducing of single wagon group. The arrival of materials is more when in supply period 2. Although there will be more wagon groups, but most of them achieve the limit of locomotive capacity. Meanwhile, they don't need to switch the unloading special lines. This will lead to efficiency improving of single wagon group.

V. CONCLUSION

A simulation model of the rail transshipment system in the hydropower development project has been proposed. Combining with characteristics of the problem and actual management methods, wagon grouping rules and para-unloading schemes are designed under the background of Yalong River Hydropower Development Company. The application of this simulation model is illustrated through doing simulation experiment in different wagon grouping rules.

The main meanings of this research are as follows:

1) The simulation model combining with material in and out is fully established. It can evaluate the rationality of the material transshipment roundly and find the bottlenecks in the operation management of the rail transshipment system. This will improve the operation efficiency of the rail transshipment system.
2) In the simulation, possible problems in the actual operation, such as the fluctuation of running time, wagon grouping, wagon unloading and running time of locomotive, are fully considered. Contrastively analyzing the simulation results and parameters concerned by decision makers under different wagon grouping rules, decision makers can analyze operation effects under different wagon grouping rules through multiple aspects and choose the appropriate wagon grouping rule, thus providing decision support to the real-time operation scheduling of the rail transshipment system.

In addition, the rationality of the distribution of storage tanks can be evaluated by collecting the real-time volumes of storage tanks. Adding the dispatching rules of transshipment vehicles into the simulation model can also evaluate the rationality of dispatching rules of transshipment vehicles.

The future direction of the research is being more focused on the complete supply chain and the coordinated operation of multiple rail transshipment systems.

REFERENCE

[1]Elena. O, Mercedes. R. Decision-making in IT service management: a simulation based approach. [J]. Decision Support Systems, 2014, 66: 36–51

[2]Luis. R, Alfonso. T. S, Magdy. H, Albert . J. Supply chain and hybrid simulation in the hierarchical enterprise[J]. International Journal of Computer Integrated Manufacturing, 2015, No.5

[3]Abbas. M; Sayed. M. S; Mohammad. M. T. Simulation of a single product supply chain model with ARENA[J]. International Journal of Industrial and Systems Engineering, 2015, No.1

[4]Salvatore. C, Mónica. L. C,Roberto . D, Jalal . A, Pablo. A. M. A simulation model of a coordinated decentralized supply chain[J]. International Transactions in Operational Research, 2015, No.4

[5]Usha. R. Performance of supply chain collaboration – a simulation study[J]. Expert Systems with Applications, 2014, No.1

[6]Roberto. C,Margherita. P, Tommaso. P, Andrea. S. Linking supply chain configuration to supply chain performance: a discrete event simulation model[J]. Simulation Modelling Practice and Theory, 2014,

[7]C. Vidalakis, J. E. Tookey, J. Sommerville. Demand uncertainty in construction supply chains: a discrete event simulation study[J]. Journal of the Operational Research Society, 2013, No.8

[8]Jung. U. M. Agent-Based Construction Supply Chain Simulator (CS^2) for Measuring the Value of Real-Time Information Sharing in Construction[J]. Journal of Management in Engineering, 2008, No.4

[9]Vidalakis. C, Tookey. J. E, Sommerville. J. Logistics simulation modelling across construction supply chains[J]. Construction Innovation, 2011, No.2

[10]Ammar. A. B, Nashwan. D. Simulation-based genetic algorithms for construction supply chain management: off-site precast concrete production as a case study[J]. OR Insight, 2012, No.3

[11]Yasser. E, Simaan. M. A, Siri. F, Yasser. M. Simphony Supply Chain Simulator: a simulation toolkit to model the supply chain of construction projects[J]. SIMULATION, 2011, No.8

[12]Navin. K, Ravi. S, Roma. M. D. Supply chain efficiency: a simulation cum DEA approach[J].The International Journal of Advanced Manufacturing Technology, 2014, No.9-12

[13]J. L. He, Y. F. Huang, D. F. Chang. Simulation-based heuristic method for container supply chain network optimization[J]. Advanced Engineering Informatics, 2015, No.3

[14]Nihar. S, Marianthi. I. Hybrid Simulation Based Optimization Framework for Centralized and Decentralized Supply Chains[J]. Industrial and Engineering Chemistry Research, 2014, No.10

[15]Y. Kong, Z. Y. Liu, X. Chen, P. Zeng, L. R. H. Li. Operations optimization of rail transshipment system in hydropower development project[J]. Journals of System Engineering, 2015, No.2 (in Chinese)

The Application Of Six Sigma Theory In The Secondary Hospital Management

Hui ZHENG[1, 2], Miaojuan TIAN[1,2,*], Li-xia FANG[3]
[1]Tianjin Key Laboratory of Integrated Design and On-line Monitoring for Light Industry & Food Machinery and Equipment, Tianjin 300222, PR China
[2]College of Industrial Engineering, Tianjin University of Science & Technology, Tianjin, China
[3]Tianjin Navigation Instruments Research Institute, Tianjin, 300222, PR China
(*m18322693589@163.com)

Abstract — **With the rapid development of social economy. The medical industry's competition is becoming increasingly fierce. How to maximize the efficiency of using existing resources, reduce operating costs and improve the service quality has become the important content of hospital management. The Six Sigma theory based on statistics theory by eliminating errors, reducing waste, and optimizing constantly of existing processes to satisfy customer's changing needs. In this paper, using the Six Sigma theory's DMAIC model, combining with the secondary hospitals displayed treatment rate low, put forward improvement assumption of hospital management, the purpose is that through a series of improvements, improving the services efficiency of hospitals, and achieving the goal of increasing the hospitals benefit.**

Keywords — **DMAIC, electronic medical record system, logistics outsourcing, program analysis, Six Sigma theory**

I INTRODUCTION

We can see healthcare industry as the third industry to analyze, because it has service characteristics. The secondary hospital provides comprehensive medical services to a number of communities, as well as part of the teaching and scientific research tasks. But now, the secondary hospital in medicine and hospitalization price weakness, in policy support and advantage of geographical location of community medical and health than the tertiary hospitals, it also limited by its own talents and facilities, etc, formed an embarrassing situation that the secondary hospital in the crevice of the primary level and the third level hospitals.

Therefore, it is necessary and urgent to study and analyze the operation of the secondary hospitals, and to find out and solve the factors that it affects the secondary hospital's development. These issue had been studied by scholars, such as the paper of "the Tianjin secondary hospital financial analysis indexes system applied research ", Li Zhichun used Tianjin 38 secondary hospitals, analyzed its financial management and operation condition, think it is necessary that to analyze the financial status of the secondary hospital and to build a secondary hospital financial analysis index, it can provide basis of studying the secondary hospital operating conditions [1]. The paper of "Application of Lean Six Sigma in the human resource management", Xv Zhen used Six Sigma's successful implementation as the basis, summed up the nine six sigma theorems for human resources management, to make human cost making a positive contribution to the organization's development [2]. There are six sigma applied to hospital operation room, human resource management and so on [3-8].

Six Sigma management philosophies achieved great success in manufacturing industry, because it unique way of management and flexible application mode. Similarly, the continuous improvement in the management thought and the process improvement on the application model are also suitable for service enterprises, especially using it in hospital management, can greatly enhance the quality of hospital medical service, and improve hospital management benefit and social satisfaction.

II THEORETICAL KNOWLEDGE

The Six Sigma management method is a statistical

Foudation item: The Project for Tianjin Philosophy and Social Science Planning(No.TJGLWT15-006);The Innovation Funds for Youth Teacher of Tianjin University of Science and Technology (No.2014CXLG17).

E. Qi et al. (eds.), *Proceedings of the 23rd International Conference on Industrial Engineering and Engineering Management 2016,* DOI 10.2991/978-94-6239-255-7_23

evaluation method, its core idea is that pursues zero defect production, reduces costs, improves production efficiency and market share, and improves customer satisfaction and loyalty to the product. The DMAIC is one of the important tools in the process improvement to the Six Sigma management. It is not only a management concept, but also a method of performance breaking. It can change ideas into actions and goals into reality, including to define, measure, analyze, improve, control, generally used for the improvement in existing processes, including manufacturing process, servicing process and operating process[9].

III PROBLEM DESCRIPTION AND IMPROVEMENT IDEAS

The two ends of the three level medical structures are rotated around the stage of the medical industry with their respective characteristics, and the secondary hospital in the middle zone face an embarrassing situation. Therefore, the secondary hospitals only from the organization itself, setting reasonable processes, improving service quality, reducing operating costs, it can survive in the fierce competition. As shown in Fig.1 the Six Sigma management's DMAIC model, this paper uses the Six Sigma management's DMAIC model as a tool for the analysis of the secondary hospital.

project	meanings
Define	Defining the key quality characteristics of the process CTQ
Measure	Measuring and collecting of the actual value of CTQ
Analyze	Analyzing and researching factors of influencing CTQ
Improve	Optimizing and improving factors of influencing CTQ
Control	Standardizing the results of been improved and maintenance

Fig.1 the Six Sigma management's DMAIC model

A Definition phase (D phase)

The secondary hospitals are county's largest comprehensive hospital, it has the complete department ,such as internal medicine, surgery, orthopedics ,facial features section, pediatrics, obstetrics and gynecology, it can diagnosis and treatment all general diseases. However, the secondary hospitals visiting rate low, social satisfaction low, less profit, the operation situation of the hospital is bad.

B Measurement phase (M phase)

Study on the secondary hospital, it concluded that process as shown in Fig.2 the flow chart of the secondary hospitals .

C Analysis phase (A phase)

Looking hospital process of the longitudinal view, it divided into: outpatient service process, the length of treatment process, access to the hospital process etc; from the lateral view of the process is mainly that paying process, taking medicine process, and inspecting process, if this be refined, there will be more branching processes. Taking the process of a hospital as an example to analyze, find that the process complicated, the department arrangement chaotic, the responsibility of medical staff is not strong, so the patient has the long time of treating, and the satisfaction degree is low. The reason for this phenomenon is wide-ranging. First, business has complete categories but the specialist advantage is not prominent, so it difficult to meet the demand; the secondary hospitals' logistics service using the form of ensuring supply, its do not the scientific cost accounting and economic benefits string; its management means is administrative management, classification is too meticulous, so it makes the management lacking vigor and vitality. Secondary, the service quality is not high, patient satisfaction low: in generally, when the patient to the hospital are usually in the physical discomfort state, in this case patients also experiences the complex treatment process, or faces hangs the wrong number, waiting for a long time, could not find a medical department. Because of this, the patient's satisfaction is low.

D Improvement phase (I phase)

3.4.1 Today successful business or organization all has strong human resources management team. Like it, with the increasingly fierce competition of the medical

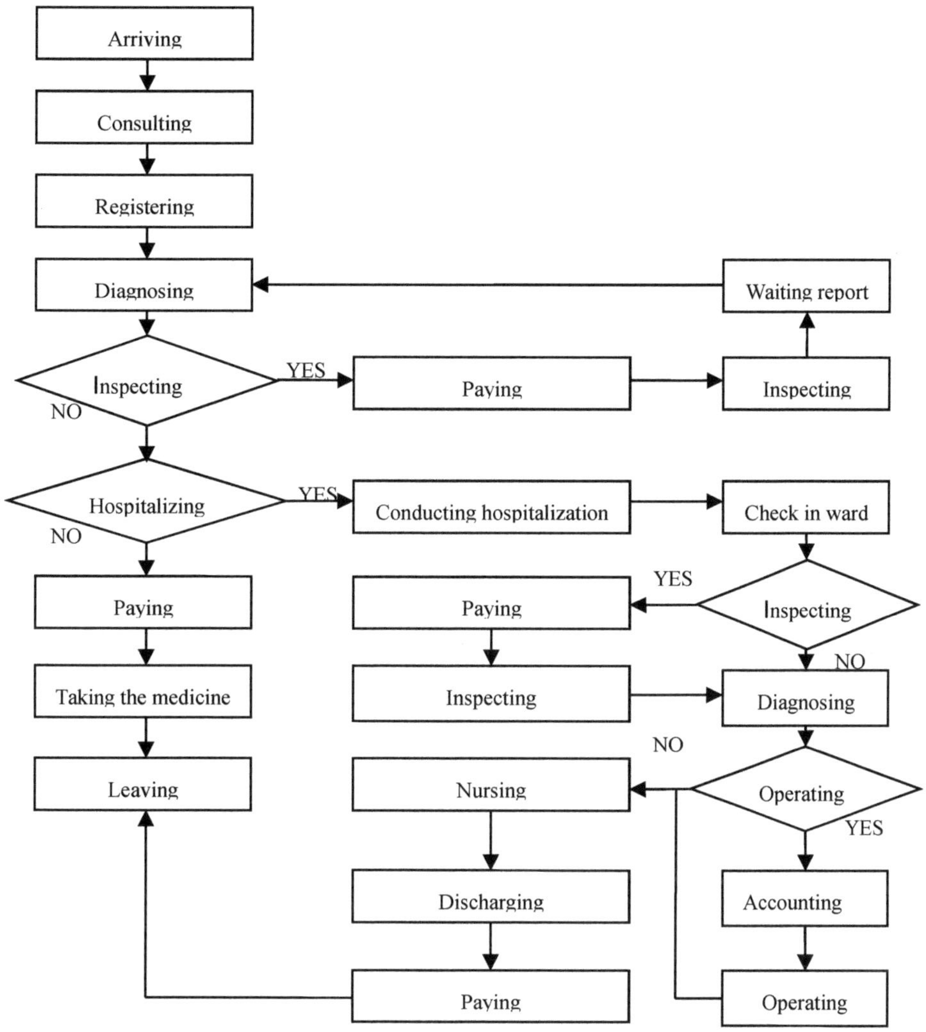

Fig.2 the current flow chart of the secondary hospitals

market, hospital wants in leading position, not only the infrastructure should be improved, the organization's human resources also to carry out creative construction, it can use cost-effective value rules of nursing staff to assess the cost [10]. In the concept of choosing and employing persons: selecting a good president, setting up the high-quality leadership team; cultivating, importing and protecting core characters; selecting middle-level cadres, especially department director, head nurse; making good use of the hospital core staff. In the management mechanism, the hospital needs to reform personnel distribution system. To establish a perfect competition system work performance appraisal system and the compensation system, these with distinctive distinguishes [11]. Such as using a clear score rating system in some positions for the staff, the work can be developed clearly defined standards and assessment records.

3.4.2 Process improvement using the "5W1H" technology and process analysis to improve the outpatient process. First, setting up the preliminary desk, arranges the professional staff to guide clinic patients, for patient with early diagnosing and preventing the wrong number. Second, promote one card's implementation, namely, establishing perfect electronic medical record system and sharing service platform to standardize the patient's medical record and make the hospital information sharing, within a certain range provide convenience for patients and doctors[12].Using their credit card to register and treat, saves the waiting time, reduces the patient's movement, and when patients finishes all medical process after the one-time payment, reducing movement time. Doing statistical work of chronic disease and maternal

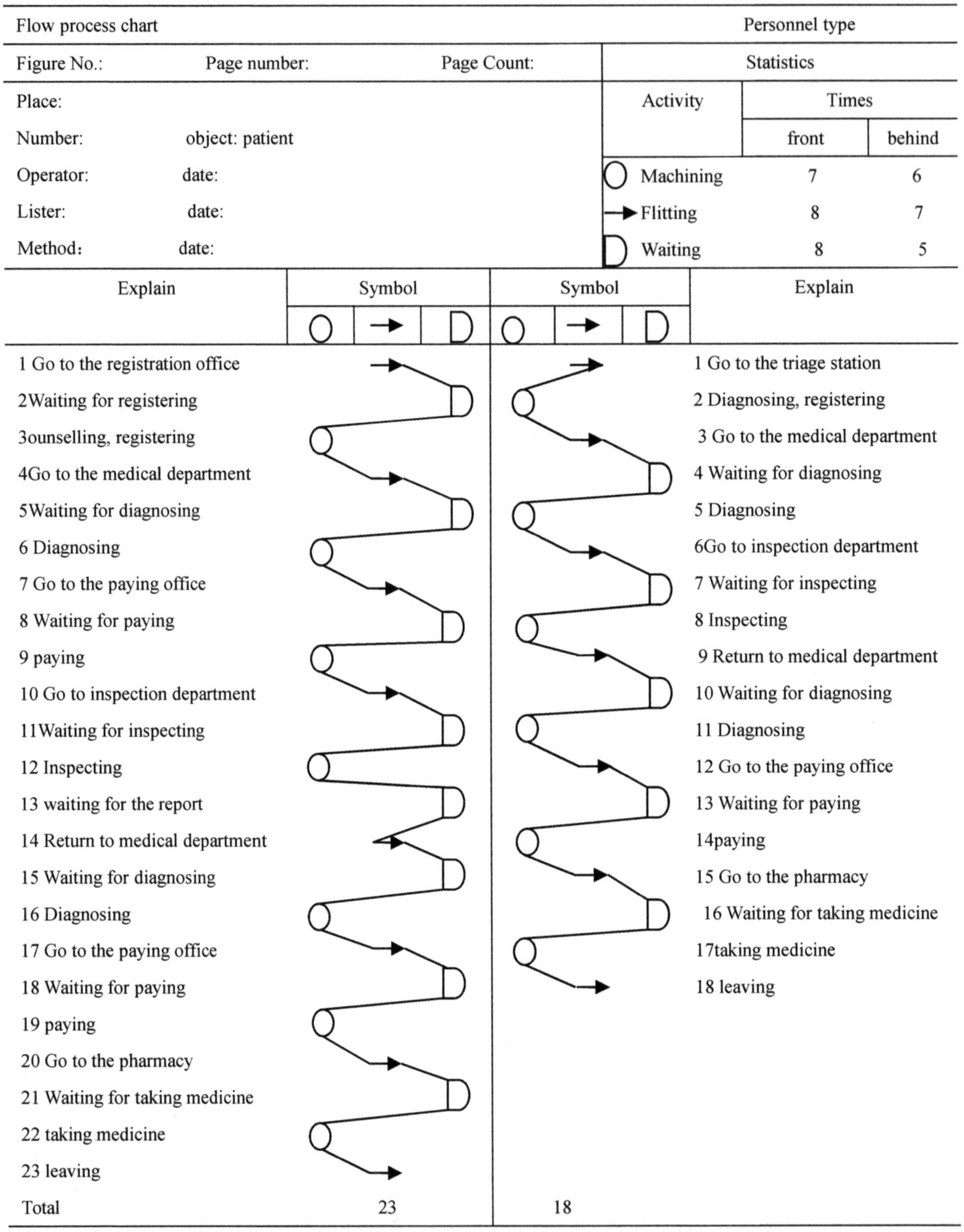

Fig.3 outpatient procedure analysis and comparison chart

medical attention. Through the analysis, it concluded that process as shown in Fig.3 outpatient procedure analysis and comparison chart.

3.4.3 Examining the hospital logistics, and evaluating of logistics costs, implementing of logistics outsourcing, logistics in the redundant links to external personnel [13-16].

E Control phase (C phase)

The main purpose is to keep the results of the control phase. Job description is the organization's development planning, operational objective, benefit goal needs, which written with a formal form to organize the internal personnel job description or post rules review, which usually include the specific content and method of the work, the project and standards of evaluation [10]. Therefore, the main work is to update the new work instructions or procedures, to monitor the whole work in real time in this stage.

IV CONCLUSION

Through the study of this subject, business logistics outsourcing makes the hospital's structure is simple and low operating costs; scientific management of staff makes service quality and service efficiency greatly improved, greatly enhance the core competitiveness of the hospital and improving measures of industrial engineering not only application in someone hospital, and similar secondary hospital can also be to improve management, improve patient satisfaction and efficiency of medical services by this method, to jump out of the medical industry three tiers architecture in secondary hospitals embarrassment.

REFERENCE

[1]Li Yuanyuan, Li Zhichun. Study on the application of financial analysis index system of two hospitals in Tianjin city (in china) [J]. China health economics, 2015,10:79-81.

[2] Xv Zhen. The application of Lean Six Sigma in the human resources management [J].China association for quality, excellence in the quality of Scientific Research Institute, mission and responsibility, quality to promote transformation and Upgrading -- the fifth session of the China academic quality and Innovation Forum in set (in china) [C]. China Quality Association, excellence in the quality of Scientific Research Institute: 2012:4.

[3] Guo Hongyan, Yao Li, Li Ang, Xie Lijuan, Xie Hong. Investigation and analysis on the status quo of nursing human resources in the secondary hospitals (in china)[J]. China nursing management, 2013,04:57-60.

[4] Li Bin, Bai Xuejuan. Investigation and analysis of the status quo of the management of the medical record department in the two level hospital (in china)[J]. China medical record, 2015,10:4-5.

[5] Ma Yuxia, Lou ChenYan, Zhou Xue, Gong Yuanyuan, Zhang Yanan. Study on the application of Six Sigma quality management plan to improve hand hygiene compliance of health care workers (in china) [J]. Chinese Journal of hospital infection, 2015,23:5506-5508.

[6] Chen Qinjuan, Zheng Lulu, Wu Jie. The application of Six Sigma in the operation room management (in china)[J]. Journal of traditional Chinese medicine, 2015,23:100-102.

[7] Xv Yan. The application of ideological motivation in hospital human resource management (in china)[J]. China health industry, 2015,16:85-87.

[8] He Jin. Study on the optimization of operation process based on Six Sigma management (in china) [D]. Third Military Medical University, 2015

[9]Wang Hui. Study on the application of Six Sigma Management in hospital management (in china) [D]. Jilin University, 2009

[10] Zhang Chunhua, Zhu Xiaoping, Liu Wenfei, Hu Fen. Study on the law of the value of the cost accounting of the two level hospital nursing service (in china) [J]. Journal of nursing science, 2013,02:22-25.

[11] Zhu Juanjuan, Zhao Changwei. Improving human resources management level and enhancing the core competitiveness of the hospital (in china)[J]. China health industry, 2015,16:91-93.

[12]Zhu Xuening, Huang Su, Li Fuxing, Qiu Mali. The application of electronic medical record system in the hospital information management system research (in china)[J]. Electronic technology and software engineering, 2015,05:192.

[13] Zhang Jingbo, Zhang Wenjie, Yang Changsong, Zhu Guobin. The hospital rear service socialization service work experience (in china) [J]. Jiangsu health service management, 2015,06:145-146.

[14]Huang Yuanyuan. Hospital performance management (in china)[J]. era, economic and trade, 2015,14:84-85.

[15]Yu Hongfeng. Hospital cost accounting problems and Countermeasures of rural health China (in china) [J]., 2016,08:29-30.

[16] Zhang Jianxin, Yang Lu, Zhao Jing, Chen Yingying. The application of job description in hospital human resource management (in china)[J]. human resources management, 2015,12:215.

Research on Parameter Optimization of ant colony algorithm based on genetic algorithm

Li-hua TAO* Peng-tao SHI Jun-feng BAI
(School of Mechanical Engineering, Chang Chun University Of Technology. Chang Chun 130012)
(*Email: taolihua2001@163.com)

Abstract-**The performance and convergence rate of the ant colony algorithm is affected by the parameters of the ant colony algorithm, and the optimal parameters of ant colony algorithm often vary from problem to problem. On the basis of analyzing the current research status of ant colony optimization, a genetic algorithm based on ant colony optimization is proposed, which carries out the combination optimization of the parameters of the ant colony algorithm through the binary chromosome coding, roulette wheel selection operation, multi-point crossover operation, single point mutation, re-insert preserving good operation and the decoding process based on ant colony algorithm. Finally, taking the TSP-Oliver30 and the TSP-20 as examples, the validity of the algorithm is verified through the simulation experiments.**

Key word- **ACO, GA, Parameter optimization, TSP**

I. INTRODUCTION

Ant colony optimization (ACO) is a kind of simulation algorithm based on the process of ant colonies' collecting food which was put forward by Italian scholars Dorigo in 1991[1]. ACO has the characteristics such as positive feedback, distribution, while it still has weakness, for instance, the slow speed of evolution, the high probability of local convergence and the poor stability. The parameter setting of ACO plays a decisive role in the optimal performance and the rate of convergence of the algorithm. The optimal parameter combination of ACO is also different for various network topologies.

Name of project fund : Research on Vehicle Regenerative Braking and stability control method based on EHB system（2015Y062）

As for the combinatorial optimization of ant colony algorithm parameters, Zhi-wei Ye and Hong-mei Xu researched the influence of the change of a single parameter on the result of convergence of ant colony algorithm. In the case of not using the best parameter combination, increasing the number of iterations and the running times can't improve the quality of the result [2]. This method ignores the coupling relationship between parameters, which couldn't achieve the simultaneous optimization of multiple parameters. Liang-jun Wang used genetic algorithm to optimize the parameters of ACO [3]. But when calculating fitness, he got the fitness only according to the best distance of one result of ant colony optimization, the result is often unable to eliminate the random interference through one operation, meanwhile, the optimization result can't guarantee the stability of the algorithm. At present, in the process of using ant colony algorithm, it is mainly based on experience and experiments to set the parameters [4, 5], while the performance and stability of the algorithm are often affected.

II .THE BASIC PRINCIPLE OF THE ANT COLONY OPTIMIZATION

Ant colony optimization is a stochastic search optimization method based on the natural world which was inspired by the foraging behavior of ant [6]. A kind of pheromone will be released in its path in the foraging process of ants. In the foraging process, the ants are more likely to choose the higher pheromone concentration sections, and then release pheromone on the road again, and a positive feedback is formed throughout the process. Eventually the ants will find a path that is most likely to be close to the food source. Most of the ants will choose this

E. Qi et al. (eds.), *Proceedings of the 23rd International Conference on Industrial Engineering and Engineering Management 2016,* DOI 10.2991/978-94-6239-255-7_24

path eventually [7].

Dorigo gave us three kinds of different models for the problem of ants releasing pheromone[8], which are the ant cycle model, the ant quantity model and the ant density model. In the ant quantity model and the ant density model, the pheromone will be updated after ants' each move between the nodes of the two positions, by using the local information. In the ant cycle model, the pheromone will be updated after all the ants complete their trips according to the global information[9].This paper mainly uses the ant cycle model to conduct the study on combinatorial optimization of the persistence of the trajectory ρ and the relative importance α and β and the initial pheromone c.

The persistence of the trajectory ρ. It reflects the attenuation degree of pheromone, whose size is directly related to the global searching ability and the convergence speed of the algorithm. When it is too large, the quantity of pheromone in the path is dominant, and it will increase the chances that the path which has been searched is searched again. The role of positive feedback is relatively weak which will slow down the convergence rate. On the contrary, if it is too small, it will lead to a consequence that some pheromones on the path disappear too fast, positive feedback function occupies the dominant position. The convergence rate of the algorithm is accelerated, but it is easy to fall into local optimum [10].

The relative importance α and β. α reflects the importance of the length of the section in guiding the ant colony search. The size of the β indicates the importance of the pheromone amount accumulated in the search process to guide the ant colony search. The greater the value of α is, the greater the probability of selecting the local shortest section is. Although the convergence rate of the algorithm is accelerated, the randomness of the search is reduced, which causes that the algorithm is easy to fall into the local optimal solution. The greater the value of the β is, the greater the likelihood that the ant chooses the traversed path is. With the weakening randomness of the search, the algorithm is easier to fall into the local optimal solution

The initial pheromone c. The amount of initial pheromone affects the accumulation process of pheromone on the path. If the amount of initial pheromone is too large, it will lead to the unobvious increment of pheromone which brings about the weakening of the guiding role of pheromone in the next generation of ants. Conversely, if the initial pheromone amount is too small, with the attenuation of pheromone, the pheromone on the path that ants never searched will disappear soon, and the algorithm will be trapped into a local optimal solution.

III.DESIGN OF GENETIC ALGORITHM BASED ON PARAMETER OPTIMIZATION

The parameter optimization of ant colony algorithm is an NP problem in combination optimization problem. Genetic Algorithm (GA) provides an effective way to solve this kind of problem. GA has the advantages of simple calculation process, strong global search capability and easy to combine with other algorithms. In this paper, we complete the combination optimization of ACO based on GA through selection, crossover, mutation, re-insertion and decoding process based on ant colony algorithm. The genetic algorithm of parameter optimization is shown in Fig.1.

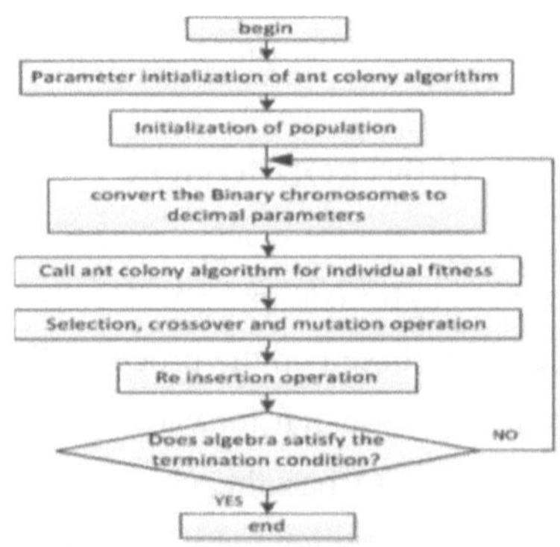

Fig. 1 Genetic algorithm flow of parameter optimization

A. Coding

The chromosomes are represented by binary coding, The chromosomes consist of ρ、α、β、c, and the binary length of each parameter is 7 bits. The total length of the chromosome is 28 bits. The binary and decimal corresponding relations of the parameters are shown in table I. If the chromosome code is 1111110 1101010 1100100 1010111, the persistence of the trajectory is 1111110, all 7 bits of binary code represent decimals. The process of converting to decimal number 0.9844 is shown in the formula (1).

$$1*2^{-1}+1*2^{-2}+1*2^{-3}+1*2^{-4}+1*2^{-5}+1*2^{-6}+0*2^{-7}=0.9844 \quad (1)$$

TABLE I .

THE RELATIONSHIP BETWEEN CHROMOSOME BINARY CODE AND DECIMAL

parameter		ρ	α	β	c
Binary code		1-7bits	8-14 bits	15-21bits	22-28 bits
decimal		(0, 1)	(0,4]	(0,4]	(0,127]
decimal code	Integer coding section		8-9 bits	15-16bits	22-28 bits
	Decimal coding section	1-7 bits	10-14bits	17-21bits	

B. Initial population

Because the chromosome is made up of binary code, the value of each gene is 0 or 1. Random alteration of genes in chromosomes is made to generate initial population.

C. Decoding based on ACO

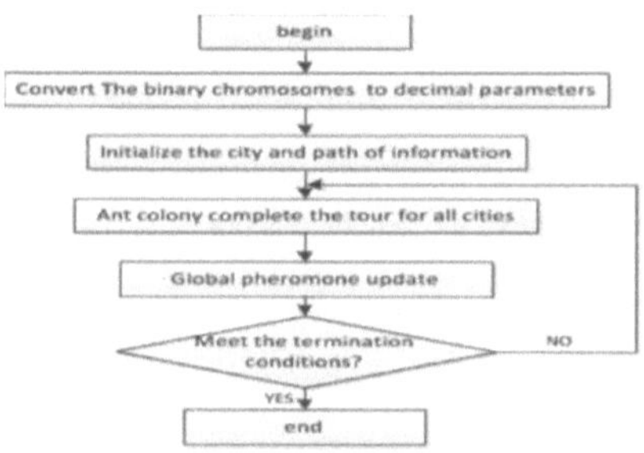

Fig. 2 Chromosome decoding based on ACO

The chromosome decoding is the conversion of individual genes to representation. The fitness value of the chromosome can be solved by decoding. Based on the traveling salesman problem, the reciprocal of the shortest path is selected as the fitness value of the chromosome in this paper. Because the ant colony algorithm is a kind of random search algorithm based on probability, the ant colony algorithm is implemented n times, and the average value of the shortest path of the chromosome is taken as the fitness value in order to reduce the influence of the randomness of the ACO algorithm on the optimization results. The chromosome decoding process based on ACO is shown in Fig.2.

D. Selection

Roulette method is used to select breeding individuals. The proportion of population reproduction (generation gap) is set to 90%, and 10% of the population of the superior individual directly as a reproductive individual, which avoids the loss of the best individual in the selection.

E. Crossover

Multi points crossover is applied. Since the chromosome is made up of four gene segments, each gene segment represents a parameter, so the double point crossover method is used for each gene segment. For each gene segment, two positions are randomly generated. The gene segments are exchanged between the two points, and two new offspring are obtained. Two paternal chromosomes ① and ② are shown in the formula (2) and (3) . After crossover, we get two offspring chromosomes ③ and ④ as shown in the formula (4) and (5).

ρ α β c

①: 1101010 1011001 1001100 1010011 (2)

②: 0101110 1010101 0101101 1010101 (3)

③: 1101110 1010001 1001100 1010101 (4)

④: 0101010 1011101 0101101 1010011 (5)

E. Mutation

The method of single point mutation is carried out. By randomly generating mutation location in each gene segments, the gene of the mutation position switches between 0 and 1 .The process of chromosome ⑤ mutating into ⑥ is shown in the formula (6) and (7).

⑤ : $11\underline{0}1110\ 10\underline{1}0001\ 110\underline{1}100\ 10101\underline{0}1$ (6)

⑥ : 1111110 1000001 1100100 1010111 (7)

F. Re-insert operation

In the process of genetic evolution, if the best individual generated offspring fitness value is less than the fitness value of the parent, the optimal individual is kept in the next generation. in order to prevent the loss of the best individual.

IV. SIMULATION EXPERIMENT

Select the city coordinates in Oliver30. Genetic algorithm initial population number is set as 15.The genetic evolution algebra is set as 100.The probability of crossover is set as 0.9. The probability of mutation is set as 0.05. The number of ants is set as 30, and ant colony iteration number is set as 30. For each individual's fitness, the algorithm is carried out for 3 times to obtain the results. The change of the average length of the shortest path with the genetic algebra is shown in Fig.3. The change of parameter ρ, α, β, c with the genetic algebra is shown in Fig. 4.

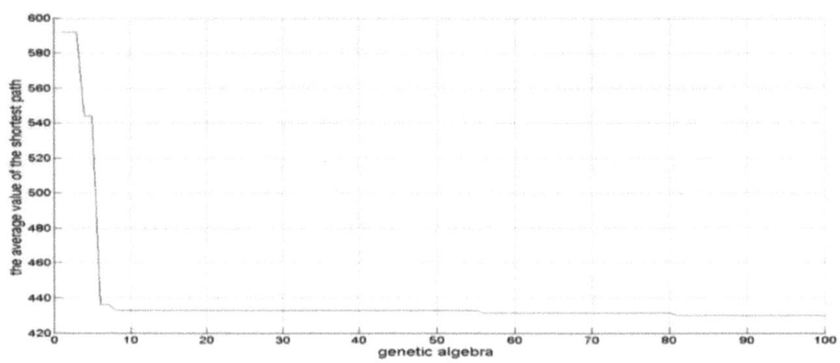

Fig. 3 The average value of the shortest path with the change of the genetic algebra

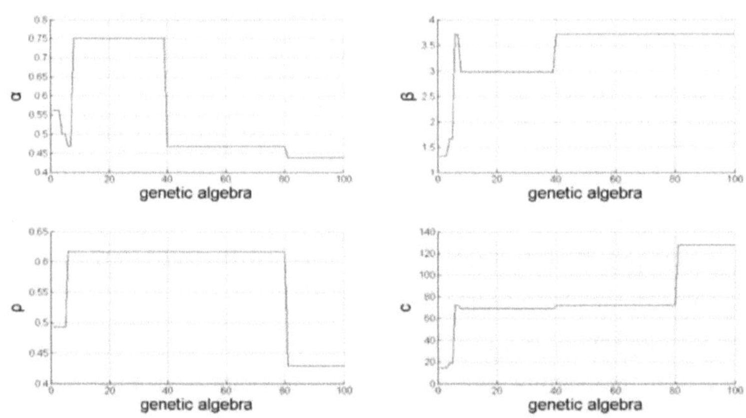

Fig. 4 Variation of ACO parameter with the genetic algebra

In Fig.3, the average shortest path decreases with the increasing of genetic algebra. After running to the 83 generation, the convergence is 430.2814. At this time, the parameters of the ant colony algorithm converge to their optimal values in Figure 4. The values of ρ, α, β, c are 0.4297, 0.4375, 3.7188, and 127 respectively. In the same way, we can get the optimal parameter combination of ant colony algorithm corresponding to the TSP- 20 problem, the values of ρ, α, β, c are 0.9129, 0.4375, 3.0313, and 65 respectively.

In order to verify the effect of ACO parameters on the performance and convergence rate of the ant colony algorithm, The Oliver30 and TSP- 20 problems are solved by using the optimized ACO and the non-optimized ACO parameters respectively. The number of ants is set as 35, and the ant colony is set as 120 times. The parameters ρ, α, β, c, that are not optimized are set as 0.61719, 0.75, 2.9688, and 69 respectively. 10 consecutive experiments are made for each question and the results are shown in Table II. The iterative procedure of the two parameter combinations in the Oliver30 problem is shown in Fig.5.

TABLE II

PERFORMANCE COMPARISON OF ANT COLONY ALGORITHM UNDER TWO COMBINATIONS OF PARAMETERS

number	Oliver30		TSP- 20	
	The optimized parameters	not optimized	The optimized parameters	not optimized
1	426.5438	431.0574	24.5263	25.1479
2	425.2667	433.8465	24.8902	26.9154
3	427.8769	430.6978	24.6422	24.7995
4	427.1752	438.1335	24.5263	25.5370
5	432.4055	438.7744	24.6422	25.8807
6	429.7853	432.0405	24.6422	26.2369
7	430.6178	430.3726	24.5263	24.6422
8	428.6337	437.4301	24.7278	25.5370
9	434.5089	430.2935	24.7995	25.0474
10	430.4807	434.6696	25.1877	25.9201
Max	434.5089	438.7744	25.1877	26.9201
Min	425.2667	430.2935	24.5263	24.6422
average	429.3295	433.7316	24.7359	25.5416

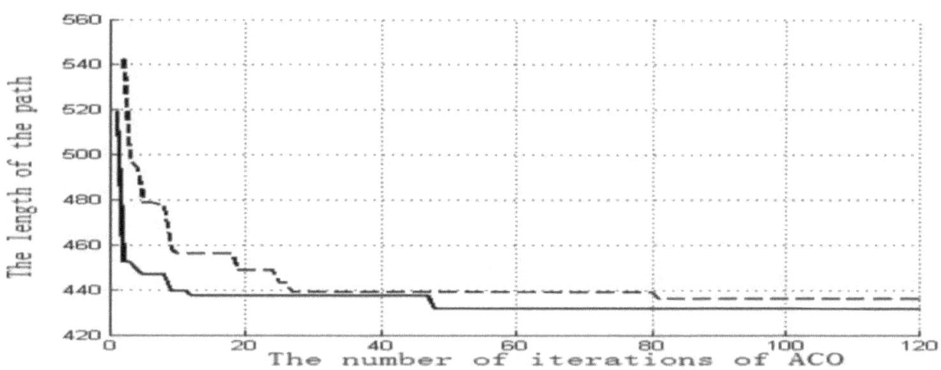

Fig.5 Oliver30 in two kinds of parameters optimization process diagram

From table II, we can see that the optimized parameters can improve the optimization ability of the algorithm, and the quality of the solution is better than that of the not optimized parameters.

In Fig.5, the solid line represents the iterative process using the optimized combination of parameters, while the dashed line represents the iterative process using a combination of random parameters. As can be seen from Fig.5, the performance of the algorithm has been greatly improved after using the combination of optimization parameters.

V. CONCLUSION

The research shows that the optimal parameter combination of the ant colony algorithm is different for various combination optimization problems. In this paper, genetic algorithm is used to optimize the parameters of ant colony algorithm. Taking the TSP problem as an example, the validity of the method is proved by comparing the experiment results. The parameter optimization method presented in this paper provides a basis for the rational allocation of the parameters of the ant colony optimization algorithm. The parameter optimization method can be used to improve the performance of the ant colony algorithm and its related algorithm.

REFERENCE

[1] M. Dorigo, L M Gambardella. Ant Colonies for the Traveling Salesman Problem. Biosystems, 1997, (43), pp. 73–81.

[2] Zhi-wei Ye, Zhao-bao Zheng. Research of parameters alpha

and beta, rho set of ant colony algorithm(Take the TSP problem as an example) [J]. Journal of Wuhan University. 2004,29（7）， pp. 597–601.

[3] Liang-jun Wang, Li-juan Sun, Ru-chuan Wang, Research of using an improved ant colony algorithm to solve TSP [A]. Journal of china institute of communications. 2004,25(10), pp. 111–116.

[4] Jun-Wang, Research on parameter setting of ant colony algorithm for solving TSP [A]. Science Technology and Engineering. 2007,7（17）, pp.4501-4503.

[5] Shi-chang Zhan, Jie-Xu, Jun-Wu, The optimal selection on the parameters of the ant colony algorithm [A].Bulletin of science and technology. 2003,19（5）, pp. 381–386.

[6] Liang-Ma. Optimization strategies from the world of insects −the ant colony algorithm[J]. Nature magazine. 1999.21（3）, pp. 161–163.

[7] Dorigo, LM Gambardella .Ant colony system : A cooperative learning approach to the traveling salesman problem [J].IEEE Transactions on Evolutionary Computation ,1997,1(1), pp. 53–66.

[8] B Bullnheimer, R F Hartl , C Strauss. Applying the ant System to the Vehicle Routing Problem. Meta-Heuristics: Advances and trends in local search paradigms for optimization. Boston : Kluwer, 1998,109–120.

[9] Sheng-wei Yu.Case analysis and application of MATLAB optimization algorithm [M]. Bei-Jing: Tsinghua university press. 2014, 09.

[10] Hong-mei Xu, Yi-bao Chen. Research of parameters set of ant colony algorithm. Journal of Shandong University of Technology(Natural Science Edition). 2008. 22（1）, pp. 9–10

Multi-objective Integrated Optimization Problem of Preventive Maintenance Planning and Flexible Job-Shop Scheduling

Zha Jing, Jin Hua*, Zhu Yi ,
College of Mechanical & Power Engineering,Three Gorges University,
Hubei Yichang, 443002, China
(*jinhua_vds@163.com)

Abstract - **In order to discuss the influence of equipment failure on the production schedule and equipment utilization rate in the pull production environment, a multi-objective integrated optimization problems of preventive maintenance planning and flexible job-shop scheduling model is established. The model is to minimize both the maximum completion time and the total equipment' s workload. In this article an improved genetic algorithm is proposed to solve the problem. The multi-objective optimization model was compared with independent maintenance model to verify its superiority, and the multi-objective improved GA can find more Pareto solutions for the decision maker. Finally the example calculation shows that the integrated optimization can find the solution with shorter completion time, eliminate waste and improve production efficiency.**

Keywords - **Preventive maintenance, Flexible job shop scheduling, Integrated optimization, Improved genetic algorithm**

I. INTRODUCTION

In order to prolong the life of the machine, as well as prevent normal production disruption caused by the machine fault, preventive maintenance of equipment in advance is essential. On the other hand, reasonable production scheduling can help enterprises quickly produce customer satisfaction products with low cost, which contains arrangement of processing machines and machining sequences .So preventive maintenance and the shop scheduling are two key problems in the production organization.

Although these two issues are often considered separately in last studies and enterprise management, there is a close connection between the two [1-2]. Fixed time maintenance planning according to the worker' s experience which not only cause equipment maintenance to premature or excessive, but also affect the utilization rate of equipment and production efficiency seriously. Production scheduling without considering machines' condition often deviates the actual scheduling and interrupts implement .Therefore, it is necessary to study the integrated optimization problem of job shop scheduling problem and equipment preventive maintenance in order to achieve the effective balance between production scheduling and equipment maintenance and to maximize the overall benefits of the enterprise.

Najib.et al [3] proposed CLSP integrated with machine preventive maintenance ,and added a new decision-making factor - out of stock costs and the reliability of the system in the model. M.Sbih et al [4] studied the single machine scheduling problem with a variety of maintenance strategies and the goal is to minimize the cost of delay .Yu Ming-rang et al[5]. Considered influence of the time on the preventive maintenance of the equipment and the single objective integration optimization was carried out for the general job shop. De Huaxu et al. [6] presented the validity of LPT algorithm to solve two parallel machine maintenance strategy under the condition of certain and the goal is to minimize the completion time. Song Wen-jia et al.[7] studied the flexible job shop and designed a weighted TOPSIS method and mixed colonial competitive algorithm to solve the problem .

In short, the previous research about the shop scheduling problem integrated with equipment preventive maintenance plan, which just consider the general job shop, or single objective optimization problem , so it is necessary to study multi-objective problem of maintenance plan and flexible job shop scheduling .This paper establish a multi-objective integrated optimization model based on the maximum completion time and the minimum total load of the machine, and then design the improved genetic algorithm to solve the problem.

II. PROBLEM DESCRIPTION

This paper studies the flexible job shop scheduling (FJSP), which is described as follows: There are n jobs and m machines in a work shop, each job has one or more processes, the processing of each work piece process route is fixed and known, each operation can choose one or more machines [8-9].In each time period, the machine can perform only a process; processing time of operation in different machine may be the same or be different. By changing each process when the choice of machines and adjust the process of machining sequence can make the whole system of certain goals become optimal. Therefore, flexible job shop scheduling problem can be divided into two sub problems: to determine the machine selected by the working procedure and the machining sequence of each process [10].

This work is partially supported by National Science Foundation of China Grand (No. 71501110) and Science and technology research project from Hubei Provincial Department of Education (No. Q20151206)

E. Qi et al. (eds.), *Proceedings of the 23rd International Conference on Industrial Engineering and Engineering Management 2016,* DOI 10.2991/978-94-6239-255-7_25

In the actual manufacture, the probability distribution of the machine's failure is always independent and diverse [11].In this paper we adopted preventive maintenance policy, that is, when the failure probability and reliability of the machine are lower than the predetermined threshold, a preventive maintenance start to execute on machine .We assumed that the time spent for a preventive maintenance is the same for each machine. After each maintenance, the failure rate of the machine is reset to zero. According to the literature [12], the probability distribution of machine failure is exponential distribution, its function shown as Eq.2.

$$p_{M_i(t)} = \begin{cases} 0 & ,t = t_{ri}^k \\ 1 - e^{-\lambda t_a}, & t_j^k \le t \le t_j^{k+1} \end{cases} \quad (1)$$

Among them $P_{M_i(t)}$ denotes the fault probability of machine in times t , t_i^k represent the machine M_i repair $k\,th$ start time. t_i^{k+1} Represent machine M_i represent repair $k+1th$ start time. λ is the parameter of the exponential distribution, $\lambda = 1/\bar{t}$, $\bar{t}$ is the damage of the machine average time (obtained by experiments). t_a Represent equipment operation time .In summary, the following main strategies of preventive maintenance of the i machine:

Step1: according to experimental or previous experience to obtain the parameters: the mean time to failure $\bar{t}$ and spent time to preventive maintenance T.

Step2: according to the failure probability Eq.2. Calculated machine fault probability $P_{M_i(t)}$ and general threshold $t_{ak} = 0.7$.

$$t_{ak} > 1 - e^{-\lambda t_a} \quad (2)$$

Step3: If $P_{M_i(t)} \ge t_{ak}$ the equipment maintenance, if $P_{M_i(t)} < t_{ak}$ equipment without maintenance.

The failure rate of the machine and the total processing time are zero when the process begins. O_{jh} and $O_{j(h+1)}$ are the two adjacent operations on machine i .After the completion of the operation O_{jh} ,the operation $O_{j(h+1)}$ starts, their processing time are t_{ijh} and $t_{ij(h+1)}$ .If $t_{ijh} \le t_{ak}$ and $t_{ij(h+1)} > t_{ak}$ the equipment maintenance execute after the operation O_{ij} complete .If the maintenance and process at the same time in the original plan, the start time and completion time of the operation should be postponed.

III. INTEGRATED OPTIMIZATION MODEL

The multi-objective integrated optimization model established as follows:

Objective function:

$$\min(\max_{1 \le j \le n}(C_j)) \quad (3)$$

$$\min(\max_{1 \le j \le m} \sum_{j=1}^{n} \sum_{h=1}^{h_j} p_{ijh} * x_{ijh}) \quad (4)$$

Constraints:

$$s_{jh} + x_{ijh} \times p_{ijh} + T * \lceil (\operatorname{sgn}(PM_i(t) - 0.7) + 1)/2 \rceil = c_{jh} \quad (5)$$

$$C_j = \max(c_{jh}) \quad (6)$$

$$s_{jh} \le s_{j(h+1)} \quad (7)$$

$$c_{jh} \le s_{j(h+1)} \quad (8)$$

$$s_{jh} + p_{ijh} \le s_{kl} + L \times (1 - y_{ijhkl}) \quad (9)$$

$$\sum_{i=1}^{m_{jh}} x_{ijh} = 1 \quad (10)$$

$$x_{ijh} = \begin{cases} 1 & O_{jh} \textit{select machine } i \\ 0 & \textit{otherwise} \end{cases} \quad (11)$$

$$y_{ijhkl} = \begin{cases} 1 & O_{jh} \textit{before } O_{kl} \\ 0 & \textit{otherwise} \end{cases} \quad (12)$$

$$s_{jh} \ge 0 \textit{ and } c_{jh} \ge 0 \quad (13)$$

The objective function (3) denotes maximum completion time of all jobs; (4)represents the minimum total load of a machine .Constraints (5) indicates that end time of the process is the sum of the start time of each process , working time and the machine preventive maintenance time;(6)says the maximum completion time of all jobs. (7) and (8) requires that the processing sequence should meet process constrains; (9) indicates the process sequence constrain on the same machine; (10) shows at the same time a machine can process an operation only; (11) and (12) indicates that the variable is a 0-1; (13) to make sure that the start time and the end time of each process must be positive.

IV. IMPROVED GENETIC ALGORITHM

The method to solve multi-objective optimization problem is simply transforming the multi-objective problem into a single objective problem, and then we can use the single objective genetic algorithm to solve the problem [13]. Methods such as goal programming, weight method, and constraint method are commonly used [14]. But there are some disadvantages above these methods:

(1)Giving only one optimal solution which can’t help policymakers to make a variety of choices.

(2)Different objectives of the measurement criteria are not the same, it is difficult to carry out non dimensional.

(3)For the weight method, when choose the weights between different targets, researchers will put into too much subjectivity.

According to the above shortcomings, we proposed the improved GA, which avoid these defects effectively. The performance show in the following aspects:

(1)Simplified the non dominated sorting operation, and reduced the complexity of the NSGA computation;

(2)Expanded the sample space, accelerate the operation of the GA and reduced the quality of the lack of solution;

(3)Artificially specified parameters were fewer.

In this paper, an improved GA is designed to solve the multi-objective optimization problem. The flow chart of the algorithm is shown in Fig.1. The main steps show as follows:

Step1: Initialize parameters, including the genetic algebra, the size of the initial population, the probability of variation, the probability of crossover and so on.

Step2: Generate the initial population, the population size is 2N.

Step3: Initialize the genetic algebra 1, namely Gen=1.

Step4: Non dominated sorting of the population 2N, Layer all individuals in the population 2N and form set Z, the individuals in layer i marked Zi

Step5:Choose the elite ,calculate the crowding distance of individuals between different lever, select the top N based on the crowing distance and hierarchy .Put the individuals in first layer into the Pareto optimal solution set ,then non dominated sorting of the new Pareto solution set and left non dominated individuals, delete the dominated individuals.

Step6: Execute cross operation among individuals of the N, product N offspring. Combine the parents' generation individuals and offspring to create a population which scale is 2N.

Step7: Execute mutation operation among the individual in the population.

Step8: Judge whether the genetic algebra is greater

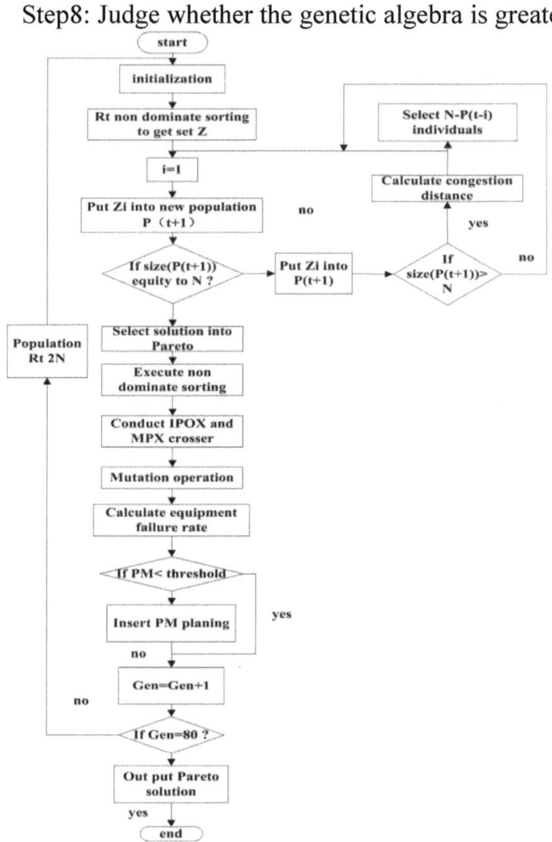

Fig. 1. Flow chart of improved GA

than the maximum. If Gen is more than 80, output the Pareto optimal solution set as the result. Otherwise return the Step4.

In this paper, we use the non dominated sorting method with the elite strategy to design and select operators. The following is a detailed description of the improvement in the improved GA algorithm.

A. Fast non dominated sorting operation

Non dominated sorting is defined as: in the multi-objective minimum value model, such as $F(x)=(f_1(x), f_2(x), \cdots f_m(x))$, x_i and x_j are the individuals in population[15]. If all the objectives $F_k(x_i)\le F_k(x_j)$ $(i,j=1,2,\cdots m)$, means x_i is non dominated by x_j as well as x_i is better than x_j .The procedure for non-dominated sorting of populations is:

(1)Calculate the amount n_j and S_i of dominated individual in population.

(2)Find the individual i which $n_i=0$, and then put these individuals into the current collection.

(3)For the individual j in Z_i , calculate its individual set S_j , update $n_j \to n_j$-1 . If $n_j=0$, individual j will put into another set H.

(4)After the completion of the population stratification, give a non-dominated sort $i_{rank}=1,2,\cdots size(layer)$ to each layer.

B. Congestion calculation and comparison

(1)Congestion calculation

In the non dominated sorting GA, the crowding degree is an essential way to keep the population diversity, the detail approach of congestion calculation is shown as follow:

a) Initialize the congestion of each point $i_d=0$:.

b) According to each objective of the model, conduct the non-dominated sorting, for the individual at the boundary define its congestion $O_d=i_d=+\infty$.

c) Calculate the congestion of each individual in the population based on the following Eq.14:

$$i_d=\sum_{j=1}^{m}(|f_j^{i+1}-f_j^{i-1}|) \tag{14}$$

Among them i_d represents congestion of points, m is the number of objective functions, f_j^{i+1} denotes the objective value j of point $i+1$, f_j^{i+1} denotes the objective value j of point $i-1$.

The sketch map for congestion calculation of points shows as Fig.2.

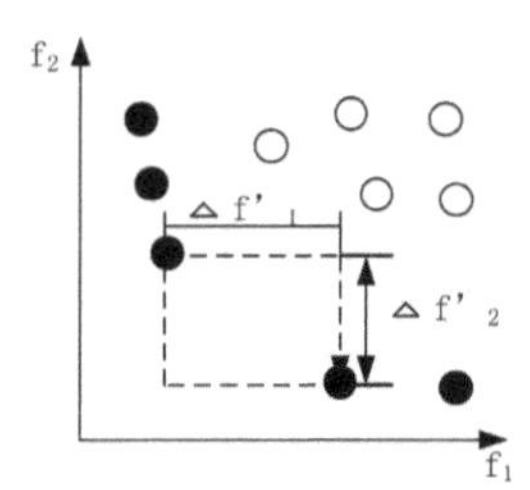

Fig.2. the point congestion

(2)Congestion comparison

After the above procedure, the rapid non dominated sorting operation and crowding degree are carried out. There are two attributes of individuals: one is congestion i_d the other is number of fast non dominated sorting i_{rank}. We can compare these two attributes to execute the Good or Bad choice between individuals i and j. If it meet any one of the conditions 1 and 2, the individual i is optimal.

Condition 1: The individuals i in the layer of non dominated is better than individual, that is $i_{rank} < j_{rank}$.

Condition 2: If the individual i and individual j are in the same layers, but is i_d bigger than j_d ,in the other words $i_{rank} = j_{rank}$ and $i_d > j_d$.

The condition 1 ensures that the priority is to choose the superior individual of the non dominated layer. Condition 2 makes sure that when the priority of the non dominated layer is the same, the larger individual is preferentially selected to cross to produce the next generation. In this algorithm the use of the elite strategy is to protect the more outstanding individuals in the process of genetic from disappearing. The sketch map for elite strategy shows as Fig.3.

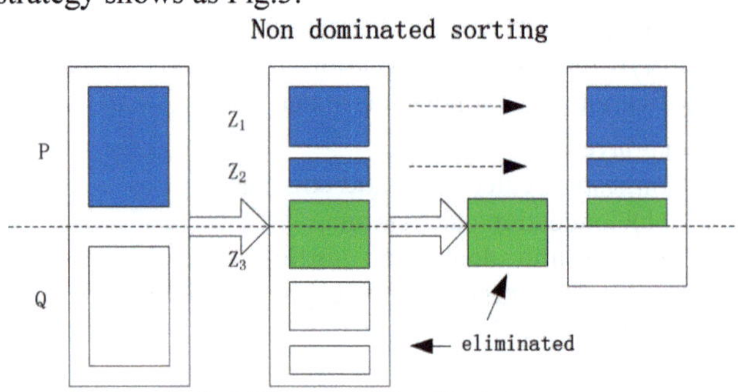

Fig.3. The steps of the elite strategy

V. EXAMPLE CALCULATION

To verify the rationality and validity of the algorithm, an instance from Z Company's production data was conducted. There are eight parts and eight machines in the company's job shop, each job need 3 or 4 procedures for processing. The T is two units time, and $\bar{t}$ is eight units time. The processing time of the jobs on the machine was given as Table .I

TABLE .I.
PROCESSING TIME OF THE JOBS

Jobs		Machines 1	2	3	4	5	6	7	8
J1	1	5	3	5	3	3	-	10	9
	2	10	-	5	8	3	9	9	6
	3	-	10	-	5	6	2	4	5
J2	1	5	7	3	9	8	-	9	-
	2	-	8	5	2	6	7	10	9
	3	-	10	-	5	6	4	1	7
	4	10	8	9	6	4	7	-	-
J3	1	10	-	-	7	6	5	2	4
	2	-	10	6	4	8	9	10	-
	3	1	4	5	6	-	10	-	7
J4	1	3	1	6	5	9	7	8	4
	2	12	11	7	8	10	5	6	9
	3	4	6	2	10	3	9	5	7
J5	1	3	6	7	8	9	-	10	-
	2	10	-	7	4	9	8	6	-
	3	-	9	8	7	4	2	7	-
	4	11	9	-	6	7	5	3	6
J6	1	6	7	1	4	6	9	-	10
	2	11	-	9	9	9	7	6	4
	3	10	5	9	10	11	-	10	-
J7	1	5	4	2	6	7	-	10	-
	2	-	9	-	9	11	9	10	5
	3	-	8	9	3	8	6	-	10
J8	1	2	8	5	9	-	4	-	10
	2	7	4	7	8	9	-	10	-
	3	9	9	-	8	5	6	7	1
	4	9	-	3	7	1	5	8	-

Notes: '-' represents the process can not be processed on the machine.

The difference between improved multi-objective GA and single objective GA is that the improved GA has a number of solution sets, rather than only one solution set. This set is called the Pareto optimal solution set. Decision makers can choose the optimal individual according to their own preferences from the Pareto optimal solution. By MATLAB on this case, the parameters were set as follow. The initial population size is 200, the maximum genetic algebra for the 80 generation, the crossover probability is 0.8, the probability of mutation is 0.05. By solving the numerical example, the number of Pareto optimal solutions are more than 200. Two representative solutions are selected for the decision maker. If delivery time is short, the decision makers will pay more attention to the goal 1, there will employ the following solution which is displayed in Table .II.

TABLE .II.
SOLUTION OF MINIMUM COMPLETING TIME

operation sequence	machine selection	operation sequence	machine selection
O{3,1}	2	O{5,3}	4
O{6,1}	5	O{8,3}	6
O{8,1}	6	O{2,2}	7
O{7,1}	3	O{2,3}	3
O{6,2}	4	O{8,3}	8
O{8,2}	7	O{1,2}	2
O{4,1}	5	O{7,2}	3
O{3,2}	7	O{5,4}	8
O{5,1}	4	O{4,4}	4
O{4,2}	1	O{2,4}	1
O{1,1}	2	O{1,3}	2
O{3,3}	6	O{7,3}	8
O{2,1}	3	O{6,3}	5
O{5,2}	1		

If the delivery time is not very tense, the decision makers will pay more attention to balance the load of equipment and reduce the maintenance cost. Under this circumstance the brains will choose solution in Table .III.

TABLE.III.
SOLUTION OF MINIMUM TOTAL EQUIPMENT LOAD

operation sequence	machine selection	operation sequence	machine selection
O{6,1}	4	O{5,3}	4
O{4,1}	5	O{2,3}	6
O{4,2}	6	O{5,4}	7
O{8,1}	3	O{7,2}	3

O{3,1}	4	O{3,2}	8
O{5,1}	7	O{4,3}	2
O{1,1}	5	O{2,4}	3
O{2,1}	7	O{7,3}	8
O{8,2}	3	O{6,3}	4
O{7,1}	1	O{3,3}	1
O{6,2}	2	O{1,3}	2
O{1,2}	6	O{8,3}	8
O{2,2}	3	O{8,4}	5
O{5,2}	1		

The results calculated by the multi-objective integrated optimization of FJSP&PM and independent maintenance are compared which was shown in Table .IV.

TABLE .IV.
COMPARISON OF INDEPENDENT AND INTEGRATED

	independent	Pareto1	Pareto2
Completing time	23	19	18*
Equipment load	107	73*	75

Notes: '*' represents the optimal solution.

The conclusion can be drawn from the above table: compared to the independent model solution, the completion time of the integrated optimization solution is greatly reduced, and the equipment load is also greatly reduced, which indicates the superiority of the integrated optimization and efficiency of improved algorithm.

VI. CONCLUSION

This article studies the multi-objective integrated optimization problem of preventive maintenance and flexible job shop scheduling. The integrated optimization model is established to minimize the maximum completion time and equipment load. Then, according to the characteristics of the integrated optimization model, a multi-objective improved genetic algorithm is designed to solve the problem. The algorithm adopts the selection operator based on elite solution retention strategy, IPOX &MPX hybrid crossover operator and mutation operator. Finally, compared to single objective optimization, the results show that improved multi-objective GA can find multiple Pareto solutions and provide a variety of choices for enterprise decision-making, which maximize the overall profit of enterprises.

REFERENCE

[1] Berrichi, A., F. Yalaoui, L. Amodeo, and M. Mezghiche, "Bi-Objective Ant Colony Optimization Approach to Optimize Production and Maintenance Scheduling," *Computers & Operations Research.*, vol. 37, no. 9, pp. 1584-1596, 2010

[2] Batun, S., and M. Azizoğlu, "Single Machine Scheduling with Preventive Maintenance," *International Journal of Production Research.*, vol. 47, no. 7, pp. 1753-1771, 2009

[3] Najid N M, Alaoui-Selsouli M, Mohafid A, "An integrated production and maintenance planning model with time windows and shortage cost," *International Journal of Production Research.*, vol. 49, no. 8, pp2265-2283, 2011

[4] Sbihi M, Varnier C, "Single-machine scheduling with periodic and flexible periodic maintenance to minimize maximum tardiness," *Computers & Industrial Engineering.*, vol. 55, pp. 830-840, 2008

[5] Yu Mingrang, Zhang Yingjie, Chen Yu, Gao Rui, Zhang Ding, "An integrated method of job shop scheduling and preventive maintenance with the adjustment time," (in Chinese), *Journal of Xi'an Jiao Tong University.*, vol. 49, no. 6, pp. 16-20, 2015

[6] Dehua Xua, Zhenmin Cheng, Yun qiang , "Makespan minimization for two parallel machines scheduling with a periodic availability constraint," *Computers & Operations Research.*, vol. 36, pp. 1809-1812, 2009

[7] Song Wenjia, Zhang Chaoyong, Shao Xinyu, Yin Yong, "Integrated optimization of equipment maintenance and job shop scheduling based on multi objective mixed colony competition algorithm," (in Chinese), *China Mechanical Engineering.*, vol. 26, no. 11, pp. 1478-1487, 2015

[8] Zhao Shikui. Bilevel Neighborhood, "Search Hybrid Algorithm for the Flexible Job Shop Scheduling Problem," (in Chinese), *Journal of Mechanical Engineering.*, vol. 51, no. 14, pp. 175-184, 2015

[9] Wu Xiuli, Zhang Zhiqiang, Du Yanhua, "Improved bacteria foraging optimization algorithm for flexible job shop scheduling problem," (in Chinese), *Computer Integrated Manufacturing Systems.*, vol. 21, no. 5, pp. 1262-1270, 2015

[10] Gao Liang, Zhang Guohui, Wang Xiaojuan, *Intelligent algorithm and its application in flexible job shop scheduling.* Wuhan: Huazhong University of Science and Technology press, 2012

[11] Hou Wenrui. "Finite time interval manufacturing system preventive maintenance strategy research opportunities," Ph.D. dissertation, Shanghai Jiao Tong University, China, 2014

[12] Liu Fanmao. "Multi equipment reliability analysis and maintenance decision making in the process of failure," Ph.D. dissertation, Huazhong University of Science and Technology, China, 2010

[13] Zhang Chaoyong, Dong Xing, Wang Xiaojuan, Li Xinyu, Liu Qiong, "Multi objective flexible job shop scheduling based on improved non dominated sorting genetic algorithm," (in Chinese), *Journal of Mechanical Engineering.*, vol. 11, pp. 156-164, 2010

[14]Cui Weiwei, Lu Zhiqiang, Pan pan, "Research on production scheduling and equipment maintenance integration based on multi-objective optimization," (in Chinese), *Computer Integrated Manufacturing System.*, vol. 20, no. 6, pp. 1398-1404, 2014

[15] Peng Jiangang, Liu Mingzhou, Zhang Xi, Zhang Mingxin, Ge Maogen, "A discrete free search algorithm based on Pareto optimization for solving multi-objective flexible job shop scheduling problem," (in Chinese), *China Mechanical Engineering.*, vol. 26, no. 5, pp. 620-626, 2015

Cost-sensitive Association Rule Modeling for Predicting Sequential Event

Hai-yan YU[1,2,3,*], Meng-li YANG[1], Jie JIAN[1]

[1] School of Economics and Management, Chongqing University of Posts and Telecommunications, Chongqing, 400065, China

[2]School of Computer Science and Engineering, University of Electronic Science and Technology of China, Chengdu, 611731, China

[3]Big Data Research Center, University of Electronic Science and Technology of China, Chengdu, 611731, China

(*yhy188@tju.edu.cn)

Abstract – To capture the behavior trends and relationships among the entities in the data, a cost-sensitive association rule model was built to achieve robust decisions. This model was trained through the imbalanced data and the reasoning loss was formulated on the concept of loss rate under uncertainty. This model distinguished the cost-sensitive losses between false negative rate and false positive rate. Considering the characteristic of ROC curve, the minimization principle of reasoning loss was obtained through a gradient descent method according to the sufficient condition for the cost-insensitive reasoning, and this approach established the mechanism of robust association rule inferences. Finally, the effectiveness of the proposed approach was validated by the simulation with real decision data sets.

Keywords – Cost-sensitive decision making, predictive customer intelligence, association rules

I. INTRODUCTION

The emergence of multi-scale decision databases presents exciting opportunities for data driven sequential event prediction. Prediction lies at the heart of customer intelligence and in this paper we propose a machine learning model for predicting entity-level sequences of activity status. Sequential event prediction (SEP) [1] has been widely implemented in guiding medical decisions, customer churn prediction, etc. Taking customer churn prediction as an example, with the rapid increasing competition in the market, the customer life cycle has been shortened, thus customer churn has become one of the important issues of the telecommunications firms. Meanwhile, the academic circles also did researches on predicting customer churn, including predicting yield loss rates in a particular industry[2] and optimizing loss rates based on various statistical models[3]. To the best of our knowledge, SEP contributes to improving the accuracy of the models on the testing samples, while those researches did not consider the sensitive decision costs of the false negatives and false positives[4].

Sequential event prediction refers to a classification method using association rules and multi stage optimization method to achieve dynamic prediction. SEP was first proposed by Rudin and Letham[5], who investigated this method to recommend case label information classification based on association rules. A derived model of this model is the evidence chain based on reasoning[6]. SEP model often represent the recommendation features of case information and the class labels assigned by the experts as apparent sequential events. Then, association rules are implemented to infer the potential outcomes. Those methods assumed that the distributions of the positive samples and negative samples were balanced data, but they did not take enough concerns on the reasoning cost underestimated by the decision risks. In details, those methods considered that the cost of correct predictions was zero and costs of reasoning errors were the same in positive[7]. However, large false negatives lead to more alarms and more costs of repairing the system, and large false positives lead to high risks in maintain the knowledge base or more efforts to improve the system discrimination ability. Despite of providing the strategies of refreshing the knowledge base on time[8], it is critical to investigate the insights of the implicit pattern underling the sequential events.

Receiver operating characteristic (ROC) curves is widely adapted to evaluate the false negative rates and false positive rates. However, the current SEP method did not performance well with ROC curves on the cost sensitive data sets[9]. Those methods often discarded some missing data in cost sensitive data set because it is not conducive to identify the potential outcomes. In this way, the distributions of the positive samples and negative samples becomes much more uneven, which declines the discrimination ability of rare class, eventually making the differences in sensitivity and specificity of the inference large. Therefore, this paper presents a cost-sensitive association rule model for predicting sequential events. This model captures the behavior trends and relationships among the entities in the data. With this model, we can achieve robust decisions under the uneven cost information. Finally, the model is applied to a real decision problem of controlling the customer loss rates.

II. MODELING APPROACH

A. Preliminary

E. Qi et al. (eds.), *Proceedings of the 23rd International Conference on Industrial Engineering and Engineering Management 2016,* DOI 10.2991/978-94-6239-255-7_26

According to the theory of association rules, the objective of SEP is to model the conditional probabilities. In this context, an association rule is a evidential chain " $a \rightarrow b$ " (or " $S \rightarrow b$ ") for information retrieval, meaning that the attribute a (or S) in the chain implies the attribute b (a) is also in the china. We define the confidence of chain as

$$\beta(a \rightarrow b) = \Pr(b \mid a) = \frac{\#(a \cup b)}{\# a}, \text{ or}$$

$$\beta(S \rightarrow a) = \Pr(a \mid S) = \frac{\#(S \cup a)}{\# S} \quad (1)$$

To infer the potential outcomes of sequential queries, the policy of finding the chains is max-confidence. For each query q, the data acquisition process has T_q periods, $t = 1,...,T_q$. In the stage t, the attribute data added to the query is $a^{(q)}(t)$. Consequently, the observations of the query is denoted as $x^{(q)}(t) = \{a_j^{(q)}\}_{j=1,...,t}$, and the complete observations of the chain is $x^{(q)}(T_q)$, abbreviated as $X^{(q)}$. Therefore, the sets of Q queries is represented as $\{X^{(q)} \mid q = 1,..,Q\}$.

B. Modeling

The tuition of SEP is a ranking model of the mapping relationship between the attributes S in the observed part of the chain and potential outcome attributes b. This ranking model is a scoring function $f(S,a)$, namely $f(x_m(t),b)$ that, given the observed part of the chain $x_m(t)$, scores each attribute according to the predicted likelihood. The predictions will be inferred by ordering the attributes in descending score. The ranking models includes two possible scoring types, the one-stage SEP model and the maximum-likelihood constrained SEP model. Their difference lies in implementing the observations or the likelihood as the information value function $\lambda_{a,b}$, which illustrates the monotonicity of $f(x_m(t),b)$. This function also demonstrates that given the observations of the chain (the current dataset or the local data), the additive feature has impact of inference value to the current feature information in the form of linear combinations.

Given the observations $x_m(t)$ in the one-stage SEP model, $x_m(t)$ ={ a_1, a_2 }, for any new additive feature b, the scoring functions is defined as

$$f(x_m(t),b;\lambda) := \lambda_{\varnothing,b} + \sum_{\tau=1}^{t} \sum_{\{a_\tau \in A \backslash \varnothing\}}^{\{a_\tau\} \subset F_{m,\tau}} \lambda_{a,b} \quad (2)$$

where $\varnothing$ is empty set, A is the feature set containing all the features in the current data, $F_{m,\tau}^{(J)}$ is the attribute or attribute set added into the feature set in the state τ. Let N be the size of the attributes, and the size of the covariates is $|A|N = N^2 + N$. Therefore, the complexity of this problem can be determined by the size of the set. For example, the attributes of ($x_m(t)$) is { a_1, a_2 }, and the scoring function is $f(\{a_1,a_2\},b;\lambda) = \lambda_{\varnothing,b} + \lambda_{\{a_1\},b} + \lambda_{\{a_2\},b}$.

The sequences of the chain $x^{(i)}$ is achieved as $\{a_1^{(i)},...,a_M^{(i)}\}$ after the process of SEP. In the historical dataset, a complete case data also involves the class label $C^{(i)}$ in spite of the features $x^{(i)}$ [10]。

C. Decision-making criteria

Given that the mapping function from the attributes $x^{(i)}(t)$ to $C^{(i)}$, $c_j \in C^{(i)} = \{-1,1\}, j = 1,2$, we exploit the impacts of the false positive rates $1 - F_0(\sigma)$ and false negative rates $F_1(\sigma)$ [11], which depicts the inference loss $L(x,c)$ from the reasoning process.

An inference loss widely used is the one defined by Arias-Nicolas[12],

$$L(x,c) = L_p(s_i,c_j) = \left\{ \left| s_i - c_j \right|^p, p \geq 1 \right\} \quad (3)$$

When we execute the SEP decision process, the decision maker consider the losses of $1 - F_0(x)$ and $F_1(x)$, which are independent with their relative distance.

$$L(x,c) = \begin{cases} 0, & s_j = c_j \\ \lambda_1, & c_1 = -1, s_1 = 1, \lambda_j > 0 \\ \lambda_2, & c_2 = 1, s_2 = -1 \end{cases} \quad (4)$$

The case $\lambda_1 = \lambda_2$ demonstrates that SEP is not cost sensitive, which this is a rare event. For a real reasoning decision problem, the costs λ_j of those errors are determined by the average value of the prior ones in the knowledge base.

$\Pr(x_i|c_j)$ is the conditional probability of the class label c_j given the observations x, and the false positive rate of *ROC* is $1 - F_0(x)$ and its false negative rate $F_1(x)$. The average loss determined by the distribution of *ROC* is

$$L_m(x|c) = \Pr_{c_1} \int_0^{\Pr_{c_2}} \lambda_1(1 - F_0(x))dx + \Pr_{c_2} \int_{\Pr_{c_2}}^1 \lambda_2 F_1(x)dx \quad (5)$$

where $\Pr_{c_j}$ is the corresponding probability of the class label c_j.

Thus, the total loss of SEP based on the min risk expectation is

$$f^* = \arg\min_f E_{X,C}[L_m(x,c)]$$

This criteria deduces the potential outcome of the cost-sensitive association rule modeling.

III. RESULTS

In intelligent decision making systems, marketing strategies are usually aided with data mining methods, i.e., predicting the churns of mobile phone customers in the Telecommunications industry. To validate the cost-sensitive decision making method, we executed the numerical experiment on a real data set. Experimental results verifies the effectiveness of the proposed method.

A. Data Source

Telecommunications Customer Management (TELCO) dataset, as the experimental data, is excerpted from the IBM Predictive Customer Intelligence database[13]. This dataset with mobile service insights industry accelerator allows decision maker to interpret the customers' mobile phone activity and to use the activity data to predict the potential chance for a customer to churn. The detailed information is depicted in Table 1.

TABLE I
DETAILS OF THE TELECO DATASET

Items	Information	Items	Number
Characteristics	Multivariate data	Instances	1310
Attribute	Integer	Attributes	8
Tasks	Classification	Training cases	1179
Missing Values	N/A	Testing cases	131

We choose the TELCO_1_0.TELCO_PEV dataset to do machine learning the proposed model. During the training stage, the attributes {CUSTOMER_ID, ESTIMATED_INCOME, SENTIMENT_SCORE, MARGIN_AMOUNT, NUMBER_OF_CLOSED_ COMPLAINTS, NUMBER_OF_MONTHS_ SINCE_ CUSTOMER_UPGRADED_THE_PLAN, NUMBER_ OF_ OPEN_COMPLAINTS, EDUCATION} are set as the observation attributes and {CHURN} as the decisive attribute. The values of the decisive attribute are taken as 0 or 1, illustrating churn or not respectively.

B. Analyze problem

In the reasoning process, the original dataset were discretized with domain knowledge first. The results after discretization are shown in TABLE II.

TABLE II
DISCRETIZATION RESULTS OF THE TELECO DATASET

Attribute	Values
CHURN	{0,1}
ESTIMATED_INCOME	(-∞, 8392]; (8392, 40014]; (40014, 64792) ; (64792, +∞)
SENTIMENT_SCORE	(-∞, 0]; (0, 1]; (1, +∞)
MARGIN_AMOUNT	①(-∞, 12000]; (12000, +∞) ②(-∞, 11000]; (11000, 14000]; (14000, 20000) ; (20000, +∞)

Second, all the redundant attributes have been remov ed with a feature selection method proposed in [14], and t he attribute weights were also been assigned with the mut ual information based ratio. Fig. 1 illustrated the weights of the selected attributes, { SENTIMENT_SCORE, MAR GIN_AMOUNT, ESTIMATED_INCOME }.

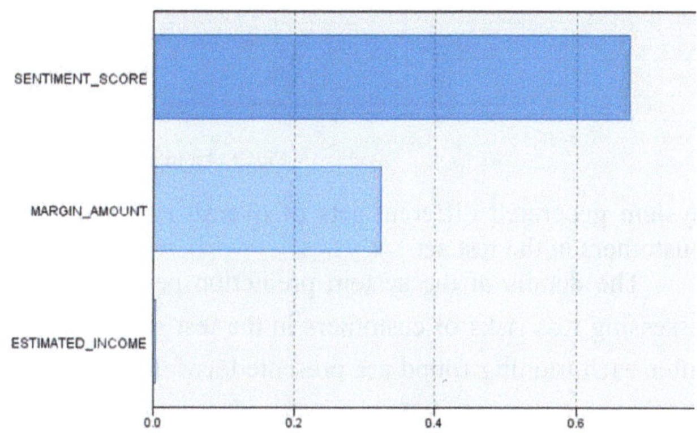

Fig. 1. Attribute weights of the experimental data set.

Fig. 2 illustrated an inference tree learned from the experimental data set.

Sequential events include SENTIMENT_SCORE, STIMATED_INCOME, MARGIN_AMOUNT and CHURN. Fig. 3 illustrated rule lists learned extracted from the inference tree.

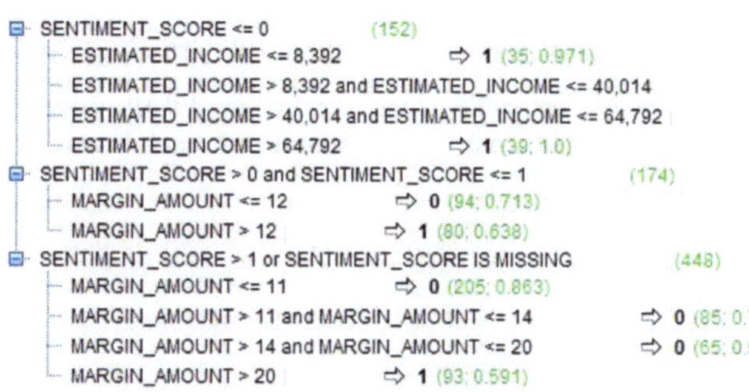

Fig. 3. Cost-sensitive association rules extracted from the CHAID tree.

C. Results

In the training process, the training module resulted in different changes to the cost-sensitive association rule model [15] for different sets of training parameters. In the test process, based on the trained cost-sensitive association rule model after different training rounds, the

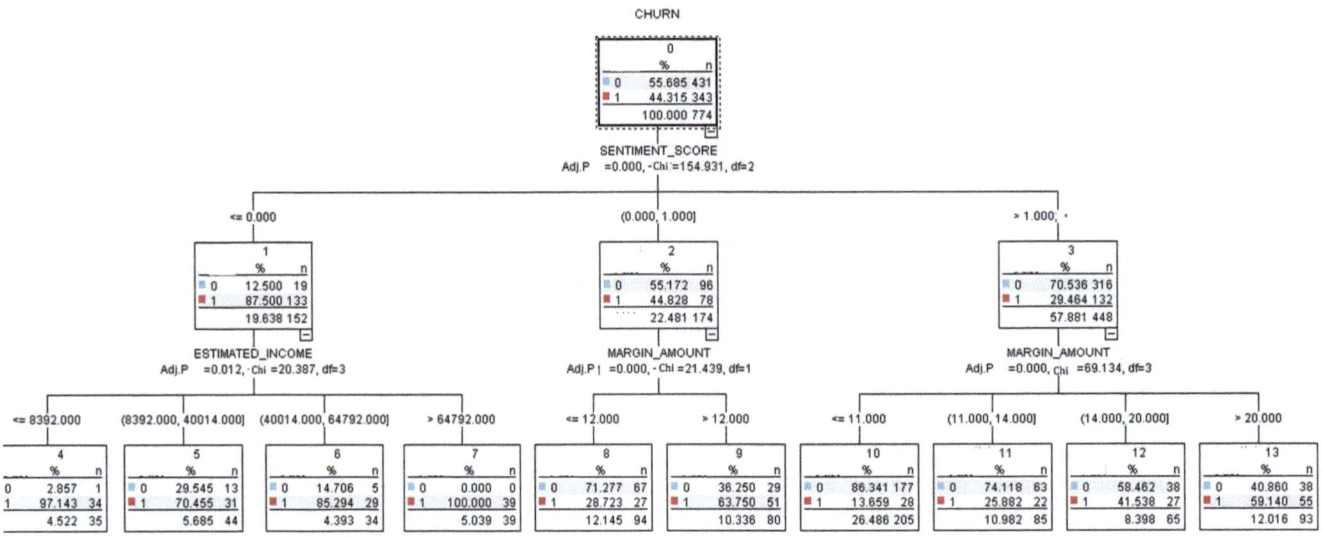

Fig. 2. An inference CHAID tree learned from the experimental data set.

system generated different sets of overall risk scores for customers in the test set.

The details of the system prediction performance in assessing loss risks of customers in the test set before and after each training round are presented as follows. We set $\lambda_1 = 1$ and $\lambda_2 = 1.5$. Before training the system using the customer data in the training set, we used the system to conduct loss risk assessment of customers in the test set based on the initial cost-sensitive association rule (CSAR) model, and generated a set of risk scores in the following discussion. After cost-sensitive association rule model training with different parameter sets, another set of risk scores was generated by the system for customers in the test set, and these scores are denoted as 'non-CSAM' scores. Based on the system generated risk scores before and after cost-sensitive association rule training, ROC curves are plotted using SPSS Modeler as shown in Fig. 4 and Fig. 5. The risk score source of each curve is described in the figure.

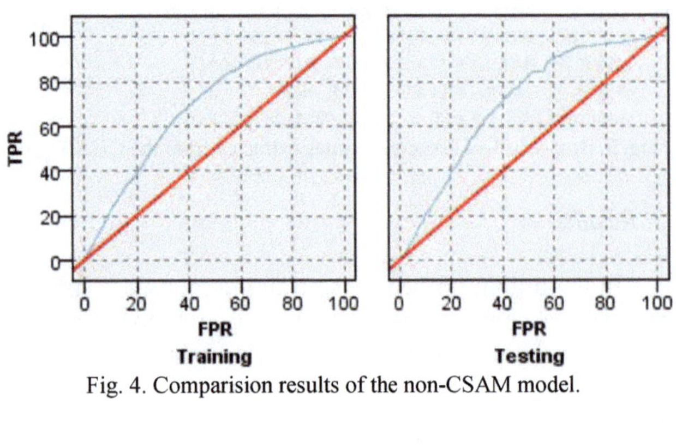

Fig. 4. Comparision results of the non-CSAM model.

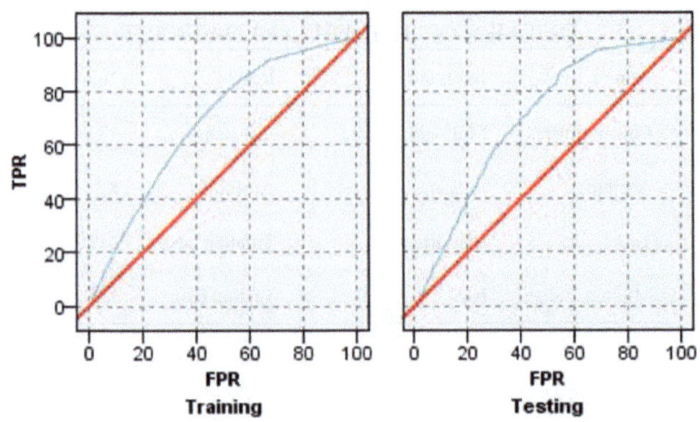

Fig. 5. Comparision results of the CSAM model.

The prediction results of corresponding AUC values on the TELECO dataset is shown as Table II.

TABLE II
PREDICTION RESULTS OF AUC VALUES

Model	Partition	Training		Testing	
non-CSAM	Correct	902	76.25%	104	80%
	False	281	23.75%	26	20%
CSAR	Correct	891	75.32%	107	82.31%
	False	292	24.68%	23	17.69

As described in Table II, the AUC of the ROC curves representing system prediction performances after training are all larger than the AUC of the ROC curve before training, and the AUC of the ROC curve after training cost-sensitive association rule model is the largest among all AUC values. However, we cannot simply conclude that training cost-sensitive association rule model is the best training which can help the system achieve the best prediction performance just based on the data as shown in Table II. To measure the significance level of the system prediction performance improvement after each training round, we used P value to compare the

AUC of the ROC curves. The results of $P<0.05$ demonstrated that the cost-sensitive association rule model is effective in predicting the potential churns and reducing the loss rate.

IV. CONCLUSION

This paper proposed a cost-sensitive association rule model for predicting sequential events. This model captured the behavior trends and relationships among the entities in the data. With this model, we achieved robust decisions under the uneven cost information. Finally, the model is applied to a real decision problem of controlling the customer loss rates in the telecommunications industry. The cost-sensitive association rules abstracted from the TELECO dataset revealed the patterns inhaled in the entity data. The results show that the proposed model is effective in predicting the potential churns and reducing the loss rate. This model can also be adapted and expended to predict the diagnosis event and hospitalization in the hierarchical diagnosis and treatment system.

ACKNOWLEDGMENT

This work was supported in part by National Natural Science Foundation of China (71571105, 71601026).

REFERENCE

[1] T. H. Mccormick, C. Rudin, D. Madigan, "Bayesian hierarchical rule modeling for predicting medical conditions", *Annals of Applied Statistics*, 2012, vol. 6, no. 2 , pp. 652-668.

[2] J. Moeyersoms, D. Martens, "Including High-Cardinality Attributes in Predictive Models a Case Study in Churn Prediction in the Energy Sector", *Decision Support Systems*, 2015, vol.72, no. 8, pp. 72-81.

[3] A. Keramati, R. Jafari-Marandi, M. Aliannejadi, I. Ahmadian, M. Mozaffari, U. Abbasi. "Improved churn prediction in telecommunication industry using data mining techniques", *Applied Soft Computing*, 2014, vol. 24, pp. 994-1012.

[4] Liping Fu, Meng Shen, Haiyan Yu, Man Xu, "Random Disturbance Reasoning Model of Decision-making System and its Anti-interference Capability", *International Journal of Industrial Engineering: Theory, Applications and Practice*, 2015, vol. 22, no. 5, pp.645-660

[5] C. Rudin, B. Letham, D. Madigan, "Learning theory analysis for association rules and sequential event prediction", *The Journal of Machine Learning Research*, 2013, vol. 14, no. 1, pp. 3441-3492.

[6] Jiang SHEN, Hai-Yan YU, Man XU, "Heterogeneous Evidence Chains Based Fusion Reasoning for Multi-attribute Group Decision Making", *Acta Automatica Sinica*, 2015, vol.41, no. 4, pp. 832-842.

[7] Man XU, Haiyan YU, and Jiang SHEN, "New Algorithm for CBR-RBR Fusion with Robust Thresholds", *Chinese Journal of Mechanical Engineering*, 2012, vol. 25, no. 6, pp. 1255- 1263

[8] X. Fang, O. L. Sheng, P. Goes, "When Is the Right Time to Refresh Knowledge Discovered from Data?", *Operations Research*, 2013, vol.61, no. 1, pp. 32-44.

[9] X. Zhang, B. Hu, "A New Strategy of Cost-Free Learning in the Class Imbalance Problem", *IEEE Transactions on Knowledge & Data Engineering*, 2014, vol. 26, no.12, pp. 2872-2885.

[10] Hai-yan YU, Jiang SHEN, Man XU, "ECs-based reasoning for group decision analysis in the mislabeled classification context", *Systems Engineering and Electronics*, 2015, vol. 37, no.11, pp. 2546-2553

[11] H. Yu, J. Shen, M. Xu, "Temporal case matching with information value maximization for predicting physiological states", *Information Sciences*, 2016, vol. 367–368, no.1, pp. 766–782.

[12] J. Arias-Nicolas, J. Martin, A. Suarez-Llorens, "Lp loss functions: a robust bayesian approach", *Statistical Papers*, 2009, vol. 50, no.3, pp. 501-509

[13] IBM. Predictive Customer Intelligence 1.1.0 product documentation. *URL: http://www-01.ibm.com/support/ docview.wss?uid=swg27046802, 2015/9/29*

[14] Man XU, Haiyan YU and Jiang SHEN, "New approach to eliminate structural redundancy in case resource pools using α mutual information", *Journal of System Engineering and Electronics*, 2013, vol. 24, no.4, pp.625-633

[15] G. Kong, D. L. Xu, R. Body, Jian-Bo Yang, Kevin Mackway-Jones, Simon Carley, "A belief rule-based decision support system for loss risk assessment of cardiac chest pain", *European Journal of Operational Research*, 2012, vol.219, no.3, pp.564-573

Study on the Improved PSO Algorithm

Used in Coal Mine Safety Resource Allocation

Jin-Feng WANG [1], Ge Zhao[1,*], Xue-QI Zhai[1],Li-Jie Feng[2]

[1]Institute of Management Engineering, Zhengzhou University, Henan, 450001;

[2] Henan Provincial Coal Seam Gas Development and Utilization c., LTD., Henan, 450016

(* 625131596@qq.com)

Abstract: **Due to the coal mine safety resource allocation is a multivariate, multiple constrained, nonlinear combination optimization problem,so the particle swarm optimization (PSO) is easy to fall into local optimal value, low precision and slow convergence speed and so on. A particle swarm optimization based on Levy (LF-PSO) is proposed. This algorithm is applied to the study of coal mine safety resource allocation, combining with the time series dynamic, complex multiple attributes and so on. first of all, to build the index of safety resources from man - machine - environment - administration - emergency; secondly, in a safe investment as the constraint conditions, the use of the nonlinear function relationship between the BP neural network is used to fit the resource security and safety level, and then to build a coal mine safety resource allocation model; finally, based on the LF-PSO algorithm for model optimization. The results show that the LF-PSO algorithm has fewer parameters and is easy to jump out of local optima. Compared with the conventional PSO algorithm, the algorithm has better convergence speed and precision.**

Key words-**particle swarm optimization, Levy flight, coal mine, resource allocation**

I. INTRODUCTION

In order to avoid the frequent accidents of coal production safety, China's coal production enterprises has successively increased safety resources investment to improve the safety level of coal mines, but the security level upgraded from high investment is very limited. At the same time, reasonable security resources are an important link to ensure the normal operation of coal production enterprises. Therefore, how to optimize the allocation of security resources to enhance the sustainable stability of coal mine safety production under the premise of safety production has become a real urgent problem to be solved in coal industry.

In recent years, the coal mine safety resource allocation whose main goal is to improve the safety level has been highly concerned, a variety of optimization methods have been appeared. Gao Yanfen etc. [1]used Constant Elasticity of Substitution(CES) production function analysis method to evaluate coal mine safety and efficiency, then used least square method to optimize the efficiency of the coal mine safety; Zhang Feiyan etc. [2]used gray system theory to analyze and optimize investment structure of coal mine safety resources. Considering the influence factors of coal mine safety, Wu etc.[3] determined the weight of the coal mine safety evaluation factor through Fuzzy-AHP method, and the safety evaluation model was built on the basis of establishing the evaluation index system; Wang Jinfeng et al.[4] constructed a maximum efficiency multiple objective optimization model to optimize the allocation of coal mine safety resources; Yu Ji'an[5] and other established a safe investment game model based on the level of coal mining technology and production costs, the interests game of safety resources investment was studied to optimize the security investment policy decisions.

In summary, the above method provide an important reference value for the study, but most of these methods have limitations. First, the traditional optimization algorithms are mostly single point linear optimization, so the results are closely related to the initial value selection, and it is easy to fall into local convergence; Second, optimization function must have continuity, and coal mine safety resource decision variables are mostly discrete variables, so the application of the conventional linear or nonlinear method to obtain the optimal solution

E. Qi et al. (eds.), *Proceedings of the 23rd International Conference on Industrial Engineering and Engineering Management 2016,* DOI 10.2991/978-94-6239-255-7_27

of security resource allocation problem will have certain errors. In order to overcome the above shortcomings, this paper puts forward the PSO algorithm based on the Levy flight mechanism based on the conventional PSO algorithm, and applies it to solve the coal mine safety resource allocation model. The empirical results show that the method comprehensively consider the influence of the global optimal value, and effectively improve the computational efficiency of the PSO algorithm, besides it is simple and accurate to find the optimal solution of the model, it can also significantly improve the safety production level of coal mine.

II. ESTABLISH SECURITY RESOURCE ALLOCATION MODEL

Analysis of the coal production enterprise's internal coupling relations, and establish the security resources index system, and then use the BP neural network to fit the functional relationship between security resources and the level of security. With the highest level of security as the objective function, the security resource input is used as the constraint condition, and the security resource allocation model is established based on this.

A. Construction of Index System

Coal mine internal coupling relationship exists, and personnel improper operation, aging equipment, management means backward and other reasons caused by the safety of coal production levels by a variety of factors, such as the level of technical personnel, equipment, management level [6]. Therefore, this paper will be divided into coal mine safety resources: personnel quality x_1, mechanical equipment x_2, environmental improvement x_3, safety management x_4 and emergency rescue x_5 and other aspects, the output for the safety level y.

B. The Establishment of Objective Function

Because of the strong nonlinear function relationship between safety index, the specific model is difficult to describe, This paper uses BP neural network model to analyze the relationship between the level of security resources and security, A multiple input, single output BP neural network model is constructed.

The security resource allocation index is used as the input, that is, the input index is $x=(x_i, x_{i+\Gamma}, \cdots, x_{i+(m-1)\Gamma})^T$, The level of security as the output, so the output $y = x_{i+1}$. Choose the typical three layer BP neural network, because of using BP neural network to fit the level of coal mine safety, the number of neurons in the input layer of the neural network is equal to the number of mine safety resources input m, the fitting effect is better [7]. So the input of BP neural network is *m*, the hidden layer is *p*, the output number is *l*, BP neural network to complete the mapping $f: R^m \rightarrow R^l$, The input of each node of the hidden layer is:

$$S_j = \sum_{i=1}^{m} w_{ij} x_i - \theta_j, \quad j = 1, 2, \cdots, p \qquad (1)$$

w_{ij} is the connection weight of the input layer value hidden layer; θ_j is the threshold of the hidden layer node.

BP neural network transfer function using Sigmoid function $f(x) = 1/(1+e^{-x})$, the output of the node of the hidden layer is

$$b_j = \frac{1}{1+\exp(-\sum_{i=1}^{m} w_{ij} x_i + \theta_j)},$$

$$j = 1, 2, \cdots, p \qquad (2)$$

Similarly, the output layer node input, output respectively:

$$L = \sum_{j=1}^{p} x_j b_j - \gamma \qquad (3)$$

$$x_{i+1} = \frac{1}{1+\exp(-\sum_{j=1}^{p} v_j b_j + \gamma)} \quad (4)$$

In the formula, v_j is the connection weight of the input layer to the hidden layer, and the γ is the threshold of the output layer.

BP neural network connection weights w_{ij}, and threshold v_j, γ can be obtained through the BP neural network training, So x_{i+1} is predictable, formula (4) that is, BP neural network prediction model.

C. Analysis of Constraint Conditions

In order to ensure the safety in production of coal mine, the enterprise is bound to increase investment in resource security, but confined to the enterprise pursuit of profit, investment will be limited in a certain range. In addition, the decision variables of enterprise security resources are non-negative, so the constraint conditions established in this paper are:

①The safety resource input of each decision variable in coal mine is restricted by the total investment, that is $\sum_{i=1}^{n} x_i \le M$;

②Considering the actual situation, the safety resources of the coal mine are put into non negative value, that is $x_1, x_2, \cdots x_n \ge 0$

D. Construction of Model

According to the establishment of the above objective function and the analysis of the constraint conditions, the model of security resource allocation is constructed:

$$Z = \max f(x) \quad (5)$$

$$s.t. \begin{cases} \sum_{i=1}^{n} x_i \le M \\ x_1, x_2, \cdots x_n \ge 0 \end{cases} \quad (6)$$

III. IMPROVEMENT OF PSO ALGORITHM

The BP neural network is used to fit the nonlinear function of the coal mine safety resource allocation. The constraint conditions are analyzed according to the actual situation, and then the model of the security resource allocation is constructed.

But the complexity of the internal relations of the coal mine, the conventional PSO algorithm is prone to "premature", convergence speed in the late evolutionary slow phenomenon. Therefore, this paper in based on the conventional PSO algorithm, Levy flight mechanism is introduced to the PSO algorithm, and proposed based on Levy flight mechanism of particle swarm optimization algorithm.

A. Introduction of PSO Algorithm

Particle swarm optimization algorithm is a kind of evolutionary swarm intelligence computation method, and it is also an iterative algorithm based on the random solution[8].

Particle swarm optimization (PSO) algorithm is based on the observation of animal collective behavior and use of individual of population information sharing, to ensure that the entire population of the activity when they encounter problems can rely on the power of the group to solve the, when an individual to find the target, the individual is the optimal solution.

PSO algorithm and other evolutionary algorithms are similar, are used to optimize the objective function.

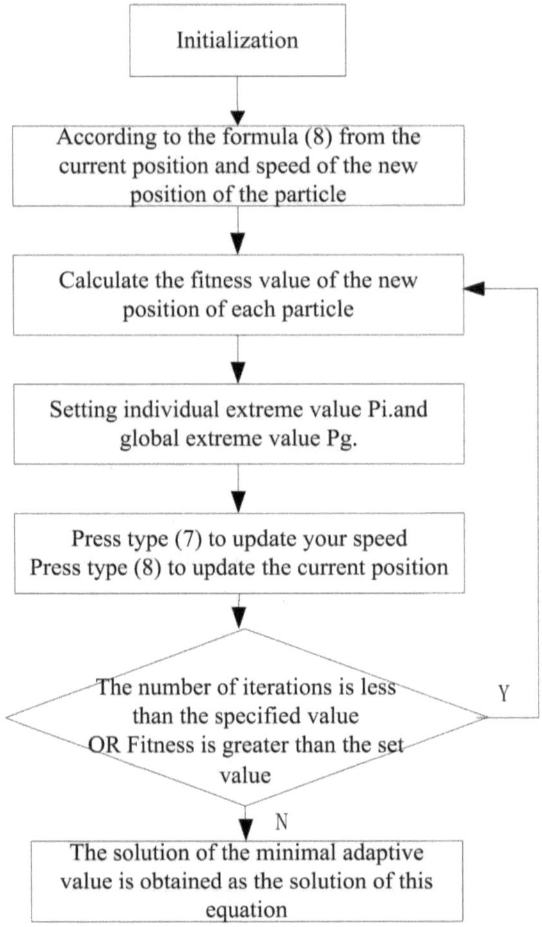

Fig. 1 solving process of PSO algorithm

PSO algorithm to solve the specific process as shown in Figure 1. According to the formula (7), (8) the velocity and position of each particle is updated[9]:

$$v(t+1)=v(t)+c1*(Pi(t)-x(t))+c2*(Pg(t)-x(t)) \quad (7)$$

$$x(t+1)=x(t)+v(t+1) \quad (8)$$

Among them, *v(t)* is the current velocity vector of the particle; *x(t)*is the position of the current particle; *Pi(t)* and *Pg(t)*are the position of the optimal solution of the particle, and the position of the optimal solution; *c1* and *c2* represent individual and group cognitive coefficients.

B. Particle Swarm Optimization Algorithm based on Levy flight mechanism

Flight Levy mechanism proposed by Pierre Levy Paul, is a walking step a to meet the Levy condition distribution of the Markov chain[10].

$$\sigma=(\Gamma(1+\beta)\times sin(3.14\times\beta/2) /(\Gamma((1+\beta)/2)\times\beta\times 2^{((\beta-1)/2)}))^{(1/\beta)};$$

$$u = rand(n,1)\times\sigma;$$
$$v = rand(n,1);$$
$$a=\alpha\times(u/abs(v)^{1/\beta}); \quad (9)$$

Where n is the dimension of the solution, Gamma function is $\Gamma(b)=\int_0^{\infty}\exp(-t)\times t^{b-1}d(t)$, $\beta=3/2$, α is step size factor.

If many of the animals in nature want to look for food in an uncertain environment, the ideal way is to use the "Levy flight" search strategy, includes two forms:

short distance carpet search and the occasional long distance raid. Short distance carpet can ensure the animals in the search for a small range of foraging, and occasionally raid with long distance and ensure the in a broader range of search. Many scholars are inspired by it, the Levy flight strategy is introduced in the evolutionary strategy, which not only improves the performance of the algorithm, but also has a significant effect.

Based on particle swarm optimization algorithm steps that, in the iterative process, the whole particle swarm are learning to the best individual, if find the optimal individual, all particles come in droves. This position update method not only greatly reduces the diversity of the particle population, and if the location is not the global optimal, it will lead to the local optimal algorithm, and then affect the convergence accuracy and speed of the particle. As far as the actual situation is concerned, in the whole search space, the global optimum of the particle population mostly exists in the vicinity of the local optimum, and the evolutionary rate of the population depends on the individual rather than the individual.

Based on the particle swarm optimization algorithm, the stochastic Levy flight mechanism is introduced to adjust the trajectory of particles. It not only helps to jump out of the local optimum, but also gets better convergence when the particle is in the vicinity of the global optimum[11]. The main effect is when the particle is moving in a small area around the local optimum, a larger step size is suddenly given to jump out of the local optimum. In D dimensional space, the particle in the t time is improved after the speed V and the position X update formula is:

$$v^{d}(t+1)=v^{d}(t)+c1\times(Pi(t)-x^{d}(t))+c2\times(Pg(t)-x^{d}(t)) \quad (10)$$

$$x^{d}(t+1)=x^{d}(t)+v^{d}(t+1)+\alpha\times sign(rand-0.5)\times levy(\lambda) \quad (11)$$

IV. MODEL SOLUTION BASED ON LF-PSO ALGORITHM

The specific solution steps of the PSO algorithm (LF-PSO) based on the flight mechanism of Levy are as follows:

①Security resource allocation model is established according to the actual situation of coal mine, initialization of each particle parameters, setting the population size, adapts the function according to the allocation of security resources, each particle randomly generated in setting down the dimensions of the initial solution and velocity value;

②According to the security resource allocation fitness equation, calculate the fitness value of each particle, then select the individual optimal value of each particle, and then compare the optimal value of each particle to determine the global optimal value of the particle;

③According to the formula (10), update the speed of each particle about security resource allocation, and its speed is limited to a certain range;

④According to the formula (11), update the location information about the security resource allocation for each particle, and the position is limited to the constraint conditions of the fitness function.

⑤When solving the security level of the fitness value does not meet the requirements or did not achieve the required number of iterations, return to step 2.On the contrary, the optimal particle position is the optimal scheme of the secure resource allocation.

V. EMPIRICAL ANALYSIS

A. Data Collection and Model Building

In combination with the actual conditions of the coal industry group Z,this paper use the BP neural network to build a nonlinear function of the relationship between the level of coal mine safety. We obtain the 40 sets of historical data in 2014 -2015 as a sample through the investigation and research, meanwhile, use the five security resource index as the input of BP neural network model and use the corresponding level of safety as output[12]. We use expert scoring method to collect the output indicators by inviting experts to rate the production safety level of Z Coal Group and the Scores are divided into 4 intervals that include excellent security level (100 ~ 90), good safety level(90 ~ 80),general security level (80 ~ 70),and worrying security level (60 ~ 70)[13].

According to the foregoing analysis, this paper selects the structure of 5-10-1 to fit the functional relationship between the security resources and the level of security [13]:

$$y = \frac{1}{1+\exp(-\sum_{j=1}^{10} v_j b_j + \gamma)} \quad (12)$$

In conclusion, the security resources allocation model is constructed for:

$$Z = \max \frac{1}{1+\exp(-\sum_{j=1}^{10} v_j b_j + \gamma)} \quad (13)$$

$$\text{s.t.} \begin{cases} \sum_{i=1}^{5} x_i \le 750 & (15) \\ x_1, x_2, \cdots x_{13} \ge 0 & (14) \end{cases}$$

Among them, the formula (13) indicated that the goal function security level is the biggest, the formula (14) indicated that the coal mine safety resources allocation total input restraint, Formula (15) indicates that the resource input of each decision variable in coal mine is not negative.

B. Solution and Result Analysis Based on LF-PSO

On the basis of the model construction, first, use of LF-PSO to write matlab program to solve the security resource allocation model.

The main code based on LF-PSO optimization is:

```
%Iterative optimization
for i=1:maxgen
    w=1-i/maxgen;
    for j=1:sizepop
        %Speed update
    V(j+1,:) = w*V(j,:) + c1*rand*(gbest(j,:) - pop(j,:)) +
c2*rand*(zbest - pop(j,:));
        V(j,find(V(j,:)>Vmax))=Vmax;
        V(j,find(V(j,:)<Vmin))=Vmin;
    %Population regeneration
    pop(j+1,:)=pop(j,:)+V(j+1,:);
        pop(j,find(pop(j,:)>popmax))=popmax;
        pop(j,find(pop(j,:)<popmin))=popmin;
```

When solving the optimal value of the coal mine safety resource allocation model, the specific parameters are set in Table I

TABLE I

THE INITIAL PARAMETERS OF THE LF-PSO SECURITY RESOURCE ALLOCATION MODEL

parameter	numerical value
Iteration number	300
Population size	25
Individual cognitive coefficient	0.5
Group cognitive coefficient	0.7
Adjusting step size factor	0.001

we can get the search results as is shown in table II.

Secondly, based on the conventional PSO and Levy based flight mechanism particle swarm optimization algorithm (LF-PSO) to solve the security resource allocation model.

TABLE II

COMPARISON OF CONVENTIONAL PSO AND LF-PSO OPTIMIZATION RESULTS

parameter	Conventional PSO	LF-PSO
Number of particles	25	25
Total number of iterations per time	300	300

The most advantage points / times	160~180	210~230
Number of particles Calculation time /s	100~2000	240~420
Safety level	77	90.2

From table 2, we can see that the improved LF-PSO algorithm proposed in this paper is superior to the conventional PSO algorithm in the model solution. According to the best position of the particle, in 2016,the best solution for the security resources allocation of the enterprise is x^T= (78.3112.5,59.6117.3,59.2) million, and the corresponding of the optimal level of security is 90.2 ,which increased by 11.68% compared with 2015. Therefore, the improved PSO algorithm is helpful to improve the security level, and the optimization results are greatly improved when solving the problem of nonlinear coal mine safety resource allocation.

It is not difficult to see that the proposed algorithm LF-PSO not only can effectively avoid the local minimum problems existing in the traditional PSO, but also can solve the conventional PSO for large-scale computational complexity and high calculation speed.

VI. CONCLUSION

Based on the conventional PSO algorithm is easy to fall into local optimum and the problem of poor accuracy, this paper proposes a PSO algorithm based on Levy flight mechanism. In the process of optimization, the algorithm of location update steps and conventional PSO algorithm updates is different, and it moves on a small scale in the process of particles movement , which given a larger time step jump to jump out of local optimum. When using this algorithm to solve the problem of coal mine safety resource allocation, we can use the BP neural network to build the security resource allocation model based on the coupling relationship between each index, and LF-PSO algorithm is used to optimize the model, so that the security resource allocation of coal production enterprise is optimized effectively. The empirical results showed that the algorithm can not only makes up for the defects of PSO algorithm, but also optimizes the optimization accuracy and convergence of the function.

REFERENCE

[1] Y.F.Gao, M.T.Jia, Z.Q.Chen,N.Li. Evaluation and optimization of safety efficiency of coal enterprises in China[J]. Journal of safety and environment, 2014,04：134-136.

[2] F.Y.Zhang, W.Meng, Y.Han. Analysis and optimization of coal mine safety resources investment structure based on Grey System Theory[J]. China mining industry,2014,01：120-124.

[3] F. D. Wu, N. L. Hu. Study on the model of safety evaluation in coal mine based on Fuzzy-AHP comprehensive evaluation method[J]. Proceedings -2011 International Conference on Mechatronic Science, Electric Engineering and Computer, 2011:1671-1674.

[4] J.F.Wang, X.Q.Zhai, L.J.Feng. Research on Optimization of coal mine production logistics efficiency oriented to safety hard constraint[J]. Management Science in China，2014，07：59-66.

[5] J.A.Yu, Z.Chen, B.Yang, J.C.Chen. Game Study on safety investment of coal mining enterprises based on utility function[J]. Chinese Soft Science，2014，01：115-125.

[6] L.Yang, C.C.Liu, L.Song, W.Sheng. Capacity evaluation of coal mine emergency rescue based on entropy weight method[J]. Chinese Soft Science，2013，11：185-192.

[7] L.F.Sun, J.B.Zhou, W.J.Lin, Z.L.Hou, F.Xu. Research on network public opinion crisis early warning based on BP neural network and genetic algorithm [J]. Intelligence Journal，2014，11：18-24.

[8] Z.F.Liu, W.Zhang, Z.L.Wang. Optimization layout of charging stations of urban electric vehicles based on quantum particle swarm optimization algorithm[J]. Chinese Journal of Electrical Engineering，2012，22：39-45+20.

[9] Z.H.Li,H.Zhu,Y.Pan. Hybrid particle swarm optimization algorithm and its application in fault diagnosis of analog circuits based on BPNN[J]. control engineering，2014，03：

450-454.

[10] Reible D, Mohanty S. A Levy flight-random walk model for bioturbation.[J]. Environmental Toxicology & Chemistry， 2002， 21(4)：875–881.

[11] Q.T.Zhang, L.Q.Fang, Y.L.Zhao. Optimization algorithm of twin group fly with Levy flight characteristics[J]. computer application，2015，05：1348-1352.

[12] B.S. Jia, B.Yin, H.Z.Wang, S.G.Li, X.W.Geng. Catastrophe theory evaluation model and its application in coal mine safety state[J]. Chinese Journal of Safety Science，2015，06：98-104.

[13] Q.F.Zhu, Z.P.Xu, L.Wang. Evaluation and comparative analysis of enterprise control activities based on fuzzy comprehensive evaluation method and BP neural network method[J]. Management Review,2013,08:113-123.

Research on the Improving Design of Laminating Machine Based on Lean Six Sigma

Fu-ying Zhang[1,*], Jian-lei Chen[1]
[1]Tianjin Key Laboratory of Integrated Design and On-line Monitoring for Light Industry & Food Machinery and Equipment, College of Mechanical Engineering, Tianjin University of Science & Technology, Tianjin 300222, PR China
(*e-mail: zhfuying@tust.edu.cn)

Abstract – **According to the quality problems of a company's carbon nanometer materials touch screen product in the processing, based on the Lean Six Sigma method, this paper analyzed the production processes and the defective rate of the big board section, and determined the offset defect as the key factors affecting the quality of the products in the process of its production. By using the experiment design, the main reason that caused offset defect was found to be the operation stability of the organic units of the laminating machine. On the basis of the working principle analyzed for the laminating machine, the defects of the structure and principle were found and a new method equipped with machine vision and automatic detection system was proposed to achieve automatic detection offset and automatic positioning. This study result would provide a feasible means for the improvement of the quality of touch screen products.**

Keywords - **Lean Six Sigma, process capability analysis, the production process of the laminating machine**

I. INTRODUCTION

The production process of carbon nanometer materials touch screen for electronic products such as mobile phones, e-books was divided into three stages: yellow light, big board section and small board section [1]. Make it have electrical conductivity by covering carbon nanotube film on the carbon nanotube (CNT) raw materials during yellow light stage. At the big board stage, after finishing the Ag-print and laser etching conductive silver lines on the bases of yellow light section, pasted it with optical film than pressurized deaeration and punch into chips. The chips of carbon nanotubes finished at big board section were continued crimping with flexible circuit board at the small board stage, and pressurized deaeration after laminating protective layer on the touch screen. The negative anomalies such as Ag- print pinhole, laser etching detonation point, laminating offset and poor crimping were often appeared in the production process of carbon nanometer materials touch screen. Because of the negative anomalies caused by laminating offset were irreparable, they will bring a great economic loss to the company. Therefore, it is urgent for enterprises to introduce a new concept of management to improve product quality, and reduce production costs.

The core of lean production is to eliminate waste, but the core of Six Sigma management is to reduce the production variation, so as to meet customer needs. The essence of Lean Six Sigma integrating two models of lean production and six sigma management [2] was to eliminate waste in the process and improve the efficiency of production. It is an effective means for improving product quality and reducing waste [3]. In this paper, the existing problems of the OCA (optical film) laminating site of the big board in the production process of carbon nanometer materials touch screen for a certain company were analyzed and improved to improve the product quality based on the lean six sigma. The results will provide reference for the application and implementation of Lean Six Sigma in the enterprise.

II. IMPROVED FRAMEWORK OF THE LAMINATING MACHINE BASED ON THE LEAN SIX SIGMA

The improvement framework of the laminating machine based on the lean six sigma method was built as shown in Fig.1, the DMAICL model was used in the Lean Six Sigma [4-5].

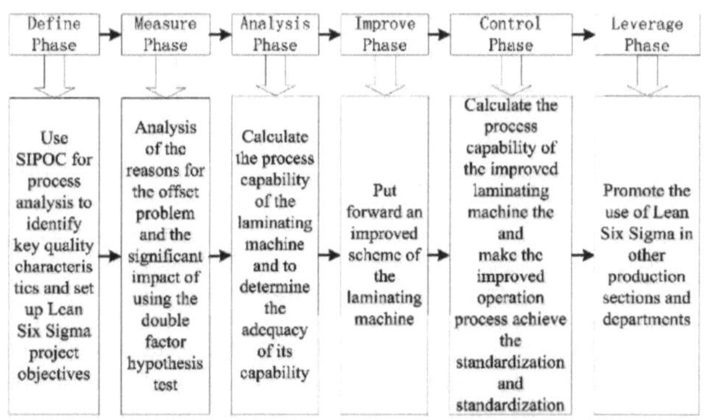

Fig.1 The improved framework of the laminating machine based on the Lean Six Sigma

The definition phase, identify the key quality characteristics and set up the goal of implementing the Lean Six Sigma on the basis of the analysis of the production process of the big board section; the Measurement phase, the Cause-Effect Diagram was used to deeply analysis the causes of the offset problem and the significant effect of various factors and to have the detection analysis on the offset; the analysis phase, according to the measurement data, calculate the process capability of laminating machine and its capability and judge whether it is enough; the Improvement stage, through the analysis of the working principle of the laminating machine, the improvement plan was put forward; the control phase, calculate the improved process capability of the laminating machine and make the improvement of the work process to achieve the

E. Qi et al. (eds.), *Proceedings of the 23rd International Conference on Industrial Engineering and Engineering Management 2016,* DOI 10.2991/978-94-6239-255-7_28

standardization and normalization; the leverage stage, promote the application of Lean Six Sigma to other production section and departments and comprehensively improve the productivity and product quality of enterprise , so as to meet the needs of users.

III. THE ANALYSIS OF THE KEY QUALITY CHARACTERISTICS AND ITS REASONS

Through the investigation, problem feedback and the analysis of product adverse phenomenon of the manufacturers of touch electrical products, it was found that the offset defect caused by the paste offset of carbon nanometer materials with optical film in laminating machine site during the big board section was the main factor affecting product quality. Therefore, in this paper, the big board section production process was regarded as improvement opportunities. The product process of big board section was shown in Fig.2, it were included of yellow light, aging, Ag-print, laser etching, open/short circuit test for big board, optical film lamination, pressurized deaeration, punch, parking. Forming a touch screen is the final goal of this product process.

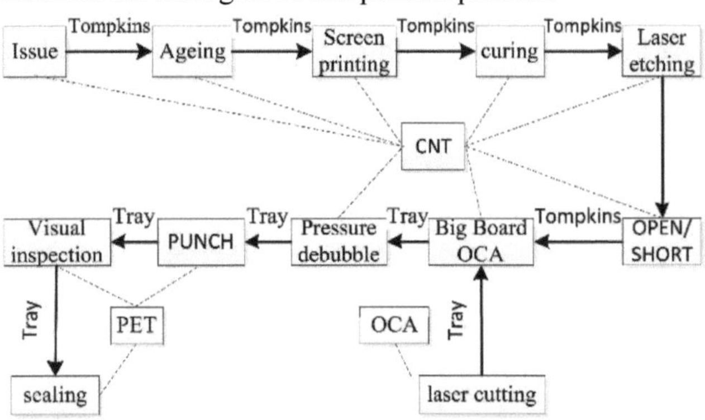

Fig. 2 The production flow chart for big board

All members of the project team was gathered to use the method of brainstorming and causal-effect diagram[6], the reasons of quality defect caused by optical film lamination offset of the big board section were found. They are mainly included of staff proficiency, temperature, humidity, the thickness of the optical film, the operating parameters of the laminating machine and process insecurity. Through the experimental design [7], the influence of the first four factors on deviation was carried out experiment. The hypothesis test results showed that staff proficiency, temperature, humidity and the thickness of the OCA was not the key quality factors, thus the key factor of causing offset was the parameters of laminating machine and process insecurity.

IV. IMPROVEMENT PROCESS CAPABILITY ANALYSIS AND WORKING PROCESS IMPROVEMENT OF LAMINATING MACHINE

The function of the optical film laminating machine was to make the carbon nanometer materials and the optical film adhered to each other. There are two ways to make attach, that is screen laminating method and the vacuum adsorption joint way [8]. The screen laminating method was used in our study company for optical film laminating; the attached diagram was shown in Fig.3. during the laminating process, first the carbon nanometer materials was placed on nylon mesh surface and the optical film was placed on a vacuum adsorption platform, then the vacuum adsorption platform rotated 180 ° , while the three markers of the up and down were all aligned, rubber roller rolling to realize the lamination of optical film and carbon nanotubes.

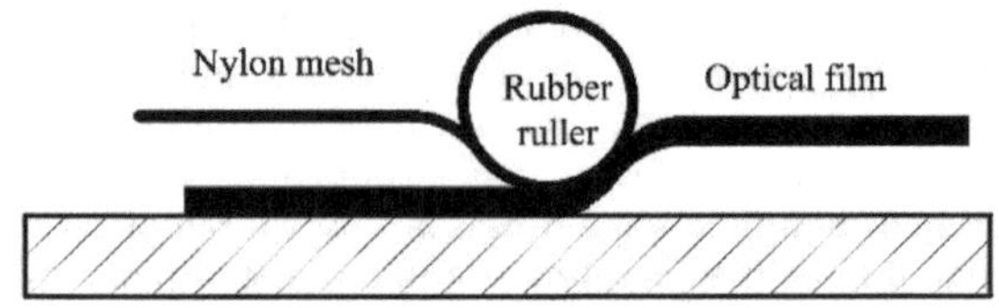

Fig. 3 The schematic diagram of sheet laminating

The working principle for the optical film laminating machine was shown in Fig.4. There are two sub-systems; one system makes the image of locating hole acquired by the CCD camera to display in the PC screen, the other system moves the adsorption platform in the X/Y direction using servo motor which was derived by control module. When the two positioning holes images in screen displayed in a required coincidence, the operator pressed the start button to start laminating, finished the lamination.

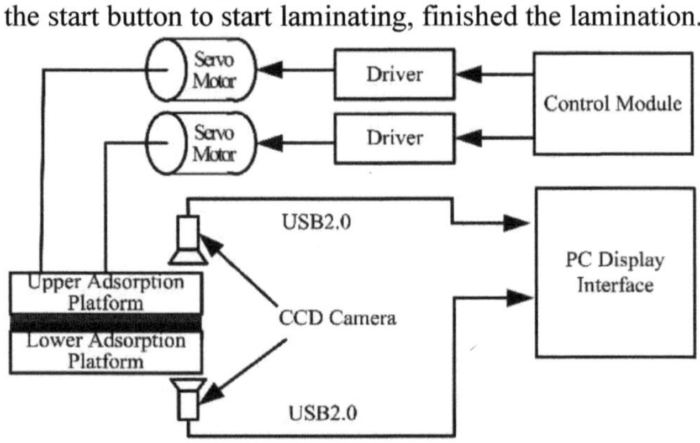

Fig. 4 The working principle of optical film laminating machine

TABLE I
THE SUMMARY TABLE OF OFFSET VALUE

Class	Groups									
	1	2	3	4	5	6	7	8	9	10
A1	0.581	0.212	0.318	0.456	0.681	0.134	0.427	0.323	0.267	0.346
B1	0.246	0.237	0.947	0.643	0.312	0.514	0.217	0.351	0.146	0.243
A2	0.172	0.149	0.264	0.315	0.247	0.248	0.243	0.134	0.482	0.461
B2	0.364	0.352	0.237	0.149	0.247	0.316	0.412	0.346	1.012	0.137
A3	0.235	0.349	0.218	0.317	0.124	0.249	0.316	0.816	0.486	0.491
B3	0.279	0.383	0.472	0.463	0.347	0.347	0.367	0.345	0.912	0.367
A4	1.143	0.413	0.269	0.349	0.425	0.431	0.465	0.364	0.367	0.491
B4	0.395	0.467	0.438	0.429	0.459	0.412	0.427	0.435	0.976	0.421
A5	0.342	0.258	0.472	0.435	0.479	0.426	0.437	0.341	0.367	1.245
B5	0.269	0.347	0.453	0.415	0.376	1.038	0.762	0.493	0.415	0.843

The product specifications required that the offset of OCA and CNT could not be more than 0.500mm. The 10 group dates got by measuring the two shifts per day, 5 days in a row, 10 randomly selected from each shift were shown in Table I. Through calculation, its process capability index Cpk = 0.16 <1, illustrated the process capability of OCA laminating machine was insufficient and need to be improved.

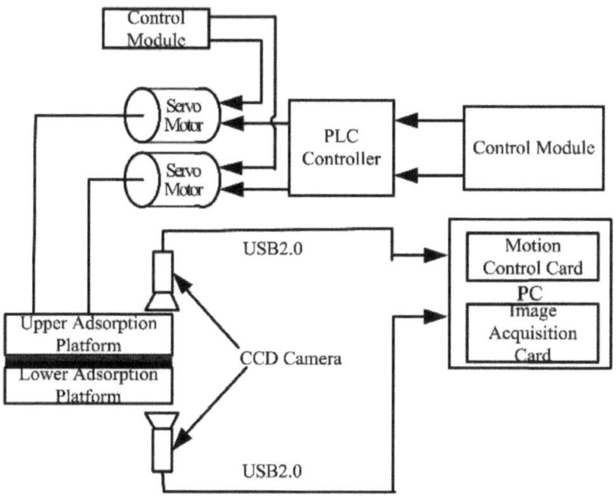

Fig. 5 The improved scheme of optical film laminating machine

The improved scheme minish the offset between the OCA and CNT by capturing the distance of two centers coordinates. Therefore, an image acquire device for the positioning hole of the CNT and OCA, and an automatic detection system were equipped in the laminating machine. After image preprocessing, the acquired image edge feature was extracted and the positioning holes of the circles were extracted, and then automatic judged the center distance of positioning hole [9, 10]. If the distances between two positioning holes within the product specifications, laminated directly, otherwise drive the up adsorption platform movement in the X/Y direction until meeting the specifications through movement control card to control PLC controller. The center coordinates for extraction as follows:

$$\left(\frac{(y_2-y_3)[(x_1-x_2)(x_1+x_2)+(y_1-y_2)(y_2-y_3)]-(x_2-x_3)(x_2+x_3)(y_1-y_2)}{2[(x_1-x_2)(y_2-y_3)-(x_2-x_3)(y_1-y_2)]},\right.$$
$$\left.\frac{(x_2-x_3)[(x_1-x_2)(x_1-x_3)+(y_1-y_2)(y_1+y_2)]-(y_2-y_3)(y_2+y_3)(x_1-x_2)}{2[(y_1-y_2)(x_2-x_3)-(y_2-y_3)(x_1-x_2)]}\right).$$

$(x_1, y_1), (x_2, y_2), (x_3, y_3)$ -- Three pixel edge points in the circle, (x_0, y_0) -- center coordinates. The improved OCA laminating machine diagram was shown in Fig.5.

The improved OCA offset was measured and the process capability was calculated. The results showed that the improved working ability index of the laminating machine Cpk=1.01>1, and the machine was in stable condition.

V. CONCLUSION

1) Based on Lean Six Sigma method, through the analysis of the production process for big board section and the scrap rate of carbon nanometer materials touch screen of a company, the offset defect was found to be the key factors affecting the quality of products.

2) Using the experimental design method, the working stability of the laminating machine was found to be the main reason caused offset defect.

3) By analyzing the working principle of laminating machine, the defects of structure and principle raised bad offset were found, and the method can realize the machine automatically detected the bad offset and automatic positioning was put forward by equipped with machine vision and automatic detection system.

4) Using the new automatic alignment principle to improve the structure of the OCA laminating machine, the improved process capability index was 1.01, and the working capability met the requirements.

REFERENCE

[1] Deng Kangfeng, Research on the Improvement of Laminating Machine Based on the Method of Lean and Six Sigma[D],Tianjin:2013.

[2] Pamfilie R,Petcu A J,Draghici M.The importance of leadership in driving a strategic Lean Six Sigma management[J].Procedia-Social and Behavioral Sciences,2012,58:187-196.

[3] Browning T R,Heath R D.Reconceptualizing the effects of lean on production costs with evidence from the F-22 program[J].Journal of Operations Management,2009,27(1):23-44.

[4] Kornfeld B J,Kara S.A Framework for Developing Portfolios of Improvements Projects in Manufacturing[J].Procedia CIRP,2013,7:377-382.

[5] Fairbanks C B.Using Six Sigma and Lean methodologies to improve OR throughput[J].Aorn Journal,2007,86(1):73-82.

[6] Lu Juan , Wang Zhouyu, Jin Guosheng. Quality Analysis and Improvement of Backlight Module Based on Pareto Diagrams and Causality Diagram, Forest engineering,2012,28(2):79-82.

[7] FlintS, Cawsey P. Use of Experimental Design and Two Stage Modeling in Calibration Generation for Variable Camshaft Timing Engines[M] . Germany: ExpertVerlag, 2003:44-49.

[8] Zhang Xiaoning, Zhai Lijun, Wang Yunzhe, Zhang Bo The Analysis of Defects Produced by the Two OCA Lamination[J], Advanced Display2011:35~37.

[9] Liu Li shuang , Zhang Yao , Lu Huiqing, A fast sub- pixel edge detection method for Images[J] Journal of Optoelectronics Laser,2005,16（8）:993-996.

[10] JIN Tao , JIA Hong zhi,The extraction method of moving circle and circle center research[J] Laser Journal.,2010, 31(4) :24-26.

A Method of Quality Evaluation for Complexity Equipment Manufacturing and Service Based on Process Area and its Reliability & Validity Analysis

Bang-jun WANG [1], Jin-feng GENG[2], Yuan-guan SHE [3]

[1.] China Aero-Polytechnology Establishment, Beijing, China;

[2.] Quality and Safety Department of Aviation Industry Corporation of China, Beijing, China;

[3.] Dongling School of Economics and Management in University of Science and Technology Beijing, Beijing, China

(E-mail: cqcwbj69@sina.com)

[1]*Abstract* - **In order to improve processes quality, an evaluation method for complexity production and service processes embedded quality requirements of production and service management was presented. Key practices and their sub-practices of production and service processes were corresponded to a set of quantitative evaluation indexes and criterions, while score results were corresponded to the quantitative evaluation of quality management maturity level instead of the traditional qualitative evaluation of quality management system compliance. To verify the practicality of the method, internal consistency reliability, construct validity and criterion related validity were analyzed using the classical test theory and correlation analysis, factor analysis and so on. The result shows, this method is feasible at the enterprise level, and its reliability & validity meet requirements.**

Keywords - **complexity production, manufacturing and service process, quality management, reliability, validity**

I. INTRODUCTION

In china, more than 200 aviation industry enterprises have established the quality management system (QMS) based on GJB 9001 standard, but the efficiency of the QMS is poor generally, documents of the QMS and processes of productions are dichotomy represented as documents mismatching documents and documents mismatching reality. The mode of quality management evaluation is extensive, enterprises mainly use certification and audit to measure the degree of compliance with the GJB 9001 standard without the subdivision of quality management maturity level. Therefore, it is needed to establish an aviation multi-level framework of quality management maturity model under “mother-son- grandson” enterprises based on the process area evaluation, according to the product life cycle and the quality control processes such as supplier control and purchasing, software and hardware development, production and service, scientific research project, test and other process areas. Enterprises judge the effectiveness of their QMS and the maturity level of each process area objectively through the evaluation quality management based on process area, find and analyze weakness, focus on quality improvement, assure the quality of products and services through the work quality, ensure the QMS operation validity and the realization of quality objectives by the process quality.

There are some researches on the evaluation of production and service quality management in literature. For example, the Department of Defense (DoD) developed a software capability maturity model (CMM), HAO Jianchun[1] established a quality management maturity model of aviation enterprise based on CMM, ZENG Jianghui[2] and WANG Bangjun[3] researched the process areas and key practices of the evaluation of aviation equipment technical service quality, Cui Liwei[4] established a model of the validity evaluation for the quality control process based on AHP method, Pei Xiaobing[5] proposed a quality evaluation method of complex product development processes based on the quality gate and CMM model.

The above researches focus on quality management system maturity evaluation, CMMI model method and tools, etc. But researches on the reliability and validity of the evaluation of the quality management of production and service processes are few. The model and index system of quality management and evaluation of production and service processes are established in this paper, which discusses the reliability & validity of the classical measure theory, the method of calculating and analyzing the reliability and validity, and gives you an empirical analysis of the reliability & validity of the evaluation method.

II. EVALUATION MODEL AND INDEX SYSTEM

A. Evaluation model

Drawing on the "four grades of quality" method of McKinsey, the continuous representation method of software development capability maturity (CMM) in Carnegie Mellon University, processes and results of Excellence Performance review, key elements of the maturity of the corresponding relationship description in ISO 9004 standard and so on[6-8], this paper established an evaluation model (Fig. 1) of production and service process quality management, based on the systematic analysis of the current situation of production and service process quality management of complex products, and the relevant standards of quality management in production and service process, and combine with the current situation of the quality management of aviation enterprise and the demand of aviation quality management.

In figure 1, the evaluation of production and service process quality management also uses the maturity model divided into four levels from low to high: basis(A), guaranteed （B）, prevention （C） and excellence

[1] Foundation item: Under the auspices of Beijing Municipal Natural Science Foundation (No. 9144028)

E. Qi et al. (eds.), *Proceedings of the 23rd International Conference on Industrial Engineering and Engineering Management 2016*, DOI 10.2991/978-94-6239-255-7_29

（D）， as shown in Table I. Guaranteed level is divided into three levels :B-1, B-2, B-3; Prevention level is divided into two levels: C-1,C-2; Excellence level is divided into two levels: D-1,D-2.

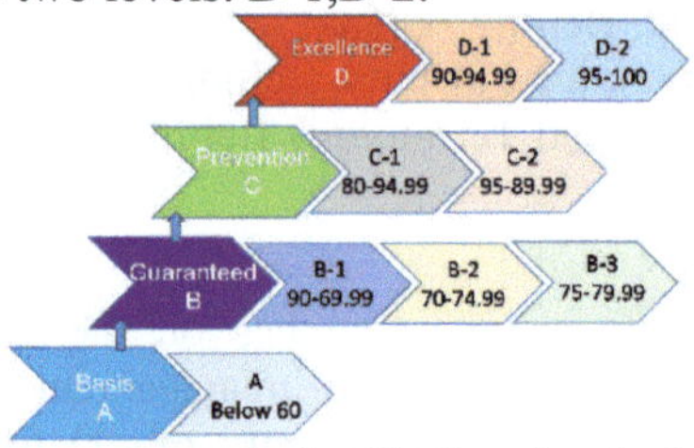

Fig. 1 The maturity grade classification of production and service process quality management evaluation

TABLE I CONNOTATION OF LEVELS FOR PRODUCTION AND SERVICE PROCESS QUALITY MANAGEMENT MATURITY

Quality management maturity	**Connotation of levels**
Basis (A)	The lowest maturity level, basis indicates that the enterprise quality assurance is in the post guarantee, passive state. In base level, success depends on individual effort, processes is disorder, confusion, less definition, and the foundation of quality management is poor.
Guaranteed （B）	The second low maturity level, guaranteed indicates the quality assurance mainly relies on process control, quality management has a certain basis, the process and product implementation of standardization, documentation management, production and service processes are approved and tailored standard operations.
Prevention （C）	The third maturity level, prevention indicates the quality management has translated from afterwards and in matter control to advance prevention, enterprise implements quantitative process control, performance prediction, based on the production and service process quality data analysis and measurement, its quality management capacity is higher.
Excellence （D）	The forth and the highest level of maturity, excellence indicates the quality management come into the innovation and quantify improvement stage, enterprise implements continuous improvement of the processes of production and service, product quality, based on the quantized feedback information, innovative thinking and methods, the quality management level and ability reached a high level.

The method of production and service process quality management maturity evaluation is “quantitative scoring” with factors of processes and results fully considered to determine the maturity level according to the conversion results in centesimal： A （below 60）， B-1 （60-69.99）, B-2（70-74.99）, B-3（75-79.99）,C-1（80-84.99）,C-2（85-89.99）,D-1（90-94.99）,D-2（95-100）.for example, the application score of an enterprise is 480 points and its evaluation score is 420 points, so its centesimal score is 420/480×100=87.50 points, and the production and service process quality management maturity level of the enterprise is C-2.

B. Evaluation index and criteria

The main objectives of the evaluation index system and evaluation criteria of the quality management of production and service processes include:(1)To strengthen the quality target system, responsibility system, assessment system, and to embed the quality requirements to workflow for high quality work to ensure high quality products;(2)To clear product realization process quality control requirements to make the "roots" of existing quality system fine, meticulous, feasible, and to evaluate the operation of the quality system effectiveness, process capability, personnel quality and the work quality; (3)To guide the self-evaluation of enterprises to analysis the weakness quantitatively and work processes detailed, to penetrate quality requirements into all aspects of the work process, to realize the quality management system compliance from qualitative description to quantitative evaluation; (4) To reduce differences between auditors for more scientific and objective evaluation results to improve the effectiveness of the existing quality management system.

The production and service process quality management evaluation index system include 7 primary indicators, and 30 secondary indicators, as shown in Table II. These indicators further development in form evaluation criteria and problems.

The evaluation criteria of production and service quality management compiles according to processes: A is basis level, the vast majority of the current aviation enterprises have passed GJB 9001 certification, their actual quality management is higher than the requirements of the level A; B is guaranteed level, focuses on the evaluation of the conformity and standardization for the existing quality management meeting requirements of national military standards and aviation industry standards; C is prevention level, focuses on the evaluation of the effectiveness of quality management and the basic ability of enterprise quality management; D is Excellence level, focuses on the evaluation of the innovation and efficiency of quality management. Key elements of each process area are based on the framework of GJB 9001B standard. From lower to higher level, the evaluation criteria requirements is specific, strict and deepen gradually, and the content of quality management and the quality of products improve gradually. The example of corresponding evaluation criteria questionnaire of some evaluation indicators are shown in Table III.

III. RELIABILITY ANALYSIS

The purpose of reliability and validity analysis for the evaluation method of production and service process is to verify the validity and the validity of the evaluation model. Reliability and validity are the two basic characteristics of the "four degree" theory framework (reliability, validity, difficulty and discrimination) for the classical measurement theory[10-12]. The former measures the reliability and consistency of the evaluation results that does not involve the correctness of evaluation results; The latter is aimed at the purpose of evaluation focusing on the effectiveness of the evaluation results. The difference between reliability and validity is errors involved. Reliability measures the influence of random errors, but validity measures systematic errors caused by variables that are independent of the purpose of the measurement. Therefore, high reliability does not mean high validity; but a high validity means a large reliability. So in the evaluation practice, reliability is a necessary condition for the validity, not a sufficient condition, an effect evaluation must be a credible evaluation, but the credibility of the evaluation may not be effective.

The evaluation index of the reliability is the reliability coefficient, which can be expressed as the ratio of the variance of the true value and the variance of the evaluation value. But in practice, due to the inability to obtain the real value and variance, people have developed a variety of based on parallel measurement indirect measurement of the reliability coefficient of assumptions, such as the test-retest reliability, split half reliability, internal consistency reliability, alternate-form reliability, scorer reliability etc. Some literatures [13-15] introduced the connotation and calculation methods of various reliability coefficients.

TABLE II PRODUCTION AND SERVICE PROCESS EVALUATION INDEX SYSTEM AND ITS WEIGHT

Primary indicators	Index code	Secondary indicators	Standard points
X0100 Production and service process planning (90)	X0101	Technological design and development	62
	X0102	Confirmation and control of production equipment, tooling and numerical control program	8
	X0103	Validation and verification of production process	20
X0200 Production process control (110)	X0201	Production management	16
	X0202	Process control	58
	X0203	Production process change and configuration management	16
	X0204	Control of facilities, equipment and environmental	20
X0300 Monitoring and measuring equipment control (30)	X0301	Measurement System Capability	4
	X0302	Monitoring and measuring equipment management system	26
X0400 Inspection and testing control (50)	X0401	Inspection and testing	34
	X0402	Control of nonconforming product	16
X0500 Delivery and service control (100)	X0501	Delivery	20
	X0502	Service	80
X0600 Production process monitoring, measurement and continuous improvement (20)	X0601	Production process monitoring and measurement	8
	X0602	Continuous improvement of production process	12
X0700 Enterprise comprehensive evaluation (100)	X0701	General requirements and planning of quality management system	5
	X0702	Management responsibilities	3
	X0703	Quality policy and target management	4
	X0704	Knowledge management	5
	X0705	Control of documents and records	6
	X0706	Human resources	8
	X0707	Customer satisfaction	4
	X0708	Internal quality audit	5
	X0709	Process monitoring and measurement	3
	X0710	Quality information management	10
	X0711	Quality cost management	10
	X0712	Continuous improvement	4
	X0713	Closed loop of quality problem and bill item management	13
	X0714	Quality reward	20
	X0715	Quality punishment	-20

TABLE III EXAMPLE OF EVALUATION CRITERION OF PARTIAL EVALUATION INDEX

Codes	Indicators	Evaluation criteria
2.3	Production process change and configuration management	
2.3.1	Control of production process change	■ Whether production processes changes identified to be subject to the approval of the customer or regulatory authority management department in accordance with the contract or regulatory requirements. ■ Are personnel authorized to approve production process changes? whether the impact of the process, production equipment, tooling and NC program changes are controlled, recorded, and approved? f necessary, whether to carry out the first inspection (identification)? ■ Are temporary changes to the production process of the document effective identification validity or effective range? ■ Are the results of the production process changed to be evaluated in order to confirm that the desired effect has been achieved and that there is no adverse effect on the quality of the product? ■ Is the establishment and implementation of the provisions of the approval process? Are the key or important features of the equipment replacement with the consent of the customer? ■ whether or not to invite customers to participate in the technical appraisal of the test and identification for important process changes?
2.3.2	Configuration identification	■ Is the technical documentation, quality control and inspection data and records of the production process and the requirements for the inspection of the configuration identified?
2.3.3	Configuration accounting	■ Have formal records and reports on the configuration of the identified technical documents, established changes and the implementation of approved changes? ■ Are the configurations of each configuration item accurately recorded in order to ensure traceability?
2.3.4	Configuration control	■ After the establishment of the configuration documents, whether or not to control the state of the configuration of all activities in accordance with the provisions of the project (Included all activities in the assessment, coordination, approval or disapproval of the configuration changes and the impact of the configuration.)

TABLE IV ORIGINAL DATA OF PRODUCTION AND SERVICE PROCESS QUALITY EVALUATION

Index code	A	B	C	D	E	F	G	H	I	J	K	L	M	N	O	P	Q	R	S	T	U	V	Mean scores of self-assessment	Mean score of on site evaluation
X0100	55.6	62.5	47.7	56.8	52.5	43.9	47.6	44	49.6	37.4	54	50.7	50.6	51.5	55.8	46	59.1	50	42.1	65.1	61	42	55.32	51.16
X0101	35.6	39.5	29.5	33.4	33	29.9	30.4	25	28.5	22.2	33.9	31.4	30	35.7	37.5	29.8	38.3	28.5	22.6	44.5	39	27	38.27	32.05
X0102	4	7	5.3	7.7	5.5	1.5	4.7	6	3.8	5.2	6.5	5.3	7.6	3.9	6.3	3	6.5	5.5	7.5	5.1	7	2	5.45	5.31
X0103	16	16	12.9	15.7	14	12.5	12.5	13	17.3	10	13.6	14	13	11.9	12	13.2	14.3	16	12	15.5	15	13	13.28	13.79
X0200	71.4	76	71.2	77.2	74	71	82.9	83	74.8	75.8	72.9	79.4	71.9	70.1	69.3	71.8	61	79	68.5	70.5	71.2	85	75.64	74
X0201	10.7	13	11	13.1	14	5.5	15.2	14	10.5	11.2	9.7	12.5	9	11.6	6.7	11.5	10.5	10	12	10	12.9	13.5	12.36	11.28
X0202	31	36.5	34.2	37.7	38	42.2	40.8	40.5	36.8	37.2	38.7	39.7	34.3	34.1	34.2	35.4	31.5	39	31.8	37.5	35.9	46	38.88	36.95
X0203	13.5	15	14.8	13.2	10.5	9.8	15.4	13	14	14.8	11.8	13.5	12.8	12.5	13.5	12.2	11.7	13.5	8.8	10.5	13	11.5	13.27	12.7
X0204	16.2	11.5	11.2	13.2	11.5	13.5	11.5	15.5	13.5	12.6	12.7	13.7	15.8	11.9	14.9	12.7	7.3	16.5	15.9	12.5	9.4	14	14.39	13.07
X0300	19.9	18	18	20.7	25	17.4	26.9	25	22	25.2	21.5	25	21	18.5	24.5	28.5	21.5	22.5	21.5	20.9	19.5	25	23.35	22.18
X0301	2	4	3.6	2.6	4	3	3.4	3	3.5	3.6	2.8	4	3.4	2	4	4	3.7	3	3.5	3.2	3.5	4	3.34	3.35
X0302	17.9	14	14.4	18.1	21	14.4	23.5	22	18.5	21.6	18.7	21	17.6	16.5	20.5	24.5	17.8	19.5	18	17.7	16	21	19.62	18.83
X0400	39.8	34	42.3	40.1	38.5	29.7	40.1	39	36.3	42.4	39.5	42.5	36.3	34.6	36.6	37.5	33.7	39.5	43.5	39.2	33.3	40	38.96	38.11
X0401	27.5	23	29.1	32.1	30.5	18.5	30.4	27	25.8	28.2	25.5	30	26.5	24.8	26.1	26.9	26.1	32	30	25.7	20.3	28	27.63	27
X0402	12.3	11	13.2	8	8	11.2	9.7	12	10.5	14.2	14	12.5	9.8	9.8	10.5	10.6	7.6	7.5	13.5	13.5	13	12	12.55	11.11
X0500	94.5	90	88.4	79.8	86	63	87.4	78	82	91.6	78.4	69.5	89.3	85.9	87.3	69.8	75	69.5	71.6	34.5	45	56	76.42	76.02
X0501	19	19	20	20	17	14	20	18	18	20	16.5	20	17	20	20	18.5	20	18.5	19	4.5	13	15	17.85	17.59
X0502	75.5	71	68.4	59.8	69	49	67.4	60	64	71.6	61.9	49.5	72.3	65.9	67.3	51.3	55	51	52.6	30	32	41	72.38	58.43
X0600	13	12.5	14	10	17.5	10	10.5	16.5	12	12.6	15	10	10.5	9.3	8	10.5	7.2	9	13.3	6	7	12	16.22	11.2
X0601	3	3.5	2.6	3	5.5	2.5	4	5	4	3.8	4.5	4	4	2.1	3	4	2.5	3	4	1	2.5	6	4.34	3.52
X0602	10	9	11.4	7	12	7.5	6.5	11.5	8	8.8	10.5	6	6.5	7.2	5	6.5	4.7	6	9.3	5	4.5	6	8.83	7.68
X0700	73	70	58.3	52.8	62.5	60.3	69.8	66	64.6	68	64.9	70.7	67.8	54.2	61.1	64.5	53.7	67.4	62	53.9	50.6	60	68.56	62.55
X0701	4	4.5	4	3	3	3.2	4.5	3	5	4.6	4.5	4	4.7	4	5	4.5	3.4	5	4.5	2.3	3.3	3	4.25	3.95
X0702	2.7	2	2.8	2.4	3	2.7	2.5	2	3	2.8	2.7	3	2.5	2.8	3	3	2.5	3	3	1.5	2	3	2.87	2.63
X0703	3.5	4	2.8	4	4	3.3	3.7	4	3.5	3.4	4	4	4	3.5	2.5	4	3.7	4	2.2	1.9	1.3	4	3.68	3.42
X0704	4.4	5	NA[a]	NA	4	NA	NA	5	NA	NA	NA	3	4	NA	NA	NA	NA	NA	NA	NA	NA	5	—	—
X0705	4.7	6	4.6	4.7	4	4.2	5	4.5	5.5	5.4	4.8	4	5.4	5.3	5.1	4.8	3	5.7	3.2	3.9	3.8	5	5.35	4.66
X0706	7.7	5	5.6	3.5	5.5	3.7	5.7	6	6	7.8	6.7	8	6.2	6	5.8	7.8	3.8	6.5	7.5	7	5.3	6	6.64	6.05
X0707	3.4	3	3.1	4	3	3.3	2.5	2	2.5	3.4	2.7	2.5	3.5	2.6	2.5	2	2.4	2.7	2.5	2.5	2.4	0	2.57	2.66
X0708	4.4	4.5	4.8	3.2	2	4	4.4	5	5	4.4	4.2	3.8	4.5	4.3	4.5	4	3.7	5	4.7	3.5	2.3	3	4.39	4.05
X0709	2.7	2.5	1.8	1.5	0.5	2.1	2	1.5	1.7	2.6	2.5	2.8	1.5	0	2.5	2.5	1.5	3	1.5	0	0.5	1	2.68	1.74
X0710	8.5	8	5	8.4	10	10	9.5	9.5	8	9.2	8.2	8.1	8.5	7.6	6	7.5	7.5	9	6.7	9.5	6.7	5	8.88	8.02
X0711	8	7.5	6.2	8.1	3.5	8.1	8	7.5	7.9	9	7	8	6	8.8	7	8	6.5	7.5	9.5	5.8	6	6	7.95	7.27
X0712	3.6	1.5	2	4	4	3	3.5	2	3	3.8	4	3.5	3.5	1.8	2.5	3	3	3.5	4	2.5	3	4	3.86	3.12
X0713	11.4	11.5	11.6	0	13	12.7	12.5	10	12.5	11.6	11.6	13	10.5	7.5	11.7	12.4	12.7	11.5	12.7	11.5	12	11	11.27	11.13
X0714	4	5	4	6	3	0	6	4	1	0	2	3	3	0	3	1	0	1	0	2	2	4	3.32	2.45
X0715	0	0	0	0	0	0	0	0	0	0	0	0	0	-2	0	0	0	0	0	0	0	0	0	-0.09

[a] *Note: NA means "not applicable".*

This paper focuses on the analysis of the internal consistency reliability of production and service process quality management evaluation. The correlation degree between each evaluation item in the evaluation scale is usually measured by Cronbach -α coefficient, and the calculation method is as follows:

Assuming there are k evaluation items, the evaluation values were x_1, x_2, …, x_k, then $x_1=t_1+e_1$, …, $x_k=t_k+e_k$, and the evaluation score is $X=x_1+x_2+\ldots+x_k$, the true total score $T=t_1+t_2+\ldots+t_k$, thus,

$$\alpha=\frac{k}{k-1}\left(1-\frac{\sum_{i=1}^{k}S_i^2}{S_X^2}\right)$$

Among them, S^2_i is the variance of the i evaluation score of all the evaluation objects; S^2_X is the variance of all the evaluation objects and all the evaluation items. Table IV is the results of the production and service process quality evaluation of 22 enterprises in china aviation industry.

We coordinate the original data of quality management evaluation in Table IV, delete the index "X0704 Knowledge management", "X0714 Quality Award", "X0715 Quality punishment" and other 3 indicators that do not carry on the analysis of reliability and validity for the main reason: (1) 15 of 22 enterprises have applied and approved for tailoring the index X0704 Knowledge management, so the missing data of the index is more;(2) "X0714 Quality Award", "X0715 Quality punishment" for the enterprise's reward and punishment, the number and level is objective data that does not subject to the subjective judgment impact of auditors on site evaluation. For the remaining 34 evaluation indicators, run SPSS software according to the evaluation data of 22 sample enterprises, select "Analysis - Scale - Reliability Analysis - Model - alpha" in turn, get the main results of the internal consistency reliability analysis including:

(1) Reliability Statistics (shown in Table V)

(2) Item-Total Statistics (Scale if item deleted) shown in Table VI

(3) ANOVA Table (shown in Table VII) etc.

Cronbach's Alpha coefficient is between 0 and 1, a higher alpha value indicates a higher internal consistency reliability evaluated. Generally speaking, the internal consistency of alpha is better than 0.8, which indicates good agreement between 0.8 and 0.6, and less than 0.6 indicates poor internal consistency. Here, we get the Cronbach's alpha coefficient is 0.675 that indicates the reliability is not very high, which may be caused by the difference in the dimension. We carry out the standard transformation of the original data, that is, after subtracting the mean and then divided by the standard deviation, the indicators are obtained after the conversion of the variables, then calculate Cronbach's Alpha based on the standard sub-index and get an Alpha coefficient of 0.798 belonging to good internal consistency of category.

Table VI shows that if item deleted the impact is not affected on the overall scale of the Cronbach's alpha reliability coefficient, It is worth noting that some items deleted result in the scale of the amount of the Cronbach's alpha reliability coefficient increased to more than the original 0.675, but the increase is very limited. Further variance analysis is carried out as shown in Table VII.

TABLE V RELIABILITY STATISTICS FOR THE QUALITY EVALUATION OF PRODUCTION AND SERVICE PROCESSES

Cronbach's Alpha coefficient	Cronbach's Alpha coefficient based on the standardization index	Number of indicators
0.675	0.798	34

TABLE VI ITEM-TOTAL STATISTICS FOR SCALE IF ITEM DELETED

Index code	Scale mean if item deleted	Scale variance if item deleted	Corrected item-total correlation	Cronbach's Alpha if item deleted	Changes of reliability
X0100	615.4409	2167.742	-0.351	0.730	decreased
X0101	634.5455	2169.727	-0.412	0.721	decreased
X0102	661.2864	1960.351	0.139	0.673	increased
X0103	652.8091	1993.470	-0.076	0.679	decreased
X0200	592.6045	1827.083	0.268	0.663	increased
X0201	655.3227	1929.214	0.245	0.669	increased
X0202	629.6455	2023.065	-0.158	0.689	decreased
X0203	653.9045	1897.412	0.558	0.661	increased
X0204	653.5318	1927.364	0.262	0.668	increased
X0300	644.4182	1905.901	0.258	0.666	increased
X0301	663.2455	1985.369	-0.021	0.676	decreased
X0302	647.7727	1904.183	0.290	0.665	increased
X0400	628.4909	1864.295	0.358	0.660	increased
X0401	639.6000	1841.036	0.450	0.655	increased
X0402	655.4909	2011.780	-0.167	0.683	decreased
X0500	590.5773	1049.887	0.694	0.595	increased
X0501	649.0091	1784.987	0.619	0.644	increased
X0502	608.1682	1185.911	0.723	0.578	increased
X0600	655.4000	1833.467	0.555	0.652	increased
X0601	663.0773	1936.671	0.458	0.668	increased
X0602	658.9227	1879.175	0.498	0.659	increased
X0700	604.0500	1576.503	0.711	0.611	increased
X0701	662.6455	1947.596	0.515	0.669	increased
X0702	663.9682	1973.143	0.300	0.674	increased
X0703	663.1773	1943.352	0.584	0.669	increased
X0705	661.9364	1949.129	0.504	0.670	increased
X0706	660.5500	1953.783	0.252	0.671	increased
X0707	663.9409	1966.122	0.260	0.673	increased
X0708	662.5455	1955.340	0.391	0.671	increased
X0709	664.8636	1942.147	0.528	0.669	increased
X0710	658.5818	1962.523	0.156	0.673	increased
X0711	659.3318	1968.344	0.122	0.674	increased
X0712	663.4773	1977.961	0.087	0.675	increased
X0713	655.4682	1997.320	-0.082	0.682	decreased

TABLE VII ANALYSIS OF VARIANCE (ANOVA) OF PRODUCTION AND SERVICE PROCESS QUALITY MANAGEMENT EVALUATION

		Sum of squares	df	Mean square	F	Sig
Between people		1225.788	21	58.371		
Within people	Between items	335254.906	33	10159.240	535.999	0.00
	Residual	13135.020	693	18.954		
	Total	348389.926	726	479.876		
Total		349615.714	747	468.026		

Grand mean=19.6059

The analysis of Table VII shows that the p value between items of the production and service process quality evaluation is close to 0, which means there is significant difference between assessment items.

IV. VALIDITY ANALYSIS

Validity refers to the accuracy of the evaluation results that is the closeness between the evaluation value and the real value. The evaluation index of validity is the validity coefficient. The commonly used method is content validity, criterion related validity, construct validity and so on. It reflects the validity of the evaluation from different perspectives. Validity analysis should follow certain sequence: firstly, evaluate content validity; secondly, extract representative latent variables (construct validity); then to assess the peer evaluation results and effective standard of correlation (criterion related validity).

Content validity analysis score items can be reflected by the content or subject. The correlation coefficient between the score and the total score is usually used to reflect the content validity. In general practice by expert judgment method, by the expert review, and use R to determine the content validity content validity

In practice content validity determine through expert judgment method by expert review and use content validity ratio R to indicate the content validity. If the correlation coefficient is not significant, in other word, if the discrimination of an item is low, the item should not be included in the evaluation scoring table. The calculation formula for content validity is:

$$R = 2\times\frac{N_e - \frac{N}{2}}{N}$$

Among them, N_e is the number of experts who consider the content to be a good evaluation; N is the total number of experts; R value varies from -1 to 1, -1 means that all people think that the content is improper, and 1 means that all people think that the content is very good.

Considering the development process of production and service quality management evaluation model, index system and evaluation standards has repeatedly in the full range of industries to solicit expert opinion, this paper does not analysis its content validity.

A. Construct validity analysis

The statistical method of construct validity analysis is commonly factor analysis, the mathematical model given in reference is $Z=AF+\varepsilon$, i.e:

$$\begin{bmatrix} Z_1 \\ Z_2 \\ \vdots \\ Z_n \end{bmatrix} = \begin{bmatrix} a_{11} & a_{12} & \cdots & a_{1m} \\ a_{21} & a_{22} & \cdots & a_{2m} \\ \vdots & \vdots & \ddots & \vdots \\ a_{p1} & a_{p2} & \cdots & a_{pm} \end{bmatrix} \begin{bmatrix} F_1 \\ F_2 \\ \vdots \\ F_m \end{bmatrix} + \begin{bmatrix} \varepsilon_1 \\ \varepsilon_2 \\ \vdots \\ \varepsilon_m \end{bmatrix}$$

Here, p is the number of evaluation items; m is the sample size; Z is the evaluation vector X composed of the evaluation results of $X_1, X_2, \ldots, X_p$ that is normalized by transformation (The standard transformation method is to subtract its mean, and then divided by the standard deviation, that is $Z_i=X_i-\mu_i$, the mean of Z_i is 0, the standard deviation of Z_i is 1), and Z is represented by the linear combination of F; F is the components extracted; ε as a special factor Z is similar to the error, the mean of ε_i is 0, andε_i is not associated with component F; A is the load matrix of relationship between the evaluation indicators and components extracted, a_{ij} is the element of A, which indicates the correlation coefficient (in psychology, it is called "load") of the i evaluation indicator and the j component extracted.

If the correlation coefficient matrix obtained from the factor analysis of construct validity shows that the correlation between the evaluation items of the same structure variable is not significant, as means the evaluation scale needs to be improved. The relationship between load matrix indicates evaluation items and components. If the same structure variable has only one component extracted, and the component load is big, then the construct validity is high. On the contrary, if there are two or more common component extracted, or the value of the load is smaller, the assessment scale or evaluation process needs to further analysis.

Run the SPSS software to carry on the factor analysis based on the data in Table IV, and get the construct validity analysis results of production and service quality management, which includes descriptive statistics, correlation coefficient matrix, the construct validity of total variance contribution rate of the production and service quality evaluation analysis (shown in Table VIII), the construct validity analysis component load matrix of production and service process quality evaluation (shown in Table IX), and factor score function coefficient matrix etc.

Table VIII shows there are 9 components with an eigenvalue greater than 1 in the correlation coefficient matrix, and the cumulative percentage of variance for all 7 components extracted has reached 74.04%.

Table IX shows that there are 21 indicators accounting for 77.78% of all 27 indicators with a load value is greater than 0.5, and the overall evaluation scale has a good construct validity. But the result also shows that the construct validity of 6 secondary indicators is not good (component load value is less than 0.5) and needs further analysis. These secondary indicators and corresponding primary indicators are 2 secondary indicators of "X0101 technological design and development" and "X0103 Validation and verification of production process" in "X0100 Production and service process planning"; " X0203Production process change and configuration management" in "X0200 Production process control";" X0602 Continuous improvement of production process" in "X0600 Production process monitoring, measurement and continuous improvement"; "X0706 Human resources" and "X0711 Quality cost management" in "X0700 Enterprise comprehensive evaluation".

TABLE VIII TOTAL VARIANCE EXPLAINED IN CONSTRUCT VALIDITY ANALYSIS OF PRODUCTION AND SERVICE

Component	Initial Eigenvalues			Sums of squared loadings			Rotation sums of squared loadings		
	Total	% of variance	cumulative %	Total	% of variance	Cumulative %	Total	% of variance	Cumulative %
1	5.794	21.459	21.459	5.794	21.459	21.459	4.689	17.368	17.368
2	3.891	14.412	35.872	3.891	14.412	35.872	3.461	12.818	30.186
3	2.927	10.839	46.711	2.927	10.839	46.711	2.554	9.458	39.644
4	2.140	7.924	54.635	2.140	7.924	54.635	2.487	9.213	48.857
5	1.996	7.392	62.028	1.996	7.392	62.028	2.423	8.972	57.829
6	1.692	6.266	68.294	1.692	6.266	68.294	2.235	8.279	66.108
7	1.550	5.742	74.036	1.550	5.742	74.036	2.141	7.928	74.036
8	1.448	5.363	79.399						

9	1.077	3.990	83.389

TABLE IX ROTATED COMPONENT MATRIX FOR 7 COMPONENTS IN CONSTRUCT VALIDITY ANALYSIS OF PRODUCTION AND SERVICE

Index code	Component 1	2	3	4	5	6	7
X0101	-0.689	-0.371	0.142	0.230	-0.109	-0.047	0.155
X0102	-0.043	-0.338	0.295	0.185	0.586	0.174	0.233
X0103	-0.260	-0.223	0.336	0.008	-0.224	-0.046	0.324
X0201	0.077	0.400	0.509	0.180	0.156	0.552	0.230
X0202	0.002	0.595	0.139	-0.202	-0.618	0.099	-0.030
X0203	0.435	-0.379	0.185	0.437	-0.293	0.356	0.248
X0204	0.510	0.028	-0.221	-0.546	-0.063	-0.037	0.215
X0301	0.016	0.477	-0.136	0.662	0.019	-0.029	-0.073
X0302	0.448	0.615	0.089	0.008	-0.013	0.013	0.325
X0401	0.518	0.261	0.405	-0.042	0.369	-0.008	0.388
X0402	-0.075	0.099	-0.677	-0.167	0.162	0.558	-0.026
X0501	0.764	-0.208	0.228	0.330	0.205	-0.121	-0.104
X0502	0.673	-0.420	0.274	0.097	0.112	0.126	-0.372
X0601	0.512	0.691	0.216	0.060	-0.009	0.169	-0.222
X0602	0.372	-0.013	0.159	-0.259	0.239	0.494	-0.584
X0701	0.729	-0.320	-0.311	0.295	-0.004	-0.152	0.114
X0702	0.649	0.276	-0.152	0.113	0.131	-0.509	-0.186
X0703	0.588	0.151	0.525	-0.068	-0.350	-0.096	-0.157
X0705	0.530	-0.285	0.088	0.078	-0.598	0.167	0.099
X0706	0.400	0.240	-0.513	-0.161	0.202	0.288	0.348
X0707	0.104	0.710	0.263	-0.282	0.283	-0.083	-0.146
X0708	0.605	-0.474	-0.372	-0.080	-0.156	0.134	0.055
X0709	0.698	-0.180	-0.197	0.068	-0.145	-0.165	0.079
X0710	0.017	-0.081	0.329	0.587	-0.185	0.035	-0.083
X0711	0.469	-0.323	-0.328	-0.275	-0.007	-0.023	0.253
X0712	0.243	0.505	0.164	-0.321	0.349	-0.376	0.148
X0713	0.027	0.292	0.585	0.286	-0.024	0.008	-0.277

B. Criterion related validity analysis

To analysis the criterion related validity, we select a direct correlation with the evaluation of the results of the independent standard used as the independent variable, according to the theory has been mastered. Then, the relationship between the evaluation results and the independent variables is analyzed, if the results indicates that the evaluation results show a strong correlation with the theory mastered, the evaluation is effective.

If X expresses the evaluation results, C indicates the criterion related, $D(X)$ and $D(C)$ are variance of X and C, cov (X, C) is the covariance of X and C, we can get the criterion related validity coefficient R by calculating the correlation coefficient of Pearson[17].

$$R = \frac{\operatorname{cov}(X,C)}{\sqrt{D(X)D(C)}} = \frac{\sum(X_i - \overline{X})(C_i - \overline{C})}{\sqrt{\sum(X_i - \overline{X})^2 \sum(C_i - \overline{C})^2}}$$

Run the SPSS software based on data in table 4 to calculate the Pearson's correlation coefficient using self-assessment score average value as the criterion, and get the criterion related validity between the mean scores of self-assessment and primary indicators. Calculation results are shown in Table X.

Table X shows, it is highly correlated between the mean score of enterprise self-assessment and the evaluation score of primary indicators, there are 6 of 7 indicates at the 0.01 level significant correlation, and 1 at the 0.05 level, which indicates that the evaluation index system has a higher criterion related validity between enterprise self-assessment and expert evaluation.

The calculation results of the criterion related validity between primary indicators and secondary indicators shown in Table XI. There are 6 of 7 primary indicators are highly correlated with their all secondary indicators by the significant test (P- value less than 0.05); There is only 1 primary indicator "X0700 comprehensive evaluation" failed in the significance test of correlation coefficients with its 4 secondary indicators ("X0702 management responsibilities", "X0707 customer satisfaction", "X0712 continuous improvement", and "X0713 Closed loop of quality problem and bill item management").

TABLE X THE CRITERION RELATED VALIDITY BETWEEN MEAN SCORES OF SELF-ASSESSMENT AND PRIMARY INDICATORS

		X0100	X0200	X0300	X0400	X0500	X0600	X0700
Pearson correlation coefficient	**Mean score of self-assessment**	0.5001^{*}	0.7616^{**}	0.5949^{**}	0.6150^{**}	0.6858^{**}	0.5804^{**}	0.9009^{**}
Significant (bilateral) P- value		0.011	0.000	0.002	0.001	0.000	0.002	0.000

TABLE XI THE CRITERION RELATED VALIDITY BETWEEN PRIMARY INDICATORS AND SECONDARY INDICATORS

		X0101	X0102	X0103		X0201	X0202	X0203	X0204		X0301	X0302
Pearson correlation coefficient	**X0100**	0.9733^{**}	0.4929^{*}	0.7792^{**}	**X0200**	0.6090^{**}	0.9172^{**}	0.5043^{**}	0.6175^{**}	**X0300**	0.4928^{*}	0.9858^{*}
Significant (bilateral) P- value		0.000	0.012	0.000		0.001	0.000	0.010	0.000		0.014	0.022

		X0401	X0402		X0501	X0502		X0601	X0602
Pearson correlation coefficient	**X0400**	0.8390^{**}	0.5047^{**}	**X0500**	0.8023^{**}	0.9654^{**}	**X0600**	0.7929^{**}	0.9176^{**}
Significant (bilateral) P- value		0.000	0.010		0.000	0.000		0.000	0.000

		X0701	X0702	X0703	X0705	X0706	X0707	X0708	X0709	X0710	X0711	X0712	X0713
Pearson correlation coefficient	**X0700**	0.5495^{**}	0.3238	0.4437^{*}	0.5417^{**}	0.5528^{**}	0.3232	0.4898^{*}	0.6562^{**}	0.4127^{*}	0.4625^{*}	0.2926	0.3393
Significant (bilateral) P- value		0.004	0.114	0.026	0.005	0.004	0.115	0.013	0.000	0.040	0.020	0.156	0.090

Therefore, there is no evidence shown that these 4 secondary indicators and their corresponding primary indicator have a highly significant correlation. However, considering these secondary indicators have nice results in the reliability analysis and construct validity, these secondary indicators should be retained in the follow-up evaluation process temporarily, but the corresponding evaluation criteria and problems should be revised to further improve the efficiency of the relevant indicators.

V. CONCLUSION

The production and service provision process is the main process of product quality formation. It is of great significance to implement the evaluation of production and service process quality management, to standardize enterprise quality management activities, and to ensure product quality of the overall promotion. For a long time, the national defense science and technology industry has attached great importance to product and process quality management evaluation methods, but the reliability and validity of the management evaluation method is not in-depth study. The innovation of this paper lies in the introduction of reliability and validity analysis method to carry out "the evaluation of the evaluation". The reliability and validity analysis results show that the reliability and validity of the evaluation model, index system and criterion established for production and service process quality management can be popularized and applied in large area. The follow-up study includes: To revise the relevant evaluation criteria, problem list, and further improve the reliability and validity of the evaluation method; To analysis the scorer reliability and validity to reduce differences between auditors.

VI. REFERENCES

[1] HAO Jianchun, YIN Taiwei, LI Bing. Research on quality management maturity model of aviation enterprise based on process area evaluation [A]. The 5th China quality academic and Innovation Forum[C].2012.

[2] Zeng Jianghui, Wang Bangjun, Wu Yueqin, et al. Study on the evaluation model for technical service quality of aviation equipment based on process area [J]. Standard Science,2011,(7): 76-78.

[3] Wang Bangjun. Management and evaluation techniques of service quality [J]. Aeronautic Standardization & Quality, 2007, (4):13-19.

[4] Cui Liwei, Ji Daqin, Xu Yuanxin. A model of the validity evaluation for the quality control process based on AHP [J]. Natural Sciences Journal of Harbin Normal University, 2011, 27(4):32-36.

[5] Method of quality evaluation based on quality gate and CMM in the development process of complex product [J]. Science and Technology Management Research, 2015, 35(20):57-62.

[6] Technical foundation of electronic information base of General Equipment Department. GJB 5000A-2008 Military software development capability maturity model[S].Beijing: The General Armament Department of PLA China, 2008.

[7] National Technical Committee 151 on quality management and quality assurance of Standardization Administration of China (SAC/TC151). GB/T 19580-2012 Criteria for performance excellence[S] . Beijing: Standards Press of China, 2012.

[8] National Technical Committee 151 on quality management and quality assurance of Standardization Administration of China (SAC/TC151). GB/T 19004-2011 Managing for the sustained success of an organization: A quality management approach [S]. Beijing: Standards Press of China,2012.

[9] Aviation Industry Corporation of China. QAVIC 09223-2012 production and severs process quality management evaluation criteria[S].

[10] Cronbach L J, Rajaratnam N, Gleser G C. Theory of generalizability: A liberalization of reliability theory [J], British Journal of Statistical Psychology, 1963, 16(2): 137-163.

[11] Spearman C E. 'General intelligence' objectively determined and measured. [J], The American Journal of Psychology, 1904, 15(2): 201-292.

[12] Cole D A. Utility of confirmatory factor analysis in test validation research [J], Journal of Consulting and Clinical Psychology, 1987, 55(4): 584-594.

[13] QIN Lei, YUAN Denghua. Applying generalizability theory to the performance assessment [J]. Psychological Science 2005, 28(3):650-651.

[14] CHEN Zhuang, LIU Yongmei, LIU Fei. An analysis method of reliability and availability based on CIMS engineering marketing [J]. Computer Integrated Manufacturing Systems, 2002,8(6):421-424.

[15] LIU Yu, WANG Bangjun, Jin Chunhua. Method and empirical study on reliability and validity analysis for bid evaluation in engineering projects [J]. The Journal of Quantitative & Technical Economics, 2009,(11):128-140.

[16] Gao Huixuan. Applied multivariate statistical analysis [M]. Beijing: Peking University Press. 2005:295-300.

[17] Teaching and research section of probability and statistics in Tongji University. Probability and statistics (3th edition) [M].Shanghai: Tongji University Press, 2004:103-110.

Analyzing "Made in China 2025" Under the Background of "Industry 4.0"

Jing Li*
School of Economy and Management, Shenyang Institute of Engineering, Shenyang, 110136, P.R.China
(*e-mail: 1653351679@qq.com)

[1]*Abstract* –**Facing the unavoidable arrival of "Industry 4.0" age, deeply understanding the strategic development plan of "Made in China 2025" from Chinese central government, revitalizing China's manufacturing industry, achieving smooth deceleration shift of our industries, achieving the implementation of the first decade action plan for building stronger country strategy, is worthy of careful thought and choices.**

Keywords - **Industry 4.0; Made in China 2025**

I. THE BASIC CONNOTATION OF "INDUSTRY 4.0" AND "MADE IN CHINA" STRATEGIC PLANNING

German "Industry 4.0" has becoming a worldwide popular term. Someone even predicted that Taobao would become obsolete in ten years, which caused quite a stir in the financial sector. At the same time, it sets off the silent fourth industrial revolution.

A. "Industry 4.0"

The concept of "Industry 4.0" is a division based on different stages of the industrial development. Industry 1.0 is the steaming engines era, industry 2.0 is the electrification era, industry 3.0 is the information era, while industry 4.0 is an era utilizing informative technology to promote industrial revolution, which is the era of intelligent. The core purpose of industry 4.0 is to improve the competitiveness ofindustry, and to occupate the initiatives in the new round of industry revolution.

The vast majority of developed countries have been whacked after 2008 worldwide financial crisis, with domestic economy dispirited. But Germany has become particularly "different". Domestic employment rate stay steady, the economy continues to set records, made the German government re-acquainted with the country's manufacturing sector, called "sky column." Subsequently, the German government announced an "upholding the future of German manufacturing: reimplementation of industrial four strategy", citing "Industry 4.0" for the first time. "Industry 4.0" utilize intelligent network platform to improve the level of manufacturing technology, a "bottom-up " production system revolution with the goal of shifting manufacturing to intelligence. It takes full advantage of Internet technology to achieve real-time production monitoring and management, greatly reducing the time and cost of technology innovation, creating new market opportunities, digging deeper potential.

To realize Industry 4.0, it is necessary to carry out a revolution inscience and technology, but the current base of technology and experience is far from enough. We therefore mustchange and renovate the key equipmentsin industries, exploring new geographic markets and to seek new solutions. If successful, this will make Germany maintain the stable development of the domestic manufacturing industry, at the same time, make fundamental changes. In the competitive world of manufacturing, Germany has a strong ability to lead the global manufacturing industry. This is not just due to Germany's "excessive" focus on industrial products, innovative research and development, but also thanks to its strict management system. Germany occupies an important positionin the world,in the field of information technology. It not only has strong production equipment and plant systems, but alsohas profound research and application in embedded systems and automation engineering and other special technology. Because of these factors, Germany occupies a very high position in the manufacturing engineering industry. That is why Germany proposes a new industrial era to suit their development: the era of Industry 4.0.

B. "Made in China 2025"

In 2013 "manufacturing power country strategy & study" consulting project was launched. At the end of 2014, the chairman of China Academy of Engineering, Zhou Ji said that the "China 2025" overall plan that China is working on, will become the overall development strategy of China's manufacturing industry. China is striving to enter the ranks of manufacturing power country by year of 2025. March 4th, theChinese People's Political Consultative Conference (CPPCC) group discussion mentioned the "Made in China 2025" plan that is under fomulation. This year's government work report pointed out to promote the industrial structure towards middle to high end. With manufacturing industry be China's leading industry, to accelerate the pace of converting from manufacturing big contry to manufacutring power country, we must absolutely implement the "Made in China", persist in reform and innovation-driven, intelligent transformation of and improve the foundation reinforcement, environmental protection and green development. Government will take

[1] 2015 the Key Project of the Social Science Foundation of Liaoning Province: The Study on the Development Strategy of Manufacturing Innovation Center in Liaoning Province Based on "Made in China 2025" (No. L15AJY011)

E. Qi et al. (eds.), *Proceedings of the 23rd International Conference on Industrial Engineering and Engineering Management 2016,* DOI 10.2991/978-94-6239-255-7_30

positive economic measures to promote the transformation of traditional industries. We should adhere to the parallel of guarantee and pressure, ease and eventually resolve the overcapacity situation. The principle of market competition is that the fittest survives. The government will support the manufacturing enterprise mergers and restructuring. To realize Industrial informatization, we must be integrated theinternet system, break the restricted area, win the opportunities, and to realize intelligent development. In a global context of industrial 4.0 concept, we are going to compete with manufacturing industries in the United States, Germany, and the develping countries in the world seize the opportunities. "Made in China 2025" will be a programmatic document guiding the transfermation of manufacturing industry to intelligent manufacturing industry in China.

At this stage, the major characteristic of China's manufacturing industry is being big but not robust. According to authoritative data as of 2014, China's manufacturing output is accounted for 20.8% of global manufacturing, industrial added value of 22.8 yuan. China has become the second largest country for global manufacturing. But there is stilla very prominent issue in China's manufacturing industry: the ability of independent innovation is relatively behind; the mechanical core components and enterprises' core technologies highly depend on foreign enterprises; the industrial structure is not sound; the industrial mechanism is not ideal. There are still large differences comparing China's manufacturing industry to the manufacturing in the United States, Japan, Germany and other developed countries.

Arrival of the new century, the development of China's thirteenth Five-Year Plan, in particular the "Made in China 2025" plan proposed, will accelerate the restructuring and development of China's manufacturing industry. China will seize this rare opportunity for development, to overtake at the corner, to catch up, and to realize the strong country dream.

II. THE DIFFERENCES BETWEEN “MADE IN CHINA 2025” AND “INDUSTRY 4.0” AS WELL AS THE “MADE IN CHINA” VISION

In the Hannover Messe, the Government of the Federal Republic of Germany, creatively proposed "Industry 4.0" strategy, aimed at improving the competitiveness of German industry, and to play a leading role in the future of the industrial revolution.

A. The Differences Between “Made in China 2025 " and " Industry 4.0 "

In 2014, "Industry 4.0" concept was highly regarded around the world as soon as it was proposed. Countries have begun to follow suit, striving to be the vanguard of the manufacturing sector.

For example, the United States so-called "re-industrialization", "manufacturing renaissance" and "Advanced Manufacturing Partnership Project"; Japan doesn’t want to fall behind, and started "the revival strategy" implementation and published the "robot new strategy” in January 23, 2015; South Korea staged a "new strategic dynamic strategy"; France also proposes "new industrial France" and so on. In order to adapt to changes in the level of science and technology in the world, at the beginning of 2014, China is also conceiving China's Industry 4.0, and in 2015 during CPPCC National Committee and NPC plenary sessions ("National Lianghui"), the Prime Minister Li Keqiang formally proposed the "Made in China 2025" Decade of Development Strategy. Li Keqiang in his government work report stressed that for China to implement the "Made in China 2025" development strategy, we must seize the opportunities, under the background of the manufacturing industry returning to Germany and America’s "Industry 4.0", accelerate industrial reform, achieve the Chinese dream of industry power country. For this we believe that the fourth industrial revolution will fire the first shot in 2015. The global industrial economy and people's lives will have dramatic changes from this.

The German Federal Ministry of Education and Research and the Federal Ministry of Economics and Technology, who first proposed "Industry 4.0", will jointly invest huge amounts of money, going to use 200 million euros as a high-tech strategic planning project investment fund. The goal is to take Germany intelligent level of manufacturing to the next level, to establish high-tech, high-performance, high-efficiency intelligent plant, to adapt to the development of resource efficiency and human factors engineering, to use network platform like the internet plus and others, to integratethe customers and business partnersin the value chain and commodities chain.

During Chinese Premier Li Keqiang’s visit to Germany October 10, 2014, the two countries jointly created "China-Germany Collaboration Brief: Bilateral Inovation", which also indicates that "Industry 4.0" will become the new direction of development cooperation between the two countries in the future. China and Germany have high complements in many fields. The perfect combination of China’s productivity and Germany’s science and technology will create the future of joint innovation and development of the world's two strongest manufacturing countries. China had earlier get rid of the label of "hundreds of millions of shirts in exchange for one plane" label. After centuries industrial development, Germany’s manufacturing is well-developed. However, it still needs the new power from informatization. As China continues to accelerate the pace of industrial upgrading , we have changed the original trade structure. In this context, the bilateral economic and trade cooperation will be deepened, innovated and developed. President Xi Jinping’s Germany visit in 2014 also makes it clear that China and Germany will joined forces together to create a new pattern of global manufacturing.

The main idea of the Industry 4.0 that we are talking about is: realizing informatization. China-Germany co-production is to rely on the informatization and realize upgrades for themselves. In the past year, how to converted the EU market into an informatized unified market has been bothered the whole Europe. The China-Germany cooperation at this time brings new opportunitiesis for the two sides, and reached a consensus. Conformist manufacturing and management is no longer suitable for the development of today's new era of information. Therefore, China and Germany can take the advantage of "Industry 4.0" platform to innovate and develop. It not only can achieve the win-win situation, but also lay the foundation for future EU-China coalition.

But we should clearly realize that the industrial development of China is not at the same level, in comparison with Germany and other European countries. The challenges still exist. China's industrialization level is still very low. Therefore the road ahead for development is still tough. For the two markets of Germany and China, if there is a uniform standard, both sides will reduce a lot of unnecessary costs and lossesin the future cooperation and exchange process.

B. “Industry 4.0” The “Made in China” vision

Every industrial Revolution drives groundbreaking changes in technology and industry, and completely changes people's way of life. The first industrial revolution achieved replacing manual work with the machinery manufacturing; the second industrial revolution made mass production a reality; the third industrial revolution completed freed people and achieved production automation. So, what challenges will the fourth industrial revolution bring to human beings?

The globalization of Industrial 4.0, to a certain extent, promotes the development of manufacturing technology in the developed countries, and further accelerates the returning trend of manufacturing enterprises. During the past two years in China, prices and human capitals have undergone tremendous changes, which makes China, the original "big country of manufacturing" face a factory outflow situation, while the low-cost India and developing countries in Southeast Asia are attracting a large number of factories. This makes the title of "Made in China" gradually replaced by “Made in Southeast Asia”. The industrial division structure in the world would change dramatically. A new round of industrial revolution will change the comparative advantage of the countries, and rebuild the structure.

From this we can conclude that: state of competitive advantage and industrial structure in countries are under changes by a new round of industrial revolution. Facing such a situation, China's manufacturing industry must adapt to new changes, develop the networking and intelligent enterprises, integrate industry and computer, create an innovative cost advantage, and enhance institutional innovation and the ability to manage changes, so that China can be at the pace of global industrial revolution, and be on a new path of development of "Chinese intelligence made".

REFERENCES

[1] Min Wang . Liaoning: Manufacturing change to " intelligence-made" [J] Observation of the northeast, 2015, 07:49 .

[2] Hui Yang. All the chiefs chatting about Industry 4.0 Opportunities and Challenges [J]. Electronic Technoligy Application, 2015,05: 5-6.

[3] Zhi Lu, Tai Li. Industrial 4.0 era manufacturing industry intelligent analysis [J]. Academic Reports of Shangqiu Vocational Technical institute, 2015,02: 32-34 .

Study on the Implementation Strategy of Liaoning Manufacturing Collaborative Innovation System Based on "Made in China 2025"

Tong-bing Ma* Jing-hong Xu
School of Economy and Management, Shenyang Institute of Engineering, Shenyang, 110136, P.R.China
(*e-mail: tbma69@126.com)

[1]*Abstract* –**"Made in China 2025" is for the whole world and China's economic and social development and changes, the overall trend of industry innovation, make long-term strategic development plan of China's development. This paper introduces the connotation of collaborative innovation method, followed by analysis of industrial engineering and management innovation, combined with the status quo and development in Liaoning manufacturing problems in the Liaoning manufacturing collaborative innovation system, based on the exploration of "Made in China 2025" Liaoning Manufacturing industry collaborative innovation model, Finally, the implementation of the development strategy based on creating synergy Shenyang equipment manufacturing industry.**

Keywords –**Liaoning Manufacturing, Collaborative Innovation, Made in China 2025, Implementation strategy**

I. INTRODUCTION

"Made in China 2025" the core guiding principle is: adhere to the type of industrialization with Chinese characteristics to achieve the equipment manufacturing industry innovation and development as the theme, industrial engineering techniques and methods as the basis, improve the quality and increase efficiency, to accelerate new generation of information technology and manufacturing depth of integration as the main line, in order to advance the main direction for the development of intelligent manufacturing, equipment manufacturing enterprises to actively reinforce basic skills, improve the overall level of integration, optimization of multi-level, multi-type innovative personnel training system, and industry the transformation and upgrading of equipment manufacturing enterprises to actively reinforce basic skills, improve the overall level of integration, optimization of multi-level, multi-type innovative Training system, to achieve industrial transformation and upgrading, innovation and development so as to realize the overall equipment manufacturing industry. In "Made in China 2025" as the guide for the equipment manufacturing industry in Shenyang, explore how to construct collaborative innovation system, proposed the implementation of the policy.

II. THE ANALYSIS OF TYPICAL INNOVATION MANAGEMENT MODEL

Depending on the subject of innovation, the complexity of the subject innovation, innovation paradigm is different, innovative management models including the independent innovation management, innovation management and joint collaborative innovation management three models, their meaning, the main innovation, management system and operational mechanism and application, They are very different occasions.

A. The Model of Independent Innovation Management

Independent innovation management means: innovation and design for a simple product of this innovative overall technical complexity, a single field, the whole innovation practice by individuals, some team members and internal company completed a certain closeness and independence sex. At the same time innovation activities clear boundary fixed, you can use corporatization, independent of the management system, the companies involved in interior design, manufacturing, human resources, sales, operations, services and other sectors. From this we can see that the company's innovation emphasis on internal company departments around centers of innovation activities, while not excluding the use of acquisitions to obtain innovations of other companies and research institutes. And this closed model in the rapidly changing global marketplace, more and more expose its flaws.

B. The Model of Independent Innovation Management

After the closure of the company's innovation stage, the industrial chain for the object, the advantages of integration and innovation resources industry chain, strengthen research cooperation, the establishment of cooperation as the core of innovation alliance groups become the main mode of innovation.

Features of cooperation as the core of joint innovation and open collaboration, innovation includes three modes: one-house innovation activities to the upstream and downstream industries enterprises to extend, to guide the industry chain enterprises in the longitudinal direction of cooperation and innovation; the second is based on the same chain enterprises to solve common

[1]2015 the Key Project of the Social Science Foundation of Liaoning Province: The Study on the Development Strategy of Manufacturing Innovation Center in Liaoning Province Based on "Made in China 2025" (No. L15AJY011)

E. Qi et al. (eds.), *Proceedings of the 23rd International Conference on Industrial Engineering and Engineering Management 2016*, DOI 10.2991/978-94-6239-255-7_31

technical problems or key technical horizontal cooperation and innovation; third is to gradually strengthen research cooperation enterprises, universities, research institutes between the three.

C. *The Model of Independent Innovation Management*

In the context of today's big science, high technology and large engineering, product / system / engineering exhibits extreme complexity, major scientific issues highlighted technical problems and bottlenecks on the original knowledge and technological innovations made high demands. Factors that affect the efficiency of innovation, technology also extends to the management level, mechanisms, marketing, finance and other aspects from, therefore, strengthen the synergy between the body and a variety of innovative elements between innovations and establish collaborative innovation model, a major innovation form.

We are talking about collaborative innovation essentially refers to innovative college resources and elements of fusion between, on the subject of innovation to break through the penalty area at the same time, more vitality to the "talent, capital, information, technology" and other innovative features of the full release . In Chen Jin standpoint, collaborative innovation is more subject to a number of factors among political research, etc. the depth of integration, by focusing on superior forces will be open, collaborate, and create equal value and innovation to improve a height. Thus we can understand: Collaborative innovation is the core, which is the biggest difference with the joint cooperation of innovation. However, we must clearly recognize that innovation alone voluntarily cooperation body is not able to solve the problem. Must be in the solid backing, the re-integration of the advantages of innovative resources by the government, the use of market functions to deploy resources, technology intermediaries and financial institutions into one, through scientific and rational organization, management and operation mechanism to coordinate market resources.

III. THE ANALYSIS OF LIAONING DEVELOPMENT AND MANUFACTURING PROBLEMS

So far, in our country, complete sets of equipment supply capacity of more than two hundred sets, Liaoning accounted for more than a third, Liaoning companies more than 40 kinds of complete sets of equipment, technical level and the market share in the domestic leading level. Liaoning province through the reform of state-owned enterprises, and gradually perfect the modern enterprise management system and methods, through equity diversification and enterprise restructuring, to promote "made in China 2025", promote the Liaoning economic vitality. Through the government guidance, take the market as the core, break through the boundaries of geography and industry, through the enterprise restructuring mergers, quickly develop a batch of world-class engineering general contracting, intelligent manufacturing enterprise group.

Is the advantage of Liaoning has a group of colleges and universities, research institutes, high knowledge and professional talent base, at the same time have a large number of national research institutes and innovative research and development ability can be used as a vanguard of scientific research in colleges and universities, is committed to research and development of Liaoning manufacturing industry. At present, compared with other domestic cities, Liaoning manufacturing talents gathered area.

But at the same time the Liaoning manufacturing also has some problems in terms of system innovation：

First, the major technical equipment still rely on imports. At present, about ninety percent of Liaoning light equipment, on seventy percent of the petrochemical industry and integrated circuit manufacturing equipment, nearly sixty percent of the numerical control equipment to import. Less than half the localization technology and capacity. Liaoning manufacturing enterprise research and development ability and manufacturing ability, compared with the international advanced level has a large gap, the foreign advanced equipment shock mounting of Liaoning equipment manufacturing.

Second, the core of key parts is still a weak link. Aircraft engine, machine tool numerical control system, electronic components manufacturing and software technology, original even basis, key parts such as problems are seriously hindering the rapid development of manufacturing industry.

Third, the enterprise independent innovation is still weak. Due to weak capacity for independent innovation in the field of equipment manufacturing, independent research and development of core technology, patent property rights, so the core of still rely on imported equipment manufacturing technology, the current will not be able to quickly get rid of the introduction -- - in the introduction of cycle of behind. And also lack the world influential brands and products, lack of international integrated core technology and competitive independent patent technology and intellectual property rights.

Fourth, enterprise overall integration capability is weak. Lack an overall design, system integration, complete sets of general contracting enterprise ability and system service ability. Under the market environment, the government only play a guiding role, the administrative department and has integrated contracting enterprise capability has not yet formed, colleges and universities, research institutes, manufacturing enterprises can't form system, still without a Liaoning manufacturing enterprises can rival Germany Siemens, General Electric, Mitsubishi, and other enterprises to become a general contractor, can provide the system service.

IV. THE ANALYSIS OF LIAONING DEVELOPMENT AND MANUFACTURING PROBLEMS

A. Relying on the "Made in China 2025", the System to Promote Collaborative Innovation

Commanding the economic development trend at home and abroad and puts forward the party central committee and state council "made in China 2025", which is based on China's international competitiveness, enhancement of national comprehensive national strength, to ensure the safety of China's national strategic height to make a long term, the major strategic deployment, its core objective is to accelerate advanced manufacturing industry innovation and development, improve the quality and enterprise benefit, realize from the manufacturer to produce world powers in China. This is built into manufacturing powerhouse is a national development strategy, is also a national systems engineering, must integrate the power of the social various aspects, integration in all aspects of the social resources, collaborative and comprehensive progress. City Liaoning province should seize the opportunity, promote the building of modern manufacturing collaborative innovation system.

B. Strengthening the Government Actively Guide, Create Synergy Innovation Environment

Over the years, technical reform and the basic construction is the direction of the main use of investment funds in China, almost no direct source of capital for innovation. Government funding basic of science and technology department is focused on the research and development of enterprises, and is not innovation, is mainly due to the innovation investment risk is big, huge investment, for a long time. So, we need to put more money on the innovation project, try to reduce all kinds of examination and approval links, and encourage financial institutions to increase the investment and financing, risk assessment and so on business development ability, the innovation of economic benefits and risks have accurate judgment.

C. Basing on Production, Alliance, Jointly Build Collaborative Innovation Ecological Environment

Relying on industry alliance, fully coordinated cooperation, alliance of industry, can form a variety of organizational form, can be either technical personnel self-employed, can also be a university-enterprise cooperation enterprises, can also be self-employed, college teachers actively encourage various forms between alliance enterprises grow rapidly. Institutions of higher learning in production has unique advantages in the league, use of the resources of teachers, personnel, equipment, and scientific research advantage, collaborative equipment manufacturing enterprises, speed up the pace of innovation. At the same time, institutions of higher learning can also be scientific research and industrialization as a quantitative assessment index, inspire innovative energy, and improve the innovation enthusiasm. The government should support and encourage the development of intermediary service enterprises, as the innovation fund, technology and service guarantee, further speed up the industrialization and marketization of scientific and technological achievements.

D. Strengthening Industrial Chain Technology Integration Innovation and Cooperation, Build Collaborative Innovation System

Manufacturing industry chain long, complex research and development and production technology, equipment, specialized tools, in their respective professional in close collaboration. Therefore, the equipment manufacturing enterprise in innovation as the innovation main body activity room, must integrate all aspects of resources, coordination of different areas, integration of upstream and downstream cooperation common development of the enterprise, from the perspective of system integration, the integration of university multidisciplinary comprehensive development characteristic, mainly owned r&d, introduction of foreign advanced technology is complementary, build integrated innovation mechanism and improve self ability to innovate. Relying on the core innovation main body ability and advantage in vertical industry chain, strengthen the communication between, in order to more realize sharing the resources and production technology, improve the ability of each enterprise in research and development and innovation, reduce duplication of procrastination, reduce costs, improve resource utilization, better adapt to the new environment in order to improve competitiveness.

V. CONCLUSION

Manufacturing industry in Liaoning, the pursue the new path of industrialization with Chinese characteristics, 2025, relying on China's manufacturing system to promote synergy innovation, realize the theme of the manufacturing industry innovation and development. To strengthen the government actively guide, on the basis of industrial engineering techniques and methods, to the quality and increase efficiency as the center, to create synergy innovation environment. Relying on the alliance of industry, to speed up the depth of a new generation of information technology and manufacturing integration as the main line, in order to promote the intelligent manufacturing as the main direction, integrated manufacturing system innovation, to create synergy innovation system and the management pattern, in order to realize the goal of Liaoning economy development, actively strengthen the infrastructure of equipment manufacturing enterprises, enhance the level of the overall integration and optimization of multi-level, multi-type innovative talent training system, building common synergy innovation of the ecological environment.

Through the government guidance, the industry developing system planning, integration of the whole industry chain, realize the industry transformation and upgrading, cultivating manufacturing culture with Chinese characteristics and the social environment, realization of Liaoning manufacturing industry from the history of the big teams.

REFERENCES

[1] Zeng Ping, Zhao Bing. Collaborative Innovation Model of SMEs under the Overall Context of Comprehensive Deepening Reform[J]. Journal of South China University of Technology (Social Science Edition) , 2015(05).1-8

[2] Tang Zhongjun, Chen Rongqiu, Xue Hong. A Innovation Model of Manufacturing Mode Based on TRIZ[J]. Management Review ,2004(10).32-38

[3] Qu Hongjian, Tuo Zhong. Synergy Innovation Model Review and Prospects[J]. Journal of Industrial Technological Economics.2013(07).133-142

Layout Analysis and Design of the Spindle Box Processing Workshop

Xiao-meng Li*
College of Information System and Management, National University of Defense Technology, Changsha, China
(*lxm2769@126.com)

Abstract - **The design of workshop layout is one of the important contents in the design of manufacturing system, and the merits of workshop layout directly influence the products quality, the productivity and economic efficiency. Because of the complexity of the workshop layout, it is sometimes difficult to meet the demand of the modern production mode simply by establishing the mathematical model. This paper optimizes the workshop layout by applying computer simulation technology, choose the FLEXSIM software for virtual simulation and dynamic layout analysis, find out the bottleneck in the workshop layout and make an improvement on it. Ultimately the production efficiency of the entire workshop is improved.**

Keywords - **FLEXSIM, simulation, spindle box, workshop layout**

Ⅰ. INTRODUCTION

Production workshop layout design is the processing that put equipment, material handling equipment, work unit, channel corridor and other layout entities in a limited production workshop.Layout design and logistics planning belongs to complex discrete event dynamic system, the traditional layouts described by programming model make a lot of simplifications, so that the model is far from the practical work conditions. Further, abstract mathematics model is difficult to adjust production plans with the current situation timely.

The integrated simulation software can solve complex system modeling problems, achieve convergence with the combination of multiple systems, meanwhile, it also provides a variety of reports and statistics, including E-Factory, FLEXSIM, Witness, etc. In this paper, the FLEXSIM software is utilized to model and simulate the facility layout of machining workshop.

A. Discrete Event System

Discrete event system is the system that state changes only at discrete time points. And these discrete time points are usually uncertain. Workshop layout design to solve the settings of each production department, section, auxiliary service department, storage facilities and work, equipment, channel and pipeline. Reasonable equipment layout plays an important role in balancing equipment capacity, keeping the balance of logistics and reducing production cost. Material handling form, and the working form of the staff is not continuous, in accordance with the characteristics of discrete events, therefore the workshop layout is a typical discrete event system.

B. Applicability Analysis of FLEXSIM Simulation Technology in Workshop Layout

FLEXSIM is a software for modeling and simulation of the discrete system, which can simulate the virtual workshop and carry out the dynamic layout. This software can avoid the limitation of the traditional workshop layout, quickly and accurately set up the 3D model of the workshop equipment, as well as simulate the actual workshop layout under different parameters to realize the optimization of workshop layout. At the same time, it also can make quick judgments about the problems that may arise in the workshop layout, and put forward the corresponding preventive measures.

Ⅱ.METHODOLOGY

A manufacturing enterprise mainly produces spindle box, the way of its layout is built on technology. There are two modes of transportation, manual handling and forklift trucks, each handling capacity of 1 units. The detailed layout of the workshop is shown in Fig. 1.

The transverse and longitudinal main channel width of original layout are 4 meters, allowing forklift

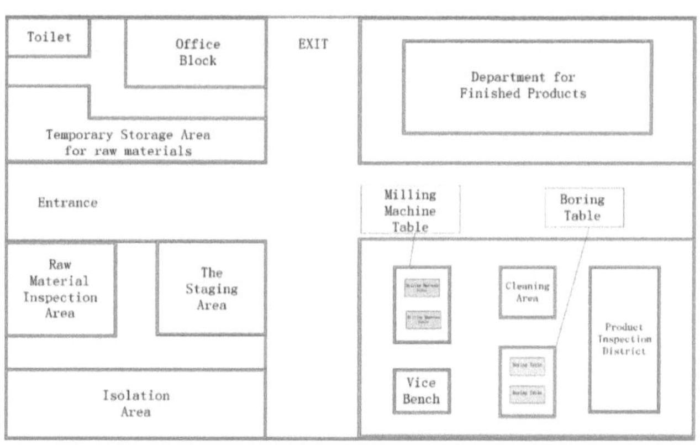

Fig. 1 The workshop layout

E. Qi et al. (eds.), *Proceedings of the 23rd International Conference on Industrial Engineering and Engineering Management 2016,* DOI 10.2991/978-94-6239-255-7_32

TABLE I MANUFACTURING PROCEDURE AND ELAPSED TIME OF EACH OPERATION

Work Number	Machining Class	Technology	elapsed time(s)
1	Milling	Rough Milling	120
2	Bench work	Drilling and Tapping	120
3	Milling	Milling bottom surface and side	120
4	Boring	Rough boring peg-hole	300
5	Boring	Finish boring main peg-hole	280
6	Cleaning		200
7	Test	Test precision	200

handling tool traveling in both directions. The channel width between regions is 2 meters, allowing forklift handling tools for one-way traffic. The distance between facilities is not less than one-third of the length of the equipment. Carrying tools are allowed to travel in the channel. Manufacturing procedure and elapsed time of each operation are shown in TABLE I.

A. Modeling and simulation

The facilities in the workshop are hereby replaced by the entities in the FLEXSIM. Use a source replaces the arrival of the raw material, a queue with capacity of 200 units in place of the temporary storage area of raw materials, a processor replaces the test of raw materials and a processor replaces the test of finishing products. For the machining area (pliers, milling, boring and other areas), there are several processors are utilized to represent operations, such as milling area needs 2 operations, with 2 processors respectively to represent the 2 milling operations.

Then connecting the various entities according to the requirements of the machining process. The output port of Queue1 in raw material temporary storage area connects the input port of Processor J, which is used to simulate the test process of raw material inspection area.

In order to facilitate memory, two forklifts are numbered. Forklift1 (transporter1) responsible for the material handling from raw material temporary storage area for raw material inspection area, forklift2 (transporter2) responsible for material handling from finished goods inspection to finished product storage area. Handlers accountable for the handling of waste which generated from machining area to a larger buffer (Queue4). Completed workshop layout of the FLEXSIM model is shown in Fig. 2.

B. The parameter setting of work units

(1) The simulation time is set at one shift, which contains 8 hours. Generator adopts the way of batch arrival. The first batch is 100 units, arrived at time zero moment and the second batch is 50 units, arrived at 14400s.

(2) Buffers in temporary storage area and isolation area have a capacity of 200. Buffer in machining area for centralized store waste has a capacity is 30. Other buffer caches used store materials between equipment have a capacity of 10.

(3) The setting of processor in FLEXSIM model is described as follows: One of the output port of the processor connects the next operation. The other connects a buffer which is responsible for receiving waste. On the temporary entity flow tab of the processor selects "by percentage (input)". Temporary entity Flow tab of the processor in a milling machine is set as follows: Selection of the output port is "according to the entity type (direct)" and select "Full" option in the input port. Finally select “polling model” in port pulled selection.

(4) The highest speed of 6 forklifts in the workshop is 2m/s, acceleration and deceleration are all $1m/s^2$, loading and unloading time is 20s, the highest

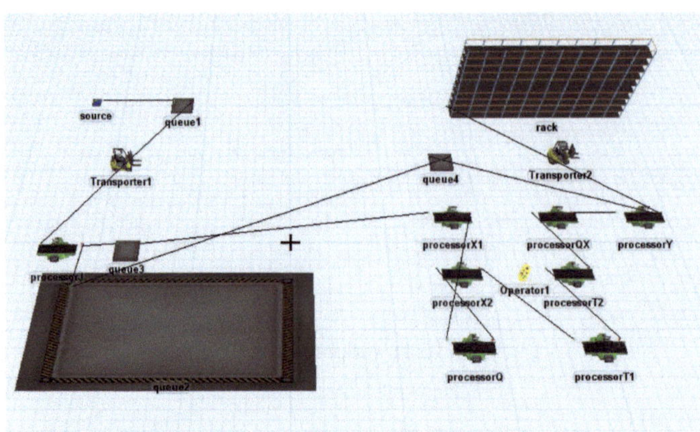

Fig. 2 The workshop layout of the FLEXSIM model

velocity of porter is 0.8m/s and its acceleration and deceleration are 0.3m/s^2, loading and unloading time is 15s.

(5) The finished product warehouse in FLEXSIM is represented by one high shelf. Due to the limited production of products within a day, the warehouse has enough capacity, so the capacity of the shelf here is not limited.

III. RESULTS

Open the statistics collection and gather statistics of the state of all entities, click on the FLEXSIM interface and select the "Reset", "compile" and "run", then the simulation model start running. In order to facilitate the observation of flowing entity, set the simulation time/real time is 80. When the simulation time is 28800s, stop running and export simulation report.

Simulation report contains the standard report and status report. Output variable of standard report includes the number of items contained in the object(Stats_content), the minimum number of items contained in the model running process(stats_content min), the maximum number of items contained in the model running process(stats_ content max), the average number of objects contained in the model running process(stats_ content avg), the shortest time stay in this object(stats_stay time min), the longest time stay in this object(stats_stay time max),the average stay time of the items in this object(stats_stay time avg), the input number of the items (stats_input) and theirs output number (stats_output). The variable of states report includes idle, processing, busy, block, collecting, releasing, travel_empty, travel_loaded, loading, unloading. Part of the generated simulation data is shown in TABLE II.

Wherein the ProcessorJ represent the processor for raw materials inspection, ProcessorX represent the processor for the milling area, ProcessorQ represent the processor of fitter zone, ProcessorT represent the processor for the boring zone, ProcessorY is the representation of finished products inspection area and ProcessorQX entity responsible for cleaning processor.

From the simulations report, we can see that ProcessorJ, ProcessorQ, ProcessorX1 are blocking and these processors exist idle time at the same time, indicating that the cause of obstruction is due to the imbalance of operations or the lack of buffer cache between processors. 70% of the processor exist WIP at the same time, Thus in this moment there are 30% of the processors are in the idle state, indicating that the continuity between the workshop process is not very well.

By making a observation for the operating process of the simulation model, we can see that the handling line in workshop has detour phenomenon, the position setting of fitter area, grinder and milling machine area, finished goods inspection area are closely connected, but the distance of fitter and milling machine area is too large with finished goods inspection area, resulting that the handling line is too long, and waste a lot of time on handling goods.

TABLE II THE SIMULATION DATA OF FLEXSIM

Object	content	contentavg	input	output	staytimeavg	idle	processing	blocked
processorJ	1	0.866168	124	123	202.867378	3855.46626	14735.4981	10217.1893
processorQX	1	0.515514	75	74	200.361576	13934.3744	14826.7566	0
processorQ	1	0.544598	107	106	148.057024	13123.6033	12713.3117	2980.73276
queue1	26	76.291831	150	124	13116.52954	0	0	0
queue3	86	41.758152	86	0		0	0	0
source	0	21.66097	0	150	1465.439921	0	0	0
processorY	0	0.286971	47	47	175.448521	20488.7869	7484.04842	0
processorT2	1	0.7913	82	81	281.061542	6004.38753	22765.9848	0
processorT1	1	0.979825	95	94	299.892934	580.436544	28189.9358	0
processorX2	0	0.683629	164	164	119.976475	9105.76296	19676.1420	0
processorX1	1	0.852093	172	171	143.598353	4262.32947	20512.7257	4042.59263
queue2	29	74.946224	150	121	12236.46901	0	0	0
queue4	10	9.350034	120	110	2339.692639	0	0	0
Transporter1	0	0.125179	245	245	14.719062	21334.392	0	0
Transporter2	0	0.061814	110	110	16.181202	25207.6229	0	0
Operator1	0	0.019581	30	30	18.211572	26812.9166	0	0
rack	47	19.278924	47	0	0	0	0	0

IV. CONCLUSION

In this paper, the FLEXSIM software is used as the main tool to make an overall modeling and simulation for the basic layout of the workshop, the logistics path and handling system.

The workshop layout is a typical discrete event system. By analyzing the characteristics of discrete events, the feasibility of FLEXSIM is confirmed and using FLEXSIM software to solve the key problems of the workshop layout design simulation, material flow form, process steps and so on.

By modeling and simulation of the workshop layout, which is based on a production workshop, the deficiency of the original layout was analyzed from the simulation report. The simulation result shows that the appropriate amount of cache capacity, the correct simulation strategy, and the accurate scheduling principle can improve the overall efficiency of the workshop production system under the existing equipment layout conditions.

REFERENCES

[1] SS Heragu. Recent models and techniques for solving the layout problem. European Journal of Operational Research, 1992, 57(2):136-144
[2] Zhang HJ. Flexsim simulation model for automatic stereowarehouse design and operation optimization. Journal of System Science and Information, 2005, Vol. 3 Issue 4: 839
[3] Flexsim User Guide Version3.0. , 2006
[4] Taucori D A, Lewis F L.A.. New Matrix Model for Discrete Event Systems: Application to Simulation. IEEE Control Systems Magazine, 1997
[5] AK Jain , VK Khare , PM Mishra. Facility Planning and Associated Problems: A Survey. Innovative Systems Design & Engineering, 2013
[6] Mark S. Daskin. Network and Discrete Location: Models, Algorithms, and Appl56ications. 2011
[7] Neghabi H, Tari F G. An optimal approach for maximizing the number of adjacencies in multi floor layout problem. International Journal of Production Research, 2015, 53(11):1-13.
[8] Chen L H, Hu D W, Xu T. Highway Freight Terminal Facilities Allocation based on Flexsim. Social and Behavioral Sciences, 2013, 96:368-381.
[9] Shan H Y, Zhang L B, Gao X Z. Studies of Balance and Simulation of Gearbox Assembly Line Based on Flexsim Software. Proceedings of 20th International Conference on Industrial Engineering and Engineering Management. Springer Berlin Heidelberg, 2013:83-90.
[10] Jia H M, Yu K C, Zhang J C. The Simulation and Optimization on the Certain Type Fuel Pump Assembly Line Balance Based on Flexsim. Applied Mechanics & Materials, 2015, 741(3):850-855.

Empirical Research on Production Leveling of Domestic Car

Zhu Guang [1,*] , Qi Ershi[2] , Xu Zeyong[3]
[1] Changchun Automobile Industry Institute, Changchun, P.R. China
[2] Economic and Management Department, Tianjing University,Tianjin, P.R. China
[3]FAW Car Co., Ltd, Changchun, P.R. China
(*fawxwz@163.com)

Abstract - **Along with consumers' growing demand for personalized and diversification in auto market, multispecies and small batch production has become the inevitable choice for automobile production enterprises. This paper analyzed FAW Car company production leveling model from monthly planning to vehicle off the assembly line. Constructed mixed production model of three kinds of cars and production transforms model of two welding lines supply one main welding line. The two models had been verified in this paper.**

Key word - **Production Leveling, Mixed Production, Production Cost**

I. FOREWORD

Automotive production line is the typical production system for standardized production on a large scale. This kind of production mode which is high in efficiency and low in cost has made great success in many countries including China. After China's accession to WTO, China automobile industry saw the protection policy for domestic national industries phased out. Auto market has entered a new stage of demand diversity that the quality requirement of consumers for cars is more and more demanding, and at the same time the requirement for personalized product is stronger and stronger. The latter posed manufacturer a new problem that how it efficiently organized many kinds of mixed flow production. If the problem failed to be solved, on one hand, a single product overcapacity would happen, and on the other hand because of larger investment the new production line would be stranded, which resulted in single product. So it is the inevitable choice for the modern automotive production line to realize many kinds of mixed flow production. The hardware condition on the production line depends on the investment in technology and product platform features. Once the production line is provided with the hardware condition, it is very important how to establish ERP production system that can realize many kinds of mixed flow production, so that it can activate the whole production line and make its logistics system efficiently. To the above problems, the level production comes into being. Implementation of the leveling is the important condition to organize many kinds of mixed flow production. Toyota's experience has verified that the leveling is one of the best conditions to realize the mixed flow production on the production line. So it is the important problem for the enterprise how to establish the leveling production model.

II. THE ESTABLISHMENT OF LEVELING PRODUCTION MODEL

A. *Determination of Mixed Flow Production*

The implementation of the mixed flow production successfully meets the need of the customer individuation and Just-in-time, and solves the regular contradiction of the auto production. In the concrete production, the pattern of manifestation on the assembly line is the production of the front vehicle and the following vehicle can be carried out in series according to different models and different brands. See the figure below (different shapes present different models and brands). Fig.1 shows traditional production mode; Fig 2 is the leveling mixed flow production model.

E. Qi et al. (eds.), *Proceedings of the 23rd International Conference on Industrial Engineering and Engineering Management 2016*, DOI 10.2991/978-94-6239-255-7_33

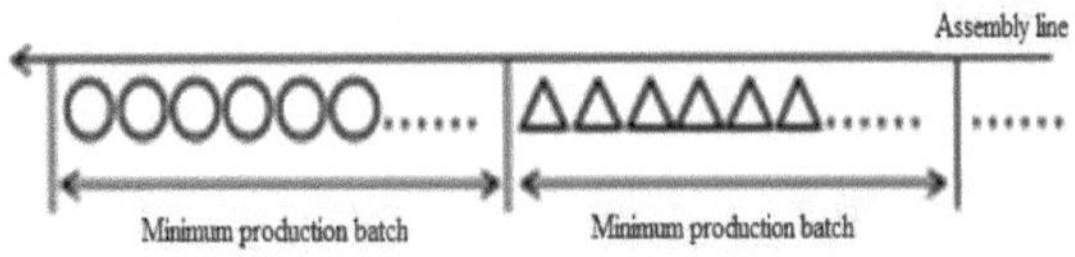

Fig.1 Traditional production model

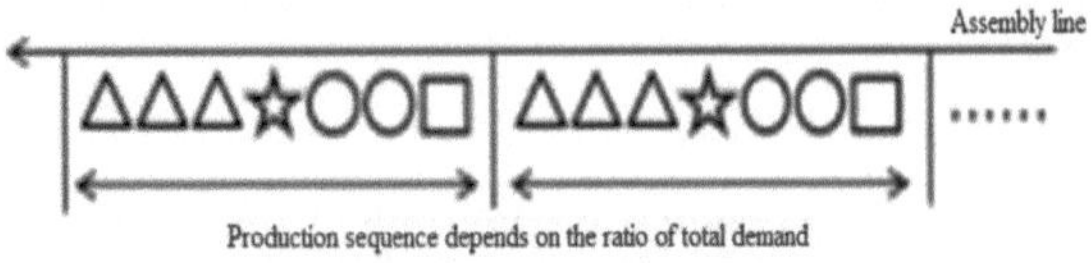

Fig.2 Leveling production model

B. *Establishment of Leveling Production Model*

The production leveling model of auto production is a comprehensive and systematic production model which consists of multiple indicating systems from monthly plan to daily operation plan, and to goods demand plan.

Take some company for instant to illustrate the production leveling model, as Fig.3 shows. Monthly production plan is the classification of large vehicle and total amount. Through BOM system, the monthly production is converted into monthly goods demand plan which then is delivered to the local and non-local suppliers. Monthly production plan is converted into weekly and daily plan for vehicle sequence, which must be complete leveling mode for the mixed flow production. The leveling plan for vehicle sequence is converted daily leveling goods demand plan through BOM system and indicating the local suppliers and local warehouse to allocate the cargo according to the vehicle sequence, and is delivered to the production line with accordance to the time series production indicators, so as to realize the leveling production.

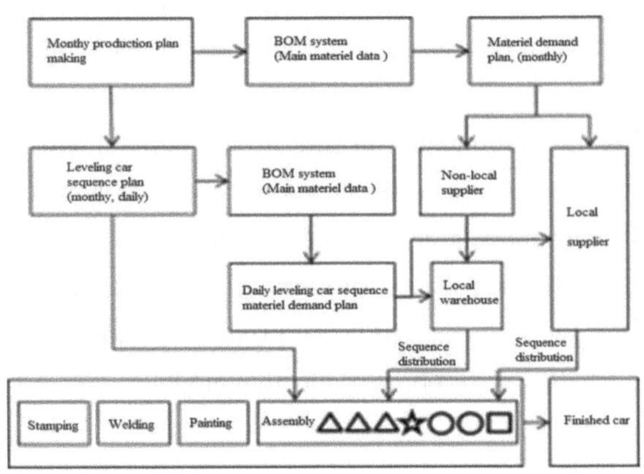

Fig.3 Leveling production model of some company

In the process of above, the key lies in the generation of monthly and daily leveling plan for the vehicle sequence. The research on its principles, logic and model is the key topic which directly relates to the achievement of production leveling.

III. LEVELING OF PRODUCTION SCHEDULING MODELFOR THREE VEHICLE TYPES

A. *The Establishment of Leveling of Production Scheduling Model for 3 Vehicle Types*

The basic problem of the mixed flow is to solve the sequence making for different vehicle types. From the point of research both at home and abroad, the scheduling problem is the key content in research on mixed flow assembly line.

A manufacturer produces 3 types of vehicles: A, B and C. The daily output is respectively Q_1, Q_2 and Q_3. The principle of production scheduling model is that the vehicle type whose reciprocal of production proportion is the minimum should be firstly produced. If the reciprocal of production proportion is the same, you should choose the vehicle type whose production proportion changes slower.

Step 1: the quantity of the production plan for all products on the mixed flow production line $Q_i (i = 1,2,3...n)$, to find the greatest common divisors q. Then to calculate the output of each product with S_i as show.

$$S_i = Q_i / q \tag{1}$$

Step 2: calculate production proportion reciprocal.

$$M_i = 1 / S_i \tag{2}$$

Step 3: make $K = 1, X_{ik} = M_i$

Step 4: According to the above two principles, choose the products to vehicle type sequence. Among X_{ik} , choose the least one, that is , when

$\min X_{ik} = X_{ik}$, product r as k number to put into production.

Step 5: if $k = \sum S_i$, get a circular flow. If $k \neq \sum S_i$, obtain numbers of times from the sequence of type of vehicle, which is equal to the product proportion in the production in the cycle options. The production proportion reciprocal of other products is updated with the following way:

$$X_{r,k+1} = X_{rk} \quad (\text{if} \quad i \neq r) \tag{3}$$

$$X_{r,k+1} = X_{rk} + M_r \quad (\text{if} \quad i = r) \tag{4}$$

Make: $k = k + 1$, feedback step 4, repeat the above process.

B .Application of Leveling of Production Scheduling Model for 3 Vehicle Types

According to three vehicle types, scheduled output are respectively 600, 300, 100. Apply the above model to determine the sequence of the production plan.

Step1: production proportion=600:300: 100=6:3:1, total output of circular flow is 10.

Step 2: calculate production proportion reciprocal. $M_A = 1/6$, $M_B = 1/3$=1/3, $M_C = 1/1$

Step 3: determine production sequence. As Table 1 shows, sequencing process of leveling of production plan for 3 vehicle types.

Step	Production of inverse ratio			Sequence
	A	B	C	
1	1/6 (*)	1/3	1/1	A
2	1/3	1/3 (*)	1/1	AB
3	1/3 (*)	2/3	1/1	ABA
4	1/2 (*)	2/3	1/1	ABAA
5	2/3	2/3 (*)	1/1	ABAAB
6	2/3 (*)	3/3	1/1	ABAABA
7	5/6 (*)	3/3	1/1	ABAABAA
8	6/6	3/3	1/1 (*)	ABAABAA C
9	6/6	3/3 (*)	2/1	ABAABAA CB
10	6/6 (*)	4/3	2/1	ABAABAA CBA

Table.1 Sequencing process of leveling of production plan for 3 vehicle types

According to Table 1, by deducing, we can obtain the sequence of the production plan: A-B-A-A-B-A-A-C-B-A, While the mixed flow batch is determined according to the actual production condition.

IV. SWITCHING MODEL OF PRDUCTIVITY FOR FOUR TYPES OF VEHICLE

A. Establishment of Switching Model of Productivity

Because each auto manufacturer has not only one type of vehicle to produce, and different types of vehicle have different characteristics in the welding workshop, many auto manufacturers use several main welding production line to produce different types of vehicle. Then through the mixed flow, they go into the adjustment line to produce. With the predictive sales-production-purchase information, the manufactures set several productivity models, and at a regular instance, switch the proportion, including the requirement of sales, forming mixed flow switching of productivity model. The workshop prepares the standard operation to preset the model. The same as the logistic set related corresponding solution to deal with the different models.

According to the requirement proportion of several types of vehicle, different types of vehicle are preset the structural behavior in the daily production plan. The daily production quantity of the different types of vehicle, the mixed flow mode in the welding workshop used as production operation indicators are described in detail, denoted by $M(Q_n(p,t), K)$, M indicates production statue in the main welding line, $Q_n(p,t)$ means n days of output p means meter, t means daily production time, K means proportion relation between variety, N means monthly vehicle order number, N_i means different varieties order number, D means whole month working day number, T_n means n model and $n+1$ means model switching time (generally once a month) , The leveling of variety of models production scheduling model is set up as follows:

Total daily output: $$\mathrm{Q_n} = \sum_{i=0}^{n} \int_0^t p_n^i dt \tag{1}$$

Type of vehicle i daily output: $$Q_n^{(i)} = Q_n \cdot K_n^i \tag{2}$$

vehicle i monthly output:

$$N^i(T)=\int_0^D Q_n^{(i)}dT$$

among which T in $[0,T_n]$ and $[T_n,D]$ sectional continuous (3)

Monthly output ： $N(T)=\sum_{i=1}^{n} N^i(T)$ (4)

Suppose 2 sets of productivity models are used this month, solve the switching time of productivity model T_n, get

$$T_n=\frac{N^i-Q_2^iD}{Q_1^i-Q_2^i}$$

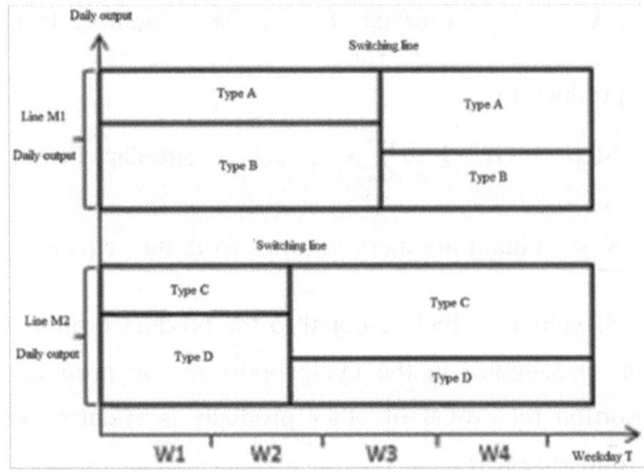

Fig.4 four models switching capacity simulation model application

B. Application of Switching Productivity Model

Take a welding workshop of a company for example, the production in the welding workshop consists of two main welding lines which supply one adjustment line. Each of main welding ling is used to produced two types of vehicle, altogether four types are produced. In certain period, according to demand fluctuations from the sales companies, the welding workshop sets the proportion of mixed flow switch to make two main welding lines M1 and M2 to produce altogether four products, and also prepares the solution to the corresponding personnel allocation and the production process. Using ERP system, the simulation is conducted to choose the solutions according to the goods need in the practical production scheduling.

The model as Fig.4 shows, abscissa means weekday, ordinate is daily output. The area of the quadrangle forms monthly output. Take production of Type A for example, on the left of vertical line is productivity model A, on the right of vertical line is productivity model B. If we adjust the position of the center vertical line, and the total monthly output has no change, the amount of production model A or model B will be different. If you move the switching line to the left, the amount of production model A will increase. If moving to the right, the amount of production model B will increase. So, productivity model is designed to dynamically adapt the sales demand change by adjusting the position of the vertical line.

For four models mixed flow productivity, conduct switching model simulation. Known to the $N+1$ as weekday, 22 days. The ordered quantity of Type A is 2000 units, the ordered quantity of Type B is 4336 units. The productivity model A: the daily output of Type A is 106 units, Type B is 182 units. The productivity model B: the daily output of Type A is 86 units, Type B is 202 units.

$M1(Q_1^1(p,t),K;Q_2^1(p,t),k)$

According to known conditions, get:

$Q_1^1(p,t)=106$

$Q_2^1(p,t)=86$

Weekday D=22 days

To calculate to obtain the productivity model switching time $T_n=5.4$ days，that is, $N+1$ month, on the 5.4th day, the production department and the logistic department begin switching between the two production models. M2 production line can be calculated in the same way, and get the corresponding result. The proportion is between production line M1and M2, that is, the production quantity of types of vehicle A+B and C+D. Through the adjustment of production meter, the two main welding lines collaborate together to the corresponding adjusting line for production, and form production scheduling solution in the ERP system to create the major plan.

V. EPILOGUE

The leveling of production will bring the enterprises great economical efficiency. Its greatest advantage is to make the enterprises meet the market demand for different varieties, different specifications of products, and also significantly reduce the inventories. It makes full use of flow line production model and production capacity to provide the necessary conditions, which is a major development for organization of flow line production.

This paper expounds the mixed flow models for three types of vehicle mixed-flowing assembly, and through some examples to verify the effectiveness. In addition, this article describes the characteristic of the leveling of production model in the welding workshop, and presents the production leveling model to produce four types of vehicle on two main welding lines. By determining how to switch between the some productivity models to meet the needs of sales companies and customers, it is used for reference for other auto manufacturers.

REFERENCE

[1] Shi Lan, Research and Application of Auto Enterprise Lean Production Management System

[D].Chongqing: Chongqing University, 2007.

[2]Xu Mingqiang, Chinese Lean Practices,[M].Beijing : Mechanical Industry Press, 2012: 127.

[3] Thomopoulos N T. Line Balancing-sequencing for Mixed-model Assembly. Management Science 1967, 14(2): 59-75.

[4] Kim Y K , Kim S J , Kim J Y. Balancing and Sequencing Mixed-model U-lines with a Co-evolutionary Algorithm. Production Planning &Control, 2000, 11 (8) 754-764.

[5] Miltenburg J. Balancing and Scheduling Mixed-model U-shaped Production lines. The International Journal of Flexible Manufacturing Systems, 2002, 14 (2) 119-151.

[6] YouYulin, Leveling Production Research, [J].Industrial Technology Economy, 1995, 14(6): 170-170.

[7] Qiershi, Industrial Engineering and Management[M].Beijing: Science Press,2011:59-64.

[8] Su Ping, Yu Zhao-qi. Multi-objective optimization for balancing problem of mixed model assembly line [J] China Mechanical Engineering, 2009, 20(19) : 2342-2347.

[9] Boysen N , Fliedner M , Scholl A . Production planning of mixed-model assembly lines: overview and extensions [J.] Production Planning and Control, 2009, 20(5) : 455-471.

[10] Zheng Min, Dong Ming. Sequencing problem for a mixed-model assembly line considering switching constraints [J] Industrial Engineering and Management, 2012, 17(2) : 85-89.

Exploration and Practice of Car Industry Lean Culture Construction

Zhu Guang [1,*] , Qi Ershi[2] , Xu Zeyong[3]
[1] Changchun Automobile Industry Institute, Changchun, P.R. China
[2] Economic and Management Department, Tianjing University,Tianjin, P.R. China
[3]FAW Car Co., Ltd, Changchun, P.R. China
(*fawxwz@163.com)

Abstract - **Lean culture has significant influences for e nterprises to promote lean production and innovate lean management methods. Based on researches of foreign adv anced enterprise culture such as Toyota, Volkswagen, For d, and BMW, this work constructed a the company lean c ulture model trough the practice of the company culture. This model, based on challenge, improvement, respect, res ponsibility and innovation, aims at zero defect, zero machi nery stop and zero inventory. Besides, safety, quality, talen t, production and manufacture were constructed based on the guidance of prevention, education, CZD, talent cultivat ion, customer orientation and technological precedence. In the end, the work also summarized lean corporate culture experience through exploring practical cases. The researc h results guide enterprises to construct and promote lean c orporate culture.**

Key words - **Lean thinking; Lean Culture; Lean Produ ction; Lean Management**

I. PURPOSES AND SIGNIFICANCES

Lean production model, a kind of advanced production management way from Japanese Tokyo Auto Company, achieved huge success in world advanced companies such as Toyota, Volkswagen, Ford and BMW. However, their successes were all supported and leaded by their unique, durable and deep-rooted culture [1-2]. As for successful enterprises, lean corporate culture is the essence and soul of lean production, the embodiment of lean thinking, inner requirement of lean management and the motivation of lean production, which is of great strategic significance to the development of enterprises.

Lean production mode, introduced to China as early as 1981, has made certain achievements in Chinese enterprises. In recent years, more and more enterprises have learned from advanced enterprises and promote lean production modes. However, few enterprises really consider lean production as a concept and culture. Thus, lean production is not deeply rooted in enterprises. There are some reasons for this phenomenon. For example, enterprises in China focus too much on lean methods, tools and techniques when promoting lean production. As a result, the essential connotation and further cultivation of lean production is ignored, resulting in slow promotion and unobvious effects of lean improvement activities. What's worse, some enterprises even give up promoting lean production [3-6]. "If you want to make trains run10 kilometers faster per hour, you only need to add certain horsepower; if you want to make the train speed doubled, you have to change tracks. Asset restructuring can temporarily increase the productivity of the company, but if there are no changes in culture, the company will not be able to maintain the development of productivity", said Jack Welch, a world famous entrepreneur [7]. Therefore, lean culture construction plays a decisive role in lean production and lean management of the enterprise. Current international automobile market is faced with complex and changeable environment with fierce and cruel competitions. Therefore, researching on lean corporate culture construction is of great significances to the survival and development of automobile enterprises in China. This work, based on the study of domestic and foreign lean culture and lean production status, established a lean corporate culture model through the practice of lean culture.

II. CONSTRUCTION OF LEAN CULTURE MODEL

Enterprise culture refers to the shared beliefs, values and assumptions among enterprise staff working for common goals [10]. Lean production requires timely manufacture and elimination of malfunction and all wastes, with purpose of zero defect and inventory [8-9]. Lean thinking emphasizes the elimination of waste and creation of values [10]. Lean management requires enterprises to reduce waste to a minimum, providing customers with satisfactory products and services. The application of lean tools and technologies brings the demand of lean management transformation. Technology innovation needs the transformation of management system as its foundation and supports. Meanwhile, lean technologies provide tools and methods for the effective implementation of lean management. In this process, lean thinking guides the innovation of enterprise culture. At the same time, culture innovation provides basic assurance for lean management. Therefore, lean tools, methods and technologies are needed in lean management to extend lean production to various fields (such as research and development, management, manufacturing, purchasing, sales). Besides, the construction of lean corporate culture plays a guiding and leading role in lean production management. Meanwhile, lean culture can be improved in lean management. When staff participate in lean production, continuous improvement, innovation and pursuit of excellence, lean culture becomes a strong core competitiveness, leading the enterprise towards excellence. The model in Figure 1 shows the principles of lean culture construction.

E. Qi et al. (eds.), *Proceedings of the 23rd International Conference on Industrial Engineering and Engineering Management 2016,* DOI 10.2991/978-94-6239-255-7_34

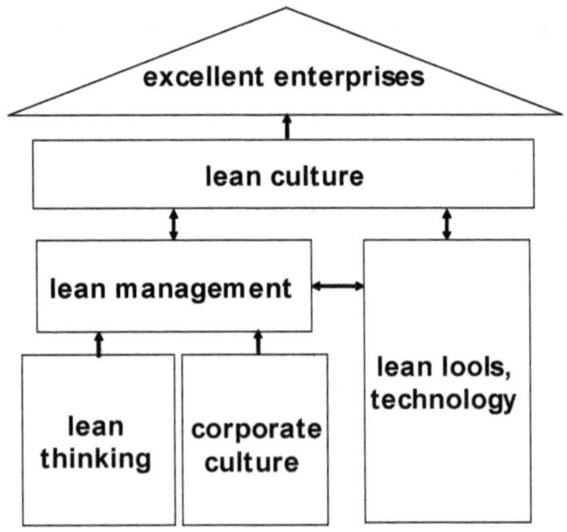

Figure 1 Principles of Lean Culture Construction

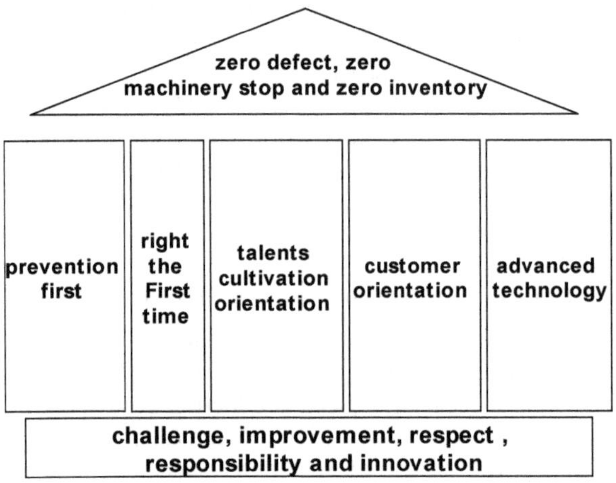

Figure 2 Lean Culture Model of the company

With the same principles of lean culture construction, different enterprises have different connotations and characteristics about lean culture. The corporate culture of Toyota is *challenge, continuous improvement, genchi genbutsu, respect for employees and team cooperation.* Volkswagen shares the corporate culture of "democracy and freedom, credibility, abidance by law, fraternity, equality, frugality, and people orientation". Ford built the corporate culture of globalization ideas, customer needs, continuously pursuit of growth and the belief—leadership is the teacher. BMW follows the culture of "production following the market", sustainable development of human beings and social role identification. Relying on their corporate cultures, those advanced automobile manufacturers have formed unique corporate business philosophy with unique characteristics and achieved success in the long-term production and fierce market competition. Outstanding corporate culture can improve internal morale, core competitiveness and the image of enterprises. As an important part of enterprise core competitiveness, it brings great power and motivation to the development and competition of enterprises. After researching the culture of foreign advanced corporations such as Toyota, Volkswagen and Mazda, the company built lean corporate culture with the ultimate goal of *Three Zeros*: zero defect, zero machinery stop and zero inventory. This model, based on challenge, improvement, respect and innovation, aims at zero defect, zero machinery stop and zero inventory. Besides, safety, quality, talent, production and manufacture were constructed based on the guidance of prevention, education, one-time pass rate, client orientation and technological precedence. Figure 2 shows the lean corporate culture model of the company. *Three Zeros* is the excellent state, lead of lean culture and reflection of lean values. The five pillars, as the core and support of lean culture, embody and enrich the content of lean culture, representing the depth and width of lean culture promotion. The five foundations are the premise and foundation of lean culture, and they are all fundamental to enterprise running. These foundations and pillars decide the realization of lean culture goals. Without them, enterprises will lose spiritual motivation and survival viability.

III. PRACTICE OF LEAN CULTURE CONSTRUCTION

Faced with fierce market competition, the company practiced lean technologies and tools based on past decades of lean production and management, thus accumulating profound Reed-Flag culture. Therefore, lean corporate culture and changes in concepts and ideas are the keys to achieve economic batch quantity of 300 thousand. Lean corporate culture is the basic guarantee for the company to achieve future development strategies, so each department of the company carried out lean culture construction. As a result, the company made obvious achievements in the past and planed a production volume of 300 thousand vehicles in 2014. The achievements of the company are illustrated from five aspects.

A. Safety culture of prevention

Potential safety hazards inevitably exist in production line, and any carelessness may cause serious threats to personnel safety and health. Thus, safety production should focus on the conversion from after process to beforehand prevention and production supervision. "Production supervision" and "beforehand prevention" can avoid "dangerous potentials" for safety production. Besides, such safety culture will respect employees and avoid safety accidents.

Anan accident happened in the company on April 27th 2011. When a worker was installing some auxiliary components for a car, his right leg was stuck between roller-bed guide wheel and the moving skidding. As a result, his toes were broken. It was found later that the major cause of this accident was the insufficient experience of process designer. When designing process auxiliary components, guide wheels were installed in the operation area. After that accident, workers passively waited security department to rectify potential safety hazards. It implies a weak safety culture guidance and inadequate capacity of employees to inspect and rectify safety problems. Therefore, the company carries out trainings about personnel safety awareness and knowledge. Then, they carried out safety team evaluation, team lead qualification examination, safety training basic education and risk predicting trainings to change employees' safety concepts. Then, "beforehand prevention" is gradually rooted in workers' heart. On July 28th, 2014, a worker identified a potential hazard around the roller bed of a grinder in the second factory. Then, he voluntarily rectified the hazard and eliminated safety accident. Those two accidents have the same safety hazard, but they were treated in different ways. After only three years, employees' attitude was changed from passively waiting for solution to autonomously identifying hazards and rectifying safety problems. Such small change reflects employees' conversion from safety for the company to safety for us.

B. *Culture of one-time pass*

In addition to prevention and zero-defect culture, the company also emphasizes the quality culture of "one-time pass".

The company created a "Quality Day" to enhance the consciousness of *Zero Defect* and encourage staff to participate in quality-culture construction activities through quality-improvement validation and zero-defect evaluation test. Since 2010, they have held about 30 rounds of "Quality Day" activities and more than 800 quality-improvement validations. About 400 excellent zero-defect employees were selected. Therefore, the company has made great success on improving quality consciousness and vehicle quality, especially in reducing defects of wrong installation and missing installation.

In 2013, the assembly workshop in first factory of the company chose "tank washing interference" issue for rectification and improvement validation. This problem happened 32 times in January 2013, affecting the accomplishment of quality goals. With "fishbone diagram" method, this workshop deeply analyzed the issue from aspects of workers, machines, materials, methods and environment. And a series of improvement measures were put forward such as establishing integrated control of quality circle and strengthening operation trainings. In August 2013, this workshop achieved *Zero Defect* for five months and effectively cultivated workers' consciousness and ability of "one-time pass", largely improving product quality.

C. *Education-oriented talent culture*

The company respects employees, considering "educating people" as an important strategy of human resource investment. It takes constructing development platform and breeding environment for talents as the most important tasks.

The company developed different career paths for general operation group and leader group. As for operation groups, they can follow management career path from worker to foreman, team lead, section foreman -and workshop supervisor. Besides, they can also choose professional career path from worker, to crew backbone, talent, expert and senior expert. General leaders can develop from secondary manager backup to secondary manager, senior manager backup and senior manager. Besides, they can also choose professional career path from talent to specialist, senior specialist and leading expert. Construction of employee career path enables every employ to find their development direction and path. Therefore, employees' development goal is consistent with enterprise development objective, effectively inspiring their initiative.

In addition, the company also built a development platform for employees, including training curriculum, trainer system and training bases. Professional trainer teams were established according to professional abilities. There are three levels of trainers: primary trainer, intermediate trainer and senior trainer. Besides, the company developed different cultivation strategies for trainers with different levels, providing teaching staff for talent cultivation. Besides, the company also developed training course for key positions to improve employees' core, general, professional abilities, thus preparing sufficient content for talent cultivation. In addition, a 7000-square training center was built as the training base. This training center includes many classrooms and 6 practicing courts such as vehicle assembly training area, mold maintenance training area, robot maintenance training area and some other courts. At present, they already established a good professional faculty team, effective curriculum, available training center and practicing courts, providing effective software and hardware supports for talent cultivation. More importantly, the company has constructed a profound talent culture through above efforts.

D. *Product culture of customer orientation*

The company is faced with diversified and differentiated needs from customers in the development process. From design to customer evaluation, they try to perfect product details and make continuous innovation for customers.

In the development of B70 model in the company, it was found that the in-front AC outlet could not be adjusted in multiple angles. The AC wind could only reach the height of passengers' face, failing to meet customers' requirements. If customers adjust the wind outlet up, the wind cannot reach passengers' shoulder. Thus, the research and development team developed strict requirements: when the outlet is adjusted to the highest angle, the wind should reach the ceiling and flow to back seats; the wind can reach passengers' face rather than shoulders. After rounds of tests and virtual researches, they finally found an improvement solution based on consideration of human-machine engineering and cost control. They decided to expand the out blade by 10° without designing new mold. In the end, the issue of cold winding blowing was solved after rigorous test and considering a balanced cost and circle requirements, which greatly improved comfortableness of air conditioner.

E. *Manufacture culture of technological precedence*

With the concept of continuous innovation, the company widely applies advanced technologies to realize a lean production mode. The applied technologies include concurrent engineering, intelligentization, virtual technology, Kanban technology, network and visualization.

Advanced virtual simulation technology is applied to the production preparation of D-Model vehicle in the company , thus identifying problems in product design and engineering design phase. It is conducive to modifying designs of product and device, as well as avoiding problems from the source. After simulation analysis, 23 items were proposed for technology assessment. And a careful production process control was conducted to avoid the influence on production of currently manufactured vehicle. Meanwhile, virtual simulation technology of digital platform is applied to conduct autonomous process planning, design and simulation analysis. Thus, rigorous virtual manufacturing ensures "one-time pass" of D-model process modification in the main welding line. The modification of machines on D-model main welding line only took seven days. After application of virtual manufacturing technology, the production is restored on schedule with only the cost for modifying an electrode holder horn.

IV. EXPERIENCE OF LEAN CULTURE

During the lean culture construction in the company, several experiences can be concluded in the process of exploration and practice:

A. *Lean culture casting the soul of lean enterprise*

During the process of lean manufacturing, the company's senior leaders actively participate to take the lead in internal lean transformation. Besides, the projects, improving the quality of personnel, three-year quality, management and structure system capacity, are developed to carry forward the red-flag spirit of being indomitable and striving unceasingly. Furthermore, to form a total improvement mechanism, the quality day, following the mass line, family visiting day, rationalization proposal, autonomous improvement and QC (quality control) group activities, are developed to mobilize every employee's enthusiasm and initiative. In this way, enterprise gradually becomes accustomed to lean management, thus forming DNA of lean culture.

B. *Lean culture developing lean talent*

The company sets the enhancement of internal quality and social image as its enterprise culture, and takes loyalty, self-improvement, learning and innovation as the spiritual foundation. It is hoped an innovative company can maximize personal quality and skill as well as strengthen team building. Furthermore, the company is devoted to the harmony between concept and action. It also uses cultural force to promote the scientific, healthy, harmonious and quick development of enterprise, making the strategic task as creating lean talents to enhance the core competitiveness of enterprises.

C. *Lean culture lean system construction*

The company regards ten elements as its starting point, including goal management, team management, total improvement, standard operation, process quality control, smoothing of production, JIT logistics, field 6S, efficient preservation and talents cultivation. Then, it constructs the HPS (Hongqi production system) to guide the improvement of whole staff and constantly increase value for customers, with the achievement of management innovation.

D. *Lean culture creating excellent enterprise*

For enterprises, lean culture is the key to ensure sustainable development and possess core competitiveness. Lean culture is very important for the company to create independent brands like red flag and Pentium, form unique and efficient structure system capacity and create excellent enterprise in market competition. The change of lean culture will promote the transformation of enterprises' production mode, thus keeping enterprises with vigorous vitality.

V. CONCLUSIONS

Exploration and practice of the lean culture construction of the company is of great significance, and conclusions in this study can be used as references for other automobile manufacturing enterprise. Toyota has devoted decades to studying the relationship between enterprise and social environment, market, and customers, as well as the relationship between enterprise internal departments. With such experience, Toyota company epurated its own enterprise culture. While learning and exploring lean production methods, Chinese enterprises should combine with the situation of China and themselves. Enterprises should firstly understand the essence of lean production, then practice and strengthen it. Therefore, the whole stuff will gradually form habits under the guidance of management concept, precipitating their unique lean culture during the ever-lasting learning and improving process. Lean production mode is the high integration and unity of enterprise internal people, technology and management concept. Lean culture can effectively promote the advancing of lean production mode, thus ensuring the effective implementation of management strategy.

REFERENCES

[1] Jeffrey Liker, Michael Hoseus, Toyota Culture: *Heart and Soul of the Toyota Way*, translated by Wang Shiquan, China Machine Press, 2011: 4-8.
[2] Hirotaka Takeuchi, Emi Osono, Norihiko Shimizu. *Contradictions that Drives Toyota's Success*, Harvard Business Review.2008, 6: 96.
[3] Qi Ershi, Enterprise Lean Culture Construction, *Science of Science and Management of S. & T.* 2008 (12): 133-136.
[4] Sasa Kimoto. Learning Toyota's Three Key Points, *Beijing: SINO Foreign Management*, 2007 (7): 92-93.
[5] Su Xuedong. We Don't Really Understand Toyota. *Beijing: SINO Foreign Management*, 2008 (3): 82-83.
[6] Liang Qiuhong, Research on Organizational Culture of Lean Production, *Logistics Technology*, 2009, 28(2): 7-10
[7] Wu Minghua, *Corporate Culture in the View of Lean Management*, Maritime China, 2006 (7): 78.
[8] Ornotaki, Toyota Production System, *China Railway Publishing House*, 2006: 25.
[9] Xaio Zhijun, et.al Lean Production JIT, *Shenzhen Haitian Publishing House*, 2003: 12-21.
[10] James P.Womack, Daniel T.Jone, Lean Thinking, translated by Shen Xi Jin, Zhang Wenjie and Li Jingsheng, audited by Li Wang. *China Machine Press*, 2011:12.

The study about intelligent manufacturing resource allocation efficiency based on mutual information criterion

Ning Kang[1,*], Jiang Shen[1], Man Xu[2]
[1]College of Management and Economics, Tianjin University, Tianjin, China
[2]TEDA college Nankai University, Tianjin, China
([*]kangning@tju.edu.cn)

Abstract - **With appearance of "Internet +" and "Industry 4.0", Intelligent manufacturing has become a hotspot. This mode realizes the optimal allocation of manufacturing resources through the network platform. At present, scholars use intelligent algorithms studying resource allocation model, however, these studies are based on assumption of definite resource environment. The study from perspective of data matching is less. This paper highlights research about efficiency of resources matching from the internet perspective, putting forward concept of "resource that is data". Combined with data redundancy issues, this paper separates many unordered resources features into redundant and nuclear attributes. It studies distribution of attribute characteristic variables in same space with information entropy, researches the redundancy among attribute features variables with mutual information and studies attributes feature selection mechanism with attributes features redundancy criterion based on mutual information. We can find the resource with robust attribute features quickly and accurately from many intelligent manufacturing resources.**

Keywords - Intelligent manufacturing; resource allocation; redundancy; mutual information

I. INTRODUCTION

Progress of science and technology provides favorable conditions for change of manufacturing industry, Networked manufacturing mode has wider use, such as networked manufacture, parallel manufacturing, cloud manufacturing. With appearance of " Internet + " , we combine internet technology method with manufacturing industry and gain a new manufacturing mode, intelligent manufacturing. According to the need of users, it organizes manufacturing resources by net and cloud manufacturing service platform, and provides all kinds of service for users.

In the Internet platform, resources are present in form of data. Core of resources allocation is the process of data integration and data matching. The efficiency of manufacturing resource allocation actually is the efficiency of matching data. The efficiency of Resource Allocation is measured by two main indicators, one is speed of resources allocation, the second is accuracy of resources allocation, namely the speed and accuracy of data matching. Many scholars carried out deep study. From perspective of resource allocation method, we often use intelligent algorithms and operations research methods to study optimal resource allocation model. Gong Guan [1] used dispersion degree of marginal product value of input elements as a measure of efficiency of resource allocation. This method is more accurate in terms of computational efficiency of resource allocation, and can be individually investigated configuration of a specific resource. Zhou Ke [2] proposed collaborative optimization configuration of resources with the standard manufacturing cell. Wang Shilong [3] constructed an optimal resources allocation model with the target of a minimum cost of time and the best quality. From perspective of manufacturing system reconfiguration, Porter [4] proposed manufacturing system integration of internal resources. Laili Y [5] proposed vertical supply chain integration. Shen Jiang [6] proposed network integration with representative of virtual enterprise based on the enterprise strategy. From perspective of usage efficiency of manufacturing system service resources, Chen Bin [7] introduced a distributed system, and proposed an allocation method with a set of scalable resources to improve the efficiency of resource use. Bertsimas [8] proposed multi-objective resource planning model based on configuration fairness. Huang Haixia [9] analyzed science and technology resources allocation efficiency from the point of input and output with data envelopment analysis method. While the above research gave deep discussions on data matching, the study to improve the efficiency of resource allocation from the perspective of excluding resources data redundancy is few.

Based on above problems, this paper highlights the research about efficiency of resources matching from internet perspective, putting forward concept of "resource that is data". Combined with data redundancy issues, starting from point of attribute characteristic angle of intelligent manufacturing resources system, this paper separates a lot of unordered resources features into the redundant attributes of resources and nuclear attribute of resources. It researches the distribution of attribute characteristic variables in same space with information entropy, researches the redundancy among the attribute characteristic variables with mutual information and studies the attribute feature selection mechanism with property features redundancy criterion based on mutual information. This method can eliminate redundancy attributes and gain optimized feature set, It also can put these features in order accord to importance. Then, we can find the resource with robust attribute features quickly and accurately from intelligent manufacturing resources.

E. Qi et al. (eds.), *Proceedings of the 23rd International Conference on Industrial Engineering and Engineering Management 2016,* DOI 10.2991/978-94-6239-255-7_35

II. METHODOLOGY

Great advance in big data technology make a lot of complex heterogeneous data in intelligent manufacturing resource platform, which carries of mass disordered property characteristic information, of course, including some of the repetitive or low-value information, namely redundant information. These redundancies will lead to low efficiency, difficult search, and even hard to get the optimal configuration path in the resource allocation process. We divided cloud resource data attribute space in redundant attribute subspace and core attribute subspace with the attribute feature selection method. We studied property value distribution of properties characteristic variables in the same state space with information entropy and redundancy of attributes characteristic variables with mutual information. And we also studied mechanism of attribute feature selection with attribute features redundancy criterion based on mutual information, eliminating redundancy to achieve optimal feature data set. Then we reduced the scope of the allocation of resources and improve the efficiency of resource allocation.

1) *Resource Properties feature selection*: In the resource data space U, We let A be a set of characteristics logical view in the set of data characteristics, called the characteristic expressed. T is a logical view of the target feature, which removed the redundant resource properties characteristic in the input feature set, called optimal feature subset. P is a pattern which builds the correlation between feature representation and the destination selecting. $\hbar(f_\lambda,C)$ is the sort function, which outputs a features sequence about the best features subset. $f_\lambda \in A$,C is a logo category set including m logo categories C_i, namely $C=\{C_i| \ i=1,2,...,m\}$,for the relevance attributes f_λ to C, we divided the property features into nuclear and redundancy attributes.

Definition 1: Set $T \subseteq A$,we call the un-omitted original relationship clusters of T core T of A, Attribute characterizing the relationship of cluster is called core attributes. They pose a collection called Core Attribute Set, referred to as $Core_T(A)$.

Definition 2:Supposing attribute $f_\lambda \in A$ but $f_\lambda \notin Core_T(A)$,namely f_λ is redundant in C, we call f_λ redundancy attribute. These attributes form a set called redundant attribute set, referred to as $Red(A)$.

After attribute feature selection, we call resource data space U' ,$\hbar(f_\lambda,C_k)$ is a homomorphism from U to U',U' is the shape of U in $\hbar(f_\lambda,C_k)$.As is shown in Fig. I

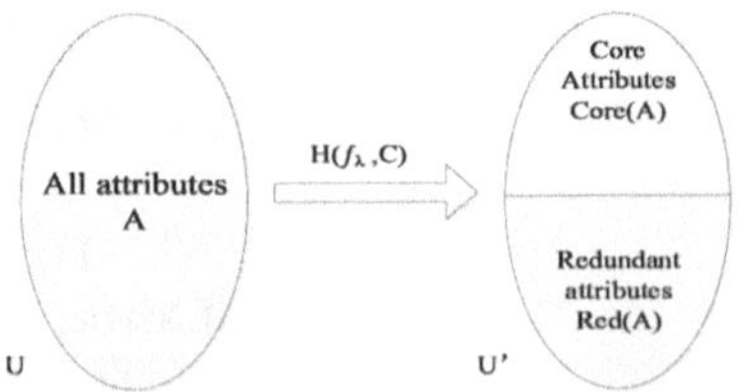

Fig. I. Attributes feature selection of U and U'

Attributes feature selection model can be described as: A is original feature representation with n features, C is logo category set with m logo categories, We select k the most effective features from A ,The optimal feature subset with a new feature sequences is called T ,meeting $\hbar(f_\lambda,C_k)$, $\lambda=1,2,...,k$, $k<n$.so, Sort function $\hbar(f_\lambda,C_k)$ of attributes Feature Selection in U is the key of the method and model of attributes feature selection.

2) *Resource attribute mutual information*: In order to describe the disorder or uncertainty of attribute value distribution of attribute variable of intelligent manufacturing Resource, we introduce the concept of information entropy.

Definition 3:In the resource data space U, A discrete variable Attribute $f_i \in A$,Its attribute value is X, $X=\{X_1,X_2,...,X\}$, The corresponding probability distribution is $p_i=P(X_i)$,We define

$$H(X)=-\sum_{i=1}^{n}P(X_i)\log\frac{1}{P(X_i)} \tag{1}$$

$H(X)$ is information entropy of f_i. When the feature is a continuous variable, it can come to similar conclusion. Information entropy represents the degree of uncertainty or disorder of attribute value X belong to resource attribute variable f_i.

Supposing another discrete attributes variable in U, $f_j \in A$, Its distribution of Attribute value is

$$Y=\begin{bmatrix} Y_1 & Y_2 & \dots & Y_m \\ q_1 & q_2 & \dots & q_m \end{bmatrix} \qquad \sum_{j=1}^{m}P(Y_j)=1 \ .$$

then, Conditional Entropy of attributes variable f_j is

$$H(Y\mid X_i)=-\sum_{j=1}^{m}P_{ij}\log Pi_j$$

Now, Conditional Entropy represents the degree of disorder attributes value distribution of f_i , when we know attributes value distribution of attribute variable f_j.

For space U and U', the whole information entropy in redundant attributes subspace and core attributes subspace remains unchanged, one increases ,another one reduces.

Information entropy represents the degree of uncertainty or disorder of attribute value X belong to f_i . Conditional Entropy represents the degree of disorder of attributes value Y of f_i ,when we know a attributes value distribution of X .But, for a accurate discussion of redundancy of f_i and f_j ,we need to know their common degree of disorder, We introduce mutual information of attributes variable.

Definition 4:Supposing we know the distribution of Attribute value X and Y belong to f_i and f_j. we define their Mutual information:

$$I(X;Y)=\sum_{f_i,f_j\in A}\sum_{f_j\neq f_i}P(f_i,f_j)\log_2\frac{P(f_i\mid f_j)}{P(f_i)P(f_j)} \quad (2)$$

Physical meaning of $I(X;Y)$ is:

(1) Mutual information reflects the information that attribute value X gets from another attribute value Y ,It reflects the common disordered degree of attribute value X and Y ,It can measure the redundancy of attribute variables f_i and f_j.

(2) $I(X;Y)$ measures general dependence between two attributes, If $I(X;Y)$ is larger, the relation between f_i and f_j is closer. When $I(X;Y)$ is close to zero, they are independent.

In resource attributes space, to solve the sorting function problem of attributes characteristics selection model between U and U', we introduce mutual information and use characteristics redundancy criterion to evaluate. We can reduce the redundancy of intelligent resource data space by construction of redundant attributes criterion with mutual information. Normalized mutual information matrix can measure mutual information of high-dimensional secondary Renyi entropy, and reduce the computational complexity.

3) *Normalized measure of mutual information*: To reduce computational complexity of information entropy and mutual information in space U, we introduce an definition of Renyi entropy. Hanel R [10] pointed out Renyi entropy can reduce non-linear computational complexity of Shannon entropy. Since normalized mutual information can effectively deal with the redundancy of attribute variables, we use normalized mutual information measure presented by Müller B[11] to estimate the mutual information of attribute variables $I(X;Y)$.

For attribute value X of continuous variables type attribute variable $f_i \in A$, Its probability density function is $f_X(x)$, Renyi differential entropy can be defined as:

$$H_{Ra}(X)=\frac{1}{1-a}\log\left[\int_x f_X(x)dx\right]$$

$a>0, a\neq 1$.when a Approaches 1,a order Renyi entropy converges to Shannon entropy.

Renyi[12] gave the definition of Renyi entropy. giving a probability density function $p(x)$, secondary Renyi entropy is

$$H_2(x)=-\log(\int p2(x)dx) \quad (3)$$

Similar to mutual information calculation based on Shannon entropy, Renyi [13] represented mutual information of secondary Renyi entropy as

$$I(X;Y)=H_{R2}(X)+H_{R2}(Y)-H_{R2}(X,Y)$$

The non-parametric estimate complexity of secondary Renyi entropy is $O(N)$, this makes Mutual information computation complexity decrease from $O(N^2)$ to $O(N)$.We can save a lot of computing work.

In space U, Normalized mutual information expression of Attribute variables f_i and f_j is

$$NI(X,Y)=\frac{H_{R2}(X)+H_{R2}(Y)}{H_{R2}(X,Y)} \quad (4)$$

Supposing that Matrix of Normalized Mutual Information of f_i and f_j is:

$$MNI=\begin{bmatrix} NI_{11} & NI_{12} & \cdots & NI_{1n} \\ NI_{21} & NI_{11} & \cdots & NI_{2n} \\ \vdots & \vdots & \ddots & \vdots \\ NI_{n1} & NI_{n2} & \cdots & NI_{nn} \end{bmatrix}$$

$I_{ij}(i,j=1,2,\ldots n)$,this represents Normalized Mutual Information of f_i and f_j, Similar to n dimensions Mutual Information of Shannon entropy, we define Mutual Information of n dimensions secondary Renyi entropy among attribute variables $f_1,f_2,\ldots,f_n$ as

$$MI(f_1,f_2,\ldots,f_n)=-\sum_{i=1}^{n}\sum_{j=1}^{n}p(NI)\log p(NI) \quad (5)$$

In the formula $p(NI)=\frac{NI}{\|MNI\|}$ $\|MNI\|$is 1- norm of normalized mutual information matrix MNI of f_i and f_j. When attribute values Distribution of f_i and f_j are same, Information Distribution of NI is uniform, n dimensions NI is biggest. When attribute values of f_i and f_j are independent, $NI_{ij}=0(i\neq j)$,n dimensions NI is smallest. $MI(f_1,f_2,\ldots,f_n)$ calculates the mutual information with two-dimensional probability density function of attribute values, not n dimensions probability density function. Computational complexity and the memory space is a linear relationship. So, the new normalized mutual information matrix will greatly reduce the complexity of mutual information calculation of the secondary Renyi entropy. Considering the values range and value distribution of characteristics f_i, it can effectively measure high-dimensional $I(f_i,S,C)$.

4) *Mutual information criterion Structure and Properties*: When U and U' are homomorphic, We can use the mutual information to build resource data attributes feature selection criterion. Pape L [14] measured redundancy brought by uncertainty of Intelligent Manufacturing Systems Resource by information entropy to improve the efficiency of resource allocation. Basis of mutual information criterion is to minimize the Bayes error. It is able to allocate resources effectively under the conditions of massive resources data defect or interference, and enhances the reliability of the optimal solution.

The guideline of resource allocation optimization is attributes feature selection of resource data set, we can analyze from the following three aspects.

① From the point of information entropy, in the same Intelligent Manufacturing System Resources, all resource data Attributes feature value is sufficiently close. The attribute feature information entropy is small, degree of disorder is low. Therefore, using this attribute feature will improve the reliability of identify of resource characteristics.

② From the perspective of mutual information, If the cloud manufacturing resource data sets belong to different categories of identity, its attributes feature value are quite different. The larger mutual information of resource attribute feature and identification category, the stronger the ability to recognize different types of identity.

③ From the perspective of redundancy, If the multi-attribute characteristic is a decision criterion, they should have a lower redundancy. Two highly related features basically reflects the resource data belong to the same category logo, repeated using of such a high correlated multi-attribute feature will cause the data set information redundancy, computational complexity of resource allocation will increase.

Supposing intelligent manufacturing resource data sample set D has $|D|$ data records, C is logo category set

including m logo categories C_i, namely, $C_i \subset C$, Each data

identification tag category has n attributes features. data records belong to logo category C_i is $|C_i|$. For data that has n attributes feature values belong to C_i, we can calculate conditional probabilities.

$$P(T|C_i)=\prod_{k=1}^{n} P(f_k|C_i)=P(f_1|C_i)\times P(f_2|C_i)\times\cdots\times P(f_n|C_i)$$

We use Bayesian error to measure results error with target feature set classification. Its lower limit is the improved *Fano* inequality [15], higher limit is one-half of conditional entropy [16].Noting N_c is the number of attributes values of logo category C_i, Inequality of Bayesian error is

$$\frac{E(c)-M(c;T)-1}{\log_2(N_c)} \le P(T|C_i)=P(g(T)\ne c)$$

$$P(g(T)\ne c)\le \frac{E(c\,|\,T)}{2}=\frac{E(c)-M(c\,;T)}{2}$$

According to number relationship between Bayesian error and $M(c;T)$, We translate Bayesian error minimization problem to the selection of the original features for $M(c;T)$ maximization, namely max M(c;T) .

So, attributes feature selection process with mutual information criterion not only can decrease information of feature set by eliminating redundancy features F, but also can get optimum collection of resources attribute characteristic by mutual information criterion.

4) *Resource attribute feature selection model*: In U and U', Mutual information criterion translates Minimization Bayes error to select optimal feature subset T which makes $MI(c;T)$ of normalized quadratic arbitrary entropy biggest. In attributes feature selection, if some candidate features has relation for logo category c, these relations are very redundant.

Normalized mutual information expressions of attribute variables f_i and f_j is $NI(X,Y)=\dfrac{H_{R2}(X)+H_{R2}(Y)}{H_{R2}(X,Y)}$.

In the feature selection process, Relation measures the relevance candidate features f_λ to logo category c, noting as $NI(c;\ f_\lambda)$.

Mutual information of n dimensions secondary Renyi entropy among $f_1,f_2,\ldots,f_n$ is

$$MI(f_1,f_2,\ldots,f_n)=-\sum_{i=1}^{n}\sum_{j=1}^{n}p(NI)\log p(NI)$$

In the feature selection process, feature's redundancy represents the redundancy between f_λ and S, and the features among S, noting as $MI(c;\ f_\lambda;\ S)$

$$MI(c\,;f_l\,|\,S)=NI(c\,;f_l)-MI(c\,;f_l;S)$$

So, the new information provided by f_λ can be measured by the difference between feature's Relation and redundancy. But, for greater complexity of the high-dimensional feature mutual information calculation, we use every feature's $MI(c;\ f_s;\ f_l)$ in the selected feature set to give approximate calculation of features' redundancy.

For secondary Renyi entropy (a=2) is

$$H_2(x)=-\log(\int p^2(x)dx)$$

For brevity, ratio between $NI(c;f_s)$ and its information entropy $H_2(f_s)$ is called Mutual Information Ratio, Represented by $\Phi_S,\Phi_S=NI(c;a_s)/H2(f_s),\ 0\le\Phi_S\le 1$. It has the following traits: When the value range is the same, The more uniform the value, the smaller its importance; When the feature value is evenly distributed, The greater the range, the smaller its importance. Thereby introducing theorem 1.

Theorem 1: For any $f_s \in S$, $f_\lambda \in F$, supposing in Ranges of information entropy $H_2(C)$,$H_2(f_s)$ and $H_2(f_\lambda)$, information is uniformly distributed. Logo Category c does not change the ratio between f_s' entropy and mutual information of f_s and f_λ. If all the features of selected feature are independent of each other, Redundancy mutual information I_r is

$$I_r=\beta\sum_{f_s\in S}\frac{NI(c;f_s)}{H_2(f_s)}NI(f_s;f_\lambda) \tag{6}$$

From theorem 1, coefficient of $NI(c;f_s)$ is mutual information of candidate feature f_λ and selected feature f_s compared to mutual information of f_s. And, $NI(c;f_s)/H_2(f_s)$ belong to[0,1]. This ratio shows mutual information of candidate feature f_λ and selected feature f_s belongs to mutual information of selected feature f_s and logo category C.

From theorem 1, when $NI(c;f_\lambda)\ \ne 0$,

$$NI(c;f_\lambda|S)=NI(c;f_\lambda)[1-\phi'(C,S)$$
$$=NI(c;f_\lambda)[1-\frac{\beta\sum_{f_s\in S}\{\phi_s NI(f_s;f_\lambda)\}}{NI(C;f_\lambda)}] \tag{7}$$

In U, relation - redundancy coefficient measures the parameter that all features in selected set S compares redundancy of C, It is expressed by r.

$$r=\begin{cases}\infty & \sum_{a_s\in S}\{\phi_s NI(f_s;f_\lambda)\}=0, or S=\varnothing\\ \dfrac{NI(c;f_\lambda)}{\beta\sum_{a_s\in S}\{\phi_s NI(f_s;f_\lambda)\}} & \sum_{a_s\in S}\{\phi_s NI(f_s;f_\lambda)\}>0\end{cases} \tag{8}$$

In the feature selection process, supposing weights of mutual information between each candidate features and selected features agreement is consistent. We use the average value as an approximation of mutual information of candidate features and all selected features. Based on correlation of features, they are divided into three categories: strong correlated features, weak correlated features and irrelevant features. Related - redundancy criterion has the following four properties:

(1) When $r=0$, mutual information between f_λ and C is 0, f_λ is irrelevant features .

(2) When $0<r<1$, redundancy between f_λ and f_s is stronger than the mutual information with C, It is redundant features.

(3) When $r>1$, mutual information between f_λ and C is stronger than the redundancy with f_s, this brings new information, It is correlated features.

(4) When $r=\infty$,we can put f_λ which makes $NI(c;f_\lambda)$ biggest in S.

Strong correlated features will seriously affect the classification performance and must be retained; Weak correlated features res not necessarily features and thus are subject to availability offs; irrelevant features are no necessary, and can be removed.

According to the above characteristics, we can gain the optimal feature set $S=\{f_j|j=1,2,..., \varphi\}$, As shown in Fig. II.

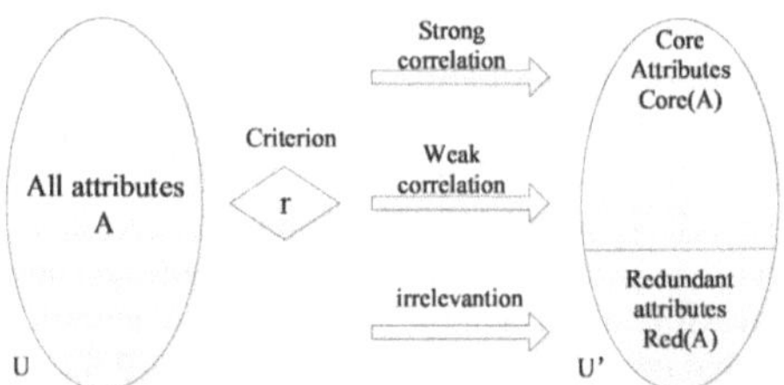

Fig. II. Attributes feature selection criterion model

The objective function combines Correlation - redundancy coefficient and features with $NI(c;f_\lambda)$ among logo categories, and constructed objective function of selecting optimal sequence characteristic. It is called correlation - redundancy coefficient feature selection model.

$$\hbar = \max\{NI(c;f_\lambda)[\frac{NI(c;f_\lambda)}{\beta\sum_{f_s\in S}\{\phi_s NI(f_s;f_\lambda)\}}-1]\} \quad (9)$$

In feature selection, if weights of f_λ and f_s are consistent, we use the mean of mutual information as Approximate estimate between f_λ and S, then $\beta=1/|S|$, $|S|$ is the number of features.

III. ALGORITHM SIMULATION ANALYSIS

A. Experimental data processing

In order to fully verify the accuracy and rationality of the model, according to the Intelligent Manufacturing Resource Classification [17] and data type in SME cloud manufacturing service platform. We use the data from Mold design and manufacturing sector in SME cloud manufacturing service platform. After checking the data input, comparison and discussion of the conclusions, we demonstrates the rationality and Science of the constructed model in this paper.

Many attributes involves mold design and manufacturing resources. Resource attributes used here consider the company, product, service, quality, cost and many other features. Various values of all attributes values are mapped to symbols. For decision attributes, selecting the resource maps to 0; otherwise 1.
Data parameters are shown in Table. I.

TABLE.I. MOLD DESIGN AND MANUFACTURING RESOURCES ATTRIBUTE PARAMETER

No.	Feature	Data description
f_1	Ownership	1:Chinese;2:Foreign;3:Joint venture
f_2	Asset size	1: (-∞,100],2: (100,500],3: (500,+∞) Unit: 10,000 Yuan
f_3	Credit rating	1:AAA;2: AA;3: A;4:B;5:C
f_4	Transport distance	1: (-∞,200];2: (200,1000];3:(1000,+∞) Unit: km
f_5	Product cost	1:<Average value;0:>Average value. Average value:2580 Yuan / Piece
f_6	Production time	1:< Average value;0:> Average value Average value:6 Days
f_7	Transportation cost	1:< Average value;0:> Average value. Average value:35 Yuan / Piece
f_8	Transportation time	1:< Average value;0:> Average value. Average value:2.5 Days
f_9	Service level	1:Excellent;1:Good;3:Moderate;4:Bad
f_{10}	Risk assessment	1:Low risk;2:Moderate Risk;3:High risk
f_{11}	Rate of qualified products	1: (0.97,1],2:(0.95,0.98],3: (0,0.95)
f_{12}	History of transactions	1:[0,10],2: (10,20],3: (20,+∞)
f_{13}	Customer Reviews	1:Five Stars;2: Four Stars;3: Three Stars;4: Two Stars;5: One Star
d	Choose	0:Yes;1:No

After the above data preprocessing, we select 200 cases of resource data from platform for experimental simulation. Before the processing results of the ten cases resource data is shown in TABLE. II.

TABLE. II. Pretreatment Results of ten cases resource data

f_1	f_2	f_3	f_4	f_5	f_6	f_7	f_8	f_9	f_{10}	f_{11}	f_{12}	f_{13}	d
1	1	3	2	0	1	1	1	2	2	3	1	4	0
3	2	4	2	1	1	0	0	1	3	2	2	3	1
1	3	2	3	0	0	1	0	3	1	2	2	5	1
2	1	3	1	1	0	0	1	2	3	3	1	3	0
1	2	2	1	0	1	1	1	4	3	1	3	2	0
1	1	5	2	0	0	0	0	2	2	1	2	3	1
2	3	4	3	1	0	0	1	1	2	1	2	1	0
1	1	3	2	0	1	1	0	1	3	3	1	4	1
1	1	4	1	0	0	0	1	3	1	2	1	2	0
3	2	5	2	1	1	1	1	2	2	2	3	3	1

B. Simulation Analysis

In the feature selection process, we get the Mutual information of candidate feature set $NI(c;f_i)$ and parameter $log_2(R^2C)$, As shown in Figure. III.

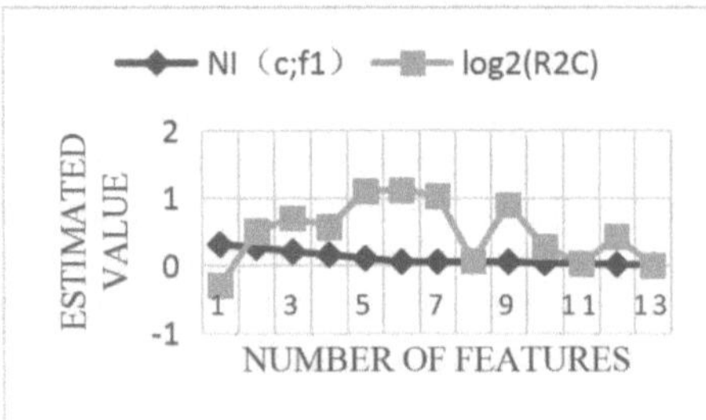

Fig. III. Mutual information Estimates of $NI(c;f_i)$ and $log_2(R^2C)$

We use the broad float attributes features selection algorithm to delate redundant attributes in space. With the formula(7), We calculate the mutual information of normalized secondary Renyi entropy $NI(c;f_i)$, Redundancy $MI(c;f_\lambda:S)$ and $MI(c;f_L|S)$, Some values are shown in Table. III.

TABLE. III. MUTUAL INFORMATION VALUE OF SECONDARY RENYI ENTROPY

Step	Feature	$NI(c;f_i)$	$MI(c;f_\lambda:S)$	$MI(c;f_L\|S)$	Time
1	f_{13}	0.217395	0	0.217395	0.040412
2	f_{12}	0.188451	0.039716	0.148735	0.068042
3	f_3	0.250012	0.104433	0.145579	0.096319
4	f_{10}	0.212886	0.116769	0.096116	0.127950
5	f_8	0.182942	0.118157	0.064785	0.152366
6	f_9	0.223176	0.170237	0.052939	0.151798
7	f_1	0.107515	0.055656	0.051859	0.206528
8	f_{11}	0.349667	0.301591	0.048076	0.237578

Based on the attribute feature selection process, we analyze mutual information metric redundancy of selected set by increasing features of F and reducing features of S. We mining attributes feature selection rule.

When we select five attribute features, redundancy between attributes features set and identified categories is minimum. Increasing or decreasing other attributes features, will increase redundancy. Therefore, this article uses feature selection method as set forth above, Optimal feature set is $\{f_{13},f_3,f_{12},f_{10},f_8\}$.

IV. CONCLUSION

The experiments carries out attributes feature selection under conditions of intelligent manufacturing resource data space with same state, and gives normalized secondary Renyi entropy mutual information criterion, which can effectively identify correlation and redundancy between attributes features indicating category. It also proposes the corresponding feature space heuristic search algorithm. By cloud resource data experiment in SME cloud manufacturing services platform, the results indicate that the model can effectively eliminate the resource attributes redundancy features.

The rapid development of big data technology makes intelligent manufacturing resource data features complex and diverse, and increases the difficulty of optimal allocation of resources. Resource attributes feature model constructed in this paper can eliminate redundant features in the resource allocation process, obtain nuclear attributes features, thereby reduces the complexity of resource features, improves the criterion of heterogeneous resources, and greatly improves the efficiency of resource allocation. It laid the foundation for the further development of intelligent manufacturing.

REFERENCES

[1] Guan Gong, Guan Liang Hu. China's manufacturing efficiency of resource allocation and total factor productivity[J]. Economic Research, 2013 (4): 4-15.

[2] Zhou Ke, Manufacturing task decomposition and manufacturing unit-level resource allocation collaborative optimization [J]. Journal of Harbin Institute of Technology, 2009, 41(11): 47-52.

[3] Wang Shilong. Manufacturing resource optimization study in cloud manufacturing environment [J]. Computer Integrated Manufacturing Systems, 2012, 18(7): 1396-1405.

[4]M.E. Porter, M.R. Kramer. "Creating Shared Value: How to Reinvent Capitalism—and Unleash a Wave of Innovation and Growth," Harvard Business Review, 63-70, 2011.

[5] Laili Y, Tao F, Zhang L. A study of optimal allocation of computing resources in cloud manufacturing systems[J]. The International Journal of Advanced Manufacturing Technology, 2012, 63(5-8): 671-690.

[6] Shen Jiang, Xu Man. Contractual relationships mechanism based on virtual enterprise strategic behavior [J]. Computer Integrated Manufacturing Systems, 2005, 11(5): 738-743.

[7] Chen Bin, Bai Xiaoying. Distributed stretchable systems research summary [J]. Computer Science, 2011,08:17-24.

[8] Bertsimas D, Gupta S, Lulli G. Dynamic resource allocation: A flexible and tractable modeling framework [J]. European Journal of Operational Research, 2014, 236(1): 14-26.

[9] Huang Haixia. Study of China's strategic emerging industries Technology Resource Allocation Efficiency based on DEA model[J]. China Soft Science, 2015 (1): 150-159.

[10] Hanel R, Thurner S, Tsallis C. On the robustness of q-expectation values and Rényi entropy [J]. arXiv preprint arXiv:0812.3855, 2008.

[11] Müller B, Zessack B, Zölzer U. Optimization of BICM mutual information and bit error rate of 8-ary signal constellations using genetic algorithms[C]//2014 IEEE 15th International Workshop on Signal Processing Advances in Wireless Communications (SPAWC). IEEE, 2014: 214-218.

[12]Renyi A. Some fundamental questions of information theory [A].In Selected Papers of Alfred Renyi [M], 10: 251-282.

[13]Renyi A. Probability Theory [M]. Budapest: Dover Publications, 2007

[14] Pape L, Dagli C. Assessing robustness in systems of systems meta-architectures [J]. Procedia Computer Science, 2013, 20: 262-269.

[15] Braun G, Pokutta S. An information diffusion Fano inequality[J]. arXiv preprint arXiv:1504.05492, 2015.

[16] Tan V Y F. Asymptotic estimates in information theory with non-vanishing error probabilities [J]. arXiv preprint arXiv:1504.02608, 2015.

[17] Tao F, Zuo Y, Da Xu L. IoT-based intelligent perception and access of manufacturing resource toward cloud manufacturing [J]. IEEE Transactions on Industrial Informatics, 2014, 10(2): 1547-1557.

Social Insurance Expenditure and Future Trend in China

Meng Lv, Hong Yin* , Xue-chao Ma

Mechanical and Electrical Engineering College, Kunming University of Science and Technology,Kunming,China

(*2473664405@qq.com)

Abstract - **Over the years, the social security system is considered to be economic "shock absorber valve" and "built-in stabilizers", which plays an important role in income redistribution. This article by searching for 2002-2012, China Statistical Yearbook relevant data, using descriptive statistics, cluster analysis, regression and time series analysis, statistical methods, development of China's social security, social security and urban and rural residents' consumption structure the impact study to obtain changes and laws of social security spending, the main constraints to predict trends in social security spending next five years.**

Keywords - **Social security, Stepwise regression, Time series analysis**

I. INTRODUCTION

The development and perfection of China's market economy need to establish a perfect social security system. Since the founding of state, the establishment and implementation of social security system for protection of people's lives, promote national economic development and social stability has played a positive role. However, with the deepening of the economic system reform and the advent of population aging, the problems and shortcomings of original social security system has been exposed. This paper analyzes the current level of social security expenditure in our country, and put forward the efficient solution.

II. METHODOLOGY

The research object of this article is social insurance expenditure that the core content of social security expenditure in China, and analysis social insurance expenditure in five aspects and effects of consumption of residents in urban and rural areas, finally obtained in social insurance expenditure accounted for major aspects of these factors and consumption of residents in urban and rural areas, and to predict the level of developing social. Thus to improve the social security expenditure, promoting stable development of China's economic and residents' consumption level of urban and rural[1].

The trend analysis of the social insurance expenditure by collecting China Statistical Yearbook, 2002 to 2012, China's urban residents consumption level of the relevant data shown in the following table:

TABLE I
CHINA'S URBAN RESIDENTS CONSUMPTION LEVEL OF THE RELEVANT DATA

Year	Social Insurance Fund expenditure (Billion)
2002	347. 15
2003	401. 64
2004	462. 74
2005	540. 08
2006	647. 74
2007	788. 78
2008	992. 51
2009	1230. 26
2010	1481. 85
2011	1805. 46
2012	2218. 16

First of all, according to the existing data using the Minitab software to draw a forecast map of social insurance expenditure for the next five years[2]:

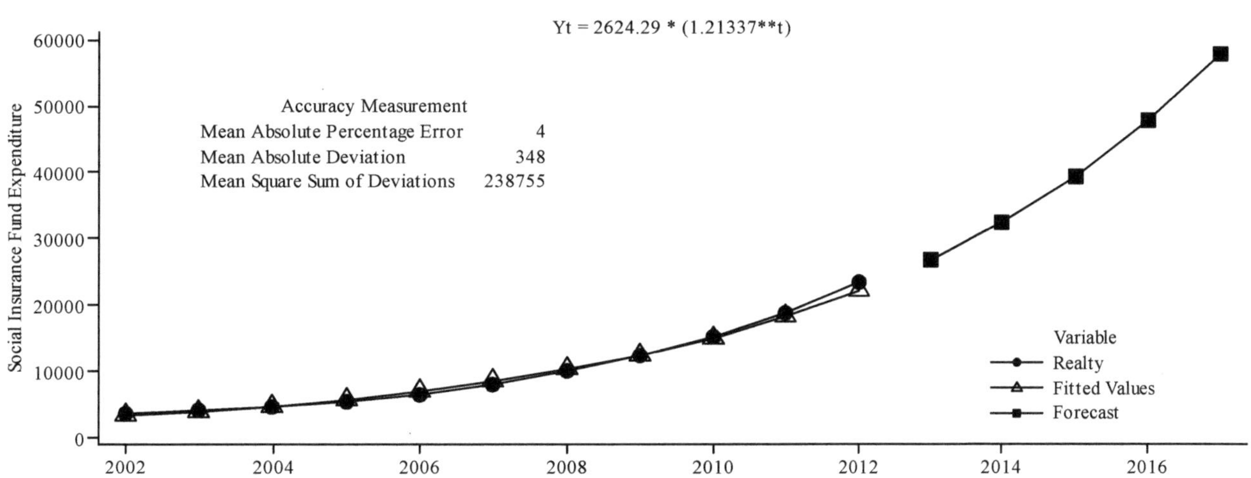

E. Qi et al. (eds.), *Proceedings of the 23rd International Conference on Industrial Engineering and Engineering Management 2016,* DOI 10.2991/978-94-6239-255-7_36

Fig 1. Trend analysis of social insurance fund expenditure

III. RESULTS

It is verified that the fitting effect of the "exponential growth" is the best in the trend analysis. Fitting trend equation:

Yt = 3249.94 * (1.22010**t)

Stepwise regression analysis of social insurance structure

Through access to the Statistical Year Book, it above the social insurance fund in our country is divided into five aspects: basic old age insurance fund, unemployment insurance fund, medical insurance fund, industrial injury insurance fund, maternity insurance fund.

TABLE II. 2002-2012 SOCIAL SECURITY DATA

Project / Year	Basic Old Age Insurance Fund	Unemployment Insurance Fund	Medical Insurance Fund	Industrial Injury Insurance Fund	Maternity Insurance Fund	Social Insurance Fund
2002	2,842.9	182.6	409.4	19.9	12.8	5308.99
2003	3,122.1	199.8	653.9	27.1	13.5	6030.00
2004	3,502.1	211.3	862.2	33.3	18.8	6510.94
2005	4,040.3	206.9	1,078.7	47.5	27.4	7182.10
2006	4,896.7	198.0	1,276.7	68.5	37.5	7942.86
2007	5,964.9	217.7	1,561.8	87.9	55.6	8696.55
2008	7,389.6	253.5	2,083.6	126.9	71.5	9997.47
2009	8,894.4	366.8	2,797.4	155.7	88.3	11242.85
2010	10,554.9	423.3	3,538.1	192.4	109.9	12264.54
2011	12,764.9	432.8	4,431.4	286.4	139.2	13471.46
2012	15,561.8	450.6	5,543.6	406.3	219.3	15160.9

TABLE III COEFFICIENTA OF STEPWISE REGRESSION ANALYSIS

Model		Non Standardized Coefficient B	Standard Error	Standard Coefficient Trial Version	t	Sig.
1	(constant)	2001.590	.602		3322.975	.000
	Basic Old Age Insurance Fund	.001	.000	.960	10.299	.000
2	(constant)	1999.068	.986		2027.662	.000
	Basic Old Age Insurance Fund	.002	.000	2.214	5.010	.001
	Industrial Injury Insurance Fund	-.034	.012	-1.269	-2.873	.021
3	(constant)	1999.789	.771		2593.147	.000
	Basic Old Age Insurance Fund	.003	.000	3.352	6.412	.000
	Industrial Injury Insurance Fund	-.051	.011	-1.872	-4.788	.002
	Unemployment Insurance Fund	-.017	.006	-.569	-2.782	.027
4	(constant)	2003.133	1.107		1809.085	.000
X_2	Basic Old Age Insurance Fund	.001	.000	1.893	3.469	.013
X_5	Industrial Injury Insurance Fund	-.075	.010	-2.763	-7.612	.000
X_3	Unemployment Insurance Fund	-.032	.006	-1.033	-5.454	.002
X_4	Medical Insurance Fund	.006	.002	2.787	3.367	.015

a. Dependent Variable: Year

The results from the stepwise regression show that the final model is:

$$Y=2003.133+0.001x_2-0.075x_5-0.032x_3+0.06x_4$$

Thus, it is concluded that, in the social insurance fund, basic old age insurance fund, industrial injury insurance fund, unemployment insurance fund and medical insurance fund are significant factors.

The stepwise regression analysis of social insurance expenditure impact on the consumption level of urban and rural residents.

TABLE Ⅳ 2002-2012 CONSUMPTION OF URBAN AND RURAL AREAS AND THE INSURANCE FUND EXPENDITURE

Year	Basic Old Age Insurance Fund	Unemployment Insurance Fund	Medical Insurance Fund	Industrial Injury Insurance Fund	Maternity Insurance Fund	Per Capita Consumption of Cities	Per Capita Consumption of Towns
2002	2,842.90	182.60	409.40	19.90	12.80	195.8	57.7
2003	3,122.10	199.80	653.90	27.10	13.50	215.1	43.0
2004	3,502.10	211.30	862.20	33.30	18.80	240.2	48.3
2005	4,040.30	206.90	1,078.70	47.50	27.40	277.8	54.5
2006	4,896.70	198.00	1,276.70	68.50	37.50	309.5	63.1
2007	5,964.90	217.70	1,561.80	87.90	55.60	357.7	74.2
2008	7,389.60	253.50	2,083.60	126.90	71.50	418.3	76.7
2009	8,894.40	366.80	2,797.40	155.70	88.30	474.2	84.1
2010	10,554.90	423.30	3,538.10	192.40	109.90	499.2	94.0
2011	12,764.90	432.80	4,431.40	286.40	139.20	581.3	122
2012	15,561.80	450.60	5,543.60	406.30	219.30	657.1	147.6

First, the urban per capita consumption expenditure to do the stepwise regression analysis, the results are as follows:

TABLE Ⅴ COEFFICIENT[A] OF STEPWISE REGRESSION ANALYSIS

	Mod el	Non Standardized Coefficient		Standard Coefficient		
		B	Standard Error	Trial Version	t	Sig.
1	(constant)	124.450	13.714		9.074	.000
	Basic Old Age Insurance Fund	.036	.002	.991	21.719	.000
2	(constant)	72.515	24.467		2.964	.018
	Basic Old Age Insurance Fund	.056	.009	1.545	6.557	.000
	Industrial Injury Insurance Fund	-.708	.297	-.562	-2.383	.044
3	(constant)	90.021	19.596		4.594	.003
X_1	Basic Old Age Insurance Fund	.078	.010	2.139	7.493	.000
X_2	Industrial Injury Insurance Fund	-1.105	.269	-.876	-4.102	.005
X_3	Unemployment Insurance Fund	-.423	.159	-.297	-2.657	.033

The result from stepwise regression show that the final model is:

$$Y=90.021+0.78x_1-1.105x_2-0.423x_3$$

The conclusion: in the structure of consumption per capita in the urban areas, the basic old age insurance, Industrial Injury Insurance Fund, unemployment insurance are significant factors. Similarly, rural per capita consumption expenditure to do a stepwise regression analysis, the final model for the:

$$Y=44.296+0.260x_2$$

Therefore, it can be concluded that: in the rural consumption structure, industrial injury insurance is a significant factor.

IV. DISCUSSION

1. The range of social pooling must be greater, can not be too narrow. To establish a social security system that covering urban and rural areas, make all types of people integrated into the coverage gradually, Implementation co-ordination and the assuring all those should be assured in urban and rural. In urban areas, should continue to improve the insurance system of

pension, medical, unemployment, work injury, maternity, and the various types of workers and flexible employment shoule be included in the coverage gradually. in rural areas, should be fully established rural minimum living standard security system, and further promote the new rural cooperative medical care .

2. The cities should strive assuring all those should be assured; The eligible poor people in backward regions should be into minimum living guarantee range and guarantee their basic living needs.

V. CONCLUSION

1. The expenditure of social insurance in our country present a upward trend.

2. The basic old-age insurance, industrial injury insurance, unemployment insurance is a significant factor in the urban per capita consumption expenditure, industrial injury insurance is a significant factor of per capita consumption expenditure in rural areas.

3. The consumption of China's social insurance fund will continue to rise within the next five years. But unemployment insurance growth will be present a downward trend, while the basic pension insurance, medical insurance, maternity insurance, industrial injury insurance these four spending will still increased year by year.

This article discusses the above, it was found that generous social security to improve efficiency, improve political stability and social cohesion of unequal support[3], but the current social security field of our country, there are still existence disadvantage to promotion resident consumption, and therefore effective measures should take to promotion the consumer expenditure with finance social security, in order to expand the consumption level of urban and rural residents, increasing consumption overall.

From the current situation in China, the main sources of social security funds are paid by enterprises, a small part by the National, individuals pay less, it does not reflect the principle of the state, collectives and individuals jointly responsible. In fact the basic security with government-sponsored becomes all insured, which is incompatible with conditions of China's low level of productivity, population, weak economic foundation and the country's overall financial strength is not strong.

ACKNOWLEDGMENT

Thank the teacher usually strict requirements, and promote the completion of this thesis gradually. Thanks for the patience guide of the laboratory students, help me to solve all kinds of questions about the process, and not only learned a lot, but also gain more friendship in the process.

REFERENCES

[1] KuangnanFang, ZiyiZhang. Study on the Influence of Social Security on the Consumption of Urban and Rural Households.[J] Statistical Research (2013)

[2] NanHong. MINITAB of Statistical Analysis Tutorial. [J]Electronic Industry Press (2008)

[3] Giorgio Bellettini, Carlotta Berti Ceroni. Social Security Expenditure and Economic Growth: An Empirical Assessment.[J] Research in Economics (2000) 54, 249–275

A Two-Level Principal-Agent Model for Schedule Risk Control of IT Outsourcing Project Based on Genetic Algorithm

Fuquan Sun, Shupeng Duan, Fuqian Lu*, Hualing Bi

(College of Information Science and Engineering, Northeastern University, Shenyang, China 110819)

(fqlu@neuq.edu.cn)

Abstract - **IT service outsourcing is an effective way to enhance the core competitiveness for many enterprises. But the schedule risk of IT outsourcing project may cause enormous economic loss to enterprise. In this paper, the principal-agent theory is used to build the model for schedule risk control of IT outsourcing project. According to the feature of problem, Genetic Algorithm (GA) is designed to solve the proposed model and the optimal plan of controlling schedule risk is given based on the simulation analysis. The simulation results demonstrate that this paper can provide the scientific quantitative proposal for the decision maker who needs to control the schedule risk of IT outsourcing project.**

Key Words - **IT Outsourcing Project, Schedule Risk, Risk Control, Principal-agent theory, Genetic Algorithm**

I.INTRODUCTION

The milestone event of IT outsourcing is that in 1989, Kodak which total sales is $ 18,000 million signed outsourcing contracts worth $ 250 million over 10 years. Since the case of Kodak that plays an important role on IT outsourcing, the industry of IT outsourcing has became mature gradually [1-6].

Although IT outsourcing has advantage in reducing cost and enhancing the core competence, the corresponding theoretical research is still in its infancy, especially the problem of controlling the schedule risk of IT outsourcing, which may bring about huge loss to company. Consequently, it is very vital to research how to control the schedule risk of IT outsourcing.

Researchers have done a lot of related research [7-12]. But most methods and models proposed in the literature only discuss the risk management issues on project itself, and ignore the cooperation between principal and agent and the distribution characteristics of the IT outsourcing activities. In recent years, principal-agent theory has been widely employed to solve the problem of risk control of IT outsourcing and good results have been achieved through these studies.

Aubert, B. A. concluded some adverse consequences of IT outsourcing based on existing literatures [13-15]. Then, he argued that the risk of IT outsourcing came from enterprises, agents and the process of IT activities and proposed corresponding risk control measures based on principal-agent theory. Bahli, B. proposed a scenario-based conceptualization of the IT outsourcing risk and applied it to the specific context of IT outsourcing using transaction cost and agency theory [16].Osei-Bryson offered a method and some mathematical models for analyzing risks and constructing incentive contracts for information system outsourcing [17]. Sanfa, C. analyzed risk factors of producer services outsourcing from the perspective of engineering and afforded managers a theoretic method to control outsourcing risks by designing the incentive and monitoring mechanism of the producer services outsourcing contract [18-19].

In this paper, we build a two-level principal-agent model considering completion probability for schedule risk control of IT outsourcing project based on the Distributed Decision Making (DDM) theory and principal-agent theory [20-21]. According to the feature of problem, GA is designed to solve the proposed model and the optimal plan of controlling schedule risk is given based on the simulation analysis. The purpose of this paper is to provide crucial decision support for the people who need to control the schedule risk of IT outsourcing project [22-23].

II.THE PRINCIPAL-AGENT MODEL CONSIDERING COMPLETION PROBABILITY

Based on the DDM theory, a two-level schedule risk control model of IT outsourcing project is built.. In the top-level, the decision maker is the principal who determines how to allocate the risk control capital among agents. The objective of top-level is to maximize the completion probability of project. In the base-level, the decision maker is the agent who determines how to allocate the risk control capital among activities of subproject. The objective of base-level is to maximize the agent's benefit.

A. Top-level

Variable Definition:

x_i:	risk control capital of subproject i
I:	the number of agents or subprojects
$P_i(x_i)$:	the completion probability of subproject i
V_i:	the upper bound of floating wage for agent i
AL_i:	the aspiration level of agent i
$\Delta T_i(x_i)$:	the duration reduction of subproject i
T_i:	the duration of subproject i before controlling schedule risk
$T_i(x_i)$:	the duration of subproject i after controlling schedule risk
$C_i(\Delta T_i(x_i))$:	the cost of developing subproject i, which is predicted by principal
C_i^{max}:	the upper bound of cost for developing subproject i
X_i^{max}:	the upper bound of risk control capital for subproject i

E. Qi et al. (eds.), *Proceedings of the 23rd International Conference on Industrial Engineering and Engineering Management 2016,* DOI 10.2991/978-94-6239-255-7_37

The top-level model can be described as follows

$$\max \prod_{i=1}^{I} P_i(x_i) \quad (1)$$

$$s.t \quad P_i(x_i)V_i - C_i(\Delta T_i(x_i)) \geq AL_i \quad (2)$$

$$P_i(x_i^*)V_i - C_i\left(\Delta T_i(x_i^*)\right) \geq P_i(x_i)V_i - C_i(\Delta T_i(x_i)) \quad (3)$$

$$C_i(\Delta T_i(x_i)) \leq C_i^{max} \quad (4)$$

$$x_i \leq X_i^{max} \quad (5)$$

$$x_i \in N \quad (6)$$

The objective of top-level shown in Eq.(1) is to maximize the completion probability of the project; Formula (2) indicates participation constraint; Formula (3) indicates incentive compatibility constraint; Formula (4) indicates the cost constraint of the subproject; Formula (5) indicates risk control capital constraint; Formula (6) indicates that the risk control capital x_i is a natural integer, which is the decision variable in the top-level model.

B. Base-level

In the base-level, the decision makers are the agents, and there are I agents. Take agent i for example, the allocated risk control capital is x_i which will be distributed into the J_i activities on the critical path.

Variable Definition:

- $C_{ij}(\Delta t_{ij}(y_{ij}))$: the cost of developing activity j
- $P_{ij}(y_{ij})$: the completion probability of activity j
- y_{ij}: the allocated risk control capital of activity j
- J_i: the number of the activities on the critical path of subproject i
- t_{ij}: the duration of activity j before controlling schedule risk
- $t_{ij}\left(y_{ij}\right)$: the duration of activity j after controlling schedule risk
- $\Delta t_{ij}\left(y_{ij}\right)$: the duration reduction of activity j

The base-level model can be described as follows

$$\max \prod_{j=1}^{J_i} P_{ij}\left(y_{ij}\right)V_i - \sum_{J=1}^{J_i} C_{ij}(\Delta t_{ij}(y_{ij})) \quad (7)$$

$$s.t. \quad \sum_{j=1}^{J_i} y_{ij} \leq x_i^* \quad (8)$$

$$\sum_{J=1}^{J_i} C_{ij}(\Delta t_{ij}(y_{ij})) \leq C_i^{max} \quad (9)$$

$$y_{ij} > 0 \quad (10)$$

Formula (7) indicates that the objective of base-level is to maximize agent's benefit; Formula (8) indicates that the sum of the risk control capital which is allocated to the activities on critical path will not be greater than the risk control capital which is allocated to subproject by principal; Formula (9) indicates the cost constraint on developing subproject. Formula (10) indicates that risk control capital y_{ij} is a positive real number, which is the decision variable in the base-level model.

III.DESIGN OF GENETIC ALGORITHM

A. Encoding Scheme

In this section, chromosome is represented by real number . Each chromosome represents a combination of risk control capital. The length of chromosome represents the number of the activities on the critical path.

B. Population Initialization

Initial population with some chromosomes is generated randomly. Punishment strategy is adopted to deal with constrains, so we don't have to judge whether the initial solution meet the constraint conditions.

C. Fitness Function

Considering the proposed constrained optimization problems, this paper set up a fitness function with penalty term to evaluate individuals. The fitness function with penalty term is given as

$$F = f\left(y_{ij}\right) - a\left(\sum_{j=1}^{J_i} y_{ij} - x_i^*\right)^+ - b\left(\sum_{J=1}^{J_i} C_{ij}(\Delta t_{ij}(y_{ij})) - C_i^{max}\right)^+ \quad (11)$$

where

$$f\left(y_{ij}\right) = V_i \prod_{j=1}^{J_i} P_{ij}\left(y_{ij}\right) - \sum_{J=1}^{J_i} C_{ij}(\Delta t_{ij}(y_{ij})) \quad (12)$$

a and b are punishment coefficients, respectively. Operation $(x)^+$ is defined as follows

$$(x)^+ = \begin{cases} x, x>0 \\ 0, else \end{cases} \quad (13)$$

D. Selection

This paper takes proportional selection strategy . For each individual, the probability of being selected is the proportion of the fitness of it to the sum of all individuals' fitness. Then, the probability of individual i being selected is given by

$$P_i = F_i / \sum_{i=1}^{NP} F_i \quad (14)$$

where F_i is the fitness of individual i and NP is the population size. Here, we adopt well-known Roulette Wheel scheme. In order to prevent the best individual in each generation from being destroyed, elite-preservation strategy is also used. That is to say, the best individual of each generation directly become one of individuals in the next generation without crossover and mutation operation.

E. Crossover

Double-point crossover genetic algorithm is adopted in this paper, which is benefit for keeping excellent individual.

F. Mutation

Here, we use non-uniform mutation . This operation is responsible for fine tuning capabilities of the system. It is applied to a single chromosome. The action of non-uniform mutation depends on the age of the population. If the chromosome x is selected for this mutation and its offspring is $x' = (x_1, \ldots, x_k', \ldots, x_n)$, then the value of x_k' is given by

$$x_k^{'}=\begin{cases} x_k+\Delta(t,x_k^U-x_k), & if\ a\geq 0.5 \\ x_k-\Delta(t,x_k-x_k^L), & if\ a<0.5 \end{cases} \quad (15)$$

where a is a random number within the interval [0,1], x_k^U and x_k^L are upper and lower bound of x_k and $\Delta(t,y)$ returns to a value in the range [0, y].

$$\Delta(t,y)=yr(1-t/T)^b \quad (16)$$

where r is a random number within the interval [0,1], T represents the maximum generation , t represents the current generation and b is the constant which represents the level of non-uniform.

G. Stopping Criterion

Use the specified precision as stopping criterion.

Ⅳ.NUMERICAL EXAMPLES

In this example, there are 1 principal and 1 agent, $I=1$. There are 5 activities on the critical path of subproject 1, J_1=5. The parameters of top-level model are set as follows.

The upper bound of floating wage V_1=5000. The risk control capital for subproject 1 is no greater than 1000, X_1^{max}=1000. The aspiration level of agent 1 is 3000. AL_1=3000. For agent 1, the cost of developing subproject 1 is no greater than 1000, C_1^{max}=1000.

The relationship between the completion probability of subproject and risk control capital is a positive correlation, namely, the higher the completion probability of subproject is, the more risk control capital is needed, which is consistent with actual situation. Before using risk control capital, the completion probability of subproject 1 is 0, that is, $P_1(0)=0$.

The relationship between the duration of project and risk control capital is negatively correlated. The duration of project will be reduced from T_1 to $T_1(x_1)$, after using risk control capital. In general, larger duration reduction of subproject corresponds to more developing cost.

The risk control capital plays a role in improving the completion probability of activity. In general, the duration of activity will be reduced from t_{1j} to $t_{1j}(y_{1j})$ under the action of risk control capital.

In this example, the initial duration of all activities on the critical path are assumed to be 100, that is, $t_{1j}(0)=\sigma_{1j}=100$, $j\in 1,\dots,5$.

The value of GA's parameters for example Ⅰ are set in table Ⅰ.

TABLE Ⅰ. The value of GA's parameters

Parameter	Value
NP	50
LC	5
NG	100
P_c	0.7
P_m	0.04

where NP is population size; LC is the length of chromosome; NG is the maximum generation; P_c is the probability of crossover; P_m is the probability of mutation.

The simulated decision-making results of example Ⅰ are shown in table Ⅱ.

TABLE Ⅱ. The simulation results

Criterion	Results
The completion probability of subproject	0.913
The duration of subproject	338.1
The completion probability of project	0.913
The duration of project	338.1
The benefit of agent	3606.1
The capital allocated to agent	978
The capital used by agent	973.7
The combination of allocated capital	257.9 205.0 189.4 167.5 154.0

We can draw some conclusions from table Ⅱ. Firstly, principal allocates 978 risk control capital which is no greater than the upper limit 1000 to agent 1. After risk control, the completion probability of the project is 0.913. The duration of the project is 338.1. The benefit of agent 1 is 3606.1. Secondly, agent 1 respectively allocates capital 257.9, 205.0, 189.4 167.5, 154.0 to the 5 activities on the critical path. The sum of allocated capital is 973.7, which is no greater than 978. Finally, for principal, the completion probability of project is improved from 0 to 0.913 and the duration of project is reduced from 500 to 338.1; for agent 1, his maximum benefit is 3606.1 that is greater than AL_1=3000. So, principal realize the objective of controlling the schedule risk of IT outsourcing project. Meanwhile, agent 1 realizes the objective of obtaining maximum benefit.

Ⅴ.THE ANALYSIS OF MODEL

Two primary model parameters are analyzed, including the upper bound of developing cost C^{max} and the upper bound of risk control capital X^{max}. The influence of C^{max} on the model is shown in table III, and the corresponding analysis graph is shown in figures 1-2. The influence of X^{max} on the model is shown in table Ⅳ, and the corresponding analysis graph is shown in figures 3-4.

TABLE III. The influence of C^{max} on the model

C^{max}	Completion Probability	Benefit of Agent
550	0.657	2775.6
580	0.698	3000.0
600	0.718	3049.1
650	0.752	3138.6
700	0.780	3244.4
750	0.820	3386.6
800	0.850	3486.6
850	0.877	3544.7
900	0.882	3553.9
950	0.896	3584.5
960	0.914	3618.0
1000	0.915	3619.1
1100	0.915	3619.0

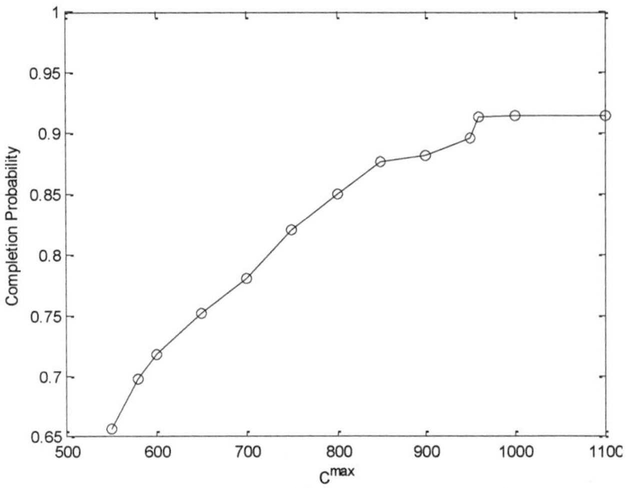

Fig. 1 The influence of C^{max} on completion probability

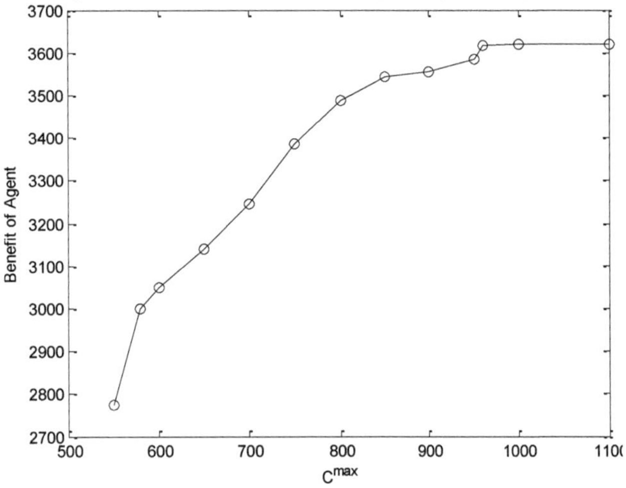

Fig. 2 The influence of C^{max} on benefit of agent

When C^{max} <580, the benefit of the agent is less than AL=3000. So, C^{max} should be greater than 580; When 580< C^{max} <960, the completion probability of subproject and the benefit of the agent will be improved with the increase of C^{max}; When C^{max} > 960, the completion probability of subproject and the benefit of agent is more or less steady. Namely, the completion probability of subproject and the benefit of the agent will be improved with the increase of developing cost until the benefit of the agent reach maximum. Then, C^{max} lose its work and no matter how it changes, agent will always spend 960 to develop his subproject. Consequently, C^{max}=1000 is reasonable.

TABLE Ⅳ. The influence of X^{max} on the model

X^{max}	Completion Probability	Benefit of Agent
600	0.649	2792.5
630	0.685	3000.0
650	0.722	3048.7
700	0.746	3118.4
750	0.801	3318.4
800	0.821	3379.4
850	0.860	3503.9
900	0.884	3576.7
950	0.909	3605.6
978	0.911	3605.2
1000	0.911	3604.5
1100	0.910	3605.1

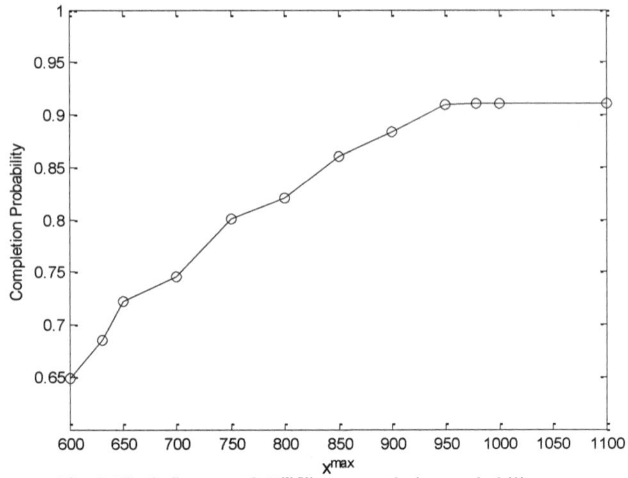

Fig. 3 The influence of X^{max} on completion probability

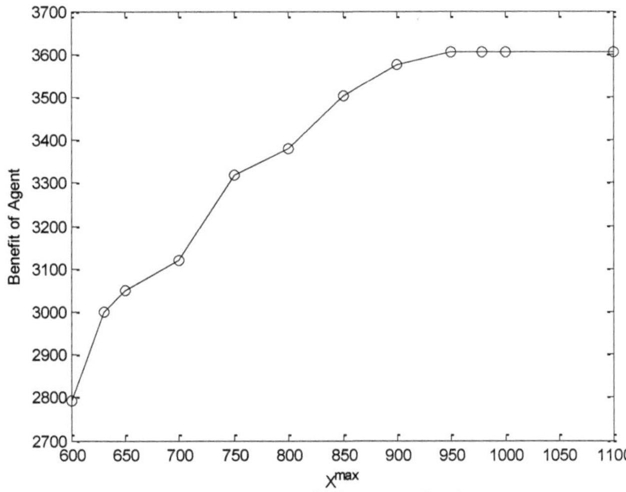

Fig. 4 The influence of X^{max} on benefit of agent

The influence of X^{max} on the model is similar to C^{max}. When X^{max} <630, the benefit of the agent is less than AL=3000. So, X^{max} should be greater than 630; When 630< C^{max} <978, the completion probability of subproject and the benefit of the agent will be improved with the increase of X^{max}; When X^{max} > 978, the completion probability of subproject and the benefit of agent is more or less steady. Namely, the completion probability of subproject and the benefit of the agent will be improved with the increase of risk control capital until the benefit of the agent reach maximum. Then, X^{max} lose its work and no matter how it changes, principal will always spend 978 to control schedule risk. Consequently, X^{max}=1000 is reasonable.

Ⅶ.CONCLUSION

This paper focus on the schedule risk of IT outsourcing project, the principal-agent theory is applied to build a two-level principal-agent schedule risk control model for IT outsourcing project, and GA is also designed to solve the resulting optimization problem. The simulation results illustrate that the model can largely improve the completion probability and reduce the duration of IT outsourcing project, which achieve the objective of effectively controlling the schedule risk of IT outsourcing project. The proposed GA can effectively solve the described problem. Consequently, the above model and algorithm can provide important decision support for controlling the schedule risk of IT outsourcing project.

ACKNOWLEDGMENTS

This work is supported by the National Science Foundation of China under Grant No. 71401027, Natural Science Foundation of Hebei Province under Grant No. G2016501086, Key Project of Science and Technology Research of Higher Education in Hebei Province under Grant No. ZD2016202.

REFERENCE

[1]. Mary C.L Shaji A.K Leslie P.W. “A review of the IT outsourcing literature: Insights for practice” Journal of Strategic Information Systems,2009,18;130-146.
[2]. Zhiping, F Weilan, S. “Identifying risk factors of IT outsourcing using interdependent information: An extended DEMATEL method ” Expert Systems with Applications, 2012, 39(3):3832-3840.
[3]. Bahli,B Rivard,S “Validating measures of information technology outsourcing risk factors” Omega, 2005, 33(2):175-187.
[4]. Kun.H “Research on risk early-warning of information technology outsourcing project based on neural network” Applied Mechanics and Materials,2013,411-414: 2400-2405.
[5]. Willcocks LP Lacity MC T Kern “Risk mitigation in IT outsourcing strategy revisited: longitudinal case research at LISA” Journal of Strategic Information Systems, 1999, 8(3):285-314.
[6]. Kliem,R “Managing the risks of offshore IT development projects” Information Systems Management,2004,21(3):22-27.
[7]. Abdullah Lili M Verner June M “Analysis and application of an outsourcing risk framework” Journal of Systems and Software,2012,85(8):1930-1952.
[8]. Alexandrova,M “Risk factors in IT outsourcing partnerships: Vendor’s perspective” Global Business Review, 2015, 16(5):747-759.
[9]. Chitrasen S Saurav D Siba S M “Risk assessment in IT outsourcing using fuzzy decision-making approach: an Indian perspective” Experts Systems with Applications, 2014,14:4010-4022.
[10]. Samadi,H Nazari-Shirkouhi,S Keramati “An identifying and analyzing risks and responses for risk management in information technology outsourcing projects under fuzzy environment ” International Journal of Information Technology & Decision Making,2014,13(6):1283-1323.
[11]. Liangjuan, Q Harris, H Nan, Z “Risk identification and conduction model for financial institution IT outsourcing in China ”Information & Management, 2010, 13(4):429-443.
[12]. Xiaoxia H. Tianyi Z. Shamsiya K “Uncertain mean-variance and mean-semivariance models for optimal project selection and schedule” Knowledge-Based Systems, 2016, 93(1): 1-11.
[13]. Earl, M. “The risks of outsourcing IT” Sloan Management Review, 1996, 37 (3): 26-32.
[14]. Boehm, B.W. “Software risk management: principals and practices” IEEE Software, 1991, 8(1):32-41.
[15]. Boehm, B.W. “Software risk management” IEEE Software, 1989, 14(3):1-19.
[16]. Bahli,B Rivard,S “The information technology outsourcing risk: A transaction cost and agency theory-based perspective” Journal of Information Technology, 2003,18(3):211-221.
[17]. Kweku-Muata Osei-Bryson Ojelanki K. Ngwenyama “Managing risks in information systems outsourcing: an approach to analyzing outsourcing risks and structure incentive contracts ” European Journal of Operational Research, 2006, 174: 245-264.
[18]. Sanfa, C Kai, C Bin, Z “Producer services outsourcing risk control based on outsourcing contract design: industrial engineering perspective” Systems Engineering Procedia, 2011, 2:308-315.
[19]. Narasimhaiah G Toni M. S “The impact of IT outsourcing on information systems success ” Information & Management, 2014, 51(3):320-335.
[20]. Christoph,S “Distributed Decision Making” Berlin: Springer-Verlag, 2003, 125-156.
[21]. Christoph,S “Distributed decision making-a unified approach” European Journal of Operational Research, 2003, 150: 237-252.
[22]. Fang C. Franck M “A simulation-based risk network model for decision support in project risk management” Decision Support Systems, 2012, 52(3):635-644.
[23]. Jerson C. Annelies M. “A multivariate approach for top-down project control using earned value management” Decision Support Systems, 2015,79:65-76.

An Empirical Analysis of Influence of Capital Allocation Efficiency on Performance in Listing Corporation

*Xu CHEN [1]

[1] School of Economic, School of Accounting, Harbin University of Commerce, Harbin，P.R.China, 150028

（*chenxu921@126.com）

Abstract: **The paper refers to value function and production function, uses 536 listing corporation's financial data of A shares from 2011 to 2014, through builds the model about influence of capital allocation efficiency on corporate performance and carries on the empirical analysis, analyzes the logic relationship and interaction mechanism between the capital factors, capital allocation efficiency and corporate performance, and it renders value advice.**

Keywords: **Capital allocation, Capital allocation efficiency, Performance**

I. FOREWORD

Industry capital allocation efficiency is determined by the enterprise's capital allocation behavior， the sensitivity of different index (profit) economic growth determines the allocation efficiency of industry hierarchy. Capital allocation process mainly includes two aspects: capital formation and capital allocation. So the capital allocation efficiency can be understood as a capital operating efficiency of the formation and allocation， those are the capital formation rate and effective allocation rate. To boost capital allocation efficiency, we are premised on the basis of the total amount of social capital is quantitative， aim at enterprise's profit， make the capital efficient flow between the various sectors， from inefficiency and low yield sectors flow to (allocate to) high efficiency and high yield sectors. When capital achieves a optimal condition， it is equal in the marginal income ratio within each of the industries. At that time， enterprise have already reached the state of Pareto Optimality.

Therefore ， this paper argues that the most important capital factors that affects the upgrading of industrial structure are human， finance， material and technology， that is the allocation of human capital， financial capital， physical capital，technological capital. Due to the industrial capital is from change of the production factors， and it is from creating value that its value form done within the enterprise， and analyzes the degree of contribution to profit in order to improve the business performance. According to the influence degree of the capital allocation efficiency to business performance ， enterprise can create better business performance which gradually tend to be the Pareto Optimality by less capital that have a reasonable allocation.

II. ESTABLISHING MODEL

A. Research hypothesis

Based on the above theoretical analysis, this paper measured that capital allocation efficiency influences performance through regression analysis. The empirical research based on the following assumptions.

Assumption 1: Human capital, financial capital, physical capital, technology capital are positively correlated with net profit.

Assumption 2: Under the total amount of business capital remains unchanged; those capital factors (human, financial, material, technology) have substitutability among them, and exist a certain substitute coefficient.

Assumption 3: The human capital profit margin is greater than the other factors of capital. Although our country's technology is in rapid development up to now, we are still a labor oriented country, so the human capital contributes more to the national economy than other factors of capital.

B. Sample selection

We choose A-share listed companies in main board market of Shanghai stock exchange as samples, and select data according to the following standard from 2011 to 2014.

(1) Due to the different according method and financial reporting format between the financial and insurance company and general listed company, we only choose non-financial and non-insurance industry as samples.

(2) In order to eliminate inconsistent problems on the market value's valuation between the B-share and the H -shares, we cut down these shares, and only select A-share listed companies.

(3) We ignore ST and PT shares of listed companies as their usually abnormal financial position or have two continuous losses accounting period.

C. Data sources

Data is taken from CCER China's economic and financial database and financial report of the Shanghai stock exchange website from 2011 to 2014. After fiction models convert to logarithms, we are opened up regression analysis by using eviews6.0 software. For that, we eliminated related samples that the variable data less than zero. As a result, we obtained research samples that included 536 A-share listed companies in the stock exchange main board of SSE. In order to expand the sample size and strengthen the persuasion of models, we

E. Qi et al. (eds.), *Proceedings of the 23rd International Conference on Industrial Engineering and Engineering Management 2016,* DOI 10.2991/978-94-6239-255-7_38

treat listed company data from 2011 to 2014 as a large cross section data. So the model is built on the basis of the cross-section data of 536 listed companies (2144 samples).

D. Variable selection and establishing model

The research focuses the fundamental situation on allocation of human capital, financial capital, physical capital and technological capital in sample companies. In order to obtain accurate and comparable data, on the choice of arguments, we choose the financial indicators to measure. We choose "pay to employees and pay for employees" in the statement of cash flow as human capital variables, it mainly reflects enterprise recognition of staff work and preservation of profession staffs. Financial capital is the result of current asset minus human capital in the balance sheet. Physical capital is based on the amount that "net non-current assets" add "construction in process" in the balance sheet. Technological capital is based on the sum amount that "intangible assets" and "development expenditure". In order to enhance the credibility of the model, we ignore the influence of time value, so the model based on 536 sample companies (2144 samples).

Research mainly draws on the cost function and production function to complete construction of the equation. In the process of research, total capital investment is changeless, capital factors can be unlimited replacement by each other, and it accords with the actual company. In fact, the production and operation process of enterprise's overall capital volatility may not very big in the short term in practice. When the overall capital is constant, increasing capital investment for one factor and other capital will decline, so the economic function is reasonable and appropriate. The main research process is to build regression model by oversee the equation and to find the relationship between the four capital factors and net profit.

Based on cobb-Douglas production function:

$$Y = f(A, K, L) = AK^{\alpha}L^{\beta} \qquad (1)$$

Where, Y stands for level of output, A stands for technology input, K stands for fixed asset input, L stands for human input。

After logarithm fetch on both sides of the model, draw out difference:

$$\Delta\ln Y = \Delta\ln A + \alpha\Delta\ln K + \beta\Delta\ln L \qquad (2)$$

In accordance with the above principles, considering the four factors of industry capital, human capital, financial capital, physical capital and technological capital have the impact on business performance, we establish the data regression model as follow:

$$\ln Y = C + \alpha*\ln L + \beta*\ln F + \gamma*\ln P + \delta*\ln T \qquad (3)$$

In the equation, explained variable Y stands for net profit, explanatory variables L stands for human capital, F stands for financial capital, P stands for physical capital, T stands for technological capital,, coefficient of explanatory variables are α,β,γ,δ , that stands for coefficient of elasticity of these capitals ,and C is constant。 In conclusion:

$$\begin{cases} \text{Profit contribution of human capital} = \dfrac{\alpha*\ln L}{\ln Y} \\ \text{Profit contribution of financial capital} = \dfrac{\beta*\ln M}{\ln Y} \\ \text{Profit contribution of physical capital} = \dfrac{\gamma*\ln P}{\ln Y} \\ \text{Profit contribution of technological capital} = \dfrac{\delta*\ln T}{\ln Y} \end{cases} \qquad (4)$$

$$\begin{cases} \text{Human capital profit margin} = \\ \dfrac{\text{Profit contribution of human capital} \times \text{net profit}}{\text{Average balance of human capital}} \times 100\% \\ \text{Financial capital profit margin} = \\ \dfrac{\text{Profit contribution of financial capital} \times \text{net profit}}{\text{Average balance of financial capital}} \times 100\% \\ \text{Physical capital profit margin} = \\ \dfrac{\text{Profit contribution of physical capital} \times \text{net profit}}{\text{Average balance of physical capital}} \times 100\% \\ \text{Technological capital profit margin} = \\ \dfrac{\text{Profit contribution of technological capital} \times \text{net profit}}{\text{Average balance of technological capital}} \times 100\% \end{cases} \qquad (5)$$

III. EMPIRICAL ANALYSIS

A. Descriptive statistical analysis of samples

This paper selects to use the data from 536 listed companies during 2011 to 2014, by eviews6.0 software using regression analysis method to measure the contribution for profit and profit margins, aim to reflect the impact on the business performance.

(1) Descriptive statistics on human capital input

Through descriptive statistical analysis in Eviews, we can see the specific situation of the human capital input, as shown in figure 1:

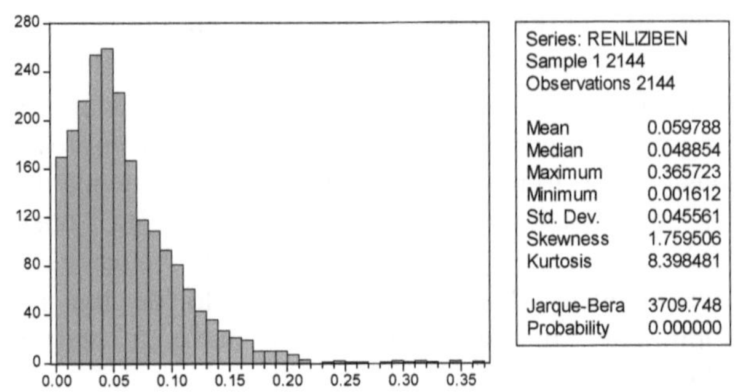

Fig 1 Situation of the human capital input

According to the analysis, average value of human capital investment is 5.98%, the median is 4.89%, the standard deviation is 0.046. Average is on the behalf of approaching value. Variance is on the behalf of discrete degree. It represents the degree of data distribution relative to averages. So 5.98% on behalf of 536 listed companies in the human capital 5.98% on average, standard deviation is 0.046 said most of the samples of human capital proportion at around 5.98%. So we can understand that human capital in the listed companies of China accounted for about 5.98%.

(2) Descriptive statistics on financial capital input

Similarly , financial capital input of sample companies, as shown in figure 2:

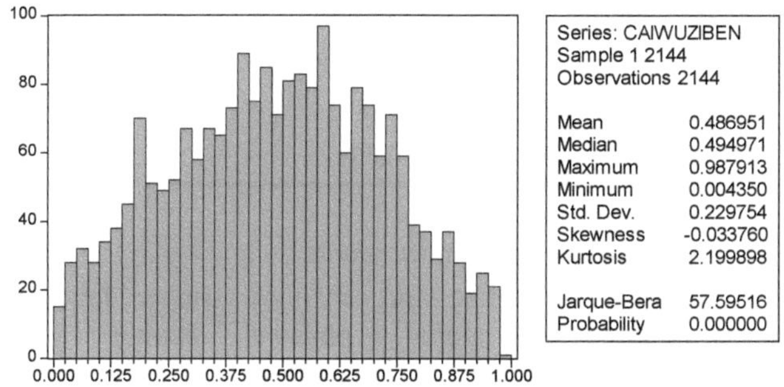

Fig 2 Situation of the financial capital input

According to the above chart, you can see that the listed company financial capital in the average is 48.70% in the total capital and the median is 49.50%, the standard deviation is 0.230. We can also learn from above: financial capital account for the proportion of total capital is concentrated in the 12.5% to 87.5%, it suggesting that samples of the company's financial capital investment situation has a certain degree of discrete.

(3) Descriptive statistics on physical capital input

Physical capital input of sample companies, as shown in figure 3:

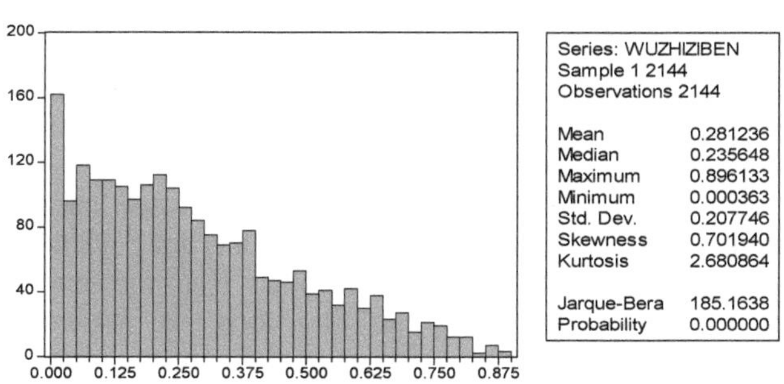

Fig 3 Situation of the physical capital input

According to analysis by Eviews software, physical capital in the listed companies accounted for the average is 28.12%, the median is 23.56%, and the standard deviation is 0.208. Proportion of physical capital presents a relatively high degree of discrete, it is concluded that physical capital in above gathered in between 12.5% and 37.5%, and the proportion of material capital investment is not concentrated.

(4) Descriptive statistics on technological capital input

We import the technological capital into the Eviews 6.0, as shown in figure 4:

Through the above it can be concluded that the technological capital invested was 5.28%, the median was 3.30%, the standard deviation was 0.071.

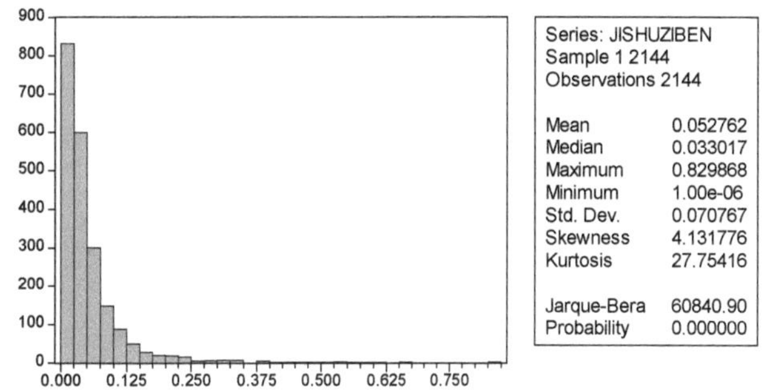

Fig 4 Situation of the technological capital input

It is said that all the data distribution relatively to the average is relatively dense; this means that the 5.28% is an overview about all data. Chinese research and development (R&D) for the ratio of the gross domestic product (GDP) is 2.05%, an increase of 0.04% over the previous year. But there is a certain gap compared with the developed countries which R&D expenditure as a share of GDP generally 3% level. It also forced the enterprise must effectively utilize technology input, strengthen the innovation level in industrial technology.

B. The establishment and calculation of regression equation

When using cross-section data as samples, the difference of other factors also can produce bigger effect in addition to the explanatory variables. Therefore, cross-section data exist certain heteroscedasticity, it will variance which produce interference is not a constant. Next we found that heteroscedasticity exists by using white test without cross to inspect the model. This paper chooses the weighted least-square method to solve this problem, and obtained the regression analysis results that capital factors have impact on the business performance, as shown in table 1and 2.

After eliminate heteroscedasticity in selection of weighted least-square method, both regression model and variable are passed the test of significance. It shows that these variables are positive correlation with net profit. R2 is 0.999, which indicates that clear influence have been made by explanations variables on operating profit, and the results are not expected vary so much. It achieves the expected effect. In a word, it can show the actual situation effectively. The final regression results are as follows format 6.

$$\ln Y = 0.379875*\ln L + 0.519238*\ln M + $$
$$(255.3541) \quad (545.3040)$$
$$0.120975*\ln A + 0.005756*\ln T - 2.096373$$
$$(204.9371) \quad (9.120878) \quad (-126.1513) \qquad (6)$$

TABLE 1 THE REGRESSION ANALYSIS RESULTS THAT CAPITAL FACTORS HAVE IMPACT ON THE BUSINESS PERFORMANCE（WLS）

Dependent Variable: LNY

Method: Least Squares

Date: 05/21/15 Time: 23:31

Sample: 1 2144

Included observations: 2144

Weighting series: W

White Heteroskedasticity-Consistent Standard Errors & Covariance

	Coefficient	Std. Error	t-Statistic	Prob.
C	-2.096373	0.016618	-126.1513	0.0000
LNL	0.379875	0.001488	255.3541	0.0000
LNM	0.519238	0.000952	545.3040	0.0000
LNA	0.120975	0.000590	204.9371	0.0000
LNT	0.005756	0.000631	9.120878	0.0000

Weighted Statistics

R-squared	0.999044	Mean dependent var	19.23430
Adjusted R-squared	0.999042	S.D. dependent var	109.0371
S.E. of regression	0.166359	Akaike info criterion	-0.747006
Sum squared reside	59.16985	Schwarz criterion	-0.733777
Log likelihood	805.4174	Hannan-Quinn criter.	-0.742166
F-statistic	558544.7	Durbin-Watson stat	1.993787
Prob(F-statistic)	0.000000		

Unweighted Statistics

R-squared	0.343964	Mean dependent var	18.92377
Adjusted R-squared	0.342737	S.D. dependent var	2.011400
S.E. of regression	1.630678	Sum squared reside	5685.178
Durbin-Watson stat	1.970436		

TABLE II THE REGRESSION ANALYSIS RESULTS THAT MODEL OF CAPITAL FACTORS HAVE IMPACT ON THE BUSINESS PERFORMANCE

Variables and constants	Coefficient	Standard deviation	T inspection value
lnL	0.379875	0.001488	255.3541
lnM	0.519238	0.000952	545.3040
lnP	0.120975	0.000590	204.9371
lnT	0.005756	0.000631	9.120878
C	-2.096373	0.016618	-126.1513
F inspection value		558544.7	
R^2		0.999	
D.W.value		1.993787	

We reorganize the calculating results, then substitute in the formula 4 and formula 5, the calculating results as shown in table 3.

TABLE III THE PROFIT CONTRIBUTION AND PROFIT MARGIN OF EACH CAPITAL FACTORS

capital factors	Profit contribution	Profit margin
Human capital	38.97%	1829.10%
Financial capital	59.40%	267.28%
Physical capital	13.31%	91.13%
Technological capital	0.57%	25.84%

C. The analysis of regression results

Through the regression analysis above, human capital, financial capital, physical capital and technological capital's profit margin is 1829.10%, 267.28%, 91.13% and 25.84%. The profit margin of the human capital is one of the biggest in these capital factors, financial capital is following it, and physical capital is at the third, technology capital is the lowest. And it is concluded that the factors of capital allocation in front case, the largest percentage is the financial capital, accounted for nearly half of the total capital, and then followed by the proportion of physical capital, human capital the proportion of 5.98%, the smallest is the technology capital ratio of 5.28%. This shows that the amount of capital factors is not proportional to the relationship with the profit margins. It is worth to note that the lowest human capital is relatively highest profit margins of the capital factors. It is enough to show that our country is a labor oriented country until now. According to the latest data of Wall Street Journal reported and the World Factbbook of CIA in 2014, the labor-intensive industry is more in China, and the employment population accounts for the proportion of total labor of three major industries is 33.6%, 30.3% and 33.6% respectively .

It is important to note that you can see human capital and technological capital accounts for 5.98% of the total capital and 5.28% respectively from the sample data. While these capital allocation is basically unchanged, but the profit margin of the human capital is the highest, and profit margin of technological capital is the lowest. Accordingly we may safely draw the conclusion that although the two kinds of capital inputs is almost the same, the return is not necessarily the same. The source of the technology can be an independent research or the imported technology from outside.

Although independent research usually exist many questions such as cost too much, or too long development cycle, we obtain core and key technology. It has very significant advantages on the sustainable development of the enterprise and industry competitiveness. On the other hand, imported technology from outside need shorter time, but spending more and is difficult to obtain the core and key technologies. So we should speed up the development of technology and the innovation of science and technology, aim to make the technological capital into productivity as soon as possible and improve the output of technological capital, and improve the efficiency of the business. Accounted for 48.70% of financial capital, profit margins are behind human capital profit margins. This is because that when the gross capital is fixed, the more financial capital, the more monetary assets of enterprise or industry. It is no change for other capital factor, this will lead to the other capital investment is insufficient. Enterprise can't create a higher market share and benefits or promote industrial development in an all-round way. The above analysis shows, the proportion of physical capital to total capital is 28.12%, achieved more than a quarter of the total capital proportion. But it is not very ideal as profit margins are third in four types of capital. Main cause of this situation is as new factors capital gets more and more attention by managers, the importance of physical capital largely ignored. That results low physical capital allocation efficiency, and the input-output ratio is low. That leads to not upgrade technology, equipment and other fixed assets cannot play its biggest utility, in a great extent, hinder the industry development and the realization of scale economics.

Through the above empirical research, it is concluded that capital allocation of Chinese listed companies is out of proportion in number. Such as too much cash slack of listed companies, less input of physical capital, both to take up a large amount of business capital. While too little investment in the human capital and technological capital, but human capital play an effective role on business performance. Technological capital, however, did not play a key role, it can be a better explanation in terms of capital allocation, there are a series of problems in capital allocation.

IV CONCLUSIONS

A. Promote enterprise's technological innovation

Technical innovation is an important condition and premise that form enterprise technology capital. Enterprises shall establish innovation system that based on industry, make the human, financial, physical, technology, and other industrial elements involve in technological change and innovation. Enterprises establish a high level of R&D team; strengthen the technology education and training of human capital. It is also important to make employment industries to upstream. Enterprise should rationally develop technological input and control R&D costs to further reduce the cost brought by the industry competition. It will have positively affected on the business performance of enterprise. Enterprise should make technological progress transit from technology import to independent innovation, focus on the development of high and new technology industries and R&D industry which is upstream industry of technological progress. Such as extending industrial chain, and expanding the scope of the economy.

B. Improve the efficiency of human capital allocation

The human capital's influence on the economic benefits goes with the upgrading of industrial structure increase. On the one hand, enterprises should strengthen the training of employees, have the attention to the cultivation of all kinds of professional staffs, and improve staff quality. Enterprise shall, on the other hand, analyze the demand for human resources. It is more reasonable allocation of human resources, more easily to realize human capital scale benefits. In the optimization of industrial structure and promote capital formation efficiency, providing more high-quality human capital for industrial development can promote the capital allocation efficiency.

C. Improve the efficiency of physical and financial capital allocation

To improve the physical capital allocation efficiency, enterprise can make relevant personnel on a regular inspection; Staff can eliminate equipment and factory in a timely manner. In this way, enterprises not only realize the normal output level, but also update the old equipment and redundant workshop. It promotes enterprise's growth and development, and use less capital factors for a higher capacity. It enhances the competitiveness of the enterprises.

To improve financial capital allocation efficiency, firstly, Listed company should focus on the corporate debt, it need maintain the liabilities in a certain range. Enterprise can regularly inspect the enterprise overall cash flow to control cash flow in a timely manner. Secondly, there should be a professional financial employee to supervise and control the accounts receivable, inventory, etc. Again, it also can optimize the distribution of financial capital through expand the scale of the enterprise. Financial capital can constantly value-added and capitalization, it will make the financial capital of enterprises gradually accumulated.

REFERENCES

[1] Chuan-Li, Wei. The application of improved genetic algorithm in the Markov state transition matrix to SME technical efficiency and scale revenue calculation methods[J]. International Journal of Future Generation Communication and Networking, 2015, 8(9):127-136.

[2] Antonio Couto and Daniel J. Graham. The contributions of technical and allocative efficiency to the economic performance of European railways[J]. Portuguese Economic Journal, 2008,7:125–153.
[3] Chen, S.M, Sun, Z., Tang, S., Wu, D.H..Government Intervention and investment efficiency: Evidence from China Original Research Article [J]. Journal of Corporate Finance, 2011, 17(2): 259-271.
[4] HOVAKIMIAN,G.. Financial constraints and investment efficiency: Internal capital allocation across the business cycle [J].Journal of Financial Intermediation， 2011 ,20(2):.175-181.
[5] CAVALLO，E.， GALINDO，A.， IZQUIERDO，A. and LEAN， J.J. The role of relative price volatility in the efficiency of investment allocation [J]. Journal Of International Money And Finance，2013， 1.33:.118-123.
[6] Oleg Badunenko ,Michael Fritsch ,Andreas Stephan. Allocative efficiency measurement revisited—Do we really need input prices?[J]. Economic Modelling,2008,25:1093-1109
[7] Thompson, H.G.,Jr.,&Garbacz,C.. Mobile, fixed line and Internet service effects on global productive efficiency. Information Economics and Policy, 2007,2:189-214.
[8]Karen， T． Negative Rebound and Disinvestment Effects in Response to an Improvement in Energy Efficiency in the UK Economy[J] ． Ecological Economics，2009，31(5)：648-666.
[9] Frei F X, Harker P T, Hunter L W. Inside the black box: what makes a bank efficient?[J]. Performance of Financial Institutions: Efficiency, Innovation, Regulation, edited by Patrick T. Harker and Stavros A. Zenios, 2000: 259-311.
[10]Chen, G. M. ; Firth, M. and Rui, . "Have China's Enterprise Reforms Led to Improved Efficiency and Profitability? "Emerging Markets Review. 2006, 7(1),:82-109.

The Game about Competition and Cooperation in Port Group

Jing WANG, *Sheng- hua SUN
Department of Economics and Management, Yanshan University, Qinhuangdao 066004, China
(E-mail address:28490698@qq.com,*850678400@qq.com)

Abstract - **Due to adjacent geographical location and hinterland overlap among ports in the port group, the port competition is fiercer and fiercer. This paper establishes a Hotelling game model to analyze the competition and cooperation among ports in the port group based on geographical location, service level and shipping distance. The research proves when service level of port enterprises is same, cooperative strategy can significantly improve the level of the port group's profit. When service level of port enterprises is different, service price of port, market share of port and port's profit are affected by the service level before and after the cooperation, the service level of port enterprise shows a trend of mutual promotion, and the port group develops into the higher service level.**

Keywords –**Competition and cooperation, game, Hotelling model, port group, service level**

I. INTRODUCTION

In recent years, due to adjacent geographical location and hinterland overlap among ports in the port group, the port competition is fiercer and fiercer. Unhealthy competition among ports is not only against the overall development of ports and port group, but also against the development of port hinterland economy. Therefore, it is a worthy concern that how to avoid unhealthy competition among ports, establish the rational relationship to enhance the competitiveness of the port group, and promote regional economy development.

With regarding to the competition and cooperation among ports, domestic and foreign scholars have done a lot of research. D W Song put forward the definition of co-operation among ports , and suggested that port should take competition and co-operation strategies [1]. Peter W.de langen, Evert-Jan Visser [2] and Evert-Jan Visser [3] analyzed the lower Mississippi port cluster and Rotterdam port cluster, proposing that the competition of port was relate to the collective behavior of the port group. E N Kaselimi, T E Notteboom and B D Borger researched that the emergence of the dedicated terminal affected competition of port group using a game theoretic approach [4]. K Wang, A K Y Nga, J S L Lam and X Fu established a game model to analyze factors and conditions affecting regional port governance in Pearl River Delta, and put forward the importance of regional port cooperation[5] . In addition , some foreign scholars also researched the competition and cooperation among ports, such as P Fu and Y Chen [6] , N Saeed and O.I Larsen [7] . Jing-han Zhang analyzed qualitatively the relationship among port group in the Yangtze River Delta, region and traffic development, presented the interest is the Fundamental factor that leads to competition of port group, and stressed the importance of port cluster from competition to cooperation[8]. Qing Gao and Jun-hai Ma developed a population model to research cooperation and competition among container terminals [9]. Hai-bo Kuang , Shu-wen Chen constructed an evaluation indexes system to analyze the competitiveness of port problem[10]. Yu-hua He researched pricing strategy of duopoly port enterprises using a Bertrand model, and the research results indicated cooperative pricing strategy can improve price and profit of two sides [11]. Xiao-jun Dai performed a simulation study for the relationship between hinterland economy and regional port system, developed a system dynamics model, and suggested port should cooperate with the rational competition [12]. Ming-zhu Yu and Jun Shan researched the two-level co-opetition problem which involved both the government strategies and the terminal operator decisions using a game theoretic approach, and proved the importance of government cooperation[13]. Xin Zhou, Mei Sha, Shi-yuan Zheng and Jian-hua Ji developed a located spatial model to research multi-port cooperative pricing strategy[14]. In addition ,some domestic scholars also researched the competition and cooperation among ports, such as Hai-bo Kuang and Wen-yuan Niu[15], Shi-yuan Zheng[16], Gang Dong[17].

Existing research mainly in the following three aspects: Firstly, an evaluation indexes system was constructed and competition and cooperation strategy was proposed; secondly, general model was established, such as population model, system dynamics model and game model; thirdly, some theory research about port group on competition and cooperation. However, some insufficient still exist. (1) There are not systematic scientific theories to direct co-operation of port group. Some scholars has researched the port competitiveness, the competition among the ports in port group, the competition among port groups and the economy relationship between port and hinterland. These theories made certain progress, but the systematic, perfect and ripe theoretical system has not formed. (2) The theory models on the competition and cooperation among ports in the port group are still deficient. Researches concentrated in two sides game and

Capital projects： National Social Science Foundation of China(13BGL006); Program of Education Department of Hebei Province, China (GH161008); Young Teacher Independent Research Program of Yanshan University (14SKA002).

E. Qi et al. (eds.), *Proceedings of the 23rd International Conference on Industrial Engineering and Engineering Management 2016,* DOI 10.2991/978-94-6239-255-7_39

the competition game, but multilateral game and cooperation game are rarely involved in the port area.

This paper establishes a multi-port Hotelling game model to research homogenization competition problem among ports which have adjacent geographical location and different service level. This paper analyzes the competition and cooperation among ports, which have the different geographical location and service level, and provides a theory reference for healthy competition and cooperation among ports in the port group.

II. PROBLEM DESCRIPTION AND ASSUMPTIONS

This paper discusses the competition and cooperation among three port enterprises, which have hinterland overlap, adjacent geographical location, and different service level. In order to scramble hinterland cargos in a certain area, there is highly homogenization competition among port enterprises. Port enterprises usually take the strategies of reducing price or improving service level to expand market share and improve profit. This paper establishes a game model among three port enterprises based on geographical location, service level and shipping distance. In order to describe the problem, the model has some assumptions as follows:

Assumption 1: Three port enterprises and consignors are completely rational in decision making. Port enterprises will maximize profit and consignors will maximize utility as the principle.

Assumption 2: Three port enterprises are not overlap, located on the x axis [0,1], and respectively distribute on the point a_1, a_2, a_3, $0<a_1<a_2<a_3<1$.

Assumption 3: Hinterland of three port enterprises is the rectangular formed by coordinates (0,0), (1,0), (0,1), (1,1). Consignors distribute evenly in the area.

Model parameters:

p_i : The service price of port i, i=1,2,3.

β : The unit land transportation cost from supplier to port.

γ : The unit shipping transportation cost from port i to destination port .

l_i :The shipping distance from port i to destination port , i=1,2,3.

S_i : The service level of port i, i=1,2,3.

$f(S_i)$: The cost of port i, let $f(S_i)=\omega S_i^2/2$, ω is cost factor ,i=1,2,3.

Q_i : The market share of port i, i=1,2,3.

m_1 : The indifference abscissa, where the utility of consignor at port1 is same to port2.

m_2 : The indifference abscissa, where the utility of consignor at port2 is same to port3.

π_i : The profit of port i, i=1,2,3.

III. MODEL ESTABLISHING AND SLOVING

The utility of consignor include land transportation cost, shipping transportation cost, port service fees and port service level. The lower the cost and the higher the port service level, the more utility. The paper assumes there are abscissa m_1 and m_2. When the abscissa of consignor's location ranges from 0 to m_1, consignor chooses port enterprise 1. When the abscissa of consignor's location ranges from m_1 to m_2, consignor chooses port enterprise 2. When the abscissa of consignor's location ranges from m_2 to 1, consignor chooses port enterprise 3. m_1 and m_2 meet the following formula :

$$m_1=\frac{p_2-p_1}{2\beta}+\frac{a_1+a_2}{2}+\frac{\gamma(l_2-l_1)}{2\beta}+\frac{S_1-S_2}{2\beta} \quad (1)$$

$$m_2=\frac{p_3-p_2}{2\beta}+\frac{a_2+a_3}{2}+\frac{\gamma(l_3-l_2)}{2\beta}+\frac{S_2-S_3}{2\beta} \quad (2)$$

The market share of port enterprise is as follows:

$$Q_1=\int_0^1\int_0^{m_1}dxdy=m_1=\frac{p_2-p_1}{2\beta}+\frac{a_1+a_2}{2}+\frac{\gamma(l_2-l_1)}{2\beta}+\frac{S_1-S_2}{2\beta} \quad (3)$$

$$Q_2=\int_0^1\int_{m_1}^{m_2}dxdy=m_2-m_1=\frac{p_3-2p_2+p_1}{2\beta}+\frac{a_3-a_1}{2}+\frac{\gamma(l_3-2l_2+l_1)}{2\beta}+\frac{2S_2-S_1-S_3}{2\beta} \quad (4)$$

$$Q_3=\int_0^1\int_{m_2}^{1}dxdy=1-m_2=\frac{p_2-p_3}{2\beta}+\frac{2-a_2-a_3}{2}+\frac{\gamma(l_2-l_3)}{2\beta}+\frac{S_3-S_2}{2\beta} \quad (5)$$

The profit function of port enterprise is as follows:

$$\pi_i(p_1,p_2,p_3)=p_iQ_i-\frac{\omega S_i^2}{2},i=1,2,3 \quad (6)$$

A. Competition Game

Three port enterprises take competitive strategies. They are completely rational and maximize profit as the principle in decision making. It is easy to prove $\frac{\partial^2\pi_i}{\partial p_i^2}<0$,(i=1,2,3). That indicates every port enterprise has a maximum profit. Let $\frac{\partial\pi_i}{\partial p_i}=0$,(i=1,2,3) and consider equation (3)~(6).The gain equilibrium solution of service price is as follow.

$$p_1^*=\frac{\beta(2+5a_1+6a_2+a_3)}{12}+\frac{\gamma(4l_2-5l_1+l_3)}{12}+\frac{5S_1-S_3-4S_2}{12} \quad (7)$$

$$p_2^*=\frac{\beta(2-a_1+a_3)}{6}+\frac{\gamma(l_1+l_3-2l_2)}{6}+\frac{2S_2-S_1-S_3}{6} \quad (8)$$

$$p_3^* = \frac{\beta(14 - a_1 - 6a_2 - 5a_3)}{12} + \frac{\gamma(l_1 + 4l_2 - 5l_3)}{12} + \frac{5S_3 - S_1 - 4S_2}{12} \quad (9)$$

Consider formula (3) ~ (5), (7) ~ (9), the market share is as follows :

$$Q_1^* = \frac{2 + 5a_1 + 6a_2 + a_3}{24} + \frac{\gamma(4l_2 - 5l_1 + l_3)}{24\beta} + \frac{5S_1 - 4S_2 - S_3}{24\beta} \quad (10)$$

$$Q_2^* = \frac{2 - a_1 + a_3}{6} + \frac{\gamma(l_1 + l_3 - 2l_2)}{6\beta} + \frac{2S_2 - S_1 - S_3}{6\beta} \quad (11)$$

$$Q_3^* = \frac{14 - a_1 - 6a_2 - 5a_3}{24} + \frac{\gamma(l_1 + 4l_2 - 5l_3)}{24\beta} + \frac{5S_3 - S_1 - 4S_2}{24\beta} \quad (12)$$

Consider formula (6) ~ (9), the port profit is as follows:

$$\pi_1^* = \frac{[\beta(2 + 5a_1 + 6a_2 + a_3) + (5S_1 - 4S_2 - S_3) + \gamma(4l_2 - 5l_1 + l_3)]^2}{288\beta} - \frac{\omega S_1^2}{2} \quad (13)$$

$$\pi_2^* = \frac{[\beta(2 - a_1 + a_3) + (2S_2 - S_1 - S_3) + \gamma(l_1 + l_3 - 2l_2)]^2}{36\beta} - \frac{\omega S_2^2}{2} \quad (14)$$

$$\pi_3^* = \frac{[\beta(14 - a_1 - 6a_2 - 5a_3) + (5S_3 - S_1 - 4S_2) + \gamma(l_1 + 4l_2 - 5l_3)]^2}{288\beta} - \frac{\omega S_3^2}{2} \quad (15)$$

B. Competition and Cooperation Game

Port enterprise 1 and port enterprise2 take cooperative strategy, and same price (p_1=p_2), at the aim of maximum their profits . Port enterprise3 take competitive strategy at the aim of maximum its profit.

The market share of port enterprise 1,2 is as follows:

$$Q_{12} = \int_0^1 \int_0^{m_2} dx dy = m_2 = \frac{p_3 - p_2}{2\beta} + \frac{a_2 + a_3}{2} + \frac{\gamma(l_3 - l_2)}{2\beta} + \frac{S_2 - S_3}{2\beta} \quad (16)$$

The market share of port enterprise 3 is as follows:

$$Q_3 = 1 - Q_{12} = \frac{P_2 - P_3}{2\beta} + \frac{2 - a_2 - a_3}{2} + \frac{\gamma(l_2 - l_3)}{2\beta} + \frac{S_3 - S_2}{2\beta} \quad (17)$$

The profit function of port enterprise 1,2 is as follows:

$$\pi_{12} = p_2 Q_{12} - \frac{\omega S_1^2}{2} - \frac{\omega S_2^2}{2} \quad (18)$$

The profit function of port enterprise 3 is as follows:

$$\pi_3 = p_3 Q_3 - \frac{\omega S_3^2}{2} \quad (19)$$

It is easy to prove that second derivatives of profit function are less than zero. That indicates port enterprises have profit maximum. Let $\frac{\partial \pi_{12}}{\partial p_2} = 0$, $\frac{\partial \pi_3}{\partial p_3} = 0$. The above two equations are simultaneously solved for equilibrium solution of service price game.

$$p_2^{**} = \frac{\beta(2 + a_2 + a_3)}{3} + \frac{\gamma(l_3 - l_2)}{3} + \frac{S_2 - S_3}{3} \quad (20)$$

$$p_3^{**} = \frac{\beta(4 - a_2 - a_3)}{3} + \frac{\gamma(l_2 - l_3)}{3} + \frac{S_3 - S_2}{3} \quad (21)$$

The market share is as follows:

$$Q_{12}^{**} = \frac{2 + a_2 + a_3}{6} + \frac{\gamma(l_3 - l_2)}{6\beta} + \frac{S_2 - S_3}{6\beta} \quad (22)$$

$$Q_3^{**} = \frac{4 - a_2 - a_3}{6} + \frac{\gamma(l_2 - l_3)}{6\beta} + \frac{S_3 - S_2}{6\beta} \quad (23)$$

The profit is as follows:

$$\pi_{12}^{**} = \frac{[\beta(2 + a_2 + a_3) + (S_2 - S_3) + \gamma(l_3 - l_2)]^2}{18\beta} - \frac{\omega S_1^2}{2} - \frac{\omega S_2^2}{2} \quad (24)$$

$$\pi_3^{**} = \frac{[\beta(4 - a_2 - a_3) + (S_3 - S_2) + \gamma(l_2 - l_3)]^2}{18\beta} - \frac{\omega S_3^2}{2} \quad (25)$$

IV. GAME ANALYSIS FOR COMPETITION AND COOPERATON OF PORT

A. The Same Service Level

Proposition 1 : Cooperative strategy can significantly improves the profit of port group, in the condition of balance.

Prove: Suppose $l_i = l_j$, (i,j=1,2,3), and $\Delta\pi_{12}^{**}$ is the difference of total profit that the port enterprise 1and 2 cooperative or not. $\Delta\pi_3^{**}$ is the profit difference of port enterprise 3 before and after port enterprise 1,2 taking cooperative strategy.

$$\Delta\pi_{12}^{**} = \pi_{12}^{**} - \pi_1^* - \pi_2^* = \frac{8 + 12a_1 + 6a_1a_3 + 28(a_3 - a_1^2) + (7a_3^2 - 5a_1^2)}{288}\beta + \frac{20a_2(a_3 - a_2) + 40a_2(1 - a_1) + 20(1 - a_1a_2)}{288}\beta \quad (26)$$

$$\Delta\pi_3^{**} = \pi_3^{**} - \pi_3^* = \frac{12(a_3 - a_1a_2) + 28(a_1 - a_2a_3) + 10(a_2 - a_1a_3)}{288}\beta + \frac{30a_2 + 60 - 20a_2^2 - 9a_3^2 - a_1^2}{288}\beta \quad (27)$$

Because of $0<a_1<a_2<a_3<1$, it is easy to prove $\Delta\pi_{12}^{**}$ and $\Delta\pi_3^{**}$ are greater than zero. That indicates cooperative strategy not only improves total profit of port enterprise 1,2 in the cooperative strategy, but also improves the port enterprise 3 in the competitive strategy and improves the port group profit. Therefore, port enterprises avoid disorderly competition and take cooperative strategy, which is conducive to the overall development of ports and port group.

Proposition 2 : The total profit difference of port enterprise 1,2 before and after their taking cooperative strategy is the increasing function of the distance between port enterprise 1 and port enterprise 2, and it is also increasing function of the distance between port enterprise 3 and port enterprise 2. The total profit difference is more sensitive to the latter.

Prove: Let $a_2 = a_1 + b$, $a_3 = a_1 + b + c$, take them into (26), and take first derivatives of b and c.

$$\frac{\partial \Delta\pi_{12}^{**}}{\partial b} = \frac{68 + 14b + 34c - 40a_1}{288}\beta > 0 \quad (28)$$

$$\frac{\partial \Delta\pi_{12}^{**}}{\partial c} = \frac{40a_1 + 34b + 14c + 28}{288}\beta > 0 \quad (29)$$

Let $\frac{c}{b} = k$, and take it into formula (26), and take first derivatives of k .

$$\frac{\partial \Delta\pi_{12}^{**}}{\partial k} = \frac{40a_1 b + 34b^2 + 14b^2 k + 28b}{288}\beta > 0 \quad (30)$$

Formula (28) and (29) indicate that $\Delta\pi_{12}^{**}$is increased with b , when c is a constant. That indicates when the geographical location of port enterprise 2,3 are stationary, the farther of the distance between port enterprise 1 and port enterprise 2, the more improvement of total profit of port enterprise 1,2 after the cooperative strategy.Similarly, $\Delta\pi_{12}^{**}$is increased with c , when the geographical location of port enterprise 1,2 are stationary.

Formula (30) indicates $\Delta\pi_{12}^{**}$is a increasing function of k .That is, which $\Delta\pi_{12}^{**}$ is more sensitive to c . In another sense, when the distance between port enterprise 1 and port enterprise 2 is less than the distance between port enterprise 2 and port enterprise 3($b < c$), the competition of port enterprise 1,2 is fiercer than the competition of port enterprise 2,3, and port enterprise 1,2 are more inclined to cooperate for higher profits.

B. The Different Service Level

Proposition 3 : Service price ,market share and profit of port enterprise are negatively related to the shipping distance between this port and destination port, and positively related to the shipping distance between the other two ports and destination port.

Prove: According to formula (13) , the relationship between profit and shipping distance is as follows:

$$\frac{\partial \pi_1^*}{\partial l_1} = \frac{-10\gamma(2 + 5a_1 + 6a_2 + a_3)}{288} + \frac{-10\gamma[(5S_1 - 4S_2 - S_3) + \gamma(4l_2 - 5l_1 + l_3)]}{288\beta} \quad (31)$$

$$\frac{\partial^2 \pi_1^*}{\partial l_1^2} = \frac{25\gamma^2}{144\beta} > 0 \quad (32)$$

According to the formula (32), π_1^* increases first and then decreases, with the increase of l_1 .And we can figure out the increasing interval, the decreasing interval and the minimum value of profit according to the formula (31).

When $\frac{\partial \pi_1^*}{\partial l_1} = 0$, π_1^* is minimized and $p_1 = 0$ according to the formula (7).

Similarly, when $p_1 < 0$, there is increasing interval of profit, but this is not reasonable. When $p_1 > 0$, there is decreasing interval of profit. In summary, π_1^* is the decreasing function of l_1 . In the same way, π_1^* is the increasing function of l_2, l_3 .

Analysis of the formula (13) and according to the formula (7) ~ (12), (20)~ (23), service price , market share and profit are decrease with the shipping distance from this port to destination port, increase with the shipping distance from the other two ports to destination port. And the profit is minimized, when service price equals zero. That is, there are usually many small ports near the hub port, competition among these ports of adjacent geographical location is more fiercer

Proposition 4 : Service price and market share of port enterprise are positively related to service level of this port, negatively related to service level of the other two port enterprises.

Prove: According to the formula (7) ~ (12), (20)~ (23), it is easy to prove proposition 4 is true. Proposition indicates port enterprise can raise service price accordingly and gain more market share, with the increase of service level. However, with the increase of service level of the other two port, service price and market share of this port enterprise are decreasing. To some degree, the service level shows a trend of mutual promotion.

Proposition 5 : Port enterprise profit is influenced by their service level, but the influence is related to geographical location of port, unit land transportation cost, unit shipping transportation cost , shipping distance and cost factor. And port enterprise profit is a decreasing function of service level of the other port enterprises.

Prove: Formula (13) example proves the relationship between profit and service level.

$$\frac{\partial \pi_1^*}{\partial S_1} = \frac{5\beta(2 + 5a_1 + 6a_2 + a_3)}{144\beta} + \frac{5[(5S_1 - 4S_2 - S_3) + \gamma(4l_2 - 5l_1 + l_3)]}{144\beta} - \omega S_1 \quad (33)$$

According to the formula (33), π_1^* is influenced by S_1, but the influence is related to a_i , β , γ , l_i and ω .The same as the proof method of proposition 3, the port enterprise profit is a decreasing function of service level of the other two port enterprises.

This proposition indicates when the other port enterprises improver service level, this port enterprise also improves service level to raise service price and gain more market share, but it also increases the cost of port and then affects this port enterprise profit. Therefore, port enterprise is not to follow other port enterprises blindly and improve service level appropriately, combining geographical location of port, land transportation cost, shipping transportation cost , shipping distance and cost factor.

V. CONCLUSION

This paper established a game model to analyze the competition and cooperation among ports in the port group. Considered the problem of the competitive strategies of three port enterprises and the cooperative strategy of two port enterprises, this paper derived equilibrium solution of service price, market share and profit. When service level of port enterprises is same, cooperative strategy not only improves total profit of port enterprises in the cooperative strategy, but also improves the port enterprise in the competitive strategy and improves the profit level of the port group. It shows that the cooperative strategy is a win-win situation. When service level of port enterprises is different, service price , market share and profit are affected by the service level before and after the cooperation, and the service level shows a trend of mutual promotion. The port group develops into the higher service level. However, the increase of port enterprise service level , that can lead to the increase of port cost and affect the port enterprise profit. Therefore, port enterprise should improve service level appropriately, combining geographical location of port, land transportation cost, shipping transportation cost , shipping distance and cost factor.

ACKNOWLEDGMENT

We wish to acknowledge the support National Social Science Foundation of China, Program of Education Department of Hebei Province and Young Teacher Independent Research Program of Yanshan University. The authors would like also to thank anonymous referees whose comments have improved the quality of the paper.

REFERENCES

[1] D. W. Song, " Regional container port competition and co-operation: the case of hong kong and south china (Periodical style)," Journal of Transport Geography. no. 10, pp. 99-11,2002.

[2]Peter, W. De Langen, Evert-Jan Visser, " Collective action regimes in seaport clusters: the case of the lower Mississippi port cluster (Periodical style),"Journal of transport geography. pp. 173-186,2005.

[3]Evert-Jan Visser, "The complementary dynamic effects of clusters and networks (Periodical style),"Industry & Innovation. vol. 2, no. 16, pp. 167-195, 2009.

[4]E. N. Kaselimi, T. E. Notteboom, B. D. Borger, "A game theoretical approach to competition between multi-user terminals: the impact of dedicated terminals (Periodical style)," Maritime Policy and Management. no. 38, pp. 395-414, 2011.

[5]K. Wang. A. K. Y. Nga , J. S. L. Lam, X.Fu, "Cooperation or competition? factors and conditions affecting regional port governance in south China (Periodical style)," Maritime Economics and Logistics. no. 14, pp. 386-408, 2012.

[6] P. Fu, Y. Chen, Analysis of competition and cooperation of ningbo-zhoushan port and shanghai port(Book style). Computer Informatics Cybernetics and Applications, Springer Netherlands,2012,PP. 1559-1567.

[7] N. Saeed, O. I. Larsen, "An application of cooperative game among container ternimals of one port(Periodical style)," Eurpean Journal of Operational Research. vol.02, pp. 393-403, Oct. 2010.

[8] Jing-han Zhang, "Research on the integration interaction of Port group in the Yangtze River Delta, region and traffic development(Periodical style)," (in Chinese), Social Sciences in Nanjing. no. 01, pp. 12–16, 2009.

[9]Qing Gao and Jun-hai Ma, "Research on competition and cooperation strategy for container terminals(Periodical style)," (in Chinese),Journal of Xidian University(Social Science Edition). no. 03, pp. 76–80, 2009.

[10]Hai-bo Kuang, Shu-wen Chen, "Research and empiric analysis on assessment model of port competitiveness based on entropy TOPSIS method(Periodical style),"(in Chinese),Science of Science and Management of S.& T. no. 10, pp. 157–162, 2007.

[11]Yu-hua He, "Duopoly pricing strategy for regional ports based on Bertrand model(Periodical style)," (in Chinese), Logistics Technology. no. 01, pp. 46–48, 2011.

[12] Xiao-jun Dai, "The port development research based on hinterland's economy and competition and cooperation among port group (Thesis or Dissertation style)," dissertation, Tsinghua University. Beijing , China,2011.

[13]Ming-zhu Yu, Jun Shan, "The multi-dimensional game about cooperation and competition in the regional port system (Periodical style)," (in Chinese),Operations Research and Management science. no. 05, pp. 93–100, 2014.

[14]Xin Zhou, Mei Sha, Shi-yuan Zheng and Jian-hua Ji, "Pricing strategy of port enterprises under cooperation condition based on located spatial model (Periodical style)," (in Chinese),Journal of Shang Hai Jiao Tong University. no. 01, pp. 125–129, 2011.

[15]Hai-bo Kuang, Wen-yuan Niu, "Research on regional port group multi-period investment decision–making optimization model(Periodical style)," (in Chinese), Systems Engineering — Theory &Practice. no. 08, pp. 1414–1421, 2010.

[16]Shi-yuan Zheng, "Research on governmental regulation on investment competition of port (Periodical style)," (in Chinese), Statistics and Decision. no. 04, pp. 59–61, 2010.

[17]Gang Dong, "Container port cooperative game based on Logit model (Periodical style),"(in Chinese), Modern Management Science. no. 03, pp. 40–42, 2011.

The factors Influencing Dwelling-Type Choices of residents in the Old Industrial District: an Investigation based on the Logistic Model

Yue WANG [1,*], Zi-wei SONG[2], Luo-di ZHANG [1]
[1] School of Management, Shenyang Jianzhu University, Shenyang ,China
[2] Department of Public Administration, Beijing City University, Beijing, China
(*wy11231027@sina.com)

Abstract - **It is high level of urbanization, more laid-off workers, and greater pressure of pension in the old industrial area. Therefore, it is considerable realistic significance to know the need of dwelling-type and solve the contradiction between supply and demand under the current condition on "getting old before rich". Applying an investigation and multinomial logistic regression model explored how factors have shaped the residential choices of dwellers. It focused on the three private-dwelling types and considered different abilities to self-care. The analysis shows that individuals appear to choose their housing with nursing care. The analysis also indicates it is better to bring in professional operation team into the old-age care project.**

Keywords - **Dwelling-type choices, logistic model, old industrial district, pension**

I. INTRODUCTION

An aging population is an issue to investigate necessarily concerned with a national and regional society, economy and population development. Under the "not rich first old" background in china, in order to meet the social demands for pension services, the government need to build social endowment service system, promote the development of pension services, guide social forces to take an active part in social endowment services, balance of supply and demand contradiction [1].

By the end of 2014, the population of the elderly people with aging 60 or above in Shenyang reached 1.5202 million weight 20.8% of the total population. The elder people with disability and semi disability accounted 220400 people weight 14.5% of the elderly population [2]. The city of Shenyang, as a typical old industrial base, it is changing in the economic structure and social structure. An ageing population has developed quickly along with the profound adjustment of the interest distribution pattern and the changing on the way of thinking. The old lifestyle and endowment concept is changing in new patterns of constructing period with concept of pension and endowment. It has important practical significance that the old industrial base of the urbanization level is high, the industrial and mining enterprises laid-off workers, social endowment pressure characteristics, scientifically satisfied the endowment living needs of residents, solve the contradiction between supply and demand of endowment real estate.

Academic circle has carried on the theoretical exploration and empirical studies with willingness to choose from different angles of pension. Research in endowment intend defined as self pension, family pension and social pension or some form of synthesis[3,4].Study of rural social endowment intend to show that the rural elderly endowment is diverse characteristics, influenced by the factors such as age, income, family structure [5-8]; Young and middle-aged peasants by cultural degree, marital status, family income and other factors influence [9]. Urban residents participate in the higher ratio of social insurance. For urban residents, the nursing research is not limited in the perspective of social security, but will focus on the pension mode (i.e., home care, community care, pension institutions) will or choice of the housing endowment buy intention [10-20].City residence is one of the most important factors in the old-age residents to buy housing。 It is also influenced by the occupation, income, education and other factors. Study on the method of analysis using Logistic or Logit model of the dependent variable for two or more regression, ordered Probit model [17] and [8] structural equation model, decision tree model [12] method of scattered. The study area is most the central and western rural areas, the North City area less research results.

The pension will mainly divide the perspective social security of residents, the prediction in the probability and the impact factors of the social insurance. The research on pension model is rare, especially old-age residence willingness, while ignoring the subdivision of the housing endowment. Choose more in the existing research of farmers as research object. More study is on the Midwest rural areas rather than in the north, especially the old industrial base, there are some urban pension problems with the highly urbanization level, the aging population of laid-off workers, the level of economic growth more backward area. Therefore, the research to the old industrial base of Shenyang City as an example, consider the elderly from life completely disability or semi disability, subdivision old-age living will, from the needs of the residents of logistic model was established. Factors affecting the analysis of the urban elderly endowment living will decision, in order to the supply side of the corresponding products in the market provide basis for decision making, balance supply and demand structural imbalance.

E. Qi et al. (eds.), *Proceedings of the 23rd International Conference on Industrial Engineering and Engineering Management 2016,* DOI 10.2991/978-94-6239-255-7_40

II. DATA AND METHODOLOGY

A. Data

In January 2015, the research data of residents in Shenyang march on questionnaire survey and interviews. Subjects are covered all ages resident population in Shenyang. Survey according to the different ages stage, adopt the method of stratified sampling. The questionnaire used by investigators access to fill out questionnaires, ensure to get the data quality reliability. It combined with the network survey and recycled effective questionnaire 329 copies (Table I).

TABLE I
STATISTIC SAMPLE BASIC INFORMATION

Variable	Category	Frequency	%
Age	20-30	45	13.68
	31-40	76	23.1
	41-50	82	24.92
	51-60	52	15.81
	More than 60	74	22.49
Degree of education	High school or below	134	40.73
	Specialty	73	22.19
	Under-graduate	110	33.43
	Master degree or above	12	3.65
Annual income	Less than 50 thousand	87	26.44
	5-10	110	33.43
	10-20	62	18.85
	20-30	51	15.50
	30-50	14	4.26
	More than 500 thousand	5	1.52
Family structure	Single	31	9.42
	Two couples	52	15.81
	Couples And children	135	41.03
	Single and children	21	6.38
	Morethan three generations	90	27.36
Occupation	Farmer	100	3.39
	Civilservice/business unit	41	12.47
	Enterprise	188	57.14

The proportion of each age stage between 14% and 25% is close to the average distribution, considering endowment living intend to the different needs of the elderly themselves simultaneously. Both situations are the children tended to give parents buy endowment housing and satisfy their future demand. Family structure is mainly combined with a husband, a wife and one child of the family with three weighted for 41.03%.The family structure including the elderly couple with a child live together; Single live with the child is the old widowed or divorced, now with a remarried family to live with other child, and such cases accounted for 6.37%; Single situation includes unmarried young people and widowed or divorced. It questions 31 people, single accounting for 9.42% of the total survey; There is only one sample that four generations live together, and three generations of 89, a total of 27.36%. Respondents is given priority to with private-sector workers, part of the rural migrant workers are still peasants in the countryside, this kind of respondents were classified into category of farmers. Respondents culture degree high school and the following accounted for 40.73%, specialized subject and undergraduate course personnel accounted for 55.62%, master's or more highly educated personnel accounts for less. Regional income to a couple family income and computing, single children don't contain revenues, total annual household income in more than 300000 RMB, more than 100000 RMB accounts for about 60%.

B. Models

Select points logistics regression model, in which y = n as a reference, compared with other category generated n - 1 non redundant logit model variables, for y = I, the prediction model is:

$$P(Y_i) = \frac{\exp(g_i)}{\sum_{k=1}^{J} \exp(g_k)} \quad (1)$$

Logit model:

$$g = log\frac{P(y=i)}{P(y=n)} = \alpha_i + \sum \beta_i^T X \quad (2)$$

$P(Y_i)$ as the probability of event occurrence,, is $X^T = (X_1, X_2 \dots X_P)$ the independent variable, $\beta_i^T = (\beta_{i0}, \beta_{i1}, \dots \beta_{ip})$. For the i the partial regression coefficient of an independent variable , α_i is regression intercepts, reference coefficient were 0[21] .

C. Variables

Combined with the current pension service provision and pension status of real estate development, according to the research purpose, the dependent variable is set to one-third variable, namely, home care, community care and institutional care (Table II).

TABLE II
VARIABLES INTERPRETATION AND ASSIGNMENT TABLE

Variable symbol	Variable name	Variable type	Variable assignment
Y	Provide for the aged living desire	Dependent variable	Home care =0; Community care (Type (not care))=1; The pension agency =2
X_1	Age	Independent variable	The actual age
X_2	Level of Education	Independent variable	High school or below =0; junior college =1; undergraduate =2; Master degree or above =3
X_3	Annual Household Income	Independent variable	Real income
X_4	Occupation	Independent variable	Farmer =0; Civil servants /institutions =1; Enterprise =2
X_5	Family structure	Independent variable	Single =0; Two couples =1; Couple and their children =2; Single and children=3; More than three generations =4

Home care housing needs are general housing market; the corresponding demand for the community care for the elderly living specifically designed to provide community-based care services, with good natural environment, real estate projects and supporting pension colorful life; the aged or disability for half disabled elderly to provide comprehensive health care services and life care apartments for the elderly. Living willingness choices influenced age, income, family structure and other factors, combined with the characteristics of the old industrial cities and to highlight the overall real estate market demand for pension forecast operability, select age, education, income, occupation, and family structure five factors as independent variables analyzed in the study area will to live pension.

III. RESULTS

A. Model test

"- 2 ll" likelihood chi-square value from the initial model of 545.153 477.993, to the final model shows that model is improved obviously. At the same time, the model test of sig. Value is, 000, significantly as a whole(Table III).

TABLE III
MODEL TEST

Model	-2 Log Likelihood	Chi-Square	df	Sig.
Interceptitem	545.153			
Final model	477.993	67.160	18	0.000

B. Results

With respect to the home care residential options, nonmedical Community Care and the type of care the aged Logistic regression results very significant difference. Compared to home care options, 10% was significantly lower level, age on retirement living community select a negative effect, but the effect is weak. In addition to the age factor, other independent variables affect the probability did not reach statistical significance sense, indicating that income, education, family structure, nature of work did not produce significant effects (Table IV).

The residents home care or community endowment pension choice. Compared to old-age home, at the 5% significance level, the independent variables are significant, indicating that age, education, household income, family structure and professional nature are affected residents between home care and institutional care options. Age and education level of the dependent variable is negative, indicating that with increasing age or level of education increased by residents choose home care increased probability, but to a lesser extent the age factors. Household income, occupation, and family structure to choose health care endowment Apartment affect forward; civil servants / institutions employees are six times the employees choose health care endowment apartment; more than three generations living under one roof with respect to family structure, single or two couples the probability of people choose to live care-type pension apartment greater influence significantly .

TABLE IV
EFFECT RESIDENTS ENDOWMENT LIVING INTEND TO LOGISTIC REGRESSION RESULTS

P_1/P_0				
independent variable	B	Standard error	Sig.	Exp(B)
Intercept item	0.173	0.646	0.789	
Age	-0.019	0.011	0.093	0.981
Level of Education	-0.211	0.188	0.262	0.810
Annual Household Income	0.016	0.011	0.137	1.016
occupation =0	-0.223	0.384	0.561	0.800
occupation =1	0.896	0.483	0.063	2.450
occupation =2	0	.	.	.
family structure =0	0.888	0.542	0.102	2.431
family structure =1	0.433	0.430	0.314	1.542
family structure =2	0.380	0.330	0.250	1.462
family structure =3	-19.848	0	0	0
family structure =4	0	.	.	.
P_2/P_0				
independent variable	B	Standard error	Sig.	Exp(B)
Intercept	-0.434	0.737	0.556	0
Age	-0.030	0.012	0.016	0.970
Level of Education	-0.738	0.228	0.001	0.478
Annual Household Income	0.031	0.012	0.014	1.031
occupation =0	1.146	0.386	0.003	3.146
occupation =1	1.805	0.501	0	6.083
occupation =2	0			
family structure =0	1.267	0.577	0.028	3.550
family structure =1	1.236	0.485	0.011	3.440
family structure =2	1.018	0.411	0.013	2.768
family structure =3	1.175	0.601	0.051	3.237
family structure =4	0	.	.	.

IV. DISCUSSION

With respect to the old-age home, the age factor has the negative effect in the choice on both pension and institutional care community, indicating that older seniors more likely traditional pension manner. Young open attitude for apartments for the elderly, the future demand for the main group pension services.

With the investigation, the lower health care endowment apartment bias, the higher education community with highly educated elderly people though educated, high degree of openness, but the impact on pension ideas that traditional culture is large, multigenerational family structure is easy to accept them, which in other parts of the Chinese aggregation (Singapore, Taiwan etc.) has been confirmed; in addition, highly educated Chinese elderly, generally relatively high and stable income, strong economic capabilities, greater choice.

The household income of elderly living choice in the community is not significantly effect, but the choice of type of care significantly affect pension apartment. Shows that in family income can afford the premise of time and energy forced other objective factors like residents of the disability, the elderly can take care of semi disability, tend to choose the professional nursing organizations; and in the case of the elderly can take care of themselves, whether less care living in a beautiful environment, rich life of retirement communities.

There are several situations in Northeast old industrial cities that civil servants and employees a stable income, good benefits, especially retirement pensions generally higher than the enterprise employees, so these groups of pension and pension apartment communities have certain preferences than their employees to purchase pension houses greater probability. Farmers care Endowment apartment selection preference is greater than the enterprise employees, which may be because the study area population movements mainly concentrated in the Northeast three province, the region's urbanization rate has been higher than the national average urbanization rate unpleasant, most of the workers in the process peasants in the countryside is still a certain amount of the contract, it is possible to obtain a more stable income supplement their income level than just a source of income for social endowment insurance of laid-off workers.

The structure of the family for more than three generations live together would shows that family members are more agree with traditional family endowment concept, and is feasible for this kind of living will, therefore, this group in the survey, chose the pension that occupy the home. Relative to the above three generations would live together family structure, single and couple with the increase of age of limited ability to self care, select the probability of medical model endowment apartment is about 3.5 times that of the three generations of family structure.

In summary of the above characteristics, single or couple life with a stable and high income, has strong preference with residents of the medical model endowment apartment.

V. CONCLUSION

The empirical analysis on home care, Logistic regression model pension and medical retirement community, development of real estate projects in the pension area should take full account of consumer demand. Designed specifically for the elderly retirement community projects, although environment, architectural design, life is easy for older people, but not enough to attract consumers; and projects with health services, care or medical services more attractive. Therefore, from the supply side point of view, in the design of the project should focus on the key elements to meet the demand side, that have medical and nursing ability to facilitate targeted medical and health management services.

China's population structure has entered a "three high" stage with high-altitude nest, aging, high disability. The formation of a trillion pension industry needs to attract significant investment into the pension agency projects, mainly insurance companies, real estate companies, medical enterprises, financial institutions, pension industrial projects should be broken down investors, developers and operators, the operational phase of the project should be introduced to professional operations team to provide old-age care.

The study is merely analysis from the perspective of individual residents living retirement wishes, preferences and their needs also in terms of the pension program product attributes, such as living environment, location, size and other factors preferences. Individual property preference for the overall prediction on the regional market demand, preferences, product attributes is to forecast market segments. In the present study, based on further investigation the old industrial area resident preferences for product attributes pension programs, predictable and market segments from a structural standpoint, provide a reference for investors, operators decision to buy government pension service or policy making in accordance with.

ACKNOWLEDGMENT

This research was supported by Social Science Foundation of Liaoning Province, China (L15BGL018) and Incubation fund of Shenyang Jianzhu University(XKHY2-85).

REFERENCES

[1] N. Liu , and Y. Chen , "Rural elderly population in China based on GIS development research of time and space" (in Chinese), Resources development and the market, vol. 28, no. 4, pp. 325-327, 2012.

[2] Liu Y, Liaoning province elderly population information and development status of aging report; [cited 2015 Dec 12]. Available from: http://www.cncaprc.gov.cn/contents/37/78149.html,2015-05-14/.

[3] P. Thomas and H. Michael, "The Dwelling-Type Choices of Older Canadians and Future Housing Demand: An Investigation Using the Aging and Social Support Survey (GSS16)" , Canadian Journal on Aging, vol. 29, no. 3, pp. 445-463, 2010.

[4] J. Park and J. Kim, " The residential location choice of the elderly in Korea: A multilevel logit model" , Journal of Rural Studies, vol. 44, pp. 261-271, 2016.
[5] JL. Hao, " The will of farmers endowment pattern choice empirical research" (in Chinese), The rural economy, no. 4, pp. 87-91, 2014.
[6] JL. Hao and JR. Jia , "In the western region of the will of farmers endowment pattern choice influence factor analysis, based on the ordered Probit model and structural equation model of empirical research" (in Chinese), Statistics and information, vol. 11, no.11, pp. 107-112, 2010.
[7] XY. Wang and C. Zhang, "Rural endowment of singleton female parents intend-based on the empirical analysis of Mianyang, Deyang city of Sichuan province survey data" (in Chinese), The rural economy, no. 3, pp. 75-78, 2013.
[8] BH. Tian , H. Huo and ZB. Zhong, "Life situation and pension intention-an empirical analysis of the influencing factors of rural elderly family endowment preference" (in Chinese), Rural economic problem, no. 2, pp. 74-85, 2012.
[9] SY. Shen, F. Li and Y. Xie, " Young and middle-aged peasants endowment intend and influence factors analysis, based on the survey data of five counties in Nanjing area " (in Chinese), Journal of rural economy, no. 11, pp. 84-89, 2009.
[10] L. Cai, " The current and future study on the relationship between the endowment of the will of endowment way" (in Chinese), Social security research, no. 2, pp. 32-42, 2014.
[11] CB. Wu, " The old mode selection factor analysis-based on logistic regression analysis tool " (in Chinese), The research world, no. 6, pp. 41-43, 2011.
[12] X. Gao, BQ. Yan and Yu.Ji, " Beijing urban residents endowment pattern selection and rationality analysis" (in Chinese), Geographical science progress, vol. 31, no.10, pp. 1274-1281, 2012.
[13] SQ. Long and XT. Feng, " Urban residents' willingness to pension and its influencing factors research, four cities of Jiangsu province the investigation and analysis of the elderly living conditions" (in Chinese), Sociological study, no. 1, pp. 97-105, 2007.
[14] M. Li, " The influence factors of community home endowment intend to research-in Beijing" (in Chinese), Population and development, vol. 20, no. 2, pp. 102-106, 2014.
[15] SJ. Chu and JJ. Sun , " Affect the urban elderly endowment intend the factors analysis" (in Chinese), Population and social, no. 2, pp. 43-46, 2010.
[16] GH. Sha and ZJ. Wang, " Urban elderly endowment intend and its influencing factors analysis, based on the empirical research in Suzhou" (in Chinese), Journal of Suzhou institute of science and technology: the social sciences, vol. 32, no. 4, pp. 24-30, 2015.
[17] Luo N. and YC. Zhang , "Urban elderly pension mode choice intention and influence factors analysis, based on: an empirical study of the Probit model" (in Chinese), Journal of xi 'an academic report of finance and economics, vol. 25, no. 4, pp. 97-102, 2012.
[18] WJ. Zhang and M. Wei, " Urban elderly institution endowment intend and influencing factors of research-in Xicheng district of Beijing" (in Chinese), Population and economic, no. 6, pp. 22-34, 2014.
[19] B. Liu, " Jinan urban elderly endowment intend and influencing factors of research-in Jinan district" (in Chinese), Journal of Shandong school of administration no. 2, pp. 69-73, 2014.
[20] QX. Kang and HM. Zhang , " Willingness to residents buy endowment property and its influencing factors" (in Chinese), Studies of finance and economics, no. 12, pp. 41-45, 2013.
[21] SQ. Bai, ZQ. Chen and Chen Q , " Based on farmers Logisfic model of ecological public welfare forest compensation standard 'willingness to accept influence factors analysis" (in Chinese), Resources development and the market, vol. 31, no. 6, pp. 744-747, 2015.

A management method of accounts receivable based on credit rating for rail equipment manufacturing industry

Kai MA, Rui MIAO*, Wan-chun YANG, Zhou PENG, Bowen SUN, Ning-yu HU
School of mechanical engineering, Shanghai Jiao Tong University, Shanghai, China
(*mrxxdeng@qq.com)

Abstract – **With the wide use of enterprises' goodwill, the amount of accounts receivable which stays is being a rising trend every period. What's more, the proportion of accounts receivable is keeping increasing in current assets. In this paper, there will be a simple summary and analysis on the emergence and the harm of accounts receivable. And an illustration on the main management of accounts receivable has been given. This paper tells the relationship between credit administration and management of accounts receivable, and it points out the inapplicability of recent popular systematic credit rating and analysis method applied to the manufacturing industry of rail equipment, also proposes a management method of accounts receivable based on credit rating.**

Keywords – **credit rating, accounts receivable, railway enterprises.**

I. CORRELATION THEORY OF ACCOUNTS RECEIVABLE

Accounts receivable is a legally enforceable claim for payment held by a business against its customer/clients for goods supplied and/or services rendered in execution of the customer's order. And it is one of the most important components in enterprises' receivables. With the development of market economy, goodwill of enterprises has been widely used. The amount of accounts receivable which stays is being a rising trend every period. What's more, the proportion of accounts receivable is keeping increasing in current assets. When most receivables occupy the operating funds, it also lower the turnover rate of business capital, increase the handling cost and bad debt loss related with accounts receivables. The opportunity cost of capital occupying has been increased, which lead to a dramatic decline of profit margin. In severe cases, it can lead to the financial deterioration and the loss of state-owned assets[1]. For manufacturing industry of rail equipment, the railway market which was monopolized before has been slightly opened at present because of the significant achievement in recent years. A large number of international famous enterprises and domestic private enterprises enter into the competition of railway market competition. Accounts receivable gets higher and higher while the scale of enterprises gets larger and larger. More competitors in this market, higher the managing risk will be faced. It is imminent to manage this kind of situation in a scientific and reasonable way.

A. The Dangers of Accounts Receivable Mismanagement

(1) Influence enterprise efficiency

As the logistics can't keep pace with the capital flow, the capital can't be regained after the products dispatched. With no capital recovering, this kind of sales revenue will definitely cause loss to the capital flow, the turn in of sale tax and income tax will gradually affect the turnover of business capital of the railway companies, and cover the actual business situation, fail the final performance indicator, reduce the capital usage rate and cause a reduction of the performance of the company[2].

(2) Bubble the operation performance of the company

Nowadays, the accounting of the company is based on accrual basis, current revenue contains all the current credit sales revenue that already occurred. Due to the existence of a large number of receivables, the increase of the corporate profit will contain bubbles caused by a lot

E. Qi et al. (eds.), *Proceedings of the 23rd International Conference on Industrial Engineering and Engineering Management 2016,* DOI 10.2991/978-94-6239-255-7_41

of charge sales, current cash inflow can't be reflected, thereby increase the risk cost of the enterprise.

(3) Influence the normal production of the enterprise

A large number of tied up capital caused by occupation sediments into the non-production processes, this will cause a fund shortage, will affect the salary, the material purchasing and investment in the security and so on, in severe cases this will affect the normal production and management of the enterprise.

(4) Increase the financial costs

With a large quantity of debt clients, the financial department will face a lot of pressure on the receivables management, and correspondingly will increase the potential risks like accounting mistakes, less awareness of the vibration of the receivables and the business management situation of those companies[3] .

B. Major Styles of Receivables Management Nowadays

(1) Pre-control

Pre-control measure is base of receivables management improvement. Measure should be taken before business to reduce the risk. Rate all your customers, perform different credit management measures. All the measures should be established in advance. In this case, bad debts could be forestalled from the very beginning. If our customer has the history of defaulting on its debt, we should be more careful about the cooperation.

(2) Mid-control: Real-time monitor customers' dynamic credit

Establish receivables early warning mechanism. Dynamically check customers' credit if their debts exceed credit line. For example, for brother company of Group Corporation, credit line could be larger and credit period could be longer. However for other company or agency, credit line would be lower, credit period would be shorter.

(3) Post-control: receivables management

The purpose of "post-control" is receivables collection and management. In this case, we can handle the issue when customer default or dishonor the receivables.

1) Make reasonable collection sequence: Lock the key collection company per collection difficulty, collection period. And make our company focus on reasonable collection.
2) Implement Responsibility System of Accounts Receivable: Sales department, contacting directly customer, takes charge of asking for accounts receivable, and undertakes relevant associative responsibility.
3) Adopt Reasonable Means: Some customers are in arrears for longer time, it is not because of their shortage of fund flow, but maybe malicious arrears or the refusal of payment. Therefore, it is necessary to adopt reasonable means or process, such as, letter notification, call & fax, sending someone for receivables and legal way.

II. STATUS QUO OF RECEIVABLES AND CREDIT MANAGEMENT FOR RAILWAY EQUIPMENT MANUFACTURERS

The accounts receivable management of railway equipment manufacturers along with receivables appearing appears. When credit quality is very high, the accounts receivable are not existed yet. Along with receivable appearing and increasing, the accounts receivable management is getting more and more important, also getting more and more professional. However, its development experiences four stages, afterwards control, intermediate control, advance control and triplicity. When the accounts receivable is just generated along with risks, most enterprises begin to pay close attention to it. This is the primary stage of accounts receivable, also afterwards control. Along with receivables increasing, most enterprises find it exist very big risks, when accounts receivable are overdue, begin to control it. Therefore, most enterprises begin to focus on intermediate control and beforehand prevention, not afterwards control. Relatively speaking, four stages of accounts receivable management get more and more difficult in turn. Especially advance control, it requires to evaluate reasonably and scientifically customers' credit. The credit is good, it is doable for business transaction, if not, to prevent from the source[4]. So, developing to present stage, the management emphasis of accounts receivable management is credit management.

Regarding Receivables and credit management of railway equipment manufacturers, some of them hand over this function to marketing department. There are existing any problems because of marketing personnel's different knowledge, sales tasks and receivables tasks. When marketing personnel are in the face of large order, some of them do not have the high control risk ability, just only for getting the large orders, but ignore credit management. Some of them are overanxious and overcautious to cause the drop in sales, the enterprise will suffer the loss. Some enterprises hand over this function to finance department. But during credit evaluation, the finance department usually only pay attention to financial data, ignoring the sales target, because of its professional characteristics. In short, if marketing department takes charge of credit evaluation, it is easy to exist credit risk. If finance department does, it is easy to lose order because of over caution. So, an organization established by marketing department, finance department and operations management department should adopt reasonable methods in receivable management. The key point of arrangement is to advance control, meanwhile jointing intermediate control, afterwards control. The difficulty of management is to evaluate the credit of customers.

III. GENERAL APPROACH AND EXISTED PROBLEM OF CUSTOMER CREDIT EVALUATION

The development of customer credit evaluation approach roughly experiences three stages. The first stage, before 1970 most finance institutions basically relied on bank experts' experiences and subjective analysis to evaluate credit risks. Main traditional credit evaluation approaches are 6C, LAPP and the five-class classification, etc. The Second stage, financial ratio comprehensive analysis method including Dupont financial analysis and Wall's proportion grading law. The third stage, based on the analysis of financial indicators credit metrics model, main including multiple discriminant analysis model and Logit model. In recent years, analytic hierarchy process (AHP) in the small and medium-sized enterprise credit evaluation has been widely used. Analytic hierarchy process (AHP) is a kind of effective treatment that is difficult to completely by the quantitative method of dealing with the complex problem[5].However, all of the above methods, for railway equipment manufacturing industry, some problems on its applicability, basically has the following several aspects. 1, railway equipment manufacturing industry of customers is relatively simple, should not be too cumbersome when introducing credit evaluation method. 2, railway equipment manufacturing industry credit management started late, the credit evaluation generally requires an objective index for weak. 3, due to the market economy transition late railway equipment manufacturing industry, in this case, the entire industry just let go of, be familiar with the relevant enterprises, enterprises had a greater influence on the nature and scale of credit. 4, because the railway equipment manufacturing industry belongs to high and new technology industry, product market position of the supplier can directly affect the management of accounts receivable, so to speak, product market position directly impact on the outcome of the credit management.

IV. SUITABLE FOR RAILWAY EQUIPMENT MANUFACTURING INDUSTRY TO BUILD SYSTEM OF ACCOUNT RECEIVABLE CREDIT EVALUATION SYSTEM

Regarding the general method of using the current credit evaluation applicability for railway equipment enterprise is not high, we proposed a relatively simple method of customer credit evaluation. Steps are as follows: first of all, according to industry characteristics determine the relevant evaluation index, and then quantitative indicators, credit evaluation score is calculated after quantitative indicators, the final evaluation to the customer after the introduction of the historical data, the modification of historical data, so as to find the optimal evaluation scheme[6].

(1) Related to the establishment of evaluation index

At present more popular abroad element method is a kind of credit analysis, the representative of which is 6 C method, 6 C, respectively, refers to the moral character, ability, capital, collateral, macroeconomic condition, continuous career development. Through expert analysis, we are according to the characteristics of the industry, to list the following elements: 1)The importance of sales revenue from clients; 2)The importance of profit from clients; 3)Credit rating based on the past records and debit paying ability of the clients; 4)Position and scale of the client in the industry; 5)Market position of our products; 6)Coincidence level of development strategies between clients and us.

According to analysis, indicators and weightings are settled as:

TABLE I

Indicator name	Indicator explanation	weightings
Average of sales percentage from recent 3 years	The importance of sales revenue from clients	20%
Average of profit percentage from recent 3 years	The importance of profit from clients	25%
Percentage of receivables over 1 year	Sales on account from clients	15%
Turnover of account receivables	Speed of money returning from clients	10%
Position and scale in the industry	Ability of debt paying	15%
Market position of the products	Market position of our products	10%
Contrast of development strategies between clients and us	Coincidence level of development strategies between clients and us	5%

(2) Regulation of index quantification for clients

1) Standardize the four indicators: average of sales percentage from recent 3 years, average of profit percentage from recent 3 years, percentage of receivables over 1 year, and turnover of account receivables, into centesimal system.

Centesimal percentage = [sample value- Min (sample value)]X100/[Max(sample value-min(sample value)]

2) Position and scale in the industry as below sheet

TABLE II

No.	Position and scale in the industry	score
1	Internal group enterprise/ head enterprise in the industry/ enterprises with registered capital over 100 million Yuan	91-100
2	External group enterprise/ leading enterprise in the industry/ enterprise with registered capital over 50 million Yuan	81-90
3	Industry-famous enterprise/ enterprise with registered capital over 30 million Yuan	71-80
4	Normal enterprise/ enterprises with registered capital over 8 million Yuan	61-70
5	others	60

Note: Enterprise customers are defined to be qualified supplier, in this case their baseline score, then reward with more score per enterprise character.

3) Product market position per table below

TABLE III

No	Product market position	Score
1	Monopoly/Share of market larger than 80%	91-100
2	Leading position/Share of market larger than 60%	81-90
3	Share of market larger than 40%	71-80
4	Weak position/Share of market larger than 20%	61-70
5	New product/Share of market less than 20%	60

4) Contact ratio between customer business and development strategy

TABLE IV

No.	Contact ratio	Score
1	More than 90%	91-100
2	More than 70%	81-90
3	More than 50%	71-80
4	More than 40%	61-70
5	Less than 40%	60

(3) Credit calculation

Credit score= three-year sales occupy ave.*20% + three-year profit occupy ave. * 25% + more than 1 year receivables occupy * 15% + receivables turnover ratio * 10% + Enterprise position in industry * 15% + Product market position *10% + Contact ratio between customer business and development strategy *5%.

(4) Credit score correction and application

After calculation, score should be corrected and verified, then the final score can be calculated. Then customers can be rated as Grade AAA, Grade AA, Grade A, Grade BBB, Grade BB, Grade B, Grade CCC, Grade CC, Grade C and Grade D.

(5) Amendments for proposal of credit rating

Since enterprises' credit is dynamic, the proposal of credit rating shall be updated periodically. Therefore by calculation, we decide to amend the proposal once every year.

V. CONCLUSION

The paper did a research on current Administration of Account Receivable and Credit Administration of China's railway equipment manufacturing industry. It analyzed the relationship between the above 2 Administrations; and pointed out that common credit rating method is not applicable for the industry. Expert marking method is adopted for a combination of qualitative analysis and quantitative analysis, against credit administration; Analysis and discussion are executed for account receivable, and a new credit rating method is proposed.

As China's rail transportation industry develops, the market is becoming more and more open. To improve enterprise competence, market share in global market, enterprises may take method of Sale on Credit. As it's more applied, however, the amount enterprises' account receivable increases, the financial risk increases as well. Hence, the administration of account receivable plays a remarkable important role during enterprises' development. With the application of customer credit rating system, an effective, reasonable and scientific management for account receivable can be reached, and will benefit enterprise development.

ACKNOWLEDGMENT

The authors gratefully acknowledge the financial support of the innovation practice program of Shanghai (IPP11022), and the national natural science foundation, China (No. 51435009).

REFERENCES

[1] Xiaodong Tang, "Explore of the enhancement of accounts receivable management in enterprises"[J], *Journal of Shanxi Economic Management College*, 2011, (03), pp.61-63

[2] Rui Yu, "Problems and solutions in the accounts receivable management in enterprises"[J], *Chinese foreign Capital*, 2011, (17), pp.31

[3] Jie Huang, "Enterprises Management strategies of accounts receivable"[J], *Chinese National Situation and Power*, 2009, (05), pp.28-29

[4] Bingyi Ma, "Analysis of the cause and solution about risks of accounts receivable"[J], *Shenzhou (Midmonth Release)*, 2013, (20), pp.275

[5] Hong Chen, "Structure and application of teaching quality assessment model in universities"[J], *Lab Research and Explore*, 2009, 26(6), pp.159-161

[6] ShuFeng Peng, Ying Tian, "Application of FAHP in the credit evaluation of small and medium-sized industrial enterprises"[J], *Journal of Chongqing Institute of Technology*, 2008, 22(1), pp.131-133

The Research of Doctors Recommendation Algorithm based on Clustering and Collaborative Filtering

Chen Wang [1,*], Man Xu [2]

[1]Department of management science and Engineering, NanKai business school, Tianjin, China

[2]Department of information resources management, NanKai business school, Tianjin, China
(*1562767687@qq.com)

Abstract - **With the continuous deepening of informatization in the medical field, the massive data in electronic medical records makes it possible to carry out intelligent medical recommendation technology based on data mining and cloud computing [1].In this paper, we proposed an algorithm for doctor assessment based on collaborative filtering to achieve the goal of personalized medical recommendation in the large data environment. On this basis, to reduce complexity and improve operational efficiency, we proposed an improved algorithm based on clustering and clustered doctors who are professional on certain illness. The result shows that the algorithm can recommend optimal doctors to patients efficiently.**

Key words - **Clustering, collaborative filtering, doctor score, personalized recommendation**

I. INTRODUCTION

For nearly 20 years, a major social problem in China is the difficulty and high cost of getting medical treatment, which is also a concern of people's livelihood. Although many new health care policies continue being proposed, the problem has not been solved fundamentally. A survey found that the most prominent difficulty of getting medical treatment for patients is to find suitable doctors. For safety and security, patients tend to pay high fees only to select big hospitals, regardless of their illnesses. But when all patients choose big hospitals, a series of problems emerge, including long queuing time, medical shortage, high fees and bad service attitudes of medical staff, etc.

Personalized medical recommendation can recommend the most suitable doctor according to each patient's interests and needs. On one hand, it will not delay the illness; on the other hand, it can solve the problems caused by too many people choosing limited number of big hospitals. For example, resource in big hospital is often in shortage while grass-roots medical facilities are idle. Compared with the other algorithms, recommendation system has a brighter future in development and application in e-commerce, which makes people feel more interested in it and start trying to find its rational uses in other areas.

Traditional methods of recommendation system include recommendation based on association rules[2], demographic[3], content[4] , collaborative filtering[5] [6][7]and on network[8], etc. The method based on association rules tries to mine the internal linkages between the items selected by the user find the items that are selected together and then use them in next recommendation. The method based on demographic and content finds similar users and projects to complete the recommendation, respectively. The method based on collaborative filtering considers users' preference and finds the correlation between users and projects, besides the items we mentioned above. The method based on network completes the recommendation by building network models. However, only one single recommendation method has limitations inevitably, which is not suitable for complex medical diagnosis.

Baofu Yu (2012) [9] proposed the medical recommended algorithm based on BP neural network and SOM neural network. The former one uses the idea of the method based on content and classifies the information by subject – put the information pages on the same subject in the same group. The latter one uses the idea of recommendation algorithm based on collaborative filtering, which matches the users' interests model with different groups of pages, and then recommends the patients with those pages which have greater values than threshold.

Peng Li and Zhongxin Yu (2013) [10] used network models for medical recommendation, which explored new ideas for our research. They proposed building a heterogeneous medical information network model to describe different parts in a medical process, such as the relationship among patients, doctors, diseases and treatment. Then classify the data and rank them so the same category of doctors and treatment method will be effective on patients. When a new patient comes, the recommendation process can be achieved by calculating the similarity between the new patient and the one in the network models.

In this paper, we try to use the most widely used collaborative filtering techniques to design a patient-oriented doctors recommendation model. Inspired by the doctor rating system in the paper of Chonglin Sun (2015) [11],we try to design a novel rating algorithm. First, cluster doctors according to the illnesses that they are professional on, then we can get the basic score using

E. Qi et al. (eds.), *Proceedings of the 23rd International Conference on Industrial Engineering and Engineering Management 2016,* DOI 10.2991/978-94-6239-255-7_42

collaborative filtering methods. The recommendation in original doctor rating system may be inaccurate and unreasonable since the same doctors are recommended all the time, and the proposed algorithm can best overcome these shortcomings. Besides, we also improve the efficiency of the algorithm.

II. MODELING

A. Doctors score algorithm based on collaborative filtering techniques

In this paper we use the following features as evaluation criteria for doctors: (1) basic information, including educational background, positional titles and the rank of hospital; (2) academic information, including the number of papers published, the number of co-workers and the academic conferences participated etc.; (3) patients' appraisal information for doctors. We need to highlight that even though we haven't taken the score of patients as a part of evaluation like England or other western countries do, in the long run, in the medical system where patients tend to become the service core the evaluation system is doomed to be an important part of future construction, so we have to take this factor into consideration. Since the information of patients score for doctors collected by hospitals is limited, we try to use the collaborative filtering techniques to predict the score for doctors. This algorithm assumes that if the academic achievements of two doctors are similar, their ability is probably at the same level and the evaluation of patients is similar.

This algorithm mainly relies on two parts of data information: (1) information for co-workers of academic papers published by the specific doctor; (2) score information of patients for the doctor. Three steps are included in this process: (1) calculating the similarity for doctors in academic achievements; (2) predicting the score for an unknown doctor with the similarity information and the score for certain doctor; (3) calculating the basic and comprehensive score based on the indexes of evaluation.

1. Completing the matrix of co-workers

We abstracted the information of co-workers in the database of published papers, and constructed the matrix for co-workers. We used n proxies for the number of doctors, and built a n*n matrix M based on the information. For doctor i and doctor j, $M_{i,j}$ stands for the frequency of their academic cooperation. This is a symmetric matrix. Significantly, since the papers' record is limited and it could hardly cover all papers published by every doctor, we need to complete the data and predict the potential information for co-workers. We took the algorithm based on singular value decomposition (SVD) as our optimal choice since it has the advantage of little disturbing for original matrix eigenvalue[12].

2. Algorithm of the basic score for doctors

When the original matrix is completed, we use person correlation formula (1) to calculate the similarity between every two doctors.

$$\mathrm{sim}(u,v) = \frac{\sum_{i \in I}(R_{u \cdot i} - \overline{R_u})(R_{v \cdot i} - \overline{R_v})}{\sqrt{\sum_{i \in I}(R_{u \cdot i} - \overline{R_u})^2}\sqrt{\sum_{i \in I}(R_{v \cdot i} - \overline{R_v})^2}} \quad (1)$$

With sim(u,v) standing for the similarity between doctor u and doctor v, I is the set of all doctors and i means each doctor in this set. $R_{u \cdot i}$ and $R_{v \cdot i}$ are the number of papers cooperated by doctor u, i and doctor v, i respectively. $\bar{R}_u$ and $\bar{R}_v$ are the average number of cooperated papers with other doctors for doctor u and v respectively.

Based on the collaborative filtering among users, we could predict the score of an unknown doctor with information of score for him or her. The algorithm is (2) as follows, and c_u is the score of doctor u given by patients.

$$c_u = \frac{\sum_{v \in I(u)} sim(u,v)(c_v - b_{u,v})}{\sum_{v \in I(u)} |sim(u,v)|} + b_{u,v} \quad (2)$$

In this equation, $b_{u,v}$ is the basic score depending on the mean value for all doctors we know.

By now we have got the patient score for every doctor, and we may start to calculate basic score with the appraisal indexes mentioned above. Specially, we need to transform the sequential data to numerical data. For example, for the variable positional rank, we may take value 1 to 4 for physician, officer physician, associate chief physician and chief physician respectively. Other sequential variable are treated in the same way, and the value of every index is between 0 and 9. After that, we give different weights to different criteria and get the basic score for doctors. The weights are derived from expert comprehensive opinion in related field with Delphi method.

3. Some problems of the current algorithm

In the process of completing the matrices, if the dimension of original matrix R increases, the calculation will be quite complex. And when we apply the algorithm to a real system, the efficiency will be rather low owing to the massive data. In another aspect, the way of filling the missing value blank with the mean value of known items is most commonly used to normalize a matrix. But this will make a sparse matrix transformed into a dense matrix, and will occupy a large storage space, so it's not feasible. With the two issues mentioned above, we proposed an improved scheme based on clustering algorithm.

B. The improved method based on clustering algorithm

Clustering is a method of gathering similar data on the premise of no artificial marks. Data with some similar features will be in the same group, that's to say, data in the same group is alike and data in different groups is different [13][14]. Collaborative filtering algorithm is based on highly similar users, so the quality of clustering algorithm will straightly influence the efficiency in general.

In this paper, we took a simple clustering algorithm and put doctors who specialize in the same illness into a group. A new patient just need to search for an appropriate doctor in this group since taking other irrelevant doctors into consideration is meaningless. For instance, there is no need for a cardiac to consult a chief physician specializes in leukemia. We should pay attention to the following aspect: (1) Clustering should be sorted by a special illness rather than a kind. For example, leukemia and Mediterranean anemia should be treated as two clusters, they can't be classified as diseases of blood simply. The more rigorously the clustering is conducted, the lower the dimension of matrices of co-workers will be, and the less complicated the algorithm will be; (2) If a doctor is good at treatment for two kinds of illnesses, then he can be classified in two groups; (3) Whether a doctor is good at a specific illness depends on his research area and the number of successful cases for this disease. And we need to give value to the professional level for calculation, and the value is between 0 and 9; (4) For common diseases like cold, it won't be considered as a kind of clustering since most doctors can treat it and the treatment effects are similar.

The improved algorithm for completing matrices is as follows:

Import: original doctors' co-worker matrix M (①M is a n*n matrix and we take the unknown value as 0 ; ② For doctor k, the paper worked by himself is recorded in matrix $M_{k,k}$.)

Export: the completed matrix PM (even though there are some elements valued at 0, we need more data information to calculate the similarity between doctors more accurately, so we need to complete the original matrix. What's more, this method is reasonable since the complement is based on the unchanged matrix eigenvalue.)

Detailed process:

(1) Find the accurate doctor clustering according to patients' illnesses, and build the original matrix M according to the corresponding co-workers' information;

(2) Calculate the mean number of co-workers mi for the rows of M;

(3) Find a non-zero value m_{ij}, and subtract it by m_i to get a matrix M' ;

(4) Decompose the matrix M' into three matrices U, S and V by SVD algorithm;

(5) Keep the part whose singular value is higher than 1 in matrix S, and delete the other row and column, then we get the simplified matrix S_k (keep the first k row and k column). Consistently we get U_k and V_k^T , U_k is a n*k matrix and V_k^T is a k*n matrix;

(6) Calculate A and B with the equations $A = U_k \times S_k^{1/2}$ and $B = S_k^{1/2} \times V_k^T$. For the missing value in M, we complete it by $PM = \overline{M_i} + A_e \times B_f$. ($\overline{M_i}$ is the mean number of papers of doctor i cooperated with other doctors, e and f is the row e in matrix A and column f in matrix B respectively.)

According to the improved algorithm, we should first cluster doctors by the treatment for illness they are good at. Thus the dimension of matrix need to be decomposed can be transformed into the number of doctors in every cluster. and we could solve the two problems mentioned above quite well. The second and third steps are aimed to normalize the matrix. The different performance between doctors may be more significant when we use a known item subtracted by the mean cooperated papers numbers. In another aspect, there will be more elements valued at 0 in the matrix, and by ignoring these zero values when storing, the second issue can be solved further. In the sixth step, we add the original mean value, which makes the prediction more accurate.

C. The calculation for doctors' comprehensive score

With the process mentioned above, we get two data parts. One is the basic score for doctors, the comprehensive character of a doctor can be measured by the score. Another part is the goodness level for specific disease in the clustering, we map it into numbers 0 to 9. The higher the value is, the higher the goodness level for the illness is. The comprehensive score for doctor i is calculated in the following equation (3):

$$S_i = a \times (\text{the goodness level for a specific illness}) + (1-a) \times (\text{basic score}) \quad (3)$$

The value of 'a' depends on the proportion of the goodness level and comprehensive character.

We can demonstrate the whole process in this way. When a new patient appears in the system, firstly we find the exact doctor cluster according to his illness. Then we filter the doctors on the basis of the severity (We classify the severity into light, serious and very serious. When the patient chooses the very serious sort, we only recommend doctors with goodness level value above 5. If the patient doesn't choose anyone, we then take the very serious sort as default), the education background of doctors, the professional position, the class of hospital, the threshold for each score and whether the doctor is appointed or not. Finally we get comprehensive scores for every doctor and recommend the doctor with highest score to the patient.

III. EXPERIMENT

A. Data sources

We test the doctor recommendation algorithm mentioned above with the information of 200 doctors in Tianjin Nankai Hospital. The data set includes the basic information of doctors, academic information and the assessment information given by the patients. We analyze the singular value decomposition part with Matlab.

B. The experimentation and analysis

1. Data clustering process

Since the variables of educational background and professional position are characters, we need to transform the character into numerical type if we want to get the basic score for doctors. As mentioned above, professional position can be mapped into numbers between 0 and 4 from low to high level, and similarly, the education background can be mapped into numbers between 1 and 3 for the undergraduates, postgraduates and doctors.

After the data is pre-processed, we can cluster the doctors according to the illness they are good at dealing with. We finally get 15 clusters, and choose 20 doctors who do well in treating pancreatic cancer to analyze.

2. Decomposition of singular value

After the data is clustered, we can calculate the similarity between doctors with the completed matrix. Taking doctor 3 for example, part of data after decomposition and filling is shown as Table I. It's clear that the completed matrix after decomposition and filling is a dense matrix. The matrix predicts the number of cooperated papers between doctor 3 and doctors in other clusters, and avoids the error caused by the incompleteness of recorded papers.

TABLE I PART OF PREDICTION DATA FOR DOCTOR 3'S COOPERATED PAPERS

Sequence number	Sequence number	Number of cooperated papers
3	2	0.6
3	3	3.0
3	5	0.7
3	6	0.5
3	12	0.6
3	13	4.0
3	16	0.2
3	18	3.0

TABLE II PART OF SIMILARITY DATA BETWEEN DOCTOR 3 AND OTHER DOCTORS

Sequence number	Similarity to doctor 3
2	0.01
5	0.01
9	0.19
13	0.82

3. Prediction of score for unknown doctors

We take doctor 3 for example as above. We calculate the similarity between doctor 3 and other doctors in the cluster respectively, finally we get part of the similarity data shown in Table II. The similarity level is higher when the value is higher.

We can calculate the score for doctor 3 according to the equation mentioned above, the result is 7.71 and we don't repeat the exact process. Similarly, we can get the score for other unknown doctors in the cluster.

4. Calculating the basic score and comprehensive score for doctors

The basic score for doctors depends on the weights of each index. By repeatedly testing different weights for basic information, academic information and patients' score, we finally find the best performance when the weights are 0.4, 0.4 and 0.2 respectively. In the same way we found that when the best proportions for basic score and goodness for specific illness are 0.4 and 0.6 respectively. The basic condition for 5 doctors with the highest score is shown in Table III, and we can make recommendation according to the score. The results show that when the patient's condition is very serious, and no other filter condition is proposed, the comprehensive level and goodness level of the recommended doctors are rather high. On the other hand, we find that this recommendation includes rather lax scope of basic ability and academic ability, which suggests the high cover of the algorithm and the goodness of recommendation effects.

C. Assessment of the experiment

The experiment above tests the validity of doctor recommendation algorithm proposed in this paper. However, there is some limitation which needs to overcome in the future research:

(1) Since the data used is quite limited, the problem of data bias may be severe. For example, the comprehensive goodness level for doctors in hospital A is higher than that in hospital B.

TABLE III DATA FOR FINAL SCORE OF DOCTORS

Sequence number	Educational background	Professional position	Papers published	Academic conferences participated	Goodness level	Patients' score	Basic score	Comprehensive score
10	3	4	15	15	8	8	16.4	11.36
1	1	4	20	8	8	8	14.8	10.72
11	2	3	20	6	8	9	14.2	10.48
14	3	3	20	3	8	8	13.2	10.08
9	3	4	10	10	7	7	12.2	9.08

(2) The set of weights is the results of multiple tests, it doesn't apply to the condition for massive data and it lacks authority.

(3) The experiment just makes some simple assessment for the recommendation results, which lacks rigorous validation.

In summary, abundant data is not available to test the algorithm for the time being, since the medical information platform and electronic medical record and health record are being constructed at present. With the accumulation of electronic medical data and deepening of relevant research, we may access to more valid data and information, so the two issues mentioned above can be solved perfectly. As for the third problem, we can solve it from the property and accuracy. We can get the former one by comparing time complexity with other algorithm when the data is sufficient, and the later one with the feedbacks from patients. So we plan to carry out the relevant follow-up research when the database is enough.

IV. CONCLUSION

With the continuous deepening of informatization and development of massive medical data, we proposed an algorithm for doctor assessment based on clustering and collaborative filtering in this background. We evaluate doctors' comprehensive quality from the views of basic information for doctors and score from patients. Besides, we took doctors' goodness for diseases into consideration to make best recommendation for specific patients, which provides strong reference for intelligent decision-making and medical data knowledge mining. Finally, we verified the validity of the algorithm by the simulation of recommending 200 doctors of Tianjin Nankai hospital. Compared with other algorithms, the proposed method has the following three advantages:

（1） We combine traditional clustering algorithm with collaborative filtering recommendation algorithm, which greatly reduces the complexity and improves the efficiency of the algorithm;

（2）We recommend the most suitable doctor according to patient's condition, which solves the problem that some methods recommend the same doctors all the time unreasonably;

（3）The complexity is further reduced when we use singular value decomposition to reduce the dimension of matrices.

In the part of processing patient's rating, we should note that regarding treating rating as an important evaluation indices of doctors should be based on standardized clinical guidelines and national payment standard. Only with these two criteria evaluation of medical services can be more objective and effective, and accredited by both doctors and patients [15]. Current grading diagnosis and treatment in China is relatively undeveloped. The doctor-patient relationship is at an uneasy state and patients could not fully understand doctors. So when calculating doctors' scores, we can't set large weight of patient's rating. But what can be sure is that with the development of patient-centered health care system, patient rating is playing a more and more important role in doctor evaluation.

ACKNOWLEDGEMENT

This work was supported in part by the National Natural Science Foundation of China (Grant 71571105).

REFERENCES

[1]Guoxia Wang, Heping Liu. Summary of personalized recommendation system[J]. Computer engineering and Applications, 2012，48(7):1-8. (Chinese)

[2]Zhenguo Ping, Jing Chen. A personalized recommendation system based on association rules[J]. Computer integrated system, 2003,9(10):1-2. (Chinese)

[3]Chen L, Nayak R, Xu Y. A Recommen-dation Method for Online Dating Networks Based on Social Relations and Demographic Information[C].In Proceedings of ASONAM, 2011, 407-411.

[4]Kompan M, Bieliková M. Content-Based News Recommendation [J/OL]. In Proceedings of EC-Web, 2010: 61-72.

[5]Ailin Deng, Yangyong Zhu, Bole Shi. Collaborative filtering recommendation algorithm based on item score prediction[J]. Journal of software, 2003, 14(9):1621-1628. (Chinese)

[6]Chuangguang Huang, Jian Yin, Jing Wang. Collaborative filtering recommendation algorithm for uncertain nearest neighbor[J]. Journal of Computer Science, 2010, 33(8):1369-1377. (Chinese)

[7]Bobadilla J, Hernando A, Ortega F, etal. A framework for collaborative filtering. recommendation systems[J]. Expert Systems with Applications, 2011, 38(12):14609-14623.

[8]Kehan Chen, Panpan Han, Jian Wu. Recommendation algorithm for heterogeneous social networks based on user clustering[J]. Journal of Computer Science, 2013,36(2):2-5. (Chinese)

[9]Baofu Yu. Research and implementation of personalized recommendation system for medical information[D]. Zhejiang: Zhejiang University, 2012:46-50. (Chinese)

[10]Peng Li, Zhongxin Yu, Ning Li. Research on intelligent medical recommendation system based on heterogeneous network analysis[N]. Chinese Journal of health information management,2013-12-10(6):1-6. (Chinese)

[11]Chonglin Sun. The research and implementation of doctor recommendation system based on multi label classification and collaborative filtering[D]. Liaoning: Liaoning University, 2015:35-36,65-66. (Chinese)

[12]Boling Wang, Zhihong Tian, Yongzheng Zhang. Singular value decomposition algorithm optimization[J]. Electronic journal，2010，(10). (Chinese)

[13]Linzhu, Jingsheng Lei, Zhongqin Bi, Jie Yang. Soft subspace clustering algorithm based on data stream[J]. Journal of software,2013,24(11). (Chinese)

[14]Qifu Yao, Guoqing Zhang. Personalized intelligent recommendation service based on URL clustering model[J]. Intelligence Journal, 2006,(7). (Chinese)

[15] Global medicine preparation. Why Taobao style "buyers" can not be used to evaluate the doctor[EB/OL].http://www.cn-healthcare.com/article/20150812/content-476828. html, 2015-08-12.(Chinese)

Resource Constrained Project Scheduling Optimization With Robust Objective Under Stochastic Duration Of Activities

Wei-bo Zheng[1,*], Yu-kang He[2]
[1]School of Management, Xi'an Jiaotong University, Xi'an, China
[1]The Key Lab of the Ministry of Education for Process Control & Efficiency Engineering, Xi'an, China
[2]School of Computer Science and Technology, Heilongjiang University, Harbin, China
(*zhengweb2011@163.com)

Abstract - **This paper involves the resource constrained project robust scheduling problem where the objective is to maximize the robustness of the baseline schedule under the constraints of resources and project deadline. On the basis of the identification of the problem, the optimization model is constructed. Then, for the NP-hardness of the problem, the two versions of variable neighborhood search (VNS), i.e., the hybrid VNS and the parallel VNS, are developed to solve the problem. To evaluate the performance of the two VNS methods, a computational experiment is conducted based on a set of data consisting of 50 instances generated randomly. The results show that the developed algorithms can solve the problem with an acceptable computational time. With the increase of the problem scale, the hybrid VNS performs better while the parallel VNS becomes worse.**

Keywords - **Optimization model, project scheduling, robustness, variable neighborhood search**

I. INTRODUCTION

In past decades, the majority of researches on project scheduling focused on the generation of the optimal schedule under deterministic environment. In reality, however, the execution of projects may be interfered by numerous uncertain factors including bad weather, equipment failure, personnel change, material supply postponement, and so on. Therefore, how to prepare a robust baseline schedule is an undoubtedly meaningful problem which is worthwhile to be studied intensively.

As for the robust project scheduling problems, Herroelen and Leus[1] developed a mathematical programming model to minimize the expected weighted deviation of activity start times so as to obtain a robust baseline schedule at first in 2004. Different from Herroelen and Leus[1] where the single objective was used to measure the performance of projects, Al-Fawzan and Haouari[3] presented a bi-criteria model to generate robust schedules with minimum makespan under uncertain activity durations. Similarly, Van de Vonder, Demeulemeester, Herroelen and Leus[2] studied the trade-off between project makespan and stability in multi-disruption project without resource constraint. They[4] extended the method in [2] to resource constrained cases and developed a resource flow dependent float factor method. Furthermore, Van de Vonder, Demeulemeester, and Herroelen[5] compared several proactive buffer insertion methods, including the virtual activity duration extension method, the starting time criticality method, and the resource flow dependent float factor method.

In recent years, more and more researchers pay their attentions to the proactive project scheduling problem. Lambrechts, Demeulemeester and Herroelen[6] analyzed the effects of resource breakdown on the activity duration extension using the probability theory and transformed the problem into the stochastic duration situation. Deblaere, Demeulemeester and Herroelen[7] presented an algorithm that integrated the proactive and reactive procedures where the optimal proactive schedule was created first for a given reactive policy and then, a heuristic iteratively modified this policy and recalculated the baseline schedule in order to increase the robustness. Hazir and Schmidt[8] addressed the integrated proactive scheduling and reactive control approach for the discrete time/resource trade-off problem. Khemakhem and Chtourou[9] developed a series of new robust measures based on the combination of slack functions and weights in which slack functions were based on free slacks. Lamas and Demeulemeester[10] developed a novel chance-constrained programming procedure to generate the robust baseline schedule.

Based on the literature reviews above, this paper investigates a special proactive project scheduling problem. In this problem, activities' durations are stochastic variables with known expected value and standard variance. The robustness of schedules is defined as the sum of the products of activities' time buffers and their weights. Under the constraints of resources and project deadline, the aim is to determine the activities' start time so as to maximize the robustness of the baseline schedules. For the NP-hardness of the problem as in [11], we develop a heuristic algorithm using variable neighborhood search which was originally proposed by Mladenović and Hansen[12] and had been successfully applied to the solution of project scheduling problems by a few researchers[13, 14]. The main contribution of this paper is the construction of the optimization model and the development of the heuristic algorithm for the studied problem. We believe that the research may have an important implication for project managers in reality.

The remainder of this paper is organized as follows. In section 2, we construct the optimization model for the problem. The two versions of variable neighborhood search are developed in section 3 and the computational experiment is presented in section 4. Finally, section 5 concludes the paper.

II. PROBLEM FORMULATION

Consider a project represented as an Activity-on-Node (AoN) network, $G=(N, A)$ where the set of nodes, N,

E. Qi et al. (eds.), *Proceedings of the 23rd International Conference on Industrial Engineering and Engineering Management 2016*, DOI 10.2991/978-94-6239-255-7_43

represents activities while the set of directed arcs, A, denotes finish-start zero-lag precedence relationships between activities. The activities are numbered from 0 to n+1. Activities 0 and n+1 are the dummy start and end activities respectively while others are non-dummy activities. There exist K types of resources and their availabilities are a_k (k=1, 2, …, K). The execution of activity i (i=0, 1, …, n+1) requires r_{ik} amounts of resource k. The duration of activity i is a stochastic variable d_i with known expected value $E(d_i)$ and standard variance $V(d_i)$. Note that r_{ik} and d_i for dummy activities (where i=0 or i= n+1) are always equivalent to zero. The project deadline is D.

Suppose that based on $E(d_i)$ the start time of activity i is arranged as s_i, thus forming a baseline schedule, $S=(s_0, s_1, \ldots, s_{n+1})$. Given an S, the time buffer which is added to activity i and represented as b_i, can be calculated by the formula $b_i = \min_{(i,j)\in A} s_j - [s_i + E(d_i)]$ where (i, j) is a direct arc in A. In order to add more time buffers to the activities with higher variability, we define the weight of activity i as $w_i = V(d_i) \big/ \sum_{i=0}^{n+1} V(d_i)$ and then the robustness of the baseline schedule is defined as $R = \sum_{i=0}^{n+1}(w_i b_i)$. Therefore, the problem needed to solve can be identified as how to determine S optimally so as to maximize R.

Let P_t be the set of the activities which are in progress at time t. Then based on the discussion above, the optimization model of the problem is constructed as follows.

$$\text{Max } R = \sum_{i=0}^{n+1}(w_i b_i) \quad (1)$$

$$\text{s.t. } s_0 = 0 \quad (2)$$

$$s_i + E(d_i) \le s_j, \text{ for } \forall (i,j) \in A; i=0, 1, \ldots, n \quad (3)$$

$$s_{n+1} \le D \quad (4)$$

$$\sum_{i\in P_t} r_{ik} \le a_k, \text{ for } t=0, 1, \ldots, D; k=1, 2, \ldots, K \quad (5)$$

$$s_i \ge 0, \text{ for } i=0,1,\ldots,n+1 \quad (6)$$

In the constructed model, the objective (1) is to maximize the robustness of the baseline schedule. Constraint (2) arranges the start time of the dummy start activity at time 0. Precedence feasibility between activities is enforced by constraints (3) while a project deadline is imposed by constraint (4). Constraints (5) are resource constraints which make the requirement not exceed its availability for every type of resources at any time over the course of the project. Constraints (6) declare the domain for the decision variables.

III. VARIABLE NEIGHBORHOOD SEARCH (VNS)

A. Solution representation

A solution for the problem is represented by the following two lists.

- Activity sequence list (ASL): This list consists of n+2 elements which define the scheduling priority of activities. In the list, the m-th (m=0, 1, …, n+1) element, i^m, indicates that activity i is placed at the m-th position and thus this activity can only be arranged when all the m–1 activities before it have been scheduled. Note that the activities' priority defined by the list must not violate the precedence feasibility constraints.
- Time deviation list (TDL): This list is an (n+2)-element list and the i-th (i=0, 1, …, n+1) element, ε_i ($\varepsilon_i \in [0, LS_i - ES_i]$), indicates how many periods of activity i's start time deviating from ES_i. Note that LS_i and ES_i are the earliest and the latest start times of activity i determined by the critical path method respectively.

B. Decoding procedure

Let s'_{i^m} denote the earliest start time of activity i for a given ASL. Based on the solution representation aforementioned, a solution can be transformed into a project schedule according to the decoding procedure described as follows.

Step 1 Input the ASL and TDL.
Step 2 Set m:=0, $s'_{i^0} := 0$, and $s_{i^0} := s'_{i^0} + \varepsilon_i$.
Step 3 Set m:=m+1. Determine s'_{i^m} using the serial schedule generation scheme as in [15] and then set $s_{i^m} := s'_{i^m} + \varepsilon_i$.
Step 4 Check whether m is less than n+1 or not. If the answer is true go to step 2; otherwise, go to the next step.
Step 5 Output the obtained schedule, i.e., $S = (s_{0^m}, s_{1^m}, \ldots, s_{(n+1)^m})$.

C. Initial solution

Given the fashion of solutions being represented, a feasible solution can be constructed in the following way.

Step 1 Without the violation of precedence feasibility constraints, arrange the activity sequence randomly and thus obtain an ASL.
Step 2 Within $[0, LS_i - ES_i]$, select ε_i randomly, hence generating a TDL.
Step 3 Transform the generated ASL and TDL into a schedule using the *decoding procedure.*
Step 4 Check whether the project deadline constraint is met or not. If the answer is true go to the next step; otherwise, go to step 1.
Step 5 Output the initial solution, i.e., the ASL and TDL obtained at last.

In order to get a good initial solution, for a given instance, $10\cdot(n+2)$ feasible solutions are generated in the above way and the best one is chosen as the initial solution.

D. Neighbourhood Structure

In the light of the solution representation, we define the following three kinds of neighbourhood structure, NS_q (q=1, 2, …, q_{max}), for the current solution. These neighbourhood structures are applied in different versions of the variable neighborhood search, which are presented below.

- Neighbourhood structure for ASL (NS-ASL): This structure is defined as the set of all the solutions in which only q pairs of elements in their ASLs are different from those in the current solution while their TDLs are identical with those of the current solution.
- Neighbourhood structure for TDL (NS-TDL): This structure is defined as the set of all the solutions in which only q elements in their TDLs differ from those in the current solution whereas their ASLs are the same with those of the current solution.
- Mixed neighbourhood structure (NS-MIX): This structure is defined as a mixed version of NS-ASL and NS-TDL, where the NS_q neighbourhood of a current solution is defined as the set of all the solutions in which q' ($0 \le q' \le q$) pairs of elements in their ASLs and $q-q'$ elements in their TDLs are different from those in the current solution.

E. Shaking

Corresponding to the three neighbourhood structures presented above, three shaking ways are proposed as follows.

- Shaking for NS-ASL (SHAKING-NS-ASL): For the current solution, without violation of precedence feasibility constraints, select q pairs of activities randomly and swap their positions in the ASL, thus obtaining a new ASL. Under the obtained ASL and the TDL which remains unchanged during the above process, check whether the project deadline constraint is satisfied or not. If the answer is true a feasible solution has been obtained; otherwise, repeat this operation until a feasible solution is obtained.
- Shaking for NS-TDL (SHAKING-NS-TDL): For the current solution, keep the ASL unchanged and choose q elements from its TDL randomly. Within their value ranges, change the values of the chosen elements to other ones randomly and thus obtain a new solution consisting of the changed TDL and the unchanged ASL. If the new solution meets the project deadline constraint we have gotten a feasible solution; otherwise, perform this operation repeatedly until getting a feasible solution.
- Shaking for NS-MIX (SHAKING-NS-MIX): For the current solution, select q' pairs of elements randomly from its ASL and choose $q-q'$ elements randomly from its TDL. Swap the positions of the elements selected from the ASL and thus obtain a new ASL. At the same time, randomly change the values of the elements selected from TDL to other possible ones within their value ranges, generating a new TDL. If the obtained solution, which consists of the new ASL and TDL, satisfies the project deadline constraint, then we accept it; otherwise, we refuse it and repeat this operation until a feasible solution is obtained.

F. Two Versions of VNS

We design two versions of VNS named the parallel VNS and hybrid VNS respectively, where different searching structures are adopted. In the parallel VNS, we use NS-ASL, NS-TDL, SHAKING-NS-ASL, and SHAKING-NS-TDL to search the desirable solution in a concurrent way. In the hybrid VNS, we use NS-MIX and SHAKING-NS-MIX to fulfill the same task in a mixed way. In these two versions of VNS, a local search with the first improvement strategy as in [12] is employed to find a local optimum solution. By using such a local search, the neighbourhood of the current solution is explored systematically and a move is made as soon as a descent direction is found. During the process, if there is no direction for descending the search terminates and returns the local optimum solution obtained. The stop condition of the two VNSs is defined as an assumed number of the feasible solutions visited, which is denoted as Num_{max}.

Let (ASL^{init}, TDL^{init}) be the initial solution, (ASL^{curr}, TDL^{curr}) the current solution, (ASL^{loca}, TDL^{loca}) the local optimum solution, and Num the number of the feasible solutions that have been visited during the searching process, and then the implementation steps of the parallel VNS can be described as follows.

Step 1 Input (ASL^{init}, TDL^{init}) and initialize q_{max} and Num_{max}. Set (ASL^{curr}, TDL^{curr}):=(ASL^{init}, TDL^{init}), q:=1, and Num:=0.

Step 2 Select a neighbourhood structure from NS-ASL and NS-TDL with equal probability. Generate a random solution in the chosen neighbourhood structure using the corresponding shaking, i.e., SHAKING-NS-ASL or SHAKING-NS-TDL, and then set Num:=Num+1.

Step 3 Starting from the solution generated in step 2, seek (ASL^{loca}, TDL^{loca}) by the local search. Record the number of the feasible solutions which visited during the process and add it to Num.

Step 4 If (ASL^{loca}, TDL^{loca}) is better than (ASL^{curr}, TDL^{curr}), set (ASL^{curr}, TDL^{curr}):= (ASL^{loca}, TDL^{loca}) and q:=1, and then go to step 2; otherwise, go to the next step.

Step 5 q:=q+1. If $q<q_{max}$ go to step 2; otherwise, go to the next step.

Step 6 If $Num<Num_{max}$ set q:=1 and then go to step 2; otherwise, output the desirable solution, i.e., the (ASL^{curr}, TDL^{curr}) obtained at last.

The implementation steps of the hybrid VNS are the same as those of the parallel VNS except for step 2 which are described as follows:

Step 2 Generate a random solution in NS-MIX using SHAKING-NS-MIX and then set Num:=Num+1.

IV. COMPUTATIONAL EXPERIMENTS

A. Setup

The two versions of VNS developed above are tested on a data set constructed by ProGen project generator as in [16]. The set consists of 50 instances and parameter setting used to generate the instances is presented in TABLE I. The values of the key parameters, including the project deadline, D, and the resource strength, RS, which determines the availability of resources based on their requirements, are set at three levels, and a full factorial experiment of the two parameters with three levels results in 3×3=9 replications for each instance and 50×3×3=450 ones overall.

TABLE I
PARAMETER SETTING USED TO GENERATE THE DATA SET

Parameter	Setting
Number of non-dummy activities, n	10, 20, 30, 40, 50
Number of instances generated under a given number of non-dummy activities	10
Number of initial and terminal activities	Randomly selected from 2, 3, and 4
Maximal number of successors and predecessors	4
Expected value of activities' duration, $E(d_i)$	Randomly selected from [10, 20]
Standard variance of activities' duration, $V(d_i)$	Randomly selected from [1, 5]
Number of resource types, K	2
Activities requirement for resource 1, r_{i1}	Randomly selected from [1, 10]
Activities requirement for resource 2, r_{i2}	Randomly selected from [1, 10]
Resource strength, RS	0.3, 0.5, 0.7
Project deadline, D	$\rho \cdot CL$, where CL is the length of the critical path of the network, and ρ is set at 1.1, 1.3, and 1.5

The following four indices are defined to evaluate the performance of the algorithms.

- G_{avg}: Average relative percent below the best solution. For a given instance, the best solution is defined as the best desirable solution found by any version of the VNS developed.
- G_{max}: Maximal relative percent below the best solution.
- T_{avg}: Average computational time in seconds of the algorithm.
- T_{max}: Maximal computational time in seconds of the algorithm.

The two versions of VNS are coded and compiled with Visual C++ 6.0 and the computational experiment is performed on a Pentium-based computer with 1.60GHz CPU and 768MB RAM. They start from the same initial solution and based on a preliminary test, Num_{max} and q_{max} are set at 10000·(n+2) and 5 respectively.

B. Computational results

The computational results are shown in TABLE II. As for the quality of the desirable solutions, the G_{avg} and the G_{max} in the table indicate that there seems no remarkable difference between the two versions of VNS. However, with the increase of n, it is clear that the G_{avg} of the parallel VNS descends while that of the hybrid VNS ascends. This implies that, compared with the parallel VNS, the hybrid VNS is more suitable for solving larger scale problems. These results may be explained as follows. In the shaking of the hybrid VNS, parameter q' is uncertain in every operation, making the search process own a higher randomness than the parallel VNS. As n increases the solution space enlarges accordingly. In such a situation, since the randomness of the hybrid VNS can make it explore more areas than the parallel VNS, the desirable solutions obtained tend to be better than those gotten by the parallel VNS.

TABLE II
COMPUTATIONAL RESULTS OF TWO VESIONS OF VNS

Parameter Value	Parallel VNS				Hybrid VNS			
	G_{avg}	G_{max}	T_{avg}	T_{max}	G_{avg}	G_{max}	T_{avg}	T_{max}
n=10	3.01	6.45	3.1	8.3	5.27	8.56	3.4	6.6
n=20	3.31	5.19	11.5	18.8	5.66	7.98	11.1	19.5
n=30	4.64	6.57	26.3	45.3	4.31	7.05	25.5	43.2
n=40	4.51	6.44	42.1	80.1	4.15	6.72	43.7	78.7
n=50	5.77	8.88	66.3	109.9	3.85	5.58	65.9	111.1
RS=0.3	3.81	6.71	32.7	109.9	4.89	8.47	31.5	103.8
RS=0.5	4.24	7.02	29.6	101.5	4.46	7.31	30.3	111.1
RS=0.7	4.56	8.88	28.1	99.7	4.58	8.56	28.2	101.4
D=1.1·CL	3.68	6.34	35.9	109.9	5.14	8.56	36.3	111.1
D=1.3·CL	4.28	8.38	29.2	103.2	4.61	6.67	29.9	107.7
D=1.5·CL	4.71	8.88	24.5	100.1	4.19	7.28	23.2	97.0

The similar conclusions can also be drawn from the results under different RS and D. When RS and D increase, the computational results of G_{avg} show that the quality of the desirable solutions generated by the parallel VNS goes down whereas that of the desirable solutions obtained by the hybrid VNS goes up. The reason for this phenomenon is identical with that aforementioned. With the increase of RS and D, the constraints of resources and project deadline are relaxed to an extent, thus making the solution space increased. This makes the hybrid VNS exert its advantage and hence obtain better desirable solutions than the parallel VNS.

Considering the computational time of the algorithms, the results of index T_{max} show that the parallel and the hybrid VNSs can solve all instances within 109.9 and 111.1 seconds respectively, which are acceptable in practice. As the scale of the problem increases, it is not surprising that the average computational time of the two VNSs, which are reflected by index T_{avg} in TABLE II, increases monotonously since they are required to visit more feasible solutions in the searching process. However, as RS and D grow the average computational times tend to decrease, which seems a little different from the conclusion in the case of various n. This is because with the increase of RS and D the generation of feasible solutions becomes easier for the loosened constraints of resources and project deadline. Therefore, under the same stop condition, namely the total number of the visited feasible solutions, the computational time of the algorithms decreases.

V. CONCLUSIONS

This paper studies the project robust scheduling problem where activities' durations are stochastic variables. Under the constraints of resources and project deadline, the objective is to generate a baseline schedule with the maximal robustness so as to deal with the influences of stochastic durations. Based on the identification of the studied problem, the optimization model is constructed and for its NP-hardness, a special heuristic algorithm, i.e., variable neighborhood search is developed. In the light of the solution representation method, the variable neighborhood search is designed as two versions, namely the parallel VNS and the hybrid VNS which adopt different searching structures to seek desirable solutions. The two VNSs are tested through a computational experiment which is conducted based on a data set generated randomly. The conclusions obtained are as follows: With the increase of problem scale, resource strength, and project deadline, the quality of the desirable solutions obtained by the hybrid VNS increases while that of the desirable solutions generated by the parallel VNS decreases. The computational time increases as the problem scale becomes larger, however, when the resource strength and the project deadline increases the computational time decreases. Note that in this research the obtained results are not compared with the optimal solutions. In the future study, we will try to develop the exact algorithm to obtain the optimal solution for the problem.

ACKNOWLEDGMENT

The research is supported by the National Natural Science Foundation of China under Grants No. 71371150.

REFERENCES

[1] W. Herroelen and R. Leus, "The construction of stable project baseline schedules (Periodical style)", *European Journal of Operational Research*, vol. 156, no. 3, pp. 550–565, Aug. 2004.

[2] S. Van de Vonder, E. Demeulemeester, W. Herroelen, and R. Leus, "The use of buffers in project management: The trade-off between stability and makespan (Periodical style)", *International Journal of Production Economics*, vol. 97, no. 2, pp. 227–240, Aug. 2005.

[3] M. Al-Fawzana and M. Haouari, "A bi-objective model for robust resource-constrained project scheduling (Periodical style)", *International Journal of Production Economics*, vol. 96, no. 2, pp. 175–187, May 2005.

[4] S. Van de Vonder, E. Demeulemeester, W. Herroelen, and R. Leus, "The trade-off between stability and makespan in resource-constrained project scheduling (Periodical style)", *International Journal of Production Research*, vol. 44, no. 2, pp. 215–236, Jan. 2006.

[5] S. Van de Vonder, E. Demeulemeester, and W. Herroelen, "Proactive heuristic procedures for robust project scheduling: An experimental analysis (Periodical style)", *European Journal of Operational Research*, vol. 189, no. 3, pp. 723–733, Sep. 2008.

[6] O. Lambrechts, E. Demeulemeester, and W. Herroelen, "Time slack-based techniques for robust project scheduling subject to resource uncertainty (Periodical style)", *Annals of Operations Research*, vol. 186, no. 1, pp. 443–464, Jun. 2011.

[7] F. Deblaere, E. Demeulemeester, and W. Herroelen, "Proactive policies for the stochastic resource-constrained project scheduling problem (Periodical style)", *European Journal of Operational Research*, vol. 214, no. 2, pp. 308–316, Oct. 2011.

[8] Ö. Hazir and K. W. Schmidt, "An integrated scheduling and control model for multi-mode projects (Periodical style)", *Flexible Services and Manufacturing Journal*, vol. 25, no. 1, pp. 230–254, Jun. 2013.

[9] M. A. Khemakhem and H. Chtourou, "Efficient robustness measures for the resource-constrained project scheduling problem (Periodical style)", *International Journal of Industrial and Systems Engineering*, vol. 14, no. 2, pp. 245–267, Jan. 2013.

[10] V. P. Lamas and E. Demeulemeester, " A purely proactive scheduling procedure for the resource-constrained project scheduling problem with stochastic activity durations (Periodical style)", *Journal of Scheduling*, on line (DOI 10.1007/s10951-015-0423-3), pp. 1–20, Mar. 2015.

[11] J. Blazewicz, J. K. Lenstra, and K. A. H. G. Rinnooy, "Scheduling subject to resource constraints: Classification and complexity (Periodical style)", *Discrete Applied Mathematics*, vol. 5, no.1, pp. 11–24, Jan. 1983.

[12] N. Mladenović and P. Hansen, "Variable neighbourhood search (Periodical style)", *Computers and Operations Research*, vol. 24, no. 11, pp. 1097–1100, Nov. 1997.

[13] K. Fleszar and K. S. Hindi, "Solving the resource-constrained project scheduling problem by a variable neighbourhood search (Periodical style)", *European Journal of Operational Research*, vol. 155, no. 2, pp. 402–413, Feb. 2004.

[14] V. Bouffard and J. A. Ferland, "Improving simulated annealing with variable neighborhood search to solve the resource-constrained scheduling problem (Periodical style)", *Journal of Scheduling*, vol. 10, no. 6, pp. 375–386, Dec. 2007.

[15] R. Kolisch, "Serial and parallel resource–constrained project scheduling methods revisited: Theory and computation (Periodical style)", *European Journal of Operational Research*, vol. 90, no. 2, pp. 320–333, Apr. 1996.

[16] R. Kolisch and A. Sprecher, "PSPLIB - A project scheduling problem library: OR software - ORSEP operations research software exchange program (Periodical style)", *European Journal of Operational Research*, vol. 96, no. 1, pp. 205–216, Jan. 1997.

Dynamic scheduling of public bicycles based on artificial bee colony algorithm

Yu-jie TIAN [1,*], Qing-hong XIE[2]
[1,2]Department of Industrial Engineering, Nanjing Tech University, Nanjing, China
([*]524914557@qq.com)

Abstract - **In order to solve the ubiquitous problems of public bicycle rental stations--"hard to borrow and return", this paper applies artificial bee colony algorithm to dynamic scheduling of public bicycles. It also established the cost minimization model , which considered time-windows, transport costs and service satisfaction. It selected some public bicycle rental stations in Nanjing Pukou as research objects to verify the model and algorithm above. And the experimental result shows that this algorithm enables the highest service satisfaction while making the transport costs of public bicycles to a minimum.**

Keywords - **Artificial bee colony algorithm, dynamic scheduling , multi-objective optimization, public bicycles**

I. INTRODUCTION

To ease the pressure on traffic and environment , and in response to the call of the green travel. Since 2007, public bicycle system had been introduced to our country.After that , the facilities were laid out in cities and towns, it brought a lot of convenience to people's daily getting around . However ,the phenomenon " imbal -ance between borrowing and return" is existing in many rental stations during the running of public bicycle system. It leads to the greatly reduced in system efficiency.This phenomenon reflects the lack of public bicycle scheduling, so that many scholars conduct research on public bicycle systems and its scheduling problems.

Lu LIU used Method of Factor Analysis to study the service satisfaction of Wuhan city public bicycle system and gave out some suggestions for its improvement[1]. Tian-wen JIANG thinks that public bicycle systems which developed in the public-private-partnership(PPP) model will make the systemic efficiency higher and the risk lower[2]. Mi-yuan SHAN puts forward that the reasonable subsidies set by government will maximize the system utility[3]. Liu HE combined demand forecasting model with scheduling model, and applied it to public bicycle scheduling to significantly improved scheduling effect[4].Hong-zhao DONG used association rules and spatial clustering method to divide scheduling areas[5], and used Multi-logit improved model to analyze the nature rental demand of public bicycles[6]. Wei ZHU found that the quality of getting around will significantly improved under the discrete choice model[7]. Ran-zi XU YE made [1]distribution method and time series a combination, and applied it to forecast demand to improve the accuracy of the system reasonable dispatch[8].Hui ZHAN studied scheduling model with fuzzy time window based on ant colony algorithm [9].Zhi-xiao PAN made an innovation on genetic algorithms in order to reduce fuel consumption[10].

Due to the advantages of artificial bee colony algorithm, such as high search precision,simple operation and strong robustness. This paper will use it to solve dynamic scheduling model of multi-objective optimization, and choose rental stations near Nanjing Tech University in Nanjing Pukou to make experimental analysis.

II. MODEL

A.Car pile ratio and dynamic scheduling strategy

Due to limited scheduling resources and different demand of different rental stations,we make scheduling strategy by monitoring the currently available public bicycle numbers of rental station i and its car pile numbers K_i to calculate "car pile ratio"α_i.Corresponding formula goes here:

$$\alpha_i = \frac{K_i}{H_i} \tag{1}$$

Then make hierarchical scheduling strategy according to α_i , corresponding formula goes here:

$$|\alpha_i - 0.5| \geq 0.4 \quad \text{level 1} \tag{2}$$

$$|\alpha_i - 0.5| \geq 0.3 \quad \text{level 2} \tag{3}$$

B.Scheduling demand

The demand of each rental station can be determined by the car pile ratio α_i , corresponding formula goes here:

$$d_i(t) = \frac{(\alpha_{\min} + \alpha_{\max})B_i}{2} - R_i(t) \quad \alpha_i(t) < \alpha_{\min} or \, \alpha_i(t) > \alpha_{\max} \tag{4}$$

$$\alpha_i(t) < \alpha_{\min} , d_i(t) < 0 ; \quad \alpha_i(t) > \alpha_{\max} , d_i(t) > 0 ;$$

$$Q_{i+1}(t) = Q_i(t) + d_i(t) \quad \alpha_i(t) < \alpha_{\min} or \, \alpha_i(t) > \alpha_{\max} \tag{5}$$

Definition:

α_{min} represents for minimum car pile ratio which is able to provide normal service ;

α_{max} represents for maximum car pile ratio which is able to provide normal service ;

$R_i(t)$ represents for the number of public bicycles in station i at time t ;

B_i represents for the number of car piles in station i ;

$Q_i(t)$ represents for the number of bicycles on the car when it arrives at station i at time t .

C.Time-window and time penalty function

[1]This work is supported by National Nutural Science Foundation of China (Grant:No.71171110).

E. Qi et al. (eds.), *Proceedings of the 23rd International Conference on Industrial Engineering and Engineering Management 2016,* DOI 10.2991/978-94-6239-255-7_44

This paper takes service satisfaction as one optimization goal , it reflected by reducing users' waiting time.When $\alpha_i(t)<\alpha_{min}$ or $\alpha_i(t)>\alpha_{max}$,time now is set as t. The degree of satisfaction will maintain at a level in duration *a*, and the satisfying service time-window is [t,t+a].When time overstep the time-window, it will produce penalty. The penalty Z_2 is total penalty of each rental station Z_i, corresponding formula goes here:

$$Z_2 = \sum_{i=0}^{N} Z_i \tag{6}$$

$$Z_i = \beta(t_i - (t+a)) \qquad t_i > t+a \tag{7}$$

Definition:

β represents for penalty factor.

To calculate the time of arrival, this paper takes transport time, service time, vehicle refueling and rest time into consideration .Specific formula is as follow:

$$t_{i+1} = \begin{cases} 0 & i=0 \\ t_i + \dfrac{s_{i,j}}{v_k} + service(i) + refuel\ \ or\ \ rest\text{-}time\,(k) & i>0 \end{cases} \tag{8}$$

Definition:

t_i represents for arrival time t of station *i*;

service(*i*) represents for service time of station *i;*

refuel _ *or* _ *rest* _ *time*(*k*) represents for vehicle refueling and rest time.

D.Transport costs

$$Z_1 = C + \sum_{j=1}^{m}\sum_{i=1}^{n}(\chi S_{ij}) \tag{9}$$

Definition:

C represents for fixed scheduling cost of a car;

χ represents for transport cost of a unit;

S_{ij} represents for distance between station *i* and station *j* .

E.Dynamic traffic condition

This paper use v_{ij} to represent for speed between station *i* and station *j* ,specific formula is as follow:

$$v_{ij} = \begin{cases} mean\ speed\ between station\ i\ \ \&\ \ station\ j, & \mathbf{i \neq j} \\ \mathbf{0} & \mathbf{i = j} \end{cases} \tag{10}$$

In this paper the measurement of speed between station *i* and station *j* is mainly depends on the number of lanes, traffic lights and motor vehicles per hundred meters. In daily life, traffic leans is varying, we assumed that there are *L* kinds of roads between i and j. Road *l* is one kind of it,we can firstly calculate the speed on Road *l* between *i* and *j* ,specific formula is as follow:

$$v_{ijl} = \frac{v\delta_l cd_l}{hd_l jc_l} \tag{11}$$

Definition:

v represents for the normal speed under ideal cond -ition;

δ_l represents for the correction factor on Road *l;*

cd_l represents for the number of lanes of Road *l*;

hd_l represents for the number of traffic lights of Road *l* ($hd_l \geq 1$);

jc_l represents for the number of motor vehicles per hundred meters on Road *l* ($jc_l \geq 1$).

After calculating speeds on *L* kinds of road between *i* and *j*, get mean speed between *i* and *j* by formula (12) , specific formula is as follow:

$$v_{ij} = \frac{\sum_{l=1}^{L} v_{ijl}}{L} \tag{12}$$

F.Objective function

$$\min Z = \omega_p Z_1 + \omega_t Z_2 \tag{13}$$

Definition:

Z represents for total costs;

Z_1 represents for transport costs;

ω_p represents for the importance of transportation costs to objective function;

Z_2 represents for time penalty;

ω_t represents for the importance of service satisfacti -on to objective function.

In addition, the establishment of objective function should take many related constraints into consideration.In order to facilitate the mathematical description of the constraints, decision variables x_{ijk} defined as follow:

$$x_{ij} = \begin{cases} 1 & (\textbf{from}\ \ \mathbf{i}\ \ \textbf{to}\ \ \mathbf{j}) \\ 0 & (\textbf{otherwise}) \end{cases} \tag{14}$$

Corresponding constraints goes here:

$$\sum_{j \in U, i \neq j} x_{ij} = 1 \tag{15}$$

$$0 \leq Q_i(t) \leq Q \tag{16}$$

$$\sum_{i \in U}\sum_{j \in U} x_{ij} - \sum_{j \in U}\sum_{i \in U} x_{ji} = 0 \tag{17}$$

Definition:

Formula (15) to ensure that each rental station only be visited once;

Formula (16) to ensure that the number of public bicycles on the car won't exceed the maximum limit Q;

Formula (17) to ensure that each distribution vehicle always starting from the dispatch center, after finishing all scheduling tasks, they are all back to the beginning.

III. ARTIFICIAL BEE COLONY ALGORITHM

Artificial Bee Colony Algorithm (abbreviation:ABC algorithm) is a intelligent optimization algorithm invented by Turkish scholars Karaboga in 2005[11]. The proposal of the algorithm is derived from bees internal division mechanism and foraging behavior. The characters of bees are converting among employed bees, onlooker bees and scout bees, they search high-quality food sources according to the principle of greedy algorithm[12]. Employed bees are used to preserve high-quality solutions, onlooker bees are helpful to improve convergence speed for solving solutions, scout bees can make the algorithm get rid of the local convergence. Now it has been applied to reverse logistics, the optimization of power system, pipeline scheduling, the design of Systems Engineering and many other fields.

A.Basic principles of ABC algorithm

While using ABC algorithm to solve models, setting colony size as ***NP***, regarding potential solutions as food sources ***i(i=1,2,...,NP)***.The fitness value ***fit*** is correspond to food sources.Setting the dimension of the problem to be solved as ***D***, its position after *t* iterations can be expressed as follow:

$$X_i^t = \left[x_{i1}^t \; x_{i2}^t \; \ldots \; x_{iD}^t \right] \qquad (18)$$

In formula (18) *t* is the current iteration number, $x_{id} \in (L_d, U_d)$, L_d and U_d are respectively represent for the upper and lower limits of search range.***d=1,2,...D***.

The initial position of food source can randomly generated by formula (19) as follow:

$$x_{id} = L_d + rand\,(0,1)(U_d - L_d) \qquad (19)$$

In the initial stage of seeking solutions,employed bees generate new food sources in defined neighborhood of food source *i* according to formula (20):

$$v_{id} = x_{id} + \varphi(x_{id} - x_{jd}) \qquad (20)$$

d is a random integer in **[1,D]**, which represents for the dimension that employed bees searched in. $j \in \{1,2,...NP\}$, $j \neq i$ means that employed bees need to choose a food source except ***i***. $\varphi \in [-1,1]$, it means a uniformly distributed random number and its changes will effect the perturbation amplitude. If employed bees find better food sources by searching around, replace the current food source ***Xi*** with a new food source ***Vi*** in accordance with greedy algorithm . If they didn't find a better one, preserve current food source ***Xi***. After finishing search tasks, employed bees fly to information sharing area to spread message.Onlooker bees will judge whether to follow employed bees after receiving information. The probability of following depends on the quality of food sources.Corresponding formula goes here:

$$p_i = fit_i \Big/ \sum_{i=1}^{NP} fit_i \qquad (21)$$

Onlooker bees commonly take roulette way to select employed bees.The usual approach is to generate a random number *r* in [0,1], when $p_i > r$, onlooker bees will generate a new food source according to formula (22).If they don't find a better food source after *try* times iterations and the times has reached to the threshold *Limit*, this food source will be abandoned and corresponding employed bee will turn into a scout bee.

Scout bees regenerate a new food source to replace X_i, corresponding formula goes here:

$$X_i^{t+1} = \begin{cases} L_d + rand\,(0,1), & trial \geq \lim it \\ X_i^t, & trial < \lim it \end{cases} \qquad (22)$$

For the new solution, we need to make an accurate evaluation about its fitness. Corresponding formula goes here:

$$fit_i = \begin{cases} 1/(1+f_i), & f_i \geq 0 \\ 1 + abs\,(f_i), & otherwise \end{cases} \qquad (23)$$

f_i represents for function value of solution in formula (23).

B.Main computing steps of ABC algorithm

Step1: Assign value to each parameter (for example: colony size NP, employed bees number En, onlooker bees number On, scout bees number Sn, maximum search time Limit, iterations Cycle, Solution dimension D).

Step2:Randomly generated NP/2 solutions, initializing the population while calculating the fitness value of each solution.

Step3: Assign 1 to variable iterations in the algorithm, namely Cycle=1.

Step4 : Through all the solutions above and have all solutions to execute Step5-Step8.

Step5 : Find new solutions in the neighborhood of the solution.

Step6 : Calculate fitness of new solution.

Step7: Compare the fitness value of new food source and former ones, and preserve higher one.

Step8: If the original solution is unable to updated, local search times will turn into limit+1.If it has been updated, local search times will turn into 0, namely limit=0.

Step9: Calculate the probabilities that onlooker bees choose to follow each employed bees.

Step10: Onlooker bees randomly choose a solution in Step9, find new solutions in the neighborhood of this solution and calculate its fitness value, the same as Step7,Step8.

Step11:To traverse all possible solutions and judge whether local search time has exceeded maximum limit time.If it has exceeded the limit, it will be abandoned and replace it with a new one.

Step12: Record current lowest-cost solution.

Step13: Each time the Cycle plus 1.

Step14: Judge the Cycle value, if it doesn't reach limit times, jump to Step4. But if it has reached limit times, the current solution is global optimal solution.

IV. EXPERIMENT AND ANALYSIS

A.Experimental data

This paper selects rental stations near Nanjing Tech University in Nanjing Pukou as study objects.So far,there are one dispatch center (located at West Bay home, abbreviation:M0) and 57 rental stations in this area.This experiment will choose 9 rental stations which satisfy level 1 dispatch condition during 17:20~17:25 at one day to study.Only one car set off at 17:25 with nothing in it, and start the "allow only once visit" scheduling work.Associated data can be found in TABLE I , TABLE II.

TABLE I
STATION DATA SHEET

Station number	Station Name	Pile quantity	Available bicycle quantity	Car pile ratio	Demand	Service Time /min
05	West station of Njut	32	3	0.09	13	4.3
08	Longhua Road 1	20	18	0.90	-8	2.7
28	Jinzhu Garden	20	2	0.10	8	2.7
29	North River Heights	20	19	0.95	-9	3.0
30	North West Bay	24	2	0.08	10	3.3
32	Sheng Jing Heights	22	21	0.95	-10	3.3
43	The Old River	14	13	0.93	-6	2.0
45	Jiangpu Forth Middle School	20	18	0.90	-8	2.7
50	Dingye Garden	38	3	0.08	16	5.3
$\alpha_{min} = 0.15$, $\alpha_{max} = 0.85$						

Other scheduling data of this dispatch center are as follow:

(1)The maximum number of pubilc bicycles for one car to load : Q=40;

(2)Fixed scheduling cost of one car: C=120yuan;

(3)`Transport cost of a unit: χ = 3.5 yuan/km;

TABLE II
DISTANCE DATA SHEET (UNIT:KM)

Station number	M 0	05	28	29	30	32	43	45	50
M 0	0	3.54	1.06	1.5	0.22	2.35	2.1	3.26	1.64
05	3.54	0	2.45	3.54	3.31	1.23	1.93	0.29	1.86
08	2.59	1.25	1.59	2.32	2.33	1.07	0.68	1.16	1.12
28	1.06	2.45	0	1.72	0.86	1.27	1.33	2.2	0.58
29	1.5	3.54	1.72	0	1.35	2.77	1.63	3.4	2.05
30	0.22	3.31	0.86	1.35	0	2.18	1.85	3.1	1.45
32	2.35	1.23	1.27	2.77	2.18	0	1.42	0.94	0.76
43	2.1	1.93	1.33	1.63	1.85	1.42	0	1.84	1.08
45	3.26	0.29	2.2	3.4	3.1	0.94	1.84	0	1.62
50	1.64	1.86	0.58	2.05	1.45	0.76	1.08	1.62	0

TABLE III
CAR SPEED BETWEEN TWO STATIONS (UNIT:M/S)

Speed Vij	M 0	05	08	28	29	30	32	43	45	50
M 0	0	2.3	1.9	2.9	1.7	2.1	2.4	1.8	2.6	3.2
05	2.3	0	3.3	2.2	1.5	1.9	2.2	2.4	1.9	1.8
08	1.9	3.3	0	2.6	3.1	2.2	2.9	1.7	2.4	3.1
28	2.9	2.2	2.6	0	3.2	2.1	1.9	1.5	2.4	2.2
29	1.7	1.5	3.1	3.2	0	1.3	2.7	2.2	2	1.9
30	2.1	1.9	2.2	2.1	1.3	0	2.8	3.4	2.5	2.6
32	2.4	2.2	2.9	1.9	2.7	2.8	0	2.7	3.1	2.9
43	1.8	2.4	1.7	1.5	2.2	3.4	2.7	0	3.1	3.4
45	2.6	1.9	2.4	2.4	2	2.5	3.1	3.1	0	2.9
50	3.2	1.8	3.1	2.2	1.9	2.6	2.9	3.4	2.9	0

(4)Transport cost of a unit: χ = 3.5 yuan/km;

(5)Average loding time of one bicycle:20 s;

(6)Weight of one public bicycle:20kg.

Other experimental parameters settings:

(1)Penalty factor:β = 3;

(2)Duration without penalty: a =10/min;

(3)Normal speed under ideal condition: v =30km/h;

(4)The importance of transportation costs to objective function: ω_p=0.75;

(5)The importance of service satisfaction to objective function: ω_t=0.25.

The speed of the car between two stations can be calculated according to formula (11) and formula (12), as shown in TABLE III.

B.Experimental results and analysis

Using Java programming language to write the model code and do experimental testing, the optimal solution of the scheme obtained after 74 iterations. The data of experimental result and Artificial thumb are show in TABLE IV.

TABLE IV COMPARATIVE DATA OF SCHEDULING

Method	Scheduling route	Total distance /km	Transport cost	Penalty /RMB	Total cost /RMB	Scheduli-ng time /min
Artificial thumb	M0--29--43--08--32--50--28--45--05--30--M0	12.24	202.8	233.7	212.1	35.9

ABC algorithm	M0--29--43--08--05--45--32--50--28--30--M0	8.71	190.5	216.7	198.4	32.67

V. CONCLUSION

This paper applied Artificial bee colony algorithm to dynamic scheduling of public bicycles.Under circumstan -ces of "allowing only once visit to each station" and "using one car" to study the scheduling. It also established "cost minimization scheduling model" with time-windows, and took transport costs and service satisfaction into consideration. This paper can provide a reference for public bicycle scheduling.

Shortcomings of this paper are as follow:

(1)It studied under circumstances of "allowing only once visit" and "using one car", which lead to many scheduling tasks can't be finished within time-windows and made penalty higher .

(2)It ignored the impact of short time.If the number of public bicycle in rental stations change in a short time, it's unable to adjust scheduling in time with only one car.

REFERENCE

[1] Lu LIU, Yang LI, Guo-hu XU , "An empirical study of Wuhan City public bicycle service satisfaction," Logistics Engineering and Management, vol. 33, no.5, pp. 116-117, May. 2011. (Chinese)

[2] Tian-wen JIANG, Sheng-huan DIAO, Run-xi ZENG, "Urban public bicycle system construction mode - comparison and practice based on low carbon perspective," Economy and Management, vol.26, no.6, pp.88-92, Jun. 2012.(Chinese)

[3] Mi-yuan SHAN, Zi-wei ZHOU, Ren-long ZHANG, Lin-feng LI, "Research on government subsidies of public bicycle based on dynamic Bayesian game and Utility theory," Statistics and Decision, no. 4, pp.47-50, Apr. 2016. (Chinese)

[4] Liu HE, Xu-hong LI, Da-wei CHEN, Jing LU, Yuan-yuan WU, "Research on the demand forecast model of public bicycle dynamic scheduling system," Wuhan University of Technology (Transportation Science & Engineering), vol. 37, no. 2, pp. 278-282, Apr. 2013. (Chinese)

[5] Hong-zhao DONG, Cai-xia SHI, Ning CHEN, "Clustering division of public bicycle scheduling regional based on association rules," Bulletin of Science and Technology, vol.29,no.9, pp.209-212, Sept. 2014. (Chinese)

[6] Hong-zhao DONG, Man-jin WU, Dong-xu LIU, Ning CHEN, "Estimation method of natural demand of urban public bicycle system," Journal of Zhejiang University (Engineering Science), vol.50, no.2, pp.265-270, Feb. 2016. (Chinese)

[7] Wei ZHU, Yu-qi WANG, De WANG, "Travel quality by public bicycles: an empirical study," Journal of Tongji University (Natural Science), vol. 42, no. 7, pp.1077-1081, Jul. 2014. (Chinese)

[8] Ran-zi XU YE, Jin SHEN, "Demand analysis of public system based on circular distribution method and time series method," Industrial Engineering Journal, vol. 17, no. 2, pp.55-63, Apr. 2014. (Chinese)

[9] Hui ZHANG, Peng-jun ZHENG, "A scheduling research of city public bicycle based on ant colony optimization," vol. 17, no. 6, pp.32-36, Nov. 2015.(Chinese)

[10]Zhi-xiao PAN, Qing-hong XIE, "Study on distribution vehicle routing optimization aimed at reducing fuel consumption," Logistics Technology, vol. 34, no. 11, pp.112-115, Jun. 2015.(Chinese)

[11]Karaboga D. An idea based on honey bee swarm for numerical optimization[R]. Technical report-TR06, Kayseri:Erciyes University, Engineering Faculty, Computer Engineering Department, 2005.

[12]Karaboga D, Basturk B, "A powerful and efficient algorithm for numerical function optimization: artificial bee colony (ABC) algorithm," Journal of Global Optimization, vol. 39, no. 3, pp.459-471, Apr. 2007.

Web service classification approach with an integrated similarity measure

Xueshan Wang[1], Fuzan Chen[1,*], Minqiang Li[1,2]
[1] College of Management and Economics, Tianjin University, Tianjin, China
[2]State Key Laboratory of Hydraulic Engineering Simulation and Safety, Tianjin University, China
(* Corresponding author: Fuzan Chen, fzchen@tju.edu.cn)

Abstract – **Service-oriented computing creates a surge of web services. With the increasing number of web services advertised via the Web, service repositories have to face a challenging task, i.e., to automatically group those Web services so as to help themselves or end-users to retrieve those services effectively and efficiently. In this paper, a new web service classification method is proposed. Firstly, a new integrated similarity measure for Web service is developed. It combines a statistic measure (TF-IDF) and a semantic similarity measure based on information content. Then, a couple of popular classifiers, e.g. radial basis function neural network (RBFN) and K-nearest neighborhoods (KNN) are used to group Web services. Experimental results on the OWLS-TC dataset indicate the proposed integrated approach outperforms the semantic method which is popular in Web service discovery.**

Keywords - **Web service, Classification, Semantic Web service**

I. INTRODUCTION

Service-oriented computing has been a new model of IT industry that attracts extensive attention. Technology companies increases the investment of cloud computing to expand the scale and promote business revenue. Service-oriented computing creates on-demand and flexible environment in which Service Oriented Computing (SOC) provides the basic framework to establish the network [1]. They encapsulate applications as services which contain standards of interoperability in different system and platform. As a consequence, the development of service-oriented computing brings the growth of the web service. Hence, the automatically management of web service is imminent.

UDDI is the core technical standard of web service which provides web service registration specification. Web services are registered in UDDI by providers. UDDI supports information sharing, web service discovery and interaction among business activities that are end-users of the repository. Web service discovery in UDDI returns a web service set which matches the key words provided by requester [2]. Web service discovery serves as a search engine in of UDDI so that web services are retrieved conveniently. Based on discovery, Web service composition is the process of combining several services into a large granularity service set according to logic rules [3]. During web service discovery and composition categorization information is used as vital parameter.

Current classification method in UDDI is manual assigned. This method is insufficient semantic supports. Because of the gap of different knowledge domain between service provider and service requester ambiguities are generated so that there would be misunderstanding of the classification information among users. On the other hand, given the tremendous growth trend of web services there has been a proliferation of published services on the Web. As a consequence the manual work is heavy workload and not flexible. Therefore, automatic classification of web service is essential in UDDI which helps to classify the services to certain field and web service providers get rid of manual assigned work. In this situation UDDI can return more accurate results to web service requester when they search web services in UDDI. With the automatically grouping service process categorization information is valuable that improves the low recall and precision in discovery and composition.

In this paper the performance of new semantic measure of web service classification is exploited. The remainder of the paper is organized as follows: Section II introduces the related works about our research. Section III states the details of the classification framework. Section IV shows experiment results and Section V concludes the paper and figures out the future work.

II. RALATED WORK

Web services are stored in UDDI register by professional service providers. Web service requesters search the required service in UDDI. The search process is only according to the key words. These key words are assigned by providers. UDDI provides catalogs of web service. At present the classification mechanism in UDDI requires web service provider must register services in "proper" category of UDDI and requester need to scan this "proper" category and find associated services [4]. However, the "proper" category is not the same meaning from the point of view of the two sides at times. Thus the manual work is arbitrary and low efficiency so that automatic grouping method of web service is necessary.

Cluster and classification are two kinds of grouping method. Cluster gathers web services with the same features into a group of which data are unlabeled [5]. Classification assigns the unknown web service into a certain domain based on the pattern mining from the labeled test set. Classification can improve the web service discovery and web service retrieve method of discovery can be applied in classification [6].

Keyword-based search is an important method of web service discovery. Vector space model (VSM) is adapted to translate web service documents into vectors. TF-IDF is used to evaluate the importance of word that belongs to the document in the VSM [1]. With the usage

E. Qi et al. (eds.), *Proceedings of the 23rd International Conference on Industrial Engineering and Engineering Management 2016,* DOI 10.2991/978-94-6239-255-7_45

of TF-IDF many similarity metrics such as cosine similarity, jaccard similarity and edit distance are used to measure the similarity between two web service documents. Utilizing TF-IDF the expected web services are found.

The VSM with TF-IDF only thinks about the similarity of morphology and neglects the similarity of semantic. The approach taking into account the semantics of classification can overcome the shortcomings of TF-IDF and improve the performance of the classifier.

Semantic similarity has been the research hotspot in web service classification. Importing semantic theory to text classification can improve performance ([7],[8]). Semantic similarity helps to get document model and calculates the distance between documents. The most crucial problem is how to define the semantic similarity. There are two methods to get semantic similarity. One is based on the semantic distance and the other is based on information content. Distance based method can be found in [9],[10]. In this method the depth of word pair, the most specific common node in the hierarchy structure, the shortest path distance between two concepts and the local density are all considered.

Resnik [11] firstly propose the information content method. He defines the information content of most specific common abstraction of word pair as the semantic similarity. After Resnik, Lin ([12]) proposes a new method. In this method the degree that the two concepts share the information content of the most specific common node decides the similarity of a word-pair. The main problem of this method is how to get the information content of the concept.

There are two way to get the information content respectively using corpus [11] and ontology. The difficulty of corpus method is that it is hard to find a suitable corpus that can provide consistent explanation with the ontology of concepts. Meanwhile the update of corpus results in the instability of information content.

The new trend of IC-based method is to combine the ontological structure with information content. Reference [13] transfers the measure tool from corpus to ontology. In this research the probability of a concept is defined by the ratio of total hyponyms and total concepts, however, the depth of word is not considered. Reference [14] thinks the semantic information of a concept in the ontology can be expressed by all the leave node of child tree. Generally the leave node in ontology always has the same probability and the probability of non-leave node is the sum of its child nodes' probability. This method exaggerates the influence of leave node but neglect the relations in the ontology. Reference [15] proposes a new method of semantic similarity that contains more links including the indirect relations. The new parameters weaken the effect of direct links and increase calculation complexity.

Based on the consideration of semantic similarity there already exists some researches focusing on the web service classification on semantic level. In [16] Heß proposes an iterative relational classification algorithm for semantically classifying web service, their operations and input and output messages. In [17] Patil presents MWSAF that classifies web services into domains according to the degree that concepts match the domain ontology. Corella and Castells develop a heuristic approach for web services in [18]. The probability that a web service belongs to a category is estimated by the inclusion - exclusion principle applied to a set of computed similarity values. A solution for achieving functional level service categorization is proposed in [19]. Under the guidance of ontology, web services are categorized by hierarchical clustering methodology.

III. METHODOLOGY

Web service classification is a kind of text classification essentially. The first step is preprocessing the documents. In our research service description part is selected as test source. Then the integrated semantic measure is used as modeling method to get the vector space model of documents. Finally well-known classifier RBFNN and KNN are used to evaluate the performance of the algorithm.

A. Document preprocess

Each term in the web service documents is regarded as a feature. Firstly documents are split into words in sequence of class. Convert the case of the words that extracted from the service description part and remove preposition, function word and auxiliary word using the stop words list.

Semantic similarity evaluates the relations on semantic level and improves the performance of text classification which has been proved. But the appearance of word is also a significant factor to describe documents. In this paper TF-IDF is combined with semantic similarity to get the VSM of documents.

B. Integrated similarity measure for Web services

Features left after mRMR are modeled in this step. X_k is the *ith* space vector model of document d_k . $d_k = X_k = (x_{1k}, x_{2k}, \cdots, x_{ik}, \cdots, x_{nk})$. x_{ik} is the weight of i - th term in k - th document. Suppose T_c is the term set of class C . For $t_i \in T_c$, $tf(t_i)$ is the term frequency of term t_i .The value of $tf(t_i)$ is calculated as follows:

$$tf(t_i) = \frac{Freq(t_i, d_k)}{Sum(d_k, term)} \quad (1)$$

Therein $Freq(t_i, d_k)$ is the frequency of t_i in d_k , $Sum(d_k, term)$ is the total terms in d_i .

The measure of $idf(t_i)$ is the inverse document frequency that measures the significance of term t_i .

$idf(t_i)$ is calculated in the following way:

$$idf(t_i) = \log \frac{|D|}{|\{D_i | t_i \in D_i\}|} \quad (2)$$

Therein $|D|$ is the document number of training set, $|\{D_i | t_i \in D_i\}|$ means the number of documents that contain term t_i.

The *tf-idf* combines the frequency and significance of term in document. In order to normalize the inherent bias of the TF measure in short documents, a higher weight is given to the *idf* value ([20]). Meanwhile the effect of length of documents is considered. Finally the *tf-idf* model is expressed by:

$$tf - idf(t_i) = \frac{tf(t_i)}{tf(t_i) + 0.5 + 1.5 \times dl/avgdl} \times idf^2(t_i) \quad (3)$$

Therein dl is the length of document. The length is equal to the number of terms. $avgdl$ is the average length of documents.

Reference [13] figures out that the similarity between two objects is related to their commonalities and difference. The similarity formula is defined following Lin's method:

$$Sim(t_i,t_j) = Sim_{Lin}(t_i,t_j) = \frac{2 \times IC(Share(t_i,t_j))}{IC(t_i) + IC(t_j)} \quad (4)$$

$Sim(t_i,t_j)$ is semantic similarity of t_i and t_j. $IC(t_i)$ is the information content of concept t_i. $Share(t_i,t_j)$ is the most specific common node of t_i and t_j in Wordnet. In this method the similarity of a word-pair is based on the degree that two concepts share the information content of the most specific common node. In formula (4) the main task is to get the information content. In this paper the information content is calculated using the ontological. In order to avoid the problems mentioned in Section II, the *is-a* link, depth and spread scale of concepts in WordNet is considered.

The vector space model of documents is created with the semantic relations of words and relations between the documents and its words.

$$sim(d_k,t_i) = \sum_{t_j \in d_k} \left(tf - idf(t_j) \times sim(t_i,t_j) \right),\ ti,tj \in d_k \text{ and } t_i \neq t_j \quad (5)$$

Therein $sim(t_i,t_j)$ is the semantic similarity of t_i and t_j. $sim(d_k,t_i)$ is the compactness of t_i and d_k. $sim(d_k,t_i)$ is decided by the semantic similarity of t_i and terms in d_k as well as the $tf - idf$ of each term. In this way the weight of term t_i is modeled as:

$$w_{ik} = tf - idf(t_i) \times sim(d_k,t_i) \quad (6)$$

Therein w_{ik} is the weight of term t_i in document d_k.

The model of document d_k can be expressed as follows:

$$W_k = (w_{1k}, w_{2k}, \cdots, w_{nk}) \quad (7)$$

Therein W_k is vector space model of *k*th document.

C. Classifier

In this paper we select two classifiers to group web services. Radial basis function neural network is a kind of artificial network with three layers structure, ie., the input layer, the hidden layer and the output layer. The input layer conveys data to the nodes in hidden layer. The weights connecting the hidden layer and the output layer determine the final output of the network. The j - th output is described as followings:

$$y_j = \sum_{j=1}^{K} w_{ij} \varphi_j(x) \quad (8)$$

$x = (x_1, x_2, \cdots, x_n)$, which is the n dimensional input vector. w_{ij} is the connecting weight of hidden node and output node. $\varphi_j(x)$ is the activation function of the hidden layer.

$$\varphi_j(x) = \exp\left(-\frac{\|x - c_j\|^2}{2\sigma_j^2}\right) \quad (9)$$

Therein $\varphi_j(x)$ is Gaussian function. c_j and σ_j are the center and the radius width of the j - th hidden node, respectively.

RBFNN classifies data based on pattern recognition. K-nearest neighboring (KNN) algorithm is another classification approach using distance measure. In this algorithm the KNN classifier finds the top k nearest neighbors of the unlabeled web service among the training data. The class of the unlabeled web service is judged by the class that the majority of the k nearest neighbors belongs to. The decision rule can be shown as followings:

$$y = \arg\max_{c_j} \sum_{x_i \in N_k(x)} C(y_i = c_i) \quad (10)$$

Therein x_i is the input vector. $N_k(x)$ is the set of the k nearest neighbors. y_i is the class label of x_i. c_j is one of the classes. C is the function that indicates whether x_i belongs to class c_i.

$$C(y_i = c_i) = \begin{cases} 1 & x_i \in c_j \\ 0 & x_i \notin c_j \end{cases} \quad (11)$$

In this paper the cosine distance is used to measure the distance the labeled and unlabeled web services.

IV. RESULTS

The web service data set in our experiment is from

the OWLS-TC 4.0. The collection contains 1082 semantic web services written in OWL-S 1.1 including nine different domains.

This section describes the results of the experiment. In order to figure out the effectiveness of the new method, experiment is carried out to compare the integrated similarity measure of web service (ISeM) with the method that only uses semantic similarity measure (SeM). Two classifiers are used to show the accuracy of web service classification-RBF and KNN, respectively. In order to show the different effects of the new method on each class, class-dependent method is used. Current class is labeled as positive class and other class is negative in the classify process. Accuracy indicates the performance of classifier on both positive and negative set, which is expressed as followings:

$$Accuracy = \frac{N_{right}}{N} \qquad (12)$$

Therein N_{right} is the sample number of instances grouped into the right class and N is the total number of samples in training set. Meanwhile, standard deviation (STD) is another indicator that reveals the stability of the algorithm.

The classification is evaluated using 10-fold cross-validations. The tables below show the average accuracy for two modeling methods of different classifier. T-test is employed on each class and overall situation to show the significance of the difference between two methods.

TABLE 1 and TABLE 2 show the comparison of two methods in different classifiers. The p-value in last row of two tables means the t-test result of the two model method that contains all the class.

The result of SeM in geography class is a litter higher than ISeM in RBF classifier. Travel class achieves the same accuracy of ISeM and SeM. However, ISeM can reach higher accuracy than SeM in other classes. In class education (p=0.049) and medical (p=0.030) the results is significant that verified by t-test. The integrated method brings the raise of economy class is 3.78%% which is the biggest of all classes. The average improvement compared ISeM with SeM in RBF is 0.91%. As for other class, although in most class ISeM can get a higher accuracy than SeM the difference is not significant (p=0.0835).

In KNN classifier the biggest raise appears in economy class that is 5.36% and the value is approximate in education. The average improvement is 1.66% that is better than RBF. The results of education (p=0.016) are significant on the level 95%. The different in communication class pass the test under the significant level 99% (p=0.007). The difference is the most significant in economy on the level 99.5% (p=0.000). The results of simulation and weapon stay almost the same in two methods. From the perspective of all class ISeM is significantly different from SeM using KNN at 95% level (p=0.0302).

TABLE 1

ACCURACY OF CLASSIFICATION WITH RBF (%)

Method / Class	ISeM	SeM
Communication	98.61±0.85	98.05±1.00
Economy	88.90±3.51	85.12±3.18
Education	85.38±2.56	82.70*±3.07
Food	98.24±1.20	98.14±0.71
Geography	97.13±0.77	97.31±0.99
Medical	98.05±0.97	97.04*±0.55
Simulation	99.44±0.61	99.26±0.64
Travel	93.99±2.02	93.99±1.60
Weapon	98.70±1.03	98.61±0.85
p-value		0.0835

*p<0.05, ** p<0.01, *** p<0.005

TABLE 2

ACCURACY OF CLASSIFICATION WITH KNN (%)

Method / Class	ISeM	SeM
Communication	98.89±0.69	97.50**±0.75
Economy	88.74±3.56	83.38***±4.51
Education	90.66±1.93	86.68*±2.97
Food	98.34±0.90	97.89±1.17
Geography	98.71±0.74	98.61±1.11
Medical	98.34±0.80	96.02±0.96
Simulation	99.17±0.65	99.17±0.87
Travel	95.11±2.21	93.81±3.58
Weapon	98.05±0.77	97.97±0.55
p-value		0.0302

*p<0.05, ** p<0.01, *** p<0.005

Further analysis of results indicates that ISeM improves notably of classes with more features which are common and low discrimination such as communication, economy and education. From the data in the tables we can see the results of class simulation and weapon are the same or approximate of the two methods in two classifiers. The reason is that feature dimension in the two classes is low and the distribution of features is uniform. The addition of TF-IDF doesn't influence the distribution of vector space model so that the results of the two classes are the same.

Experiment results in this part show that using the integrated semantic measure to establish the vector space model of web services improves the accuracy of web service classification in different algorithm. The standard deviation of ISeM and SeM in two classifier is between 0.40% - 4.51% which is within acceptable range. The standard deviation value indicates that the two algorithms using in this paper is stable.

V. CONCLUSION AND FUTURE WORK

Facing the fact that the number of web services is growing fast, this paper attempts to propose web service classification algorithm in UDDI. An integrated method for web service semantic measure is presented. This method combined TF-IDF and semantic similarity that aims at eliminating drawbacks of the single method. Based on the new semantic similarity measure, a framework of web service classification is proposed. Experiment results using OWL-S TC dataset prove that the integrated method is efficient to improve the classification accuracy using RBF and KNN method than the conventional approaches in web services discovery.

In this paper we proposed a new method to calculate semantic similarity applied all words that appearance in web service documents. In further research the feature selection will be explored to improve the performance.

ACKNOWLEDGMENT

The work was supported by General Program of the National Natural Science Foundation of China (No.71101103) and National Natural Science Fund for Distinguished Young Scholars of China (No. 70925005).

REFERENCES

[1] Crasso M, Zunino A, Campo M. A Survey of Approaches to Web Service Discovery in Service-Oriented Architectures [J]. Journal of Database Management, 2011, 22(1):102-132.

[2] Nayak R, Lee B. Web Service Discovery with additional Semantics and Clustering [C]// Web Intelligence, IEEE/WIC/ACM International Conference on. IEEE, 2007:555-558.

[3] Bayati S, Bahreininejad A, Nejad A F, et al. Improving Semantic Web Services Composition Performance, Using Data Mining Techniques [J]. Journal of Algorithms & Computational Technology, 2010, 4(4):409-423.

[4] Liu M, Shen W, Hao Q, et al. A fuzzy matchmaking approach for Semantic Web Services with application to collaborative material selection [J]. Computers in Industry, 2012, 63(3):193-209.

[5] Skoutas D, Sacharidis D, Simitsis A, et al. Ranking and Clustering Web Services Using Multicriteria Dominance Relationships [J]. IEEE Transactions on Services Computing, 2010, 3(3):163-177.

[6] Farrag T A, Saleh A I, Ali H A. Semantic web services matchmaking: Semantic distance-based approach [J]. Computers & Electrical Engineering, 2012, 39(2):497-511.

[7] Luo Q, Chen E, Xiong H. A semantic term weighting scheme for text categorization.[J]. Expert Systems with Applications, 2011, 38(10):12708-12716.

[8] Nasir J A, Varlamis I, Karim A, et al. Semantic smoothing for text clustering [J]. Knowledge-Based Systems, 2013, 54(4):216–229.

[9] Wu Z, Palmer M. Verb Semantics and Lexical Selection [J]. Proceedings of Annual Meeting on Association for Computational Linguistics, 1994:133--138.

[10]Li Y, Bandar Z A, Mclean D. An Approach for Measuring Semantic Similarity between Words Using Multiple Information Sources [J]. IEEE Transactions on Knowledge & Data Engineering, 2003, 15(4):871-882.

[11]P. Resnik, Using information content to evaluate semantic similarity in a taxonomy, in: Proc. of the International Joint Conferences on Artificial Intelligence (IJCAI 1995), vol. 1, Montreal, Canada, 20 August 1995, pp. 448–453.

[12]Lin, Dekang. An Information-Theoretic Definition of Similarity [C]// Proceedings of the Fifteenth International Conference on Machine Learning. Morgan Kaufmann Publishers Inc., 1998:296--304.

[13]Seco N, Veale T, Hayes J. An Intrinsic Information Content Metric for Semantic Similarity in WordNet.[C]// Proceedings of the 16th Eureopean Conference on Artificial Intelligence, ECAI'2004, including Prestigious Applicants of Intelligent Systems, PAIS 2004, Valencia, Spain, August 22-27, 2004. 2004:1089-1090.

[14]Sánchez D, Batet M, Isern D. Ontology-based information content computation [J]. Knowledge-Based Systems, 2011, 24(2):297-303.

[15]Taieb M A H, Aouicha M B, Hamadou A B. A new semantic relatedness measurement using WordNet features [J]. Knowledge & Information Systems, 2014, 41(2):467-497.

[16]Heß A, Johnston E, Kushmerick N. ASSAM: A Tool for Semi-automatically Annotating Semantic Web Services [C]// Proceedings of The Semantic Web - ISWC 2004: Third International Semantic Web Conference, Hiroshima, Japan, November 7-11, 2004. 2004:320-334.

[17]Patil A A, Oundhakar S A, Sheth A P, et al. Meteor-s web service annotation framework [C]// Proceedings of the 13th international conference on World Wide Web, WWW 2004, New York, NY, USA, May 17-20, 2004. 2004: 553-562.

[18]Corella M Á, Castells P. Semi-automatic Semantic-Based Web Service Classification[C]// Business Process Management Workshops, BPM 2006 International Workshops, BPD, BPI, ENEI, GPWW, DPM, semantics4ws, Vienna, Austria, September 4-7, 2006, Proceedings. 2006:459-470.

[19]Adam N. Semantics-Based Automated Service Discovery [J]. IEEE Transactions on Services Computing, 2012, 5(2):260-275

[20]Segev A, Toch E. Context-Based Matching and Ranking of Web Services for Composition [J]. Services Computing IEEE Transactions on, 2009, 2(3):210-222.

How to Optimize Production and Carbon Resources under Cap and Trade

Meirong Zhou, Ming Zhou*, Yanchun Pan, Zhimin Chen, Mei Wen, Lei Li
College of Management, Shenzhen University, Shenzhen, P.R. China
(*mzhou@szu.edu.cn)

Abstract - **With the increase of human carbon emissions global climate warming is becoming more and more significant. Natural disasters, climate change and extreme weather events become more frequent, these environmenta problems have caused serious influence to human life. China has established regional carbon emissions trading markets at seven pilot cities which are based on cap and trade Participating enterprises are restricted in total carbon emissions through initial allocation of emission quotas. They can purchase emission quotas to satisfy their excessive emission needs via the carbon emissions trading market Also, they can conduct self-purification to reduce carbon emission level to satisfy their excessive emission needs. And they can sell the surplus emission quotas to gain revenue via the carbon emission trade market. Under cap and trade enterprises must balance their economic and environmenta, benefits. Based on such background, we have established a multi-period optimization model of integrated decision to help enterprises make production and carbon resource allocation decisions. It is found that government should combine industry self-purification characteristics and emission characteristics to allocate emission quotas; the fluctuations of carbon price should be considered when to set punishment force; also, self-purification, carbon trade and punishment should be balanced when to allocate carbon resources.**

Keywords - **Cap and trade, production planning, carbon resource allocation**

I. INTRODUCTION

In recent years, the global warming has become an increasingly serious problem. The effects of global climate warming are the ecological and social changes caused by greenhouse gas emissions. There is a scientific consensus that climate change is occurring, and that human activities are the primary driver[1][2]. Evidence of climate change includes the instrumental temperature record, rising sea levels, and decreased snow cover in the Northern Hemisphere. To reduce greenhouse gas emissions (i.e. CO2 equivalent emissions), different mechanisms have been designed and implemented Among them cap and trade has been one of the widely applied programs based on eco-economic theory[3] Known as emission quotas trading, cap and trade is a market-based program for controlling CO2e emissions Under such a program, emission generating companies (e.g. manufacturers) in a region are allocated emission allowances to offset their emission needs. Initial allowance can be grandfathered or auctioned through a regulatory agency, and then be traded later among the generating companies through transactions at a regional carbon emission trade market[4]. The total of emission allowances issued is restricted by a pre-determined cap, which is gradually tightened over time. There are many established regional carbon emission trade market systems around the world, such as European Union Emission Trade System (EU-ETS) and Regional Greenhouse Gas Initiative (RGGI). In China, the government has promised to reduce the intensity of carbon emissions per unit of GDP by 40 to 45 percent compared with the level of 2005 by 2020 at the World Climate Conference in Copenhagen in 2009[5], and is aggressively pushing for a compulsory carbon emission reduction program that combines administrative penalty and market trade of emission allowances[6]. Shenzhen has taken the lead in the establishment of carbon emission trade market in June 2013, and there are 635 large industrial enterprises to be included. This has inspired and provided a valuable opportunity for the proposed research.

Under cap and trade, the manufacturing enterprises will face more severe challenge. They can not only consider the cost of the traditional production, but also consider the carbon resource allocation and its cost, when making the production plans. The increasing production of an enterprise will increase sales revenue, but at the same time will lead to the increase of carbon emissions, which will produce the high cost of carbon emissions. Although the decreasing production of an enterprise can avoid the high cost of carbon emissions, sales revenue will decrease at the same time. That is to say, the production plans and carbon resource decision-making are related to each other, and have influence on each other. There is the dynamic trade-off phenomenon among them, and the optimal production simply determined by market demand may not be optimal[7][8][9]. In addition to the production plans, an enterprise should make carbon resource allocation decisions according to the different situation of carbon resources. When there is surplus carbon quotas, an enterprise can sell them for revenue or retain them; When there is insufficient carbon quotas, which means excessive emissions, an enterprise can choose to buy carbon quotas from the carbon emission trade market, do a self-purification or pay the fines to offset the excessive emissions[10][11][12]. But there are different risks between carbon trading and self-purification: there exists such as financial, technical and organizational risks in self-purification, which may lead

E. Qi et al. (eds.), *Proceedings of the 23rd International Conference on Industrial Engineering and Engineering Management 2016,* DOI 10.2991/978-94-6239-255-7_46

to problems such as high investment, slow effect and long period; and the research of carbon emission trade market shows that the changes of the carbon price, carbon yield, and market risks and volatility are not subject to mean reversion process, and carbon emission trade market has greater risks than the stock market[1]. So, under cap and trade, if enterprises want to maximize their own profits, they must balance their economic and environmental benefits. That is to say that they must arrange the production plan and configure the carbon resources, which has very important sense to enterprises and government. Based on such background, we have established a multi-period optimization model of integrated decision to help enterprises make production and carbon resource allocation decisions.

Although the literature seems abundant, there is clearly a lack of research to address a basic modeling problem for production plans and carbon resource allocation decisions under cap and trade conditions.

A. There is a lack of carbon resource allocation in traditional production decisions.

Most of the existing researches focus on the traditional production operations of an enterprise, and a few involve carbon emission reduction, which add the costs of carbon emissions to the traditional production decision-making of an enterprise, but only involve the carbon trade, which do not involve the self-purification of an enterprise, and only involve the one-period decision-making of an enterprise. Therefore, a multi-period decision model which helps an enterprise make a decision has a significant sense.

B. The self-purification is not considered as a medium- and long-term decision variable.

Either the existing researches do not involve the self-purification decisions, or they assume that the self-purification decisions and the carbon trade decisions are the complementary continuous variables in the same planning scale. According to the research of participating enterprises in Shenzhen, we find that an enterprise do a self-purification in the form of a program, the self-purification is served as a medium- and long-term decision variables.

C. The time value of the costs of a self-purification program and fines is omitted

The purpose of this research is to develop a multi-period optimization model of integrated decision to help enterprises make production and carbon resource allocation decisions. The study focuses on enterprises' decisions for production plans and carbon resource allocation decisions under cap and trade conditions. In our research, the self-purification will be done in the form of a program, and we do not assume that the self-purification decisions and the carbon trade decisions are the complementary continuous variables in the same planning scale. Also, we take into consideration the time value of money, and that makes the model be more close to reality. In our model, carbon trade decision is a short-term decision, and the self-purification is a medium- and long-term decision. When doing trade-offs, an enterprise must consider the time value of the costs of a self-purification program. At the same time, because of the government check time, an enterprise also take into consideration the time value of fines. In Section 2, we propose a conceptual modeling framework that captures essential characteristics of a multi-period optimization model of integrated decision about production plans and carbon resource allocation decisions under cap and trade conditions. Section 3 shows the limited results of testing a baseline optimization model. Finally, Section 4 concludes the report.

II. A MATH PROGRAMMING MODEL

Under cap and trade conditions, the government will give an enterprise some carbon quotas for free at the beginning of the performance period (usually one year), and these carbon quotas can be served as a kind of resources of the enterprise, which can be retained to offset its own excessive emissions, and also can be sold to earn revenues in the carbon emission trade market. If carbon emissions are expected to more than a given carbon quotas, an enterprise can purchase carbon quotas in the carbon emission trade market or can conduct self-purification programs to offset its own excessive emissions. At the end of the performance period, the government will check an enterprise's carbon emissions. When the government finds that an enterprise has excessive emissions, it will punish the enterprise. Therefore, an enterprise will face trade-offs about production and carbon resource allocation. When producing more products, sale revenues will increase, but carbon emissions will increase at the same time, which can lead to more costs of carbon emissions, and therefore an enterprise need to make trade-offs between purchasing carbon quotas and do self-purification programs. When producing less products, sale revenues will decrease, carbon emissions will decrease at the same time, which can lead to less costs of carbon emissions, and therefore it may bring carbon yields. Therefore, how to arrange the production, the carbon trade and the self-purification is important for an enterprise, which can make the total profits maximize. In this paper, we build a multi-period optimization model of nonlinear integer programming model, which integrates three types of decisions about production, self-purification and carbon trade, to analyze the production and carbon resource allocation decisions under cap and trade conditions, and to solve the problems about how many products to produce in one period, when to do a carbon trade, how many to purchase or sell, and when to do self-purification programs.

Our goal is to maximize the total profits. Before building the model, we give these assumptions:

1) The product demands are predictable, and there is no out of stock.

2) The carbon quotas are sold or purchased according to the actual carbon quota demands of an enterprise, which means the speculation of carbon quotas is not considered.

3) The money that is used to conduct a self-purification program comes from a bank and will be returned when the self-purification program completes.

The conceptual model of this research is as follow:

The objective:

Total profits = total sale revenues + total revenues of carbon quotas sold - total production costs - total inventory costs - total costs of purchasing carbon - total costs of conducting self-purification programs + total fines

The constraints:

1) There is no out of stock.

2) The carbon quotas sold is no more than that possessed.

3) The purchasing carbon quotas is no more than excessive emissions evaluated.

4) Each self-purification program can be conducted only once.

5) A self-purification program must be finished when the government does a check.

To build our model, the following notations are proposed:

The decision variables:

$q(t)$: Production quantity of the *t*th period with the unit *piece.*

$x_s(t)$: Carbon quota quantity sold of the *t*th period with the unit *ton.*

$x_p(t)$: Carbon quota quantity purchased of the *t*th period with the unit *ton.*

$y_j(t)$: Whether to start the self-purification program *j* in the *t*th period, 1 means yes and 0 means no.

z_T: Excessive emission quantity in performance period T with the unit *ton.*

I_t: Initial product inventory of the *t*th period with the unit *piece.*

The parameters:

T: Performance period with the unit *month.*

Δt: Duration to check with the unit *month.*

p_t: Product price of the *t*th period with the unit *yuan per piece.*

d_t: Product demands of the *t*th period with the unit *piece.*

v_t: Production costs per unit of the *t*th period with the unit *yuan per piece.*

h_t: Inventory costs per unit of the *t*th period with the unit *yuan per piece.*

cp_t: Carbon quota price of the *t*th period with the unit *yuan per ton.*

k: Risk preference coefficient.

g_j: Carbon emission reduction quantity of self-purification program *j* with the unit *ton.*

cy_j: costs per unit of self-purification program *j* with the unit *yuan per ton.*

b_j: Duration of self-purification program *j* to complete with the unit *month.*

r_t: Monthly loan interest rate of the *t*th period.

cz: Penalty costs per unit with the unit *yuan per ton.*

Q: Free carbon quota quantity from the government in performance period T with the unit *ton.*

δ: Carbon emission quantity per unit product of the last performance period with the unit *ton per ton.*

The model is as follow:

$$\text{Max.} \sum_{t=1}^{T} p_t \times d_t + \sum_{t=1}^{T} cp_t \times x_s(t) - \sum_{t=1}^{T} v_t \times q(t) - \sum_{t=1}^{T} h_t \times I_{t+1}$$
$$-\sum_{t=1}^{T} cp_t \times x_p(t) - \sum_{j=1}^{n}\sum_{t=1}^{T} y_j(t) \times g_j \times cy_j \times (1+r_t)^{b_j} - cz \times z_T$$

s.t.:

1) $q(t) + I_t \geq d_t$.

2) $I_1 = m$ (*m* is a parameter).

3) $I_{t+1} = q(t) + I_t - d_t$.

4) $\sum_{t=1}^{i} x_s(t) \leq Q + \sum_{t=1}^{i-1} x_p(t)$.

5) $\sum_{t=1}^{T} x_p(t) \leq k \times (\delta \times \sum_{t=1}^{T} q(t) - Q)$.

6) $min\{x_s(t),\ x_p(t)\} = 0$.

7) $\sum_{t=1}^{T} y_j(t) \leq 1, j = 1,2,\cdots\cdots,n$.

8) $(t + b_j) \times y_j(t) \leq T + \Delta t$.

9) $z_T = max\{0,\ \delta \times \sum_{t=1}^{T} q(t) + \sum_{t=1}^{T} x_s(t)$ $-Q - \sum_{t=1}^{T} x_p(t) - \sum_{j=1}^{n}\sum_{t=1}^{T} y_j(t) \times g_j\}$.

10) $q(t) \geq 0$.

11) $x_s(t) \geq 0$.

12) $x_p(t) \geq 0$.

13) $y_j(t) = 0\ or\ 1$.

III. LIMITED RESULTS OF TESTING A BASELINE OPTIMIZATION MODEL

In all controlled industries, papermaking industry is easier in the data collection, which can not only find the related data of the industry in the national bureau of statistics, but also find the annual report of papermaking industry from the association of papermaking industry. At the same time, the papermaking industry produces large carbon emissions, and therefore face more pressure about carbon mission reduction. So, It has an urgent research need about carbon mission reduction to help itself make decisions under cap and trade conditions. Thus, one enterprise of papermaking industry is selected as a model example.

The production scale of top 30 in the papermaking enterprises that China papermaking association released in 2014, is 500 to 1000 thousand tons in the majority (as shown in figure 1). So, to analyze this kind of enterprises' decisions is the most representative. Therefore, we select a papermaking enterprise, whose annual production is 500 to 1000 thousand tons, to analyze the optimal decision.

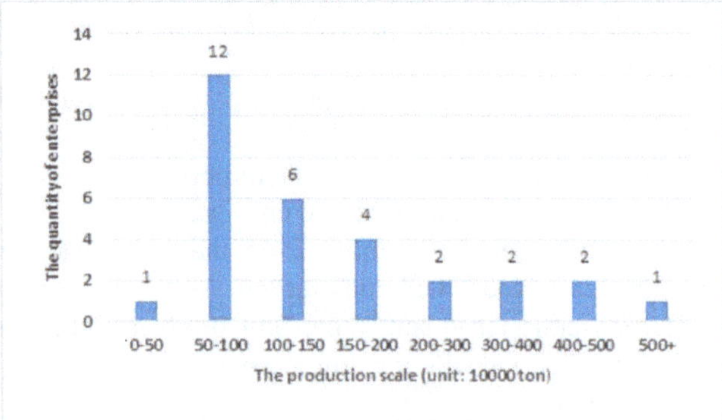

Fig. 1. The annual production distribution of top 30 of China's papermaking industry in 2014.

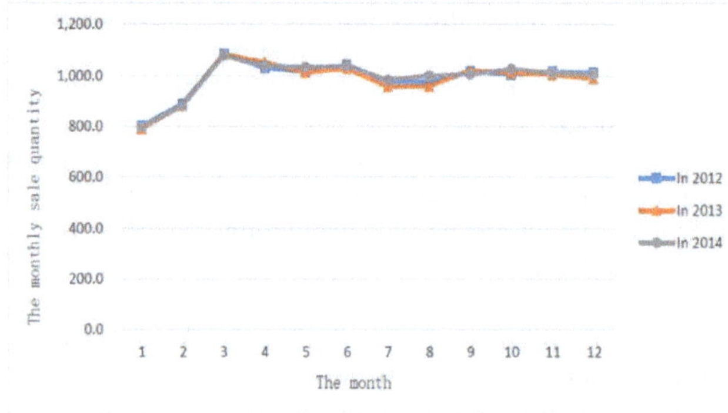

Fig. 2. The monthly sale trend of papermaking industry.

A. *The production-related parameter data*

Zhejiang Jingxing Paper JSC Ltd. is selected as the model example. It is listed in the Shenzhen Stock Exchange and its stock code is 002067. Its annual report in 2013 and 2014 shows that base paper production is 988.4 thousand tons, sale quantity is 965.3 thousand tons, the main business revenue is about 2.988 billion yuan and the costs of raw materials is about 1.376 billion yuan in 2013; base paper production is 1025.6 thousand tons, sale quantity is 972.5 thousand tons, the main business revenue is about 2.868 billion yuan and the costs of raw materials is about 1.307 billion yuan in 2014. From calculation, we can get that the average price of its base paper is 3095 yuan per ton in 2013 and 2949 yuan per ton in 2014, and the average production costs of its base paper is 1392 yuan per ton in 2013 and 1274 yuan per ton in 2014.

According to the national bureau of statistics website data about papermaking industry, we can get the past month sale quantity trends in papermaking industry as shown in figure 3. From figure 2, we can find the annual demand characteristics of papermaking industry are that the monthly demands of January, February, July and August are lower, and little difference among the other monthly demands, that is to say the other monthly demands are relatively stable. Therefore, according to the enterprise research and analysis, the papermaking enterprise product price is within the scope of 2800 to 3200 yuan per ton, the production costs is within the scope of 900 to 1150 yuan per ton, the inventory costs is within the scope of the 80 to 120 yuan per ton, and monthly demands is within the scope of the 78000 to 88000 ton. When getting random values within these scopes, we can get the production-related parameter data that the model need, as shown in table 1.

TABLE I
THE PRODUCTION-RELATED PARAMETER DATA

parameter \ period	1	2	3	4	5	6	7	8	9	10	11	12
p_t	2800	2800	3000	2800	2900	3000	3200	3000	3100	3000	2900	3000
v_t	900	1000	1100	1100	1000	1150	1000	1100	1100	1150	1000	1050
h_t	80	80	80	90	90	100	100	100	100	120	130	120
d_t	78000	79000	81000	83000	85000	88000	86000	85000	83000	84000	85000	83000

B. *The self-purification-program-related parameter data*

The pollution of papermaking industry is extremely serious, which has not only large carbon emissions, but also dust, water pollution and other problems. Since China is a little late in carbon emission reduction, a few enterprises would actively do a self-purification in fact. In academia, there are a few scholars to estimate the costs of self-purification of papermaking industry, such as Wang et al.[13]. They think that the costs of self-purification of Guangdong's papermaking industry is less than 20 yuan per ton through the GD-CGE model. But it is based on the macro-field mode, that is to say it is based on the GDP of Guangdong Province, economic growth, input and output data, and it is the pulp and paper industry is the opportunity costs of self-purification of Guangdong's papermaking industry, not the unit cost of actual self-purification of Guangdong's papermaking industry. Therefore, we will reference other related industry data to get the self-purification-program-related parameter data for the model. Through investigation and research of enterprises, we find that doing a self-purification program such as changing boilers, biogas recovery and power generation, can cut carbon emissions in 22000 to 40000 tons, unit cost of self-purification is in 48 to 70 yuan per ton, and the duration of a self-purification program is in 5 to 8 months. We will consider five self-purification programs, and get the parameter random values within these scopes, as shown in table 2.

TABLE II
THE SELF-PURIFICATION-PROGRAM-RELATED PARAMETER DATA

parameter \ program	1	2	3	4	5
g_j	22000	35000	40000	38000	28000
cy_j	70	62	70	52	48
b_j	5	6	6	7	8

C. The evaluation of carbon quota prices

Through the study of literature, we adopt the carbon quota price forecast function from Zhu et al.[14] and Tian et al.[15] to forecast the carbon quota prices that are used in the model. The carbon quota price forecast function is $x_t = 0.1273 + 0.993649x_{t-1} + \varepsilon_t$, where x_t means the carbon quota price of the *t*th period with the unit *yuan per ton*, ε_t is random error term of the *t*th period. The first period carbon quota price comes from the average carbon quota price 70.99 yuan per ton of January 2014 of Shenzhen carbon emissions exchange.

D. The macro-policy-related parameter data

The macro-policy-related parameters such as the government carbon verification time, unit penalty cost, etc. in the model are as shown in table 3.

TABLE III
THE MACRO-POLICY-RELATED PARAMETER DATA

parameter	value	basis
Δt	1 (unit: month)	Beijing: 10 March-5 April Shanghai: 31 March-30 April Guangzhou: 15 March-30 April Shenzhen: 31 March-30 April
cz	188.10=3*annual average carbon quota price (unit: yuan per ton)	Beijing: 3 to 5 times of the average carbon quota price of last six months Chongqing: 3 to 5 times of the average carbon quota price of last month Shenzhen: 3 times of the average carbon quota price of last six months
r_t	$r_1 = r_2 = r_3 = 0.00320$ $r_4 = r_5 = r_6 = 0.00425$ $r_7 = r_8 = r_9 = 0.00467$ $r_{10} = r_{11} = r_{12} = 0.00500$ (unit: yuan per month)	Central bank released the short-term loan interest rates
δ	1.5 (unit: ton per ton)	Chen et al. (2014)[16]

In this research, we want to know the production and carbon resource allocation decisions of an enterprise, so we let the performance period T be 12 months. And we assume that an enterprise has no initial product inventory, which means $I_1 = m = 0$, and the risk preference coefficient is 1, which means $k = 1$. Also, we assume that the free carbon quota quantity from the government in performance period T accounts for the proportion of 80% of the actual carbon emissions of an enterprise, the annual production is 1 million tons, and the carbon emission quantity per unit product in the performance period is 1.5 ton per ton. Thus, the total carbon emissions of the enterprise in the performance period is 1.5 million tons, and the free carbon quota quantity from the government is 1.2 million tons.

When we let LINGO solve the model, the result is as shown in table 4.

TABLE IV
THE RESULT OF MODEL EXAMPLE

decision variable	value
total profit	1906552000 yuan
$q(t)$	$q(1)$ =238000 tons, $q(2)$ =0 tons, $q(3)$ =0 tons, $q(4)$ =83000 tons, $q(5)$ =173000 tons, $q(6)$ =0 tons, $q(7)$ =171000 tons, $q(8)$ =0 tons, $q(9)$ =83000 tons, $q(10)$ =84000 tons, $q(11)$ =85000 tons, $q(12)$ =83000 tons
$x_s(t)$	$x_s(1)$ =0 tons, $x_s(2)$ =0 tons, $x_s(3)$ =163000 tons, $x_s(4)$ =0 tons, $x_s(5)$ =0 tons, $x_s(6)$ =0 tons, $x_s(7)$ =0 tons, $x_s(8)$ =0 tons, $x_s(9)$ =0 tons, $x_s(10)$ =0 tons, $x_s(11)$ =0 tons, $x_s(12)$ =0 tons
$x_p(t)$	$x_p(1)$ =0 tons, $x_p(2)$ =0 tons, $x_p(3)$ =0 tons, $x_p(4)$ =0 tons, $x_p(5)$ =0 tons, $x_p(6)$ =0 tons, $x_p(7)$ =0 tons, $x_p(8)$ =0 tons, $x_p(9)$ =0 tons, $x_p(10)$ =300000 tons, $x_p(11)$ =0 tons, $x_p(12)$ =0 tons
$y_j(t)$	$y_1(1)$ =1, $y_2(1)$ =1, $y_3(1)$ =1, $y_4(1)$ =1, $y_5(1)$ =1
z_T	0 tons
I_t	I_1 =0 tons, I_2 =160000 tons, I_3 =81000 tons, I_4 =0 tons, I_5 =0 tons, I_6 =88000 tons, I_7 =0 tons, I_8 =85000 tons, I_9 =0 tons, I_{10} =0 tons, I_{11} =0 tons, I_{12} =0 tons, I_{13} =0 tons

Limited experiments were conducted to analyze the sensitivity of related model parameters, and will be reported in a full paper under preparation. In this particular case, we find that total profits are sensitive to product price and production costs per unit, and are not sensitive to inventory costs per unit, costs per unit of self-purification program, risk preference coefficient, carbon quota quantity from the government and carbon emission quantity per unit product of the last performance period. It is found that government should combine industry self-purification characteristics and emission characteristics to allocate emission quotas; the fluctuations of carbon price should be considered when to set punishment force; also, self-purification, carbon trade and punishment should be balanced when to allocate carbon resources. The detailed results of sensitivity analysis are omitted in this paper due to the limit of space.

IV. CONCLUSION

Under cap and trade, if enterprises want to maximize their own profits, they must balance their economic and environmental benefits. That is to say that they must arrange the production plan and configure the carbon resources, which has very important sense to enterprises and government. This paper reports an on-going research that aims at building a multi-period optimization model of integrated decision to help enterprises make production and carbon resource allocation decisions. In the future research, more uncertainties such as the fluctuation of product demands and carbon price can be considered to make the model more realistic.

ACKNOWLEDGMENT

This research is partially funded by the Natural Science Foundation of China (NSFC) under grant codes 71572114, 71272089 and 71472126.

REFERENCES

[1] Q. Zhang, X. Min, Y. Chen, 2011. Reflection on the Development of Low-Carbon Cities from the Perspective of Systems Engineering[J]. Systems Engineering, 29(1): 1-7.

[2] IPCC, 2007. Climate Change 2007: The Physical Science Basis. Cambridge University Press, UK., pp. 1-18. http://www.ipcc.ch/pdf/assessment-report/ar4/wg1/ar4-wg1-spm.pdf.

[3] J. H. Dales, 1968. Pollution. Toronto, University of Toronto Press.

[4] Environmental Protection Agency (EPA). 2007. http://www.epa.gov/epahome. Retrieved December12, 2007.

[5] W.J. Yi, L.L. Zou, J. Guo, K. Wang and Y.M. Wei, 2011. How can China reach its CO2 intensity reduction targets by 2020? A regional allocation based on equity and development[J]. Energy Policy, 39(5): 2407-2415.

[6] EN (Enigin News), 2010. China Set to Use Iron Hand to Force Energy Efficiency. http://www.enigin.com/about/enigin-plc/news/-/enigin-news.

[7] Mathijn J. Retel Helmrich, Raf Jans, Wilco van den Heuvel and Albert P.M. Wagelmans, 2015. The economic lot-sizing problem with an emission capacity constraint[J]. European Journal of Operational Research, 241(1): 50-62.

[8] Nabil Absi, Stéphane Dauzère-Pérès, Safia Kedad-Sidhoumb, Bernard Penzc and Christophe Rapine, 2013. Lot-sizing with carbon emission constraints[J]. European Journal of Operational Research, 227(1): 55-61.

[9] Xi Chen, Saif Benjaafar, Adel Elomri, 2013. The carbon-constrained EOQ[J]. Operations Research Letters, 41(2): 172-179.

[10] M. Zhou, Y.C. Pan, Z.M. Chen and X. Chen, 2013. Optimizing environment resource planning under Cap-and-Trade[C]. CIE43 Proceedings, October, 2013, Hong Kong.

[11] B.Z. Zhu and Y.M. Wei, 2013. Carbon price forecasting with a novel hybrid ARIMA and least squares support vector machines methodology[J]. Omega, 41(3): 517-524.

[12] F. Zhang and D. Li, 2007. Environmental and Natural Resources Economics[M]. Shanghai: Shanghai People's Press.

[13] P. Wang, H.C. Dai and D.Q. Zhao, 2014. Policy evaluation of carbon quotas trade of Guangdong Province based on GD-CGE model[J]. Journal of Environmental Sciences, 34(11): 2925-2931.

[14] B.Z. Zhu and Y.M. Wei, 2011. International carbon market price forecasting based on GMDH-PSO-LSSVM[J]. System Engineering Theory and Practice, 31(12): 2264-2271.

[15] K. Tian, L.Q. Qiu and M. Zeng, 2012. Power grid planning model based on dynamic carbon emission price[J]. Journal of Chinese Electrical Engineering Science, 32(4): 57-64.

The Development About Social Network Analysis And The Review About Its Application On Engineering

Hua PAN[1], Fang ZHENG[1*], Yong-kui LI[2]
[1]Shanghai University of Electric Power, College of Economics and Management, 200090, China;
[2]Tongji University, College of Economics and Management, 200092, China
(*Email: 987020124@qq.com.)

Abstract-**Social network analysis as a method, its application has become more and more widely used. Firstly, this paper reviews the origin and the maturity about this method; Secondly, it also summarizes its application in all fields, and especially summarizes its application on engineering from three aspects; Finally, based on the previous literature, this article provided summary and put forward the problem about the study and research prospects, I hope this study can provided some reference for the development about the field of engineering in social network analysis in the future.**

Key-**Social network analysis, Engineering application, Research prospects**

Ⅰ. INTRODUCTION

Social network analysis as a clear specific area or a research method and has become more and more widely used， although it's only a matter of ten or twenty years, the development of this idea of social network analysis has been in the west for seventy or eighty years[1]. Through looking up the literature which is from China, we can find that it's main application focus on the knowledge management & project performance & the internal dimension relationship between social network and the simple application about social network analysis software, but rarely used in dynamic network & areas of the economy & e-commerce, particularly in the engineering field. This paper will comb the literature which used the method of social network analysis, and the key point is this method's application in engineering field.

Ⅱ.The development of social network analysis

A. The basic concept of social network

Network concept is one of the defining paradigms in modern society [2]. It includes three sources, in a nutshell, Firstly, it is to be affected by the development of the theory of physics, such as Kurt Lu who applied the concept of network to the study of social interaction, Secondly, it is influenced by mathematical methods and it begins on "social network diagram" which is put forward by Jacob Moreno in 1933, Thirdly, researchers who use the method of anthropology to study organization issues, such as the famous "Hawthorne experiment" which is the first experiment and use the concept of "community figure" or "social network diagram " to studying.

B. The origin of social network analysis

The development of social network analysis thoughts can be traced back to the earliest classical sociologists, some people's social structure point, such as Emile Durkheim & Georg Simmer and so on, Emile pointed at its own book"social division of labor" in which people's combine way of social relations from one from to another of change. "from of sociology" which was advocated by Georg has become one of the directional sources of social metrology.

The second origin is the book of 《Human Group》 which is written by George homans who studied the structure function of small group in his own book, and especially analyzed the structure of community and the position of the individual in the group, the same as the study of small group and interpersonal relationship.

The third origin is the study of "Manchester school", they began to use social network as an analysis concept.

E. Qi et al. (eds.), *Proceedings of the 23rd International Conference on Industrial Engineering and Engineering Management 2016,* DOI 10.2991/978-94-6239-255-7_47

They making an important contribute to the development of social network analysis.

III.The Application of Social Network Analysis

Social network analysis has developed rapidly in recent ten or twenty years, which has been widely applied to many fields, such as sociology & political science & anthropology & psychology & organizational management & mass communication and social policy research and other fields [3]. The following is divided into two parts to summarize its application, one is the engineering field, and the other is in other field.

A. *The application of social network analysis in other fields*

Carrington and other researchers generally proposed in1990 or so, interest in social networks and use of the wide-ranging collection of social network methodology began to grow rapidly[4].Wasserman Stanley and other researchers put forward social network analysis as a kind of important technology and research field in modern sociology, it focuses on the relationship between social entities and patterns, and the meaning of these relationships[5]. Garton and colleagues summed up the previous literature, listed four units of social network analysis: relations & ties & multiplicity and composition[6]. Pastor-Satorras, R and Mikolajczyk, R put forward in the field of science, social network can help study the propaganda of endemics or epidemics [7-8], and then be widely used to study of HIV. David, Fu-ren Lin and Chun-hung Chen discuss from the study about how to embed the function of social network analysis into a virtual team system [9]. Svenson, P and Yang, C.C applying social network analysis method to hunt and map the terrorist network respectively, and their work has important actual meaning to prevent the occurrence of terrorist activities [10-11]. Xiao-tao Yao & you-min Xi believes that the theory of social network can provide a new perspective for the practice of management under the new environment [12]. At the same time, Zhong-lin Fei and jin-gan Wang put forward the importance of putting social network analysis into the field of management research, and think of it as a method of management research and perspective[13]. Nowadays, the development trend of social network analysis in foreign countries has been concentrated in complex network & dynamic network and multiple networks, but our country's literatures which focus on these aspects are less than foreign countries. Xiao Liu first proposed a international trade complex meta-network model by extending meta-network model to the complex meta-network model, and then proposed a dynamic network framework, named the International Trade Complex Dynamic Metanet-work Model[14]. Lei Feng & ya-bing Zha and other people proposed a dynamic network organization formation algorithm which based on the viewpoint of complex networks [15]. Although there are a lot of applications of social network analysis in other fields, due to space limitations, here no longer to detail. Below this article will give a detailed overview of the application of social network analysis in the field of engineering.

B. *The application of social network analysis in the field of engineering*

In this paper, the application of social network analysis in the field of engineering is divided into three aspects to introduce. The first aspect is the method of social network analysis is used to study the large scale projects organization controlling mechanism; The second aspect is the method of social network analysis is used to study the project stakeholders relations; The third aspect is the method of social network analysis is used to study the effect of project organization for safety management.

C. *The application of social network analysis method in the study of the total control of large scale projects organization*

Chinowsky P and Koerner M think that social network analysis is an important theory & method and tool for studying large scale projects organization [16-17]. Yong-kui Li & Yun Le and other researchers summed up eight kinds of social network elements based on the theory of social network analysis[18]. Hua Pan and Yong-kui Li put forward interrelated conception & study program and construction of network model etc from the angle of top and framework design [19]. Yu-jie LU & Yong-kui Li and other people use social network analysis to study investigates the evolution of the organizational network of a megaproject in China, and then proposes corresponding governance strategies[20]. The application of social network analysis method in the study of the project stakeholders' relations.

Based on the theory of social network analysis, Qi-ping Shen & jing Yang and others established the management model of the system[21]. Rong-gui Ding & Fang Liu and other people taking an construction project of China as an example and with the method of social network analysis and builds a social network analysis model of this project governance analyzes the interaction between network embedded [22]. Based on the value of social network analysis for project governance research, Xing-zhi Liu and Yan-wei Wang proposed how to construct network model for the relationship between project governance stakeholders [23].

D. The application of social network analysis method in the study of the effect of project organization for safety management

When collecting the literature, this paper found that from the perspective of organization using the method of social network analysis to study the effect of project organization for safety management of literature is less, Jinkyun Park and others using the method of social network analysis to investigate the characteristics of human cluster who belongs to nuclear power plant, and pointed out that in the project management if or not establish effective communication security mechanism has a decisive influence to reduce the risk of accident probability [24]. Yan-chi Li through the method of social network analysis to explore emotional and work relationship between subway construction project departments of all staff and finally he draw a conclusion that the factors which affecting safety management of subway construction [25]. Then Xing-zhi Liu study the risk of project governance based on social network analysis perspective and from the viewpoint of manageability to analyze the relationship between project management risks and put forward effective countermeasures [26]. Guang-pu Lei using the method of social network analysis to build a social network model of the sub-project safety communication and identify the relationships between the structural characteristics of communication network and safety climate [27]. I am also from the perspective of organizational and using the method of social network analysis to study the security management of transmission and substation project.

Ⅳ. CONCLUSION

In this paper, we first introduce the origin of the method of social network analysis, then introduces the application of this method in other fields (it is not including engineering field) , and divided three aspects to emphasizes the application of this method in engineering field. Summary of the above literature, although the application of this method is less, these literatures also illustrates the feasibility of social network analysis method in the relevant field of research.

Ⅴ.Existing problems and research prospects

Through the way of comprehensive comparison of the literature, we can find that a large number of research which studied by domestic and foreign scholars on the social network analysis which provided a basis for the study of this paper, social network analysis as a research paradigm, it is more and more widely used and it is considered by many fields as a new research method and perspective, the reason why the method of social network analysis become maturity is that caused by the logic evolution of scientific method itself. But, unfortunately, there is few research about the method in engineering field, it mainly confined to the above of three aspects, and it is less on the application of safety management, and it lack of relevant aspects of dynamic study of model building about the project, and that domestic scholars study the method of its application is less than foreign scholars. Therefore, based on the summary of previous literature, this paper summing up the literature and I am hoping to provide some reference for the future application of social network analysis in the field of engineering.

In the future research, I hope the study can be able to discuss and evaluate the project safety management from the perspective of model building and modeling tools, and finally come up with the corresponding suggestions and strategies.

ACKNOWLEDGMENT

This study is supported by the National Natural Science Foundation of China (project numbers : 71471136) .The views represented in this article are those of the individual authors only.

REFERENCES

[1] Tian-kui Jing. Social Network Analysis: theory & method and application[M]. Beijing: Beijing Normal University Publishing, 2009. (Chinese)

[2] Feng-bin Wang , chao-wei Zhu , Social Network and organization [M]. Beijing: China Renmin University Press, 2003.(Chinese)

[3] Min Xiang. Study on Knowledge Communication Network of Management Subject [D]. Zhejing University, 2012. (Chinese)

[4] Carrington, Peter J., John Scott, and Stanley Wasserman (Eds) .Models and Methods in Social Network Analysis. New York: Cambridge University Press, 2005.

[5] Wasserman, Stanley, and Faust, Katherine. Social Network Analysis : Methods and Applications. Cambridge: Cambridge University Press, 1994.

[6] Garton, L, Haythornthwaite, C. and Wellman, B."Studying Online Social Networks", Journal of Computer Mediated Communication (3:1) , 1997.

[7] Pastor-Satorras, R, Vespignani, A, Epidemic spreading in scale-free networks, In: Physical Review Letters, Vol.86, Number14, April 2nd 2011, pp.3200-3203.

[8] Mikolajczyk, R.T, Kretzschmar, M, Collecting social contact data in the context of disease transmission: Prospective and retrospective study designs, In: Social Networks, Vol, 30, Issue 2, May 2008, pp.127-135.

[9] Fu-ren Lin, Chun-hung Chen. Developing and Evaluating the Social Network Analysis System for Virtual Teams in Cyber Communities. Department of Information Management National Sun Yat-sen University. Proceedings of the 37th Hawaii International Conference on System Sciences, 2004, pp.1-8.

[10] Svenson, P., Svensson, P., Tullberg, H., Social Network Analysis And Information Fusion For Anti-Terrorism, In: Proceedings of the Conference on Civil and Military Readiness 2006 (CIMI 2006), Enköping, Sweden, 16-18 May 2006, Paper S3.1. Försvarets Materielverk, Stockholm, 2006.

[11] Yang, C.C.; Ng, T.D., Terrorism and Crime Related Weblog Social Network: Link, Content Analysis and Information Visualization, In: Intelligence and Security Informatics, 2007 IEEE, vol., n., pp.55-58, 23-24, May 2007.

[12] Xiao-tao Yao & you-min Xi. Management Research and Social Network Analysis [J]. Modern Management Science, 2008 (6) .pp:19-21. (Chinese)

[13] Zhong-lin Fei & jin-gan Wang. Social Network Analysis: A study of management research methods and perspective [J]. Science and Technology Management, 2010 (24) .pp: 217-219. (Chinese)

[14] Xiao Liu. Complex Dynamic Meta-network Model and its Applications in International Trade [J]. South China University of Technology, 2011. (Chinese)

[15] Lei Feng, ya-bing Zha, Ji-wen Hu, Quan-jun Yin. Research of Combat Simulation Organization Modeling Based on Complex Networks [J]. Journal of System Simulation, 2012 (4) pp. 882-886. (Chinese)

[16] Clarkson M. A stakeholder framework for analyzing and evaluating corporate social performance [J]. Academy of Management Review, 1995, 20(1): pp. 92-117.

[17] Mitchell A, Wood D. Toward a theory of stakeholder identification and salience: Defining the principle of whom and what really counts [J]. Academy of Management Review, 1997, 22(4) pp. 853-886.

[18] Yong-kui Li, yun Le, yu-jie Lu. Large Scale Projects Organization Controlling Mechanisms and Empirical Study Based on SNA [J]. Journal of Tongji University (natural science edition), 2011(11): pp. 1716-1719.

[19] Hua Pan, Yong-kui Li. Study on Large and Complex Engineering Projects Organization Based on Social Network Analysis Perspective [J]. Science and Technology Management Research, 2013(20): pp. 215-217. (Chinese)

[20] Yu-jie Lu, Yong-kui Li. Organizational Network Evolution and Governance Strategies in Megaprojects [J]. Construction Economics and Building, 2015(3): pp. 19-33.

[21] Qi-ping Shen, Jing Yang. Construction project stakeholders management framework study [J]. Journal of Engineering Management, 2010(4). (Chinese)

[22] Rong-gui Ding , Fang Liu, Tao Sun, Hua Sun. The Study on Project Governance Based on Social Network Analysis-An Example of Large Construction Project Supervision [J]. China Soft Science, 2010(6): pp. 132-140. (Chinese)

[22] Xing-zhi Liu, Yan-wei Wang, Wei Wei. Analysis of Project Governance Relationship Network and Response Strategy Based on SNA [J]. East China Economic Management, 2011(6): pp. 124-129. (Chinese)

[23] Jinkyun Park. The use of a social network analysis technique to investigate the characteristics of crew communications in nuclear power plants- A feasibility study [J]. Reliability Engineering and System Safety, 2011, 96; pp. 1275-1291.

[24] Yanchi Li. Study of safety management in subway construction based on social network analysis [D]. Hua zhong University of Science and Technology, 2011. (Chinese)

[25] Xing-zhi Liu. Research on the Method of Social Network Risk Analysis in Project Governance [D]. Shandong University, 2011. (Chinese)

[26] Guang-pu Lei. A Study of Safety Communication in High-Rise Project Subcontractors Based on Social Network Analysis [D]. Tsinghua University, 2012. (Chinese)

Gallbladder Diagnosis and Importance Analysis based on Bayesian Network

Zhi-qiang CAI[1,*], Peng GUO[1], Shen LI[1], Long-long CONG[2], Zhi-min GENG[2]
[1]Ministry of Education Key Laboratory of Contemporary Design and Integrated Manufacturing Technology, School of Mechatronics, Northwestern Polytechnical University, Xi'an 710072, Shaanxi, China
[2]First Affiliated Hospital, Medical College, Xi'an Jiaotong University, Xi'an 710061, Shaanxi, China
(*e-mail: caizhiqiang@nwpu.edu.cn)

Abstract - **In recent years, diagnostic and prognostic methods of cancer have become one of the research focuses in clinical medicine with the increased risk of cancer. However, how to excavate the more information from the data, how to establish the different math model, and how to verify the diagnosis further are of great significance for doctor to improve the diagnostic accuracy and reduce the diagnostic complexity. This paper collects and teases the gallbladder cancer data from the first affiliated hospital of medical college of Xi'an Jiaotong University, China, which is used to be excavated except the known tumor markers. Then, the optimal data model is set up using some typical Bayesian networks. Finally, we improve the warning factors of tumor diagnosis from the known information by putting importance calculation method into the analysis of gallbladder diagnosis.**

Keywords - ***Bayesian network, Diagnosis model, Gallbladder cancer, Importance measure***

I. INTRODUCTION

Cancer, which is called malignant tumor in medical terms, is a disease caused by the disorder of cell proliferation mechanism [1]. Now, gallbladder cancer (GBC), which leads to death finally no matter what method is used, has a trend of increase seriously. The main reason is that gallbladder cancer doesn't have the specific clinical symptoms, which means it is hard to be detected at earlier stage. So, how to improve the accuracy of gallbladder diagnosis is becoming the medical problem to be resolved at present [2].

In recent years, with the development of computer technologies, the puzzles with respect to Bayesian statistics and posteriori distribution have been better solved. Meanwhile, Bayesian statistics successfully apply to economics, sociology and some other fields, which makes the theory and application of Bayesian statistics develop rapidly [3]. In medical field, the foreign scholars have solved some medical problems that are hard to be settled in classic statistics by using the Bayesian statistics [4]. In china, the research standard of the theory and application of Bayesian statistics is lower than western developed country, especially in medical application.

In 1986, Bayesian method was applied to the field of medical industry, such as lethal dose of 50% (LD50) experiment, crossover trial, and the use of historical information in bioequivalence study [5]. After 1998, the application scope and depth in medical field with respect to Bayesian statistics has been evidently increased. Si et al. [6] brought forward a new Bayesian networks to build a diagnosis and prognosis model for breast cancer. Ko et al. [7] had used the mixture Bayesian networks to infer the gene pathways. Loghmanpour et al [8] established a new risk stratification model based on Bayesian network to predict the short-term and long-term LVAD mortality. So Bayesian network is becoming one of the most significant methods in medical field, which means a lot for us to solve the medical problems.

By using the gallbladder data collected from the first affiliated hospital of medical college of Xi'an Jiaotong University, this paper introduces two kinds of commonly used Bayesian network (BN), including Naïve Bayes model (NB) and Tree Augmented Naïve bayes (TAN), to build the analysis optimal models. Then, the influences between pathogenic factors and pathologic results are calculated based on the theory of importance. The diagnostic accuracy and efficiency of gallbladder cancer will be great improving.

II. METHODOLOGY

A. Bayesian Networks

Bayesian network, combining the graph theory and probability theory, is a kind of probability network, which can be understood in two aspects of quantitative and qualitative. In the quantitative aspect, BN displays the dependency relationship between variables and its father node. While in the qualitative aspect, BN describes the relationship of dependent and independent during the variables by using the directed acyclic graph.

Duda and Hart first put forward a kind of Naïve Bayes classifier based on bayes formula in 1973 [11]. In NB model, every attribute variable puts target variable as the single father node, which means that all of the attributes are conditionally independent to the target variable. Fig.1 is a picture of NB structure model. Assumed that a given target variable C, every attribute variable X_i is conditionally independent to each others, as shown in (1).

$$P(X_1=x_1,\dots,X_n=x_n|C=c_l)=\prod_{i=1}^{n}P(X_i=x_i|C=c_l) \quad (1)$$

Then, the posterior probability distribution of the target variable, which is the diagnostic result, can be calculated by using Bayes theorem, as shown in (2) and realizes the classification of the target variable, when the values of the attribute variables are given.

E. Qi et al. (eds.), *Proceedings of the 23rd International Conference on Industrial Engineering and Engineering Management 2016,* DOI 10.2991/978-94-6239-255-7_48

$$P(C=c_l|X_1=x_1,...,X_n=x_n) = \frac{P(C=c_l)P(X_1=x_1,...,X_n=x_n|C=c_l)}{P(X_1=x_1,...,X_n=x_n)} = \frac{P(C=c_l)\prod_{i=1}^{n}P(X_i=x_i|C=c_l)}{P(X_1=x_1,...,X_n=x_n)} \quad (2)$$

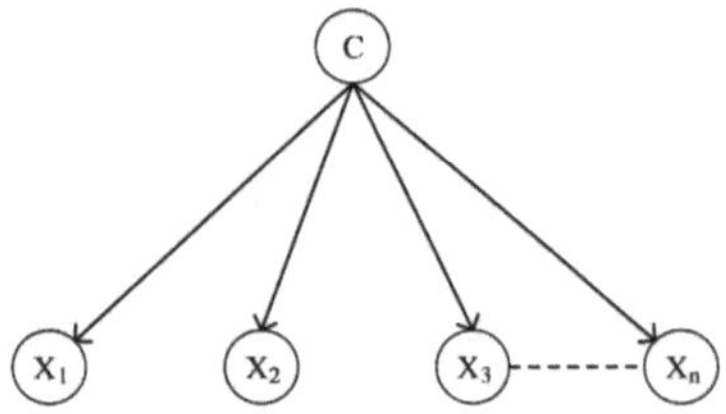

Fig. 1 Naïve Bayes structure model

Tree Augmented Naïve bayes classifier [12] is the extension of NB classifier and better than NB. TAN classifier reduces the hypothesis of any attributes that are independent to each other in NB classifier, by the dependent relationship of the attributes. What's more, it allows the attribute nodes to be connected by directed edge, as shown in Fig. 2. The joint arc from X_i to X_j shows that the influences between X_i and target node C also depend on the value of X_j. TAN classifier is a kind of BN, which views the target variable as root node, and every attribute variable views the target variable and at most one attribute variable as father nodes.

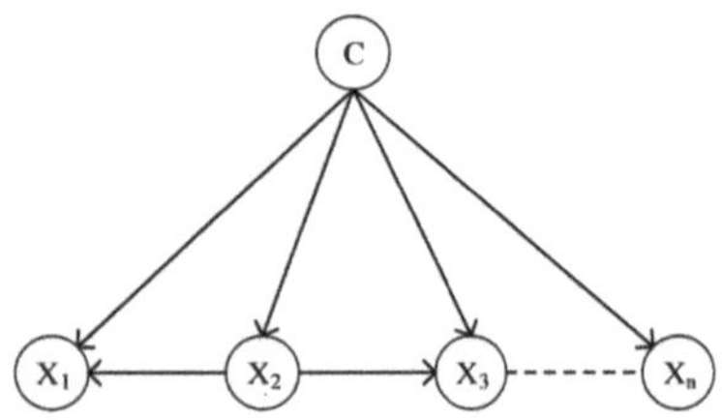

Fig.2 Tree Augmented Naïve bayes structure model

B. Importance measures

Importance measure is a very useful tool to reflect the importance of each unit in system, which is deeply studied by scholars at home and abroad [13-14]. They proposed some analytical method, such as Birnbaum (BM) importance, structural importance and other importance, and proposed some computational formulas with respect to the whole system reliability influenced by the unit reliability from the different side.

Birnbaum importance describes how the variation of one component affects the functioning of the whole system, as shown in (3), and R(·) expresses the reliability function of systems or components [15]. What's more, Birnbaum importance can be also expressed in (4) on the condition that F(·) expresses the failure function of systems or components. In the view of probability distribution of the system states, (4) can be transformed into (5).

$$I(BM)_{C_i}^{S} = \frac{\partial R(S)}{\partial R(C_i)} \quad (3)$$

$$I(BM)_{C_i}^{S} = \frac{\partial R(S=0)}{\partial R(C_i=0)} = \frac{\partial(1-F(S))}{\partial(1-F(C_i))} = \frac{\partial F(S)}{\partial F(C_i)} \quad (4)$$

$$I(BM)_{C_i}^{S} = \frac{\partial R(S)}{\partial R(C_i)} = P(S=0|C_i=0) - P(S=0|C_i=1) \quad (5)$$

III. DIAGNOSIS BN MODEL

A. The establishment of the optimal BN Model

This research collects 908 cases of data from the first affiliated hospital of medical college of Xi'an Jiaotong University, china. Because of the artificial objective reasons and faults in the process of data mining and preprocessing, many attributes of the clinical cases are critical missing. So, we study one target variable (Disease Category) and 23 kinds of attribute variables whose value is not missing seriously, including gender, age, occupation, blood type, weight, height, BMI, smoking history, reproductive number, menarche age, menstrual age at the last time, menopausal age, menstrual period (year), cholecystic polypus, gallstone, gallstone time, gallstone size, hypertension, diabetes, coronary heart disease, hepatitis B, family history of cancer, and chronic cholecystitis, as show in TABLE I. Then, computer generates a file named of dataset.txt after preprocessing the numerical variable.

TABLE I
THE DESCRIPTION OF VARIABLES

ID	Variable	Type
1	Disease Category	Discrete
2	Gender	Discrete
3	Age	Continue
4	Occupation	Discrete
5	Blood Type	Discrete
6	Weight	Continue
7	Height	Continue
8	BMI	Continue
9	Smoking History	Discrete
10	Reproductive Number	Discrete
11	Menarche Age	Continue
12	Menstrual Age at last time	Continue
13	Menopausal Age	Continue
14	Menstrual Period (year)	Continue
15	Cholecystic Polypus	Discrete
16	Gallstone	Discrete
17	Gallstone Time	Continue
18	Gallstone Size	Continue
19	Hypertension	Discrete
20	Diabetes	Discrete
21	Coronary Heart Disease	Discrete
22	Hepatitis B	Discrete
23	Family History of Cancer	Discrete
24	Chronic Cholecystitis	Discrete

Via using Bayesialab software [16], all the standalone nodes in dataset can be generated to set up the NB model and TAN model of the gallbladder diagnosis with the learning function of the software. Then, by the modeling method without the target node, factors, which are conditionally dependent to target node, can be screened out to make up the optimal model. And after screening, the optimal BN model is established, including the attribute nodes of gender, age, smoking history, chronic cholecystitis, gallstone and gallstone size. As is shown in Fig.3, (a), (b) respectively represent the NB model and TAN model of the optimal models.

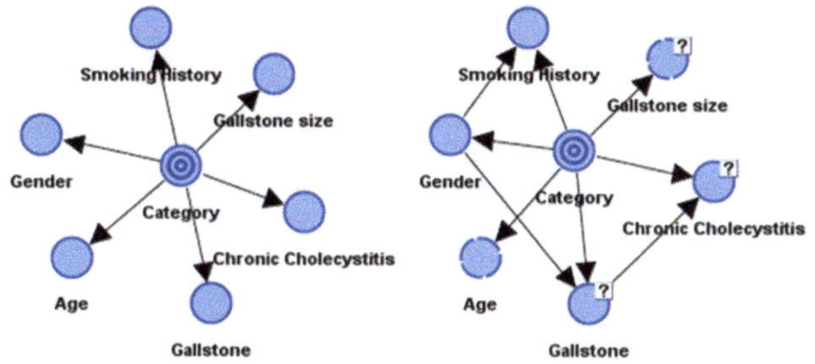

(a) NB (b) TAN

Fig. 3 The establishment of optimal models

B. Accuracy of the optimal models

In order to make sure the accuracy of the optimal models, the established BN models must be evaluated via the classification accuracy which is a standard used to evaluate classifier. Confusion matrix is one of the basic tools to assess the reliability of classifier, as shown in TABLE II and TABLE III, and G.C, H.C, H.H respectively represent the gallbladder cancer, hepatic cyst and hepatic hemangioma. The classification accuracy of BN model classifier of gallbladder diagnosis, based on confusion matrix calculation, is shown in TABLE IV.

TABLE II
THE CONFUSION MATRIX OF NB MODEL

	G.C (802)	H.C (54)	H.H (52)
G.C (805)	799	4	3
H.C (32)	0	25	7
H.H (71)	4	25	42

TABLE III
THE CONFUSION MATRIX OF TAN MODEL

	G.C (802)	H.C (54)	H.H (52)
G.C (803)	799	2	2
H.C (50)	0	31	14
H.H (55)	3	21	36

TABLE IV
ACCURACY OF THE OPTIMAL MODEL

Optimal Model	Accuracy
Naïve Bayes Model	95.26%
Tree Augmented Naïve-bayes Model	95.37%

C. The diagnostic method of the optimal model

Disease category, including gallbladder cancer, hepatic cyst and hepatic hemangioma, can be diagnosed on the condition of the known attribute variables by using the optimal BN models. Concretely speaking, the first step of diagnostic process is to set up the optimal BN model for gallbladder diagnosis based on the dataset, as shown in Fig.3 (a). The second step is to obtain the information of some new patients, including gender, age, smoking history, chronic cholecystitis, gallstone and gallstone size, as shown in TABLE V. The third step is to import this information into the optimal model to get the posterior probability of the disease category, as shown in Fig. 4. The results of new patient's diagnosis, based on the optimal NB model, will be obtained by means of the three steps.

TABLE V
THE BASIC ATTRIBUTE INFORMATION OF NEW PATIENTS

Attribute information	Patient 1	Patient 2
Gender	Female	Female
Age	38	63
Smoking history	False	False
Chronic cholecystitis	False	False
Gallstone	True	False
Gallstone size	30	0

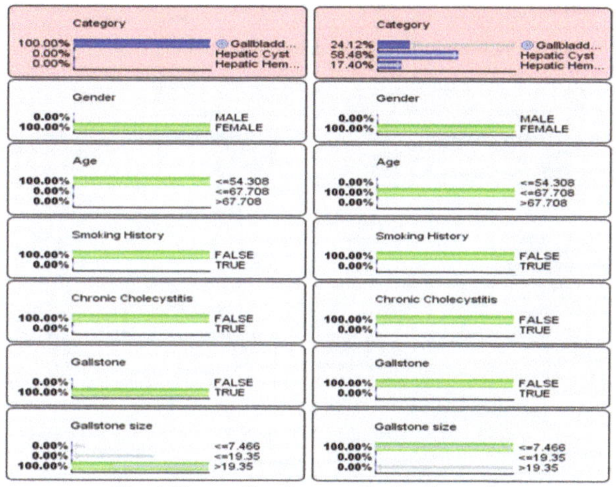

Fig. 4 The results of new patient's diagnosis

IV. IMPORTANCE ANALYSIS OF GBC

Importance measure is an extremely useful tool to reflect the importance of the every attribute information in the process of gallbladder diagnosis. Considering the optimal NB model has nearly the same accuracy with TAN model and simpler structure, the established optimal NB model, concerning gallbladder diagnosis, is selected out to calculate the importance based on the Importance Theory. What's more, the larger the importance absolute value is, the more important the attribute variables are.

Prior probability of the optimal NB model should be calculated at the start. And then, posterior probability of the target node is calculated on the condition that every state of the attribute variable is given. Fig.5 to Fig.10 is the prior probability, posterior probability of each factor.

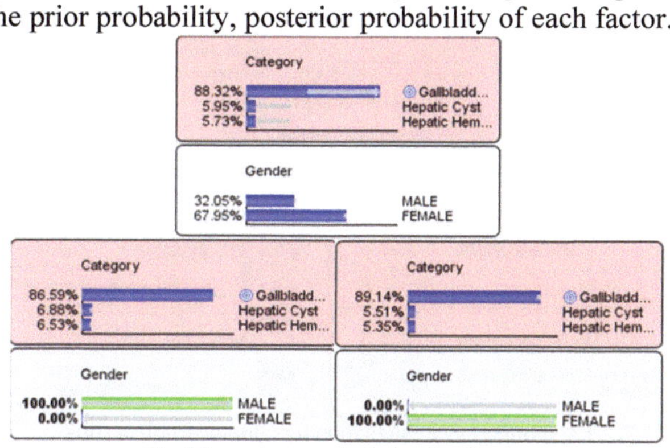

Fig.5 Importance calculation for Gender

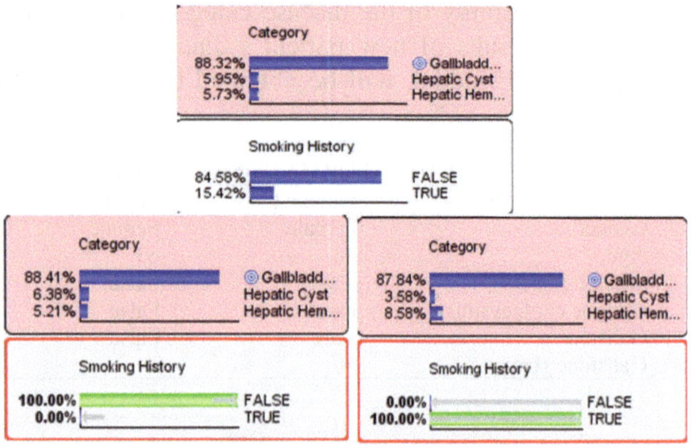

Fig.6 Importance calculation for Smoking History

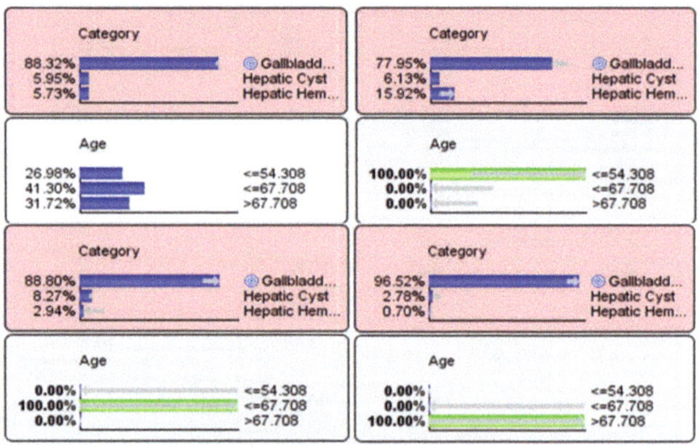

Fig.7 Importance calculation for Age

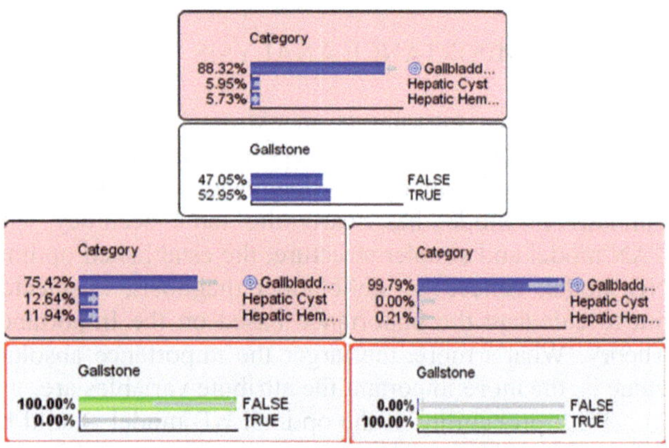

Fig.8 Importance calculation for Gallstone

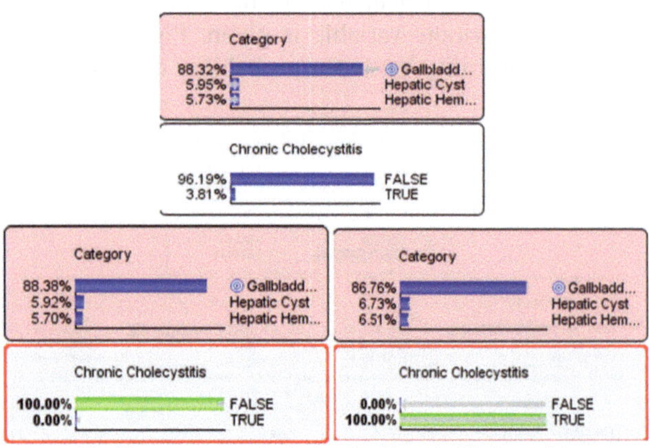

Fig.9 Importance calculation for Chronic Cholecystitis

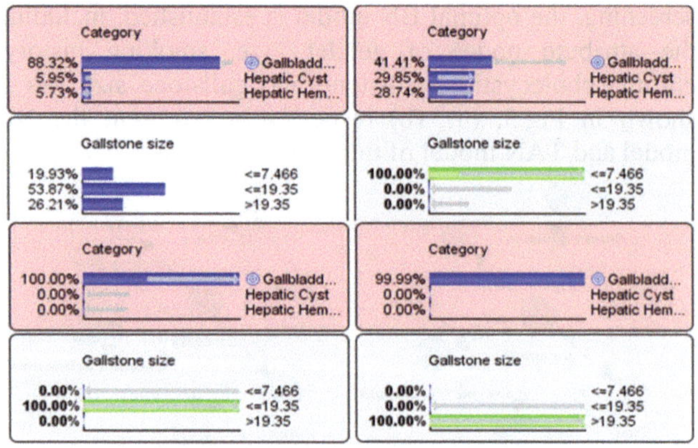

Fig.10 Importance calculation for Gallstone Size

BM importance values can be calculated, according to formula (5), as is shown in TABLE VI. According to the results of importance in TABLE VI, the BM importance values of the gender, age, smoking history, chronic cholecystitis, gallstone and gallstone size are 0.026, 0.186, 0.006, 0.016, 0.244 and 0.586, respectively. Besides, according to the importance measures theory, the larger the BM importance absolute value is, the more important the attribute variables are. In the final results, the absolute values of gallstone and gallstone size, which are 0.244 and 0.586 respectively, are much larger than others. So, in all of the diagnostic factors, doctors should pay more attention to the gallstone and gallstone size.

V. CONCLUSION

This paper introduces the BN and importance measures for gallbladder diagnosis. With the dataset gathered from the first affiliated hospital of medical college of Xi'an Jiaotong University, China, we have constructed the practical BN models by putting the disease category factor as target node. To analyze the importance of different factors, we calculate BM importance values and obtain ultimate results. The final conclusion is that the gallstone and gallstone size must be particularly attentive in the process of gallbladder diagnosis.

For the future research, we plan to collect more clinical records of gallbladder cancer, and build the diagnostic and prognostic models with some other BN learning algorithms, which may increase the effectiveness and efficiency of decision making for doctors.

ACKNOWLEDGMENT

The authors gratefully acknowledge the financial supports for this research from the National Natural Science Foundation of China (Grant Nos. 71471147), the Basic Research Project of Natural Science in Shaanxi Province (No. 2015JQ7273) and the 111 Project (No. B13044).

TABLE VI
Factor distribution and BM importance

Node	State	Prior Distribution	Posterior	BM
Pathological Result	Gallbladder cancer	0.8832		
	Hepatic Cyst	0.0595		
	Hepatic Hemangioma (benign)	0.0573		
Gender	Male	0.3205	0.8659	0.026
	Female	0.6795	0.8914	
Age	<=54.308	0.2698	0.7795	
	<=67.708	0.4130	0.888	0.108
	>67.708	0.3172	0.9652	0.186
Smoking History	TRUE	0.1542	0.8784	0.006
	FALSE	0.8458	0.8841	
Chronic Cholecystitis	TRUE	0.0381	0.8676	0.016
	FALSE	0.9619	0.8838	
Gallstone	TRUE	0.5295	0.9979	-0.244
	FALSE	0.4705	0.7542	
Gallstone Size	<=7.466	0.1992	0.4142	
	<=19.35	0.5387	1	0.586
	>19.35	0.2621	0.9999	0.586

REFERENCES

[1] S. Misra, A. Chaturvedi, N. C. Misra, I. D. Sharma, "Carcinoma of the gallbladder," *The lancet oncology*, vol. 4, no. 3, pp.167-176, 2003

[2] R. Kanthan, J. L. Senger, S. Ahmed, S. C. Kanthan, "Gallbladder cancer in the 21st century," *Journal of oncology*, vol. 2015, 2015.

[3] B. Efron, "Bayesian, frequentist and scientist," *Journal of the American Statistical Association,* vol. 100, no. 469, pp.1-5, 2005.

[4] Mr. Bayes, M. Price, "An essay towards solving a problem in the doctrine of chances. by the late rev. mr. bayes, frs communicated by mr. price, in a letter to john canton, amfrs," *Philosophical Transactions (1683-1775),* vol. 53, pp. 370-418, 1763.

[5] A. Racinc, G. P. Grieve, H. Fluhler, A. F. M. Smith, "Bayesian methods in practice : experiences in the pharmaceutical industry(with Discussion)," *Applied Statistics*, vol. 35, no. 2, pp. 93-150, 1986.

[6] S. B. Si, G.M. Liu, Z. Q. Cai, P. Xia, "Using Bayesian networks to built a diagnosis and prognosis model for breast cancer" *Industrial Engineering and Engineering Management (IE&EM), 2011 IEEE 18Th International Conference on.* pp. 1795-1799, 2011.

[7] Y. Ko, C. X. Zhai, S. Rodriguez-Zas, "Inference of gene pathways using mixture Bayesian networks." *BMC systems biology*, vol. 3, no. 12, pp: 1-16, 2009.

[8] N. A. Loghmanpour, M. K. Kanwar, M. J. Druzdzel, R. L. Benza, S. Murali, J. F. Antaki, "A new Bayesian network-based risk stratification model for prediction of short-term and long-term LVAD mortality". *ASAIO Journal*, vol. 61, no. 3, pp. 313-323, 2015.

[9] N. Friedman, D. Geiger, M. Goldszmidt, "Bayesian network classifiers," *Machine Learning,* vol. 29, pp. 131-163, 1997.

[10] S. B. Kim, K. S. Han, H. C. Rim, S. H. Myaeng, "Some effective techniques for naive bayes text classification," *Knowledge and Data Engineering, IEEE Transactions on,* vol. 18, no. 11, pp. 1457-1466 , 2006.

[11] R. O. Duda, P. E. Hart, N. J. Nilsson, "Subjective Bayesian methods for rule-based inference systems," *Proceedings of the June 7-10, 1976, national computer conference and exposition.* ACM, pp. 1075-1082, 1976.

[12] F. Pernkopf, J Bilmes, "Discriminative versus generative parameter and structure learning of Bayesian network classifiers," *Proceedings of the 22nd international conference on Machine learning.* ACM, pp. 657-664, 2005.

[13]J. E. Ramirez-Marquez, D. W. Coit, "Composite importance measures for multi-state systems with multi-state components," *Reliability, IEEE Transactions on,* vol. 54, no. 3, pp. 517-529, 2005.

[14] S. F. Song, Z. Z. Lu, W. Li, L. J. Cui, "The uncertainty importance measures of the structural system in view of mixed uncertain variables," *Fuzzy Sets and Systems*, vol. 243, pp. 25-35, 2014.

[15] S. B. Si, L. Yang, Z. Q. Cai, "A New and Efficient Computation Method of IM(Integrated Importance Measures) for Components in Binary Coherent Systems," *journal of Northwestern Polytechnical University,* vol. 29, no. 6, pp. 939-947, 2011.

[16] S. Conrady, L. Jouffe, "Bayesian Networks and BayesiaLab: A Practical Introduction for Researchers," *Bayesia USA*, 2015.

The Effect of Dynamic Capabilities on International Performance of Chinese Universities

Dan Xiang*
School of Foreign Languages, Northwestern Polytechnical University, Xi'an, China
(*xiangd@nwpu.edu.cn)

Abstract –**This paper seeks to apply the dynamic capabilities framework to examine the internationalization of higher education, specifically, the effect of dynamic capabilities on university's international performance. An empirical data analysis has been employed, and hypotheses are tested on Chinese 985 and 211 universities. The results indicate that dynamic capabilities have a significant positive effect on the university's international performance. In particular, learning mechanisms has the most prominent positive effect on university's international performance. Universities must be fully aware of the importance of dynamic capabilities when dealing with the rapidly changing environment, and should develop the university's international dynamic capabilities especially their learning mechanisms to enhance their international performance.**

Keywords – **Dynamic capabilities, higher education, international performance**

I. INTRODUCTION

Internationalization of higher education is a trend across the globe, which is also an inevitable path for Chinese universities. The external environment for Chinese universities are changing rapidly. The global market, the Chinese government and the universities have become the three main drives for the international development of Chinese universities. In particular, the Chinese government has incorporated more functions for the universities' international development. In 2010, it issued *The Guideline of National Mid-Term and Long-Term Education Reform and Development (2010-2020)*, requiring universities to learn from their oversea counterparts' new concept and experiences to educate talented students with international perspectives and competitiveness. In 2013, the designing of *New Silk Road Initiative* brought the requirements further, from one way of education import to a two-way model with both education import and export. In 2015, the government proposed the *First-class University and First-class Discipline Initiative*, with the purpose of building world's leading universities and disciplines by solving the current problems in the solidification of universities' identities, repeated infrastructure and equipment investment, and the lack of international competitiveness.

Chinese universities consider the process of internationalization as a core contributory factor in the university's development. They have undertaking considerable international activities in student's exchange, research collaboration, joint educational programs, etc. With the expansion of their international activities, many of them have realized the significance of strategic management, for instance, the strategic planning and its implementation in their internationalization process, rather than to take the international activities in a random way. Since Tianjin University issued its strategic planning for internationalization in 2010, many Chinese universities started to make their own international strategy planning and completed their first step of strategic management in their international development. However, because little research has been done on the strategic management in the internationalization of higher education, problems occurred during the implementation of the strategic planning. Many university found it hard to carry out those international strategic planning because of the changing environment which made it difficult to apply their static planning to the real changing world.

This paper chooses dynamic capabilities framework as a possible approach to deal with the problems of a static strategic planning in a changing environment. It examines the effect of dynamic capabilities in the internationalization of higher education with the purpose of finding possible solutions to enhance the international performance.

II. LITERATURE REVIEW

A. Internationalization of higher education

Internationalization of higher education is not a new concept and it appeared with the born of the first universities in the world, for instance, the University of Bologna in 1158 and University of Paris in 1180. They were the world's knowledge centers in the Middle Ages attracting students and teachers from across the European continent. The internationalization of universities experienced five main periods before the globalization, the Middle Ages, the Reformation period, the Industrial Revolution period, the World War II period, and the Cold War period [1]. However, the most prominent one is in 21st century, when the nation, university, and the global market become the three main drives of the internationalization of higher education [2].

Many research has been done on the internationalization of higher education. Most of them focus on its concept [3, 4], elements effecting university's internationalization [5, 6], drives of university's

E. Qi et al. (eds.), *Proceedings of the 23rd International Conference on Industrial Engineering and Engineering Management 2016,* DOI 10.2991/978-94-6239-255-7_49

internationalization [7], the evaluation system [2], comparative study of higher education [8], the path for internationalization [9], as well as the outlook of internationalization [10]. But little research has been done on the strategic management of the internationalization of higher education.

B. *University's international performance*

The research on university's international performance often appears in the form of evaluation system of a university's internationalization at a certain point. Organizations and researchers, for instance, the OECD, the International Association of Universities of UNESCO, the American Council on Education, and the Canadian research Jane Knight, started the study of the evaluation system in the mid of 1990's. The most accepted one is Jane Knight's system which can match different situations in different countries with proper adjustments.

In her system, Knight (2001) proposed eight dimensions that should be taken into consideration in the evaluation of a university's internationalization, including the external international environment, university's international strategies and policies, organizational structure and supporting infrastructure for international development, international academic curriculum development and international educational programs for students, university's international links and network, international research collaboration, international human resources management, as well as international contracts and services [2].

A widely recognized evaluation system among Chinese university is the one developed by the Southwest Jiaotong University. It issued the first Internationalization Ranking for Chinese University in 2013 measured by international exchange, joint educational programs, international research collaboration, cultural exchanges, and supporting structures, infrastructure, and investment for internationalization.

As for the definition of performance, there are three different views. Campbell (1991) regarded performance as behavior, which should be separated from result [11]. Kane (1996) considered performance as result, which should be closely connected to the objectives of an organization [12]. Brumbrach (1988) took both behavior and result into consideration as the components of performance, and found that behavior was one of the prerequisites for result [13].

Since this paper is going to examine the correlation between dynamic performance and university's international capabilities, performance will be regarded as result while dynamic capabilities as behavior. We could see from the evaluation system mentioned above that the international resources, international capabilities, and the universities international performance are all included, which can be divided into two categories, behavior and results. In this paper, the international performance will only examine the results.

C. *Dynamic capabilities framework*

Dynamic capabilities framework is a concept in strategic management, which is to answer the question of how firms or organizations achieve and sustain their competitive advantage in a rapidly changing environment [14]. Dynamic capabilities framework has become the effective and promising approach for firms to gain competitive advantages in the regimes of rapid changes, where the core capabilities framework, the used dominant paradigm, failed to build competitive advantages in a rapidly changing environment.

Many studies have been done in applying dynamic capabilities framework to firm performance. Makadok (2001) studied the mechanisms of dynamic capabilities on firm performance, and the result showed that there was a direct cause-effect relation between dynamic capabilities and firm performance [15]. Hitt et al. (2001) found that dynamic capabilities improved the effectiveness and efficiency of the firms' reaction to the changing environment and thus enhanced the firm performance [16]. Zollo and Winter (2002) also found that without dynamic capabilities the competitive advantages of a firm would be temporary, and it would be impossible for firms to create sustainable growth in a changing environment [17].

Dynamic capabilities framework has been applied widely to improve firm performance but not yet in the field of higher education nor in the international development of higher education. This paper is going to explore whether there is any positive effect of dynamic capabilities on university's international performance.

III. THEORETICAL HYPOTHESES

A. *Description of variables*

(1) Description of dynamic capabilities

This study looks into the effects of dynamic capabilities on university's international performances. Therefore, the dependent variable would be university's international performance, while the independent variable would be the dynamic capabilities.

As for dynamic capabilities, it was divided into four dimensions based on the studies of Caloghirou et al. (2004), Teece (2007), and Barreto (2010), including international environment detecting capabilities, learning mechanisms, mediating and integrating capabilities, as well as restructuring and innovating capabilities [18, 14, 19]. The description of the four dimensions of dynamic capabilities in this paper is shown in Table I.

TABLE I
DESCRIPTION OF DYNAMIC CAPABILITIES

Variable	Symbol	Description
International environment	IEDC	The capability to detect the changing of the international environment which may

detecting capabilities		affect the international development of the university
Learning mechanisms	LM	The mechanisms for the university to learn knowledge from the external environment and for the flow of knowledge inside the organization
Mediating and integrating capabilities	M&IC	The capability to mediate the organization's internal sectors with the changing external international and domestic environment, as well as the capabilities to integrate resources among those sectors related to the university's international development
Restructuring and innovating capabilities	R&IC	The capabilities to restructure the organization in terms of international development to meet the changing environment and the ability of innovation

(2) Description of university's international performance

As for the university's international performance, it was divided into three dimensions based on the studies of Knight (2001), Yang (2011) and Ding (2014), including international reputation, schooling internationalization and discipline internationalization [2, 20, 21]. The description of university international performance in this paper is shown in table II.

TABLE II
DESCRIPTION OF UNIVERSITY'S INTERNATIONAL PERFORMANCE

Dimension	Symbol	Description
International reputation	IR	University's intentional reputation, measured by the peers' comments, the satisfactory of oversea teachers and students, internationally influential research outcome, and numbers of faculty members who play important role in internationally renowned organizations and journals
Schooling internationalization	SI	University's international schooling capabilities, for instance, the number of English courses, the number of overseas students, etc.
Discipline internationalization	DI	The international development of the disciplines, for instance, the total citation of the papers, international patents, etc.

B. Hypothesis

The paper applies the dynamic capabilities framework to university's internationalization, and specifically it examines the effect of dynamic capabilities on university's international performance. 12 hypotheses will be proposed in this paper.

(1) The effect of international environment detecting capabilities on university's international performance

Environment detecting capabilities is the ability to detect the changing environment and the ability to rapidly figure out the demand of their customers to collect and analyze the competitor developing trends and market information in the industry [22, 23].

The detecting of opportunities and threatens from the external environment is an important prerequisite for an organization to make their new strategy or to adjust their strategy to fit the changing environment for the purpose of an improved performance [24, 25, 26].

As for university's international development, the capabilities of detecting the international environment would assist the university to analyze the external changes and to make appropriate adjustments or changes in its strategies to improve its international performance. Therefore, the following four hypotheses have been proposed.

H1 International environment detecting capabilities have significant positive effects on university's international performance

H1a International environment detecting capabilities have significant positive effects on international reputation

H1b International environment detecting capabilities have significant positive effects on the internationalization of schooling

H1c International environment detecting capabilities have significant positive effects on the internationalization of discipline

(2) The effect of learning mechanisms on university's international performance

Caloghirou et al. (2004) regarded organizational learning capability as one of the three sub-capabilities of dynamic capabilities, related to effective team work, training programs, and other learning process [27]. Teece (2007) found that learning mechanisms and knowledge management was of great significance in the improvement of the performance of an organization, where learning, sharing knowledge and the integration of knowledge was the fundamental for the formation of dynamic capabilities [14].

In terms of university's international development, learning mechanisms could assist the university to learn knowledge from the external environment on a regular basis and to keep a smooth flow of the newly learned knowledge among the internal sectors and departments. Therefore, the following four hypotheses have been proposed.

H2 Learning mechanisms has significant positive effects on university's international performance

H2a Learning mechanisms has significant positive effects on international reputation

H2b Learning mechanisms has significant positive effects on the internationalization of schooling

H2c Learning mechanisms has significant positive effects on the internationalization of discipline

(3) The effect of mediating and integrating capabilities on university's international performance

Eisenhardt & Martin (2000) found that dynamic capabilities were the ability of mediating and integrating resources and capabilities to guarantee that the firm's competitiveness could keep pace with the changing environment [26]. Cao et al. (2009) found that mediating and integrating capabilities were integrated parts of dynamic capabilities, which were to mediating the collaboration and the distribution of task among the internal sectors with the external changing environment [28].

As for a university, it is significant to mediate its internal relationship and resources among sectors, in particular in terms of international development which is not possible to be done in one sector but needs collaboration among the administrative sectors such as undergraduate office, graduate school, international office, human resources office and campus service, as well as the support and collaboration from academic schools.

Therefore, the following four hypotheses have been proposed.

H3 Mediating and integrating capabilities have significant positive effects on university's international performance

H3a Mediating and integrating capabilities have significant positive effects on international reputation

H3b Mediating and integrating capabilities have significant positive effects on the internationalization of schooling

H3c Mediating and integrating capabilities have significant positive effects on the internationalization of discipline

(4) The effect of restructuring and innovating capabilities on university's international performance

Caloghirou et al. (2004) investigated the restructuring and innovating capabilities and found that restructuring the firm's asset was a must-have process to meet the changing environment [27]. Teece (2007) regarded restructuring and innovating capabilities as a key process in promoting the firm's efficiency and maintaining the firm's long-term interest with the development of the firm, the changing environment and the upgrade of technology [14].

In this paper, the innovation and change of a university's international resources, technology, as well as the processes and convention of the university's organizational structure are all considered as the processes of restructure and innovation. The following four hypotheses have been proposed.

H4 Restructuring and innovating capabilities have significant positive effects on university's international performance

H4a Restructuring and innovating capabilities have significant positive effects on international reputation

H4b Restructuring and innovating capabilities have significant positive effects on the internationalization of schooling

H4c Restructuring and innovating capabilities have significant positive effects on the internationalization of discipline

IV. EMPRICAL ANALYSIS AND RESULTS

A. Data Collection

The sample universities selected were Chinese 985 and 211 universities, who have a relatively better international development after several rounds of investment by Chinese government for their international development, and could reflect the variables selected in this paper.

The questionnaires were distributed during the period of October 2015 and December 2015 to their administrative departments closely linked to the university's international development, for instance, the international offices, graduate schools, undergraduate schools, human resources offices, and science and technology administrative departments. Questionnaires were also distributed to those faculty members and students who were actively engaged in international activities. The number of universities investigates was 38 and the total number of questionnaires distributed was 776, among which 390 were distributed in person on campus and 386 via e-mail. The valid questionnaires collected were 438, with 299 from the former distribution way and 139 from the later one. The approach of *t* test was applied to figure out the possibility of merging the two sets of data collected, and the result showed that they could be merged as one data set.

B. Descriptive statistics of variables

TABLE III
DESCRIPTIVE STATISTICS OF INTERNATIONAL DYNAMIC CAPABILITIES

variable	N	Min	Max	Mean	Std. dev
IEDC1	438	1	5	3.59	1.041
IEDC2	438	1	5	3.73	.958
IEDC3	438	1	5	3.69	.963
IEDC4	438	1	5	3.50	1.025
LM1	438	1	5	3.50	1.067
LM2	438	1	5	3.85	.904
LM3	438	1	5	3.57	1.025
LM4	438	1	5	3.43	.991
LM5	438	1	5	3.68	1.184
M&I1	438	1	5	3.62	1.054
M&I2	438	1	5	3.42	1.131
M&I3	438	1	5	3.41	1.065
M&I4	438	1	5	3.76	.979
R&I1	438	1	5	3.61	1.070
R&I2	438	1	5	3.47	1.056
R&I3	438	1	5	3.52	1.147
R&I4	438	1	5	3.47	1.060
R&I5	438	1	5	3.62	1.014

TABLE IV
DESCRIPTIVE STATISTICS OF UNIVERSITY'S INTERNATIONAL PERFORMANCE

variable	N	Min	Max	Mean	Std. dev
IR1	438	1	5	3.32	1.084
IR2	438	1	5	3.52	1.111
IR3	438	1	5	3.63	1.061
IR4	438	1	5	3.04	1.172
IR5	438	1	5	3.06	1.198
SI1	438	1	5	3.29	1.094
SI2	438	1	5	3.21	1.114
SI3	438	1	5	2.91	1.237
SI4	438	1	5	3.18	1.127
SI5	438	1	5	3.05	1.182
DI1	438	1	5	3.36	1.068
DI2	438	1	5	3.20	1.150
DI3	438	1	5	2.94	1.177
DI4	438	1	5	2.95	1.196

C. *Correlation coefficient analysis*

Pearson correlation coefficient analysis was applied to analyze the relation between dynamic capabilities and the university's international performance. The results show that there is a significant correlation between dynamic capabilities and university's international performance, which has preliminarily verified the impact of dynamic capabilities on university's international performance as shown in table V.

TABLE V
PEARSON CORRELATION COEFFICIENT ANALYSIS

		Dynamic capabilities	University's international performance
Dynamic capabilities	Pearson correlation coefficient	1	.607**
	Sig.		.000
	N	438	438
University's international performance	Pearson correlation coefficient	.607**	1
	Sig.)	.000	
	N	438	438

D. *The impacts of dynamic capabilities on university's international performance*

The effects of dynamic capabilities on university's international performance was analyzed, and the results were shown in table VI.

TABLE VI
THE IMPACTS OF DYNAMIC CAPABILITIES ON UNIVERSITY'S INTERNATIONAL PERFORMANCE

	std. regression coefficient	t value	sig.
IEDC-->UIP	0.484	11.549	0.000
IEDC --> IR	0.433	10.044	0.000
IEDC -->SI	0.460	10.806	0.000
IEDC -->DI	0.402	9.115	0.000
LM -->UIP	0.573	14.584	0.000
LM --> IR	0.503	12.166	0.000
LM --> SI	0.555	13.945	0.000
LM --> DI	0.472	11.179	0.000
M&IC -->UIP	0.527	12.950	0.000
M&IC--> IR	0.464	10.933	0.000
M&IC--> SI	0.522	12.770	0.000
M&IC --> DI	0.421	9.690	0.000
R&IC -->UIP	0.527	12.936	0.000
R&IC --> IR	0.457	10.731	0.000
R&IC --> SI	0.539	13.345	0.000
R&IC --> DI	0.407	9.315	0.000

From table VI, the results show that the all the four dimensions of dynamic capabilities have significant impacts on university's international performance. In other words, all the 16 hypotheses were verified.

It can be seen that there are differences among the impacts of international environment detecting capabilities, learning mechanisms, mediating and integrating capabilities, and restructuring and innovating capabilities on the university's international performance, which are 0.484, 0.573, 0.527, and 0.527 respectively. The impact of learning mechanisms on university's international performance is the highest among the four dimensions of dynamic capabilities, followed by the impacts of mediating and integrating capabilities, and restructuring and innovating capabilities, while the impact of international environment detecting capabilities is the least one among the four. In addition, we could see from table VI that dynamic capabilities have the highest impact on the university's schooling internationalization compared with international reputation and discipline internationalization.

V. CONCLUSION

The findings indicate that dynamic capabilities increase university's international performance, and that learning mechanisms have the highest positive effects on university's international performance. In addition, dynamic capabilities have the highest effects on the university's schooling internationalization.

It can be concluded that dynamic capabilities have played an important role in promoting the university's international performance. With this study, it has made it clear that it is because of the lack of dynamic capabilities that have caused the problems during the process of implementing the university's international strategy. Therefore, university leaders must be fully aware of the importance of dynamic capabilities and be able to develop dynamic capabilities from the four dimensions as developed in this paper if they are to deal with the rapidly changing environment they are facing. In addition, learning mechanisms have played the most important part in promoting the university's international performance. Therefore, it takes the priority when the university is developing their dynamic capabilities.

ACKNOWLEDGMENT

The author thanks the sponsor and financial support of Higher Educational Research Fund of Northwestern Polytechnical University (2015GJZ02).

REFERENCES

[1] Huang, F. Internationalisation of higher education in the era of globalisation: What have been its implications in China and Japan? [J]. Higher Education Management and Policy, 2007, 19 (1): 1-15.
[2] Knight, J. Monitoring the quality and progress of internationalization [J]. Journal of Studies in International Education, 2001(5): 228-243.
[3] de Wit, H. On the Definition of International Education[J].European Association forInternational Education Newsletter, 1993(11): 7-10.
[4] Knight, J. Updated internationalization definition[J]. International Higher Education, 2003(33): 2-3.
[5] Harari, M. Internationalization of Higher Education, Effecting Institutional Change in the Curriculum and Campus. Long Beach, California: Center for International Education, California State University, 1989.
[6] Knight, J. Internationalization Remodeled: Rationales, Strategies and Approaches [J]. Journal for Studies in International Education, 2004, 8(1): 5-31.
[7] Altbach, P.G., Knight, J. The Internationalization of Higher Education: Motivations and Realities [J]. Journal of Studies in International Education, 2007(11): 290-307.
[8] Altbach, P.G. Trends in Comparative Education [J]. Comparative Education Review, 1991, 35(3): 491-507.
[9] Davies, J.D. University strategies for internationalisation in different institution and cultural settings: A conceptual framework, in Policy and policy implantation in internationalization of higher education, 1995: 3-18.
[10] Liu, Y.H., Wang, Q.L. University International Consortium: A New Choice of Chinese University Internationalization Strategies [J]. Journal of Beijing Institute of Technology (Social Sciences Edition), 2012(6): 150-153.
[11] Campbell, J.L. and Lindberg, L.N. The Evolution of Governance Regimes. pp. 319-355 in Governance of the American Economy [M]. New York: Cambridge University Press, 1991.
[12] Kane, M.B., Mitchell, R. Implementing performance assessment: Promises, problems, and challenges [M]. Mahwah, NJ: Lawrence Erlbaum Associates, 1996.
[13] Brumbrach. Performance management [M]. London: The Cronwell Press, 1988:15.
[14] Teece, D.J. Explicating dynamic capabilities: the nature and microfoundations of (sustainable) enterprise performance [J]. Strategic Management Journal, 2007, 28(13): 1319-1350.
[15] Makadok, R. Toward a synthesis of the resource-based and dynamic-capability views of rent creation [J]. Strategic Management Journal, 2001, 22(5): 387–401.
[16] Hitt, M.A., Bierman, Shimizu, K., Kochhar, R. Direct and Moderating effects of human capital on strategy and performance in professional service firms: a resource-based perspective. Academy of Management journal, 2001, 44(l): 13-28.
[17] Zollo, M., Winter, S. Deliberate learning and the evolution of dynamic capabilities[J]. Organization Science, 2002, 13(3): 339-351.
[18] Caloghirou, Y., Protogerou, A., Spanos, Y., Papagiannakis, L. Industry-Versus Firm-specific Effects on Performance: Contrasting SMEs and Large-sized Firms[J]. European Management Journal, 2004, 22(2): 231-243.
[19] Barreto, I. Dynamic Capabilities: A Review of Past Research and an Agenda for the Future[J]. Journal of Management, Jan 2010, 36(1): 256-280.
[20] Yang, F.L. A Study on the Development and Management of Internationalization of Higher Education [D]. Tianjin: Tianjing University, 2011: 1-122.
[21] Ding, S.C. The Study on Motivation, Pattern and Performance of Internationalization about Research Universities in China [D]. Anhui: University of Science and Technology of China, 2014: 1-133.
[22] Galunic, D.C., Rodan, S. Resource recombination in the firm: Knowledge structures and the potential for Schumpeterian innovation[J]. Strategic Management Journal, 1998, 19(12): 1193-1201.
[23] Kogut, B., Zander, U. What firms do? Coordination, identity, and learning [J]. Organization Science, 1996, 7(5): 502-518.
[24] Helfat, C.E., Eisenhardt, K.M. Inter-temporal Economies of Scope, Organizational Modularity, and the Dynamics of Diversification[J]. Strategic Management Journal, 2004, 25(13): 1217–1232.
[25] Itami, H., Roehl, T. Mobilizing Invisible Assets[M]. Harvard University Press, Cambridge, MA, 1987.
[26] Eisenhardt, K., Martin, J. Dynamic capabilities: what are they?[J]. Strategic Management Journal, 2000, 21(10-11): 1105–1121.
[27] Caloghirou, Y., Protogerou, A., Spanos, Y., Papagiannakis, L. Industry-Versus Firm-specific Effects on Performance: Contrasting SMEs and Large-sized Firms[J]. European Management Journal, 2004, 22(2): 231-243.
[28] Cao, H.J., Zhao, J.B., Wang, Y.H. Dimensions of dynamic capabilities and path analysis: an empirical research based on China enterprise [J].Study in Science of Science, 2009, 27(1): 36-44.

Researching The Role Of Online Movie Reviews Based On User Classification

Yin-ling GUO, Xian-guo ZHANG*, Zhi-peng SHEN, MESSAN Barthelemy
College of Computer Science, Inner Mongolia University, Hohhot, China
(*E-mail address: cszxg@imu.edu.cn)

Abstract - **This paper examines the relationship between the online reviews with movie box office under the environment of panel data. According to users' behavior, we divide the user reviews into two categories: average user reviews and critic reviews, then compare the influence of different types of reviews on the box office. We also analyze the relationship between box office and factors, including movie rank, release days, count of reviews and scores. Our results show that average user reviews have an obviously positive influence on box office, while critic reviews have no effect. The count of average user reviews can be helpful for predicting the movie revenues, rather than critic reviews. Thus the movie rank and release days will provide the reference for consumers to evaluate the quality of movies in daily life. The article points out a new direction for further research on online reviews.**

Keywords - **Average user reviews, box office, critic reviews, movies, online reviews**

I. INTRODUCTION

Domestic or international, there are a lot of literature about online reviews, which from different angles, using different methods, in the view of the different phenomenon, theory support, based on various data sets, etc., and researches about movies online reviews are more than count.

This paper tries to draw lessons from previous research, exploring the relationship during various influence factors of the movie. In the daily recommendations, people usually choose the movie according to the score, the box office and the rank, which is considered to represent the quality. For some patriotic movies with lower score but good quality, it may be caused by a lack of advance publicity or late without lots of reviews on diffusion, finally leads the low revenue. However, there are also traditional culture theme movies with low score and high box office. Therefore, it is important to research online movie reviews.

There are different accounts about researches on online reviews, someone put the user rating as endogenous variables to measure in the study, the results show that there is no obvious effect on movies[1]. The other research has shown that critic rating is the predictor of movies. It also shows that online reviews from consumers are more reference value than critic[2]. In view of many more contradict points, we take the online movie reviews as the research lines, to consider the internal relations with the box office. Our works focus as follows, first we draw the user reviews is divided into average user reviews and critic review, and compare their different effects on box office. Then, with the limitation of data, in the study, we contrast the role of two reviews just using 1-week data and research the relationship between user reviews and other influencing factors using 4-week data. Last, grouping by the movie total score, we discuss the relationship among movie rank, score and revenues.

There are some contributions as the following, firstly, with classifying and processing user reviews, which is a kind of inspiration for future research, and it can be directly applied to other areas. Secondly, the research shows that, average user reviews are more influence than critic reviews to the box office, indicating that people tend to be more acceptable with who has similar experiences, in order to reduce the purchase risk, so businesses can pay more attention to this kind of reviews, rather than hiring critic painstakingly to do excessive publicity. Thirdly, our experiments verify the relationship during the rank, score, release days and box office, indicating that, when assessing the quality of the movie, we should carry comprehensive consideration with above multiple indexes.

II. LITERATURE REVIEW

With the advent of the era in Internet and the rapid development of e-commerce industry, more and more products are shifted online from offline, forming new trading patterns, which combines with the online and offline. As online transaction products are mostly unable to understand the quality in advance, which leads more and more customers rely on other people's reviews to sense the quality of the products. The online reviews involve a wide range of areas, including movies, books, TV shows, DVD, videos and foods, catering and other supplies and services. Here we select the movie as object, and try to make some comparisons with the previous studies.

In the movie industry, the online reviews become one of the key factors for the final success of box office. Many previous studies on the impact of online reviews on product

E. Qi et al. (eds.), *Proceedings of the 23rd International Conference on Industrial Engineering and Engineering Management 2016,* DOI 10.2991/978-94-6239-255-7_50

sales with the empirical analysis, they have shown that, the count of online reviews has significant positive correlation with product sales, and they influence each other[1]. Some studies have shown that there is a causal relationship between reviews and sales. The average grade of online reviews is decreasing over the time[3]. There is prediction effect of user reviews, which does not affect the box office performance, and there is no correlation with the early box office, only related to the accumulative box office[2]. Researches of box-office revenues, respectively select weekly and daily revenue as the research object, in the opening week, there is great volatile among daily box office and the number of online reviews. Here, we choose the unit to study the influence during various factors.

With the different versions about the influence of online user reviews on consumers, someone considers that review information from critic is more influence on consumers[4], and another think it is more significant influence on himself from other user reviews, because it was easier to form a consensus among users[5,6]. Most literature of online reviews, they may only use the user reviews, and less one divide them into average user reviews and critic review, without explicitly pointed out that the difference between the two, this article specifically indicate user review is the ordinary users, then respectively obtain both reviews information from the professional website.

Considered that the cross-sectional data modeling could not capture the individual differences and also unfavorable to reflect the dynamic change of the influence of online reviews on box office revenue over time, therefore cannot reflect the real relationship during variables. Different from Liu[7] and many others, here we use the panel data to give detailed exploration for the dual relationship between online reviews and box office revenue.

III. HYPOTHESES

A. Hypotheses of user reviews and critic review

Many researches have shown that person often attaches importance to the opinions of others while ignores his own information in a large extent[8], which conforms to the herd mentality and authority effect of psychology. About purchasing behavior, the study verifies that, nearly half of persons say that they would refer to the reviews of other before consumption, and more than half think that consumers reviews are more valuable than critic[2]. In general, in order to reduce the purchase risk, consumers will search for the similar scene. They believe that critic reviews are too official, as the environmental difference, the credibility of the reviews is to be confirmed. However, some researches indicate that the higher credibility of the reviewers, the greater impact of online reviews on consumer purchase decisions[9],especially the professional degree of critics[10]. Consumers can get more professional knowledge from the critic review, as well as determining the feasible solution, it can be known that consumer has a strong dependence on critic review. Even, studies show that critic review has positive effects on movie and can be used to predict, especially in the early release days[11,12,13]. So we can make the following hypotheses:

H1a. Average user review has obviously positive influence on box office revenues.

H1b. Critic review has obviously positive influence on box office revenues.

H1c. The influence of average user review is greater than critic review on box office revenues.

With analyzing data of the first week, we find that the count of critic reviews will achieve a peak in the movie release, which is same with the box office, we guess that critic review can be used to predict its box office. The count of user reviews is very consistent with box office, thus we hypothesize:

H1d. Average user review and critic review both can predict the box office revenues.

B. Other hypotheses

Previous researches have verified some important variables related to the movie, in addition to the count of online reviews, which is positively correlated with box office, there is also significantly positive correlation with screen theaters and box office, the box office is decreasing with time, that is, release days have a negative correlation relationship with box office[14,15]. According to the daily box office performance, we can observe that box office will increase to the peak at weekend, thus we guess the holiday effect should be similar. As the life experience, movie's daily rank, score, count of reviews, gross to date will affect the box office at the next day, however, there is also a certain rule of movies rank, score and count of reviews, in order to improve the explanatory power for box office, we also take these variables into the model of preliminary inspection, then we derive the following hypothesis:

H2. Whether it is a weekend or holiday both has significantly positive influence on box office revenues.

H3a. The cumulative gross has a positive influence on box office revenues.

H3b. The movie rank and the count of release days both have negative influence on box office revenues.

H3c. The count of user review and score both have positive influence on box office revenues.

IV.RESEARCH METHODOLOGY

A. Data

The data for this study was collected from three sources. First, Movieslens datasets, is to get more than 300 movies released between June 2014 and July 2015.Second, the BoxOfficeMojo.com, is to get the movie box office information. Third, the database IMDB (www.imdb.com) is for movie reviews. After screening for many times, and eliminating some serious lack of movie box office and reviews, finally 66 movies are selected as the sample.

For each movie, by the title and release date, more information are collected from Mojo.com, such as daily gross, theaters, average revenue per theater, daily rank, weekly rank, daily revenues, gross to date etc., and the information is collected from the IMDB, each user review's ID, release date, overall grade, and detailed reviews. Original user score is from 0 to 10, and the critic score is from 0 to 100, to ensure the accuracy of data, we keep the two types the same as the later.

From the perspective of users, who is divided into average user and critic. The revenues are represented as daily gross and cumulative gross. According to the life cycle theory, general movie screening cycle is from 8 to 10 weeks, most researches

TABLE I

VARIABLES, DESCRIPTIONS AND MEASURES

Variable	*Description and measure*
$gross_{it}$	Daily gross for movie *i* in day *t* (in thousands, US dollars)
$grosstodate_{it}$	Cumulative gross for movie *i* until day *t* (in thousands, US dollars)
$userscore_{it}$	Average user score for movie *i* in day *t*
$usercount_{it}$	The count of average user reviews for movie *i* in day *t*
$criticscore_{it}$	Critic score for movie *i* in day *t*
$criticcount_{it}$	The count of critic reviews for movie *i* in day *t*
age_i	The release days for movie *i* ,equals with t
$rank_{it}$	The movie rank for movie *i* in day *t*
$weekend_{it}$ $holiday_{it}$	A dummy variable indicating if day *t* is a weekend or holiday(codes as 1 if the day is weekend or holiday,0 otherwise)

TABLE II

WEEK 1 DESCRIPTIVE STATISTICS

Variable	*Obs*	*Mean*	*Std. Dev.*	*Min*	*Max*
$gross_{it}$	462	888.84	1102.5	4.43	8442.45
$grosstodate_{it}$	462	4547.54	4524.99	5.15	29621.17
$usercount_{it}$	462	15.63	20.88	0	178
$userscore_{it}$	462	66.16	17.02	0	100
$criticcount_{it}$	462	4.98	6.07	0	29
$criticscore_{it}$	462	50.08	26.96	0	100
$rank_{it}$	460	2.03	1.28	1	8
age_i	462	4	2	1	7

TABLE III

WEEK 4 DESCRIPTIVE STATISTICS

Variable	*Obs*	*Mean*	*Std. Dev.*	*Min*	*Max*
$gross_{it}$	1848	428.483	687.62	4.43	8442.45
$grosstodate_{it}$	1848	8781.34	8291.11	5.15	57253.83
$usercount_{it}$	1848	8.01	13.21	0	178
$userscore_{it}$	1848	59.01	24.96	0	100
$rank_{it}$	1842	4.35	3.16	1	16
age_i	1848	14.5	8.08	1	28

select 8 weeks[7] to research the dynamic effect of the online reviews. Literature[1] takes the first two weeks as the object, to capture dynamic relationship between online reviews and the box office. Combining with the actual situation, some movie with good quality and reputation will be relatively release for longer time, and some with poor reputation, its release time may be less than 8 weeks, we finally select 4 weeks as the screening cycle. The table I provide the descriptions and measurement of the key variables, the table II and table III present descriptive statistics respectively for the 1 week and 4 week data.

B. Model

As the above hypotheses and the data, in order to verify the relationship between the two kinds of reviews, we establish the initial panel data model as equation (1) shows.

$$\begin{aligned} gross_{it} &= \alpha_1 grosstodate_{it} + \alpha_2 usercount_{it} \\ &+ \alpha_3 userscore_{it} + \alpha_4 criticcount_{it} \\ &+ \alpha_5 criticcore_{it} + \alpha_6 rank_{it} + \alpha_7 age_i \\ &+ \alpha_8 weekend_{it} + \alpha_9 holiday_{it} + \varepsilon_{it} \\ &+ \theta_i + \sigma_t \end{aligned} \tag{1}$$

Here, *i* represents the movie, according to the hypotheses,

H1a acknowledges that user reviews has obviously positive influence on box-office revenues, thus we expect $\alpha_2, \alpha_3 > 0$. H1b suggests that critic reviews has obviously positive influence on the revenues, and we expect $\alpha_4, \alpha_5 > 0$. H3a indicates that the gross-to-date has a positive effect on the revenues, we can expect $\alpha_1 > 0$. H3b says that movie rank and count of release days both negatively influence on the revenues, so we expect $\alpha_6, \alpha_7 < 0$. Both weekend and holidays have significantly positive effects on the revenues and we expect $\alpha_8, \alpha_9 > 0$. θ_i, σ_t represent the idiosyncratic characteristics associated with movie and time, ε_{it} represents other implicit factors.

Due to the data of critic reviews is relatively small, for more in-depth research about the relationship between movie rank, release days, count of user reviews, user score and box office revenues, we will directly use the average user reviews as research object, then the model is adjusted as equation (2).

$$\begin{aligned} gross_{it} &= \beta_1 grosstodate_{it} + \beta_2 usercount_{it} \\ &+ \beta_3 userscore_{it} + \beta_4 rank_{it} + \beta_5 age_i \\ &+ \beta_6 weekend_{it} + \beta_7 holiday_{it} + \varpi_{it} \\ &+ \varphi_i + \rho_t \end{aligned} \tag{2}$$

The settings of variables and coefficients in equation (2) is same with equation (1), then we can expect $\beta_1, \beta_2, \beta_3 > 0$, $\beta_4, \beta_5 < 0$, $\beta_6, \beta_7 > 0$, φ_i, ρ_t represent the idiosyncratic characteristics associated with movie and time, ϖ_{it} represents other implicit factors.

For model setting, through the Hausaman inspection, which is more suitable for fixed effects model, thus we use the model, in order to eliminate the impact of the multicollinearity, we can use generalized least squares (GLS) based on panel data to estimate the parameters of model[16]. It is generally considered GLS estimator as the best linear unbiased estimator, which is automatically considered with correlated disturbances in the model.

V. RESULTS AND DISCUSSIONS

With the particularity of experimental data, we directly use the model (1) to carry comparison analysis for two types of reviews with the data of first week, and use the model (2) to make further detailed analysis of the relationship among the variables. Table IV shows the correlation analysis of the main variables in the first week, and Table V is pairwise correlation analysis of the variables for data of four weeks. All of the estimation results of the models are presented in Table VI.

The table IV shows that the number of user review has significantly positive correlation with box office revenues, and critic reviews has not obvious relationship with the box office (0.49), therefore, it could be known that the influence of user reviews is greater than the critic. Movie rank and release days both negatively correlate relationship with the revenues(-0.35 and -0.45), combining with the fact that, the movie revenues will decrease along the days, especially at the beginning, then fall rapidly, which may due to the reducing of the count of reviews, but their overall relationship is not very obvious. The table V indicates that the relationship between the count and the revenues increased (from 0.49 to 0.51), it may be because of reducing the explanatory variables. However, the relationship between user score and movie rank changes into a negative relationship from the positive (from 0.04 to -0.22), which may exist certain influence between overall user score and movie rank.

The estimation results of our model are shown in table VI. Through the GLS analysis, it can be seen that, in the model (1), the cumulative gross is obviously positive correlation with daily gross (0.11), supports H3a. The movie rank and release days present obviously negative relationship with the box office (-81.08 and -184.30), which fits the actual situation and supports H3b. The count of user reviews has significant positive influence on the box office (12.51), does not the score, and partly supports H1a and H1d. As different from the user reviews, neither the score nor the count of critic reviews has the significant impact, and does not support H1b or H1d. It possibly explains that most consumers tend to believe the user with a similar experience and avoid of the purchase risk. By their coefficient, we can see the influence of user reviews is greater than the critic, which supports H1c. Surprisingly, not like the weekend, the holiday has not the significant influence, partly supports H2.

In the model (2), in order to avoid the individual differences, it considers the whole sample. The movie rank has no obvious correlation with the box office. In addition, the count still shows significant positive correlation with the daily gross, and the user score has no obvious impact, which has nothing with the critic reviews. The rank becomes no influence on the revenues (6.62). It can be seen that, with film screening and the spread of the movie reviews, we no longer blindly rely on the rank to understand the movie.

According to the movie ratings, we divide the model (2) into three groups, to study the various influence on box office. While group 1, group 2, group 3, respectively represent the high, medium and low rating, the estimated result as shown in table VI, for three groups, the count of user reviews all has obvious positive influence (8.59, 28.45, 20.86).About the user score, group 1 has no significant effect on the revenues, while group 2

TABLE IV

WEEK 1 CORRELATION ANALYSIS

variable	$gross_{it}$	$grosstodate_{it}$	$usercount_{it}$	$userscore_{it}$	$criticcount_{it}$	$criticscore_{it}$	$rank_{it}$	age_i
$gross_{it}$	1							
$grosstodate_{it}$	0.37	1						
$usercount_{it}$	0.49	0.21	1					
$userscore_{it}$	0.11	0.02	0.13	1				
$criticcount_{it}$	0.02	0.03	0.07	0.06	1			
$criticscore_{it}$	0.05	0.12	0.12	0.15	0.30	1		
$rank_{it}$	-0.35	-0.41	-0.23	0.04	-0.01	-0.07	1	
age_i	-0.45	0.32	-0.24	-0.12	-0.05	-0.03	0.07	1

TABLE V

WEEK 4 CORRELATION ANALYSIS

variable	$gross_{it}$	$grosstodate_{it}$	$usercount_{it}$	$userscore_{it}$	$rank_{it}$	age_i
$gross_{it}$	1					
$grosstodate_{it}$	0.16	1				
$usercount_{it}$	0.51	0.03	1			
$userscore_{it}$	0.17	0.07	0.20	1		
$rank_{it}$	-0.41	-0.30	-0.28	-0.22	1	
age_i	-0.46	0.33	-0.31	-0.21	0.59	1

TABLE VI

THE PARAMETERS ESTIMATION RESULTS

Variable	Model(1)	Model(2)	Model(2)		
			group 1	group 2	group 3
$grosstodate_{it}$	0.11(0.01)***	0.03(0.002)***	0.02(0.01)***	0.03(0.002)***	0.02(0.01)***
$rank_{it}$	-80.08(29.67)***	6.62(5.91)	-14.75(20.61)	4.89(6.94)	-0.59(11.14)
age_i	-184.30(27.63)***	-37.09(2.44)***	-52.29(6.03)***	-31.01(3.01)***	-31.38(4.54)***
$usercount_{it}$	12.44(1.75)***	18.52(0.92)***	8.59(1.64)***	28.45(1.49)***	20.86(1.86)***
$userscore_{it}$	0.65(2.03)	-0.71(0.48)	2.18(2.05)	-1.57(0.55)***	-1.36(0.77)*
$criticcount_{it}$	-3.99(5.73)				
$criticscore_{it}$	-2.11(1.31)				
$weekend_{it}$	578.40(105.80)***	328.10(23.94)***	547.40(62.20)***	268.30(27.99)***	319.50(40.50)***
$holiday_{it}$	5.85(156.00)	33.97(54.55)	85.25(147.60)	32.14(67.56)	57.66(74.01)
observations	460	1841	391	1200	250
number of i	66	66	14	43	9

Notes: Standard errors in parentheses, *** p<0.01, ** p<0.05, * p<0.1.

has obviously negative effect (-1.57), and group 3 has less obvious negative effect (-1.36), which partial supports H3c. For the most medium rating movies, user score significantly affects the box office. For the high rating movies, as its high quality, the revenues have nothing with the score, which is mainly affected by the count of reviews. While the user score has a certain role for that movies with low rating. Thus, there is no fixed relation between user score and box office, as the imbalance of grouping that can cause the certain deviation, which suggests it is not accurate to only refer to user score in evaluating the movie. In addition, there is no significant negative correlation relationship between movie rank with the revenues, then we know the rank still cannot represent the quality attributes of the movies. All in all, when evaluating a movie, it should be comprehensively considered such factors, including the count of user reviews, score, and release days, etc.

VI.CONCLUSION

Based on the present researches, our model has investigated the impact of movies online reviews on the box office. We selected movie industry as the research context, and used the panel data and combined with the classical statistical analysis method, to respectively verify the influence of the user and critic review. The results show that user review has the obvious effect on box office revenues,but the critic review has no impact on the revenues. In addition, we have further verified the relationship between box office and movie rank, user score, the count of user reviews, etc. which provides the reference to measure the quality of movies, that is to say when you assess the quality of the movie, you should carry comprehensive consideration with multiple indexes.

The methods of classifying user reviews, can be directly applied to other areas for more researches. As average user review is more significant influence than critic review on the box office, so the businesses can pay more attention to this kind of reviews, and to guarantee the quality to earn more and high praise from the public.

The present study has several limitations. The one, with data limitation, we cannot give a reasonable explanation for the abnormal movie, as well as we need enough more data for verifying the role of the critic. The other, during the study, we find there is a cross process about movie rank and score along the time, we can investigate the relationship of the two, and control the formation of the intersection, analyze influence factors in the future.

ACKNOWLEDGMENT

The authors would like to thank Lirong Chen for her guidance on the earlier version of this paper.

REFERENCES

[1] W.D, Bin G, Andrew B. Do online reviews matter? - An empirical investigation of panel data, Decision Support Systems, vol. 45, pp. 1007–1016, 2008.

[2] J. Eliashberg, S.M Shugan, Film critics: influencers or predictors, Journal of Marketing, vol. 61, pp. 68–78, 1997.

[3] X. Li, L.M. Hitt, Self-selection and information role of online product reviews, Information Systems Research, vol. 19, no. 4, pp. 456-474, 2008.

[4] D.H. McKnight, V. Choudhury, C. Kacmar, Developing and validating trust measures for e-commerce, an integrative typology, Information Systems Research, vol. 13, no. 3, pp. 334–359, 2002.

[5] W.W. Moe, M. Trusov, The value of social dynamics in online product ratings forums, Journal of Marketing Research, vol. 48, no. 3, pp. 444–456, 2011.

[6] F. Zhu, X. Zhang, Impact of online consumer reviews on sales, the moderating role of product and consumer characteristics, Journal of Marketing, vol. 74, no. 2, pp. 133–148. 2010.

[7] Yong Liu. Word-of-mouth for movies: its dynamics and impact on box office revenue, Journal of Marketing, vol. 70, no. 3, pp. 74-89, 2006.

[8] A. Banerjee, A simple model of herd behavior, Quarterly Journal of Economics, vol. 110, no. 3, pp. 797–817, 1992.

[9] Seneca S, Nantel, The influence of online product recommendations on consumers' online choices, Journal of Retailing, vol. 80, no. 2, pp. 159-169, 2004.

[10] Gilly,Mary C.,John L. Graham,Mary Finley Wolfinbarger and Laura J.Yale, A dyadic study of interpersonal information search, Journal of the Academy of Marketing Science, vol. 26, no. 2, pp. 83-100, 1988.

[11] Reinstein, David A., Christopher M. Snyder. The influence of critic reviews on consumer demand for experience goods, A case study of movie critics, Journal of Industrial Economics, vol. 53, no. 1, pp. 27-50, 2005.

[12] Basuroy, Suman, Subimal Chatterjee, S. Abraham Ravid. How critical are critical reviews? The box office effects of movies critics, star power, and budgets, Journal of Marketing, vol. 67, no. 4, pp. 103-117, 2003.

[13] Xiaoquan Zhang, Chrysanthos Dellarocas. The lord of the ratings, how a movie's fate is influenced by reviews, Proceedings of the 27th International Conference on Information Systems, 2006.

[14] Duan, W., B. G, and A. Whinston. The dynamics of online word-of-mouth and product sales-an empirical investigation of the movie industry, Journal of Retailing, vol. 84, no. 2, pp. 233–242, 2008.

[15] Godes, D,D. Mayzlin, Using online conversations to study word-of-mouth communication, Marketing Science, vol. 23, pp. 545-560, 2004.

[16] Yuanyuan Hao, Peng Zou, Yijun Li, Qiang Ye, An empirical study on the impact of online reviews sentimental orientation on sale based on movie panel data(in Chinese), Management Review, vol. 21, no. 10, pp. 95-103, 2009.

Analysis and Simulation Application for Enterprise Populations M&A Based on Predator-prey Model

Jing-hong Xu [1,*], Tong-bing Ma [2], Yun-long Yi [3]
[1, 2]School of Economics & Management, Shenyang Institute of Engineering, Shenyang, China
[3] Science and Technology Department, Shenyang Institute of Engineering, Shenyang, China
([1]jinghong_xu2003@163.com,)

Abstract - **In the circumstances of globalization, the raging trend of transnational merger and acquisition is irresistible, and trends promoted in becoming lager for enterprises are pronounced increasingly. Meanwhile, the professional and subtle-oriented small and medium size firms have developed vigorously, which agree upon the scale-oriented developing trend for enterprises in becoming lager. Thus, many enterprises consider merger and acquisition as a strategic option when oriented towards building competitiveness. Learning from the predator-prey model in ecology, and combined with numerical solution and phase diagram, a model with three typical enterprise populations is studied in this paper. The stability of solutions is investigated and the phase diagrams are described by theoretic analysis and numerical simulation.**

Keywords - **Predator-prey model, enterprise population, M&A, stability**

I. INTRODUCTION

As market is increasingly competitive, customer needs are becoming more diverse, personalized and protean, which leads to shortened life cycle of products and the difficulty of product development. All these factors increase uncertainty of commercialization of innovative products, cost of technology development and production, while promoting penetration of different technologies. Under such grim conditions, it is difficult for companies of different sizes to occupy the market for a long time. Since the beginning of this year, many small- and medium-sized enterprises (SMEs) with advantage factors have disappeared in the fierce competition because of fund shortage and lack of talents. Some large enterprises focusing on core businesses were stuck into a passive situation where overall interests and long-term development are impaired, because they ignore coordination with other companies and failed to adapt to the market fluctuation.

Given above circumstances, some enterprises acquire some existing enterprises to obtain more strategic opportunities and expand their influence on the particular industry cluster, which can reduce a competitor and bring a new partner, access to more information, less efforts on research, less time delay on plant construction and profitability improvement of the enterprises. Mergers and acquisitions (M&A) will have an impact in many business processes, such as production, market and human resources. If we can enhance the synergic effects, it enables enterprises to decentralize risks, and benefit from economies of scale and new competitive advantages.

This paper reports an exploratory study on the process of M&A from an ecological perspective. Since the 1970s, ecology has been integrated into many disciplines and popularized to some degree, which are due to its systematic description of objects and good simulation of system complexity. Nowadays, many ecological applications expand the traditional business management theory, through the use of ecological concepts, theories and methods. Practice has proved that both enterprises and market have certain characteristics of life. There are a lot of similarities between cycles of an enterprise and life, which both include birth, growth, reproduction, aging and demise. Some SMEs that are bought by powerful enterprises are able to recover from the brink of bankruptcy and start a new cycle,. Besides, the market that manipulates the survival and development of a company starts to display oscillation instead of eco-systemic stability. Therefore, the biological ecosystem might can better describe of the life characteristics of enterprises and market. The oscillation of the market follows a co-evolutionary process which either evolutes or degenerates to adapt to environment. If we include individual companies into corresponding levels in the ecosystems and establish a hierarchy form of individual enterprises - enterprise population - industrial clusters - economic system, it will create a sustainable development of the entire hierarchy and promote the healthy growth of economic systems.

In ecology, biological population is a group of individuals which occupy a certain space and interact with each other through individual dispersal. In terms of systems, both ecosystem and socio-economic system follow the law of system evolution. Thus, individual firms with certain characteristics can be defined as an enterprise population. This enterprise population ecological theory focuses mainly on the same industry or enterprise generating products with alternative functions. Daniel C. Ai Siti and Porter (1997) believe that system concept of industrial ecology could help enterprises find ways to increase value or reduce costs in the internal as well as between upstream and downstream supply chains. Additionally, the industrial ecological thinking could help enterprises to optimize their resources, and enhance the competitiveness.

If two or more populations live in a natural environment, relationship between them could be competitive, interdependent or predator-prey. Similarly, in a particular geographic location, enterprises and organizations from the same or related industries form a tight aggregate within an industry cluster, which constitutes multiple paralleled or up-down stream enterprise populations. In a certain amount of time and

E. Qi et al. (eds.), *Proceedings of the 23rd International Conference on Industrial Engineering and Engineering Management 2016,* DOI 10.2991/978-94-6239-255-7_51

space, since a variety of environmental resources are limited, including raw materials, labor, capital and customers, when business needs exceed the actual amount of resources, the M&A which is similar to predator-prey would occur. Therefore, this paper intends to draw the predator-prey theory into the M&A activities of enterprise populations within the industry cluster, from the perspective of population dynamics and ecology. Based on the Volterra model in mathematical ecology, numerical solution diagram and phase diagram analysis methods are used to construct three typical M&A models, followed by the discussion of stability of equilibrium point.

II. MODEL CONSTRUCTION

A. Assumptions:

In this paper, numerical methods are used to establish following three prey-predator systems. Assuming that there are large-sized, medium-sized and small-sized enterprise populations.

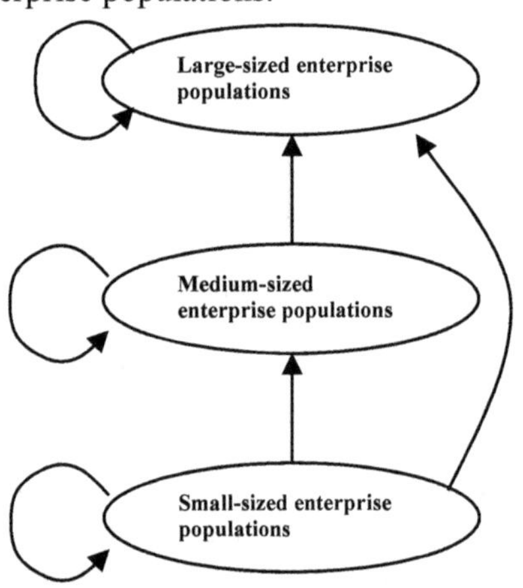

Fig. 1. Conceptual model

Assumption 1: density functions of large-, medium-, and small-sized enterprise populations are x（t）、 y（t）、 z（t）respectively;

Assumption 2: They are continuously differentiable functions;

Assumption 3: The industry cluster has adequate natural resources internally, but is constrained by the external economic boundaries, so the number of firms that the cluster can accommodate falls into a finite interval;

Assumption 4: Various populations survive independently and grow exponentially.

B. Analysis of the stability of equilibrium in the model

In this model, small-sized enterprise populations are regarded to be prey populations which grow exponentially. The plunder of small-sized enterprise populations by medium- and large-sized enterprise populations will lead to reduction in their growth rate. The large-sized enterprise populations could also plunder medium-sized enterprise populations. Medium- and large-sized enterprise populations have stronger independent survival ability than small enterprise populations. Mortality rate of predator populations could be decreased with the presence of prey populations. The potential function of how the predator population could rely on prey population is defined as multiplication of the predator's digestion coefficient and plunder coefficient of predator.

Establishment of the mathematical models of above three populations:

$$\begin{cases} \dot{x}(t) = x(r_1 - by - dz) = f_1(x,y,z) \\ \dot{y}(t) = y(-r_2 + mbx - cz) = f_2(x,y,z) \\ \dot{z}(t) = z(-r_3 + ncy + ldx) = f_3(x,y,z) \end{cases} \quad (1)$$

Expand f_1 f_2, f_3 as Taylor series at the point $p_0(x_0, y_0, z_0)$ and system（1`）is generated

$$\begin{cases} \dot{x}=f_{1x}(x_0,y_0,z_0)(x-x_0)+f_{1y}(x_0,y_0,z_0)(y-y_0)+f_{1z}(x_0,y_0,z_0)(z-z_0) \\ \dot{y}=f_{2x}(x_0,y_0,z_0)(x-x_0)+f_{2y}(x_0,y_0,z_0)(y-y_0)+f_{2z}(x_0,y_0,z_0)(z-z_0) \\ \dot{z}=f_{3x}(x_0,y_0,z_0)(x-x_0)+f_{3y}(x_0,y_0,z_0)(y-y_0)+f_{3z}(x_0,y_0,z_0)(z-z_0) \end{cases} \quad (1`)$$

According to the qualitative theory of differential equation, nonlinear system 1 and linear system 1 have uniform stability with same equilibrium points. Thus, introduce

$$p = f_{1x}(x_0,y_0,z_0) + f_{2y}(x_0,y_0,z_0) + f_{3z}(x_0,y_0,z_0)$$

$$q = \begin{vmatrix} f_{1x}(x_0,y_0,z_0) & f_{1y}(x_0,y_0,z_0) & f_{1z}(x_0,y_0,z_0) \\ f_{2x}(x_0,y_0,z_0) & f_{2y}(x_0,y_0,z_0) & f_{2z}(x_0,y_0,z_0) \\ f_{3x}(x_0,y_0,z_0) & f_{3y}(x_0,y_0,z_0) & f_{3z}(x_0,y_0,z_0) \end{vmatrix} \quad (2)$$

Five equilibrium solutions of system 1 are generated:

$$(0,0,0,),(\frac{r_3}{dr},0,\frac{r_1}{d}),(0,\frac{r_3}{nc},-\frac{r_2}{c}),(\frac{r_2}{bm},\frac{r_1}{b},0)$$

$$(\frac{-bc+ncr_1+ndr_2}{bd(-l+mn)},\frac{r_1cl+dr_2l-mbr_3}{bc(-l+mn)},\frac{dr_2l-mbr_3+mncr_1}{dc(-l+mn)})$$

For point $(\frac{r_3}{dr},0,\frac{r_1}{d})$, there is

$$p = -r_2 + \frac{mbr_3 - clr_1}{dl}$$

$$q = -r_1 r_3\left(-r_2 + \frac{mbr_3 - clr_1}{dl}\right)$$

when

$$-r_2 + \frac{mbr_3 - clr_1}{dl} < 0 \Rightarrow r_2 dl > (mbr_3 - clr_1)$$

$$\Rightarrow p < 0 \, and \, q < 0$$

Therefore $(\frac{r_3}{dr},0,\frac{r_1}{d})$ is the equilibrium point.

Other points can be deduced by the same method.

III. NUMERICAL SOLUTION DIAGRAM AND PHASE DIAGRAM ANALYSIS

A. Choose parameters

$r_1 = 1$, $r_2 = 0.5$, $r_3 = 0.6$, $b = 0.1$, $d = 0.2$

$m = 0.2$, $c = 0.2$, $n = 0.1$, $l = 0.1$

Initial value is (100 , 20 , 6) , and corresponding numerical solution diagram and phase diagram are produced by using the results from 4th-5th Runge-Kutta method.

B. Numerical solution diagram simulation and analysis

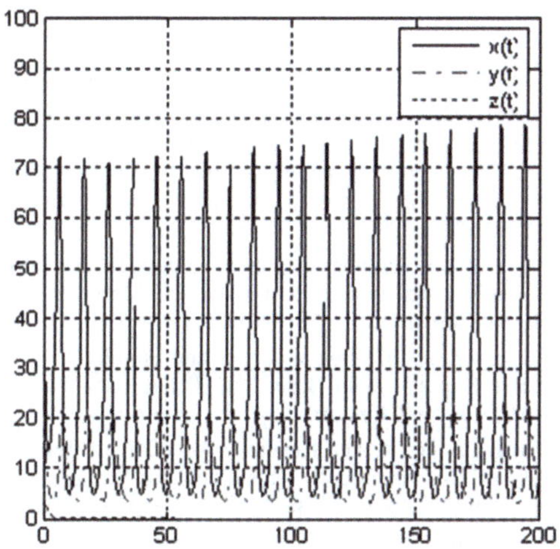

Fig.2. Numerical solution diagram

Based on the characteristics of Volterra model, balanced state, i.e. stable equilibrium point should be observed in majority of predator-prey system, instead of periodic oscillations. The numerical solution diagram (Fig 2) and related calculation results demonstrate different fluctuation characteristics of these three populations. Small- and medium-sized enterprise populations show a significant oscillation. Due to the initial limited number, and many factors needed to be coordinated and relatively long operation time, large-sized enterprise populations show reduced number and may tend to reach equilibrium state, rather than a significant oscillation within limited simulation cycle. In consistent with the simulation, actual formation of an industrial cluster or a large-sized enterprise population within a cluster takes a long time. In addition, only a few large enterprises could be dominant in a representative cluster, such as Shanghai Automotive Industry Corporation in Shanghai automobile industry cluster. Therefore, in the process of numerical simulation, that the number of enterprises in large enterprise populations is on the decline and even maintain constant is possible. In contrast to large enterprises, SMEs have a sensitive response to pressure outside, which is due to their flexible organizational structures. In view of internal constrain mechanism of enterprise management theory, disturbed large enterprises need a long period of recovery and oscillation because of complete system, while small- and medium-sized enterprises need a relatively short period of recovery and oscillation cycle can be easily observed.

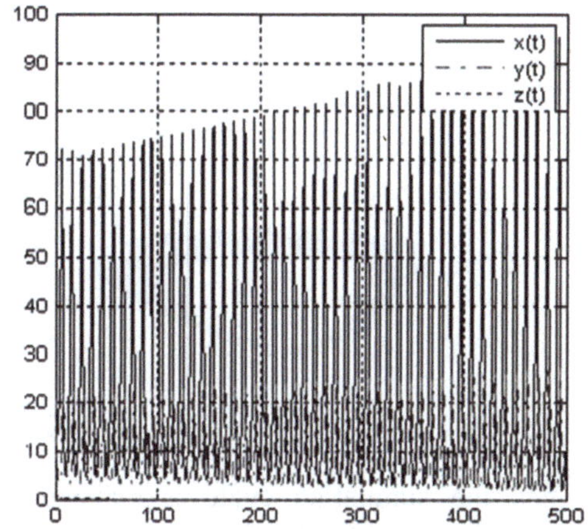

Fig.3. Numerical solution diagram after change of time parameters

When the simulation cycle is extended, we may find out that, with the presence of M&A, the number of small enterprises is significantly rising; the number of medium enterprises also shows upward but not obvious trend; and the number of large enterprises demonstrates a steady trend. Based on the practice of developed countries, SMEs should be actively developed in areas especially where large enterprises are concentrated, which can enhance regional economic prosperity. The development of SMEs in many industrial clusters and many regions of China is not satisfactory, since a lot of SMEs had been shut down. However, from the above simulation process, it is probable that the number of SMEs will have a steady rise.

C. Phase diagram simulation and analysis

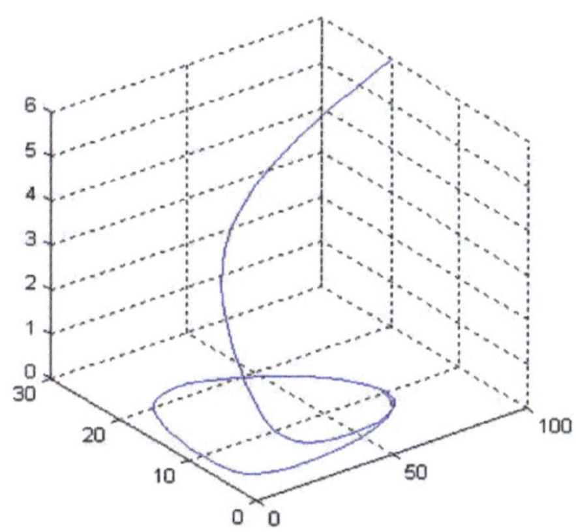

Fig.4. Phase diagram

The phase trajectory is closed and its structure is unstable, so once it leaves a closed trajectory, it will enter another closed trajectory without recovery. In a periodic balanced eco-system, internal constraint would make system recover if it deviates from the periodic orbits. As is shown in the three-dimensional phase diagram of fig 4, the system locates in a closed trajectory in a certain time, indicating the prey-predator relationship among large, medium and small enterprises could exist for a long period.

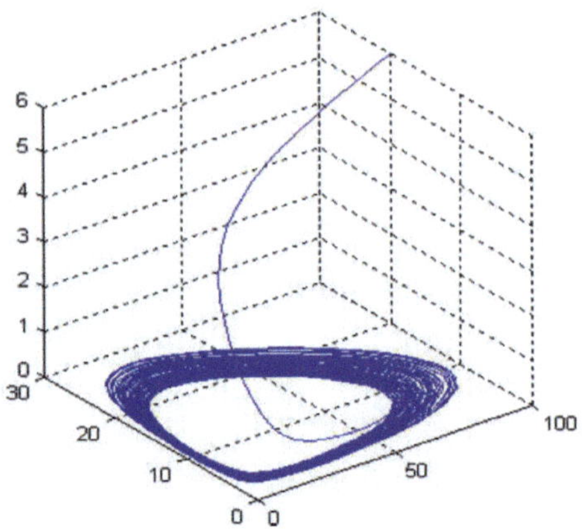

Fig.5. Phase diagram after change of time parameters

When the simulation cycle is extended, we can observe a clear triangle-like limit cycle, suggesting that this is a non-asymptotically stable system with mutual restraint among three populations and periodic oscillations. This point confirms that oscillation period exists even though simulation of large enterprises shown in fig 2 & 3 does not display it. From the point of ecology, some scholars believe that the long-standing balanced eco-system with periodic changes should be stable in structure, and internal constraint mechanism could make the system restore its amplitude or cycle if it is disturbed and deviated from the original orbit. This indicates that the M&A mode is structurally stable to some extent in a long term.

IV. CONCULTIONS AND PERSPECTIVES

From the ecological perspective, the M&A model of three enterprise populations within an industrial cluster is investigated based on the prey-predator theory. We have drawn two conclusions on the stability and trend of M&A process among three populations.

Firstly, under certain assumptions, three enterprise populations display periodic oscillation. In a long term, the number of large enterprises tends to be relatively stable, while the number of SMEs shows a growth trend. This model is consistent with the actual situation of economic development, so that we are able to use ecological theory to study similar issues.

Secondly, since the M&A activity is not the only way for the synergic growth of various enterprise populations, some conclusions are not evident due to the assumptions. Therefore, after the enlargement of time parameters, we generate a solid conclusion that M&A model is stable and three enterprise populations restrain mutually and develop synergically.

In summary, it is meaningful to apply prey-predator model to the study about the effect of M&A activities on evolutionary process of enterprise population ecosystem, which helps us better understand the sustainable development of enterprise populations and industrial cluster.

ACKNOWLEDGMENT

1、This article has been funded by Society Science Union of Shenyang "The research on the key industries cohesion and coordination development in Shenyang economic zone". Project grant number: SYSK2016-16-20

2、This article has been funded by Social Science Foundation of Liaoning province "The study on the development strategy of manufacturing innovation center in Liaoning province based on " Made in China 2025 Strategy" ". Project grant number: L15AJY011

REFERENCES

[1] Gossain Sanjvi, Cracking the collaboration code, Joumal of Busniess Srtaetgy, Nov/Dec-2002, Vol.23, Issue6, pp. 20-26

[2] Mareo Iansiti, Roy Levien, The New Operational Dynamics of Business Eeosystems: Implications for Policy, Operations and Technology Strategy, Havrard Business School working Paper Series, No.03-030, 2002.

[3] Martin V. E. E., Technology gap among suppliers and Pop&Morn Stores as a key distribution cost issue. PICMET 2008, Portland International Conference, 2008(7): pp.201-206.

[4] Xu Jinghong, Zhang Cuihua, Analysis and Simulation Application for Reverse Supply Chain Based on TP and Retailer Take-back, ICEE2010, Vol. (In Chinese).

[5] Iansiti. M.&Levien. R. The Keystone Advantage : What the New Dynamics of Business Ecosystems Mean for Strategy Innovation and Sustainability. Harvard Business School press, 2004: 255.

[6] Li Lan, Chen Linjue, Xing Zhihong,Numerical Analysis of Equilibrium Solution and Phase Diagram of some Three Species Ecological Systems. J. Journal of Jiamusi University(Natural Science Edition), 2007, Vol25, No.3.(In Chinese).

[7] Shen Yunhong, Wang Hengshan, Liu Yi, Research on Harmonious Symbiosis Mechanism of Chinese Small and Medium-sized Enterprises, Proceedings of 2005 International Conference on Management Science Engineering, 2005:1091-11394. (In Chinese).

[8] Zou Lei, Sun Changxiong, Analysis on mechanism of symbiotic behabior enterprises based on the population mutually beneficial model. J. Jounal of Harbin Engineering University, 2009, Vol30, No12, pp.1446-1450. (In Chinese).

[9] Audretsch D.Innovative, Clusters and the Industrial Life Cycle. J. Review of Industrial Organization, 1996(11), pp.253-272

[10] Miao Yongwei, A Study of Stability in Three-species Ecological System. J. Journal of Mathematics for Technology, 2002, Vol18, No4, pp.20-23. (In Chinese).

[11] Sun C, linY, HanM, Stability and Hopfbifurcate-on for an epidemic disease modelwith Delay. J. Cha-os, Solitons and Fractals, 2006, 30, pp. 204-216. (In Chinese)

[12] Chen L L, Lin Z S, Dynamic study on the non-allee and allee system of two-competitive species. JNanjing Normal University(NaturalScience Edition),Chinese Series, 2008: (2), pp.124-129

[13] S.Q.Liu, J.H.Zhang, Coexistence and stability of predator –prey model with BD functional response and stage structure.J. Math. Anal. Appl, 2008,(1): pp.446-460.

[14] Z.G.Lin, M.Pedensen, Stability in diffusive food-chain model with Michaelis-Menten function response.J. Nonliner Anal, 2004(57), pp.421-433

[15] Zhang Qipeng, Yang Guozeng, Global stability of a positive equilibrium point on two-species competition model, Journal of Nanyang Normal University, Vol8, No.3, 2009, pp.22-23(In Chinese)

Research on Mechanism of Early Warning of Health Management Based on Cloud Computing and Big Data

You-cheng SHAN[1,2,*], Chao LV[1], Qiu-ye ZHANG[1], Xin-yu TIAN[3]

[1] School of Management and Technology, Zhejiang Technical Institute of Economics, Hangzhou 310018, China
[2] School of Management and Economics, Tianjin University, Tianjin 300072, China
[3] School of Education, Tianjin University, Tianjin 300072, China
(shanyoucheng@163.com)

Abstract - **This study introduces the meaning of cloud computing and big data, analyses the research status of health management in domestic and foreign. Also, it expounds the data mechanism and key technologies, and then discusses the transmission process of monitoring data. In addition, it mentions a new pattern uses warning data cloud platform as a carrier which can give all kinds of early warning services to the families, communities, hospitals and other users in health management system.**

Keywords - **Cloud computing; Big data; Health management; Data early warning**

As new generation of information technology, cloud computing and big data has already become one of the strategic emerging industries that the government selected. It is an important work emphasis for the fields of health management and information technology of the competent departments to promote the development of cloud computing and big data. Currently, the relationship between cloud computing and big data is like different sides of the coin, they are inseparably intertwined. Both of the two applications received unprecedented attention in several of industries in our country.

I. THE MEANING OF CLOUD COMPUTING

A. Concept of cloud computing

Cloud computing[1-3] is a rapidly developing field of technology, its essence and connotation in the course of future technological developments will continue to be revised and enriched. United States University of California, Berkeley, published a report on cloud computing, cloud computing refers to both the application provided as a service over the Internet, and provide these services in the data center hardware and software, and these are called data center hardware and software. IBM believes that cloud computing is a computing style, which is based on public or private network services, and software and processing power delivery. Google believes that cloud computing is closer to a kind of cloud, first in order to optimize the performance of its search engine, after the development of infrastructure and it as a service (including word processing, map, photo manipulation, etc) for use by customers.

B. Features of the cloud computing technique

A special point of the cloud computing is that it creates a flexible organizing scheme which integrates all kinds of recourses quickly so that it meets needs of different business. This scheme fits for not only business module but also for underlying hardware. Also, it re-defines the way the computer resource on the upper layer is used, the way the services are provided and the coordination process the socialized production is made. Meanwhile it provides us with a new view to deal with the "big issue" that the internet brings and to improve the service mode. In other words, the cloud computing brings a organizing scheme which enable us to organize resources for serving, organize techniques for doing and organize process for adapting.

The cloud computing technique integrates massive hardware software by using virtual technique. Then based on data transition and load balancing, it forms a "system" in the sense of a standard concept. The feature of this system is that it is scattered physically and clustered logically, which is what we call "distributed integration". For one thing, the logical integration of massive resource means that we are able to meet "important clients" needs and deal with "big data". For another, it means we are able to provide multiple services[4] on one single platform to meet needs of all kinds of clients, for example IaaS, PaaS, SaaS and XaaS.

II. THE MEANING OF BIG DATA

A. Concept of big data

Big data[5-7], refers to a term for data sets that are so large or complex that traditional data processing applications are inadequate. It can not be achieved in a reasonable time to capture, manage, process, and organize to become more active to help business decision-making information purposes. Gartner thinks that big data is high volume, high velocity, and/or high variety information assets that require new forms of processing to enable enhanced decision making, insight discovery and process optimization. Big data represents the Information assets characterized by such a High Volume, Velocity and Variety to require specific Technology and Analytical Methods for its transformation into Value. If we compared the big data to an industry, then the key to realize the benefits is to improve the processing and handling capabilities of the data and add value through processing and handling of data.

B. Features of the big data technique

E. Qi et al. (eds.), *Proceedings of the 23rd International Conference on Industrial Engineering and Engineering Management 2016*, DOI 10.2991/978-94-6239-255-7_52

In the Internet era, big data can be described by the following characteristics: volume, velocity, value and variety.

(1) Volume. Enterprises are facing massive growth in data volumes. According to a recent IDC report forecasts, global data volume will expand 50 times by 2020. At present, the scale of the big data is a constantly changing indicator, a single data set size ranging from dozens of TB to PB.

(2) Velocity. In the era of high-speed network, it creates real-time data streams through the software of performance-optimized high-speed computer processors and servers. Enterprises not only need to know how to quickly create data, but also to know how fast the processing, analysis and returned to the user, in order to meet their real-time requirements.

(3) Value. The plenty of the statistics include lots of irrelevant information. They have huge value while it seems useless at normal time. They can be analyzed deeply in the analysis and prediction of future tend and mode, deeply complicated analysis with robot learning and artificial intelligence.

(4) Variety. This helps people who analyze it to effectively use the resulting insight. Information generation algorithms must detect and address invisible issues such as machine degradation, component wear, etc. on the factory floor.

III. THE CURRENT SITUATION OF RESEARCHING IN DOMESTIC AND FOREIGN

A. The Current Situation of Researching in Foreign

In the middle age in 20 century, with the development of countries , in order to deal with non-infective chronic disease from bad life mode and behavior mode, sorts of health management service have been risen except current clinical care mode and prophylactic medicine mode. With many years development, foreign health has been developed to nowadays comprehensive health evaluation health risk and social health acceleration, and becoming continuous behaviors and process of prophylaxes in health, curing with science, preventing disability in recovery, from solo health examination and life guidance.

In the West, health management plan has become an important part for the health care system, and has been proved that it can effectively reduce personal risk and reduce medical costs. According to the health management experience, Through effective prevention and intervention, participants in health management services as prescribed regular medication chance increased by 50%, doctors carve out a more effective drugs and treatment increases the chance by 60% so that the participants overall risk of illness by 50%.

B. The Current Situation of Researching in Domestic

In China, the health management is a new concept. Our specialized health management organizations began in 2001, a professional health management division in October 2005 formally incorporated into the health industry-specific range of occupations. Since then, the health management industry in terms of academic exchange, service delivery, risk management and business cooperation, carry out various forms of exploring practice. According to health management services, domestic health management organizations can be divided into four categories: First Class healthy bodies; Second class medical services intermediary agencies; Third, health management service establishments; Fourth, community health service class institutions.

In October 2013, the State Council issued “Opinions on promoting the healthy development of service industry”. By 2020, the opinion pointed out that, basically covers the entire life cycle of content-rich, well-structured system of health services, health services, to create a number of well-known brand of industrial clusters and a virtuous circle, and a certain international competitiveness, basically meet the majority of people's health service needs. According to the China Industry Research Network released “2016-2022 China's health management industry status quo and development trend forecasting research analysis report” shows that the next five years, Chinese health management industry market size will accelerate growth, expected to grow from 106.9 billion in 2015 to 203.9 billion in 2020, annual growth of 11.34 percent from 2015 to 2020 grew 14.82%.

This article will cloud computing and big data while the introduction of health management, the use of classification, clustering, association rules, regression analysis, variance analysis, WEB mining and other methods to analyze vast amounts of data generated by health management, health risk factors research data Warning mechanism, domestic few outside experts and academics discussed in this regard, so research in this area to improve the health of residents, improve the family, the community, the hospital three health management system, and promote urban health infrastructure and residents' health and safety management, has a promising future

IV. THE WARNING DATA MECHANISM AND KEY TECHNOLOGY

A. Warning Data

Warning data covers management engineering, artificial intelligence, information technology systematic application of cross-disciplinary, integrated cloud computing, big data, scientific management, intelligent control, data mining, knowledge discovery, statistics and other related theories and key technologies. Its core modules are data acquisition, early warning model, association rule, and prevention and control of risk.

(1) Data acquisition. Based on cloud computing and big data, analysis the risk factors[8] affecting the health of the state of health by monitoring the data collection, use scientific management, intelligent control and other key technology export health risk solutions.

(2) Early warning model. Based on discovery knowledge and mathematical statistics, use the domain related knowledge of early warning to build early warning model[9], solve the problem that warning knowledge acquires difficulties.

(3) Association rule. Based on amounts of data in health management, through data excavate, find association rules between health hazard data and warning reference value by quantization evaluation index to identify and evaluate interventions in health risk management.

(4) Prevention and control of risk. Warning data to the cloud platform as the carrier, through data processing, proofreading the accuracy of health risk prevention and management control method, verify the completeness of health management and early warning mechanism.

B. Monitoring Data Transmission Process

Monitoring data collection is the most important part of warning data, the data transmission speed and quality is a key indicator of the efficiency of the system. As in Fig.1, the core module of the data transmission flow warning system monitoring data is the data interface middleware, third-party interface software and data collection for client and so on.

(1) Data interface middleware. By the system to provide technical support unit data interface middleware (Dll dynamic library) and call instructions, completed by the monitoring unit of the monitoring device on their own real-time data upload work.

(2) Third-party interface software. Various monitoring institutions themselves or entrust a specialized unit to write interface software to realize the monitoring equipment real-time data upload work.

(3) Data collection for client. Self-developed or entrust a specialized unit development and provide technical support.

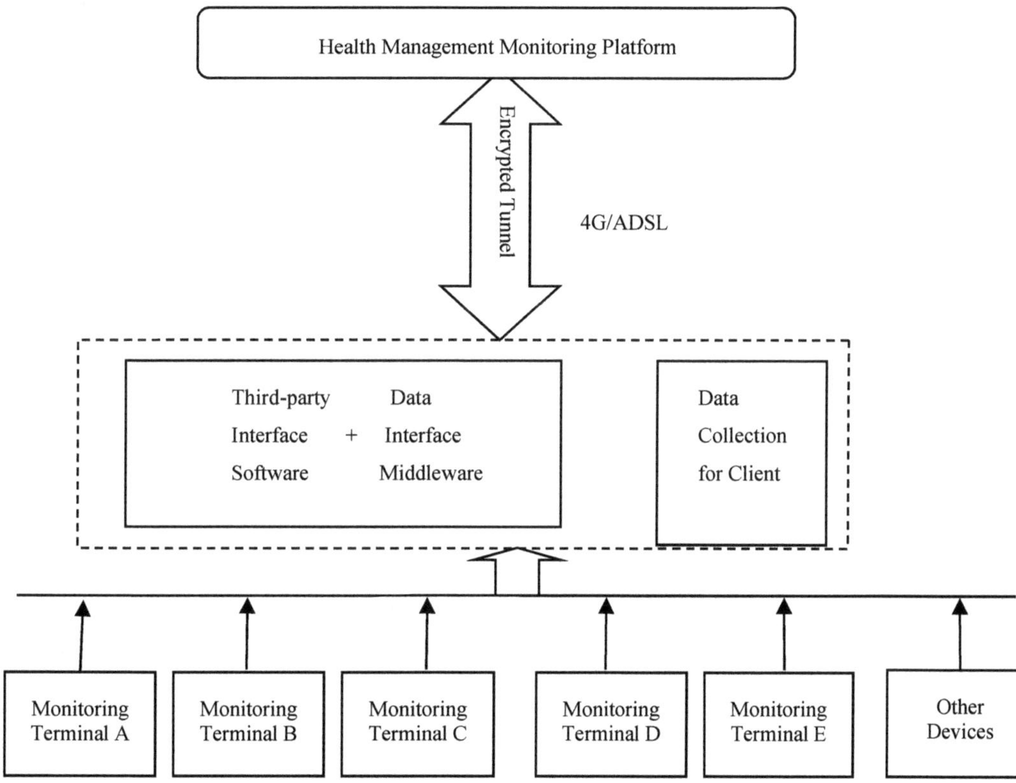

Fig. 1. The core module of early warning system for monitoring data transmission process

C. Warning Data Cloud Platform

The essence of it is to offer warning services to the families, communities, hospitals and other users in health management system. As in Fig. 2, the core module can be divided into User Access (UA), Infrastructure as a Service (IaaS), Platform as a Service (PaaS), Software as a Service (SaaS), and Service Management (SM).

(1) User Access (UA). Offering visit connector to the warning cloud service of all levels, it's the backup service for the cloud service to make customer convenient. It provides the basic parts of visit service, warning services and service catalogue. It belongs to the private cloud service.

(2) Infrastructure as a Service (IaaS). Using the virtualization technology to hide the complex physical resource, it provides the basic parts of saving service, server service, internet service and hardware resource.

(3) Platform as a Service (PaaS). Offering better service for IaaS, customer can construct their own application; it provides the basic parts of software developing, data base service and middleware service.

(4) Software as a Service (SaaS). Offering software applying service, it provides the basic parts of warning data capacity and warning resource customized system, monitoring data transmission process optimization system, risk prevention and control site management system, early warning and emergency response system, customer information control system, daily management system and smart guarantee system.

(5) Service Management (SM).Offering management function of the warning cloud service, it provides the basic parts of IT service management and business management. It belongs to the public cloud service.

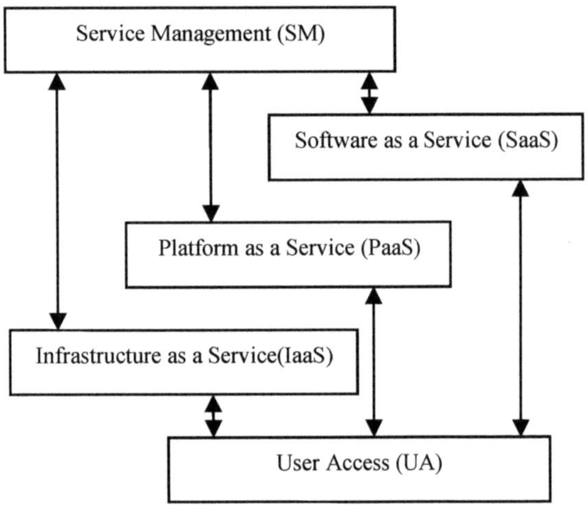

Fig. 2. Warning Data Cloud Platform core module

V. SUMMARY

Nowadays, it is the strategic chance for the development of cloud computing technology. It's urgent to accelerate the reform and the upgrading of the traditional early warning technology, use cloud computing and big data to improve the warning data ability of health management. With the background of cloud computing and big data in our context, it discerns and excavates health risk factor, analysis the rule and cause of health risk factor, research health risk data and control criteria to determine the risk state of health management, choose intervention measures to prevent health risks for residents for the performance evaluation. It also identify and deal with the mass of data that generated by health management, optimist the monitoring data transmission process, build warning data cloud services platform and explore the new path of warning data to provide effective solutions of health management for families, communities and hospitals.

REFERENCES

[1] Chang J H, Lai C F, Wang M S. A fair scheduler using cloud computing for digital TV program recommendation system[J], telecommunication systems, 2015, 60(1): 55-66.

[2] Tsai J L, Lo N W. A Privacy-Aware Authentication Scheme for Distributed Mobile Cloud Computing Services[J], ieee systems journal, 2015, 9(3): 805-815.

[3] Alkhanak E N, Lee S P, Khan S U R. Cost-aware challenges for workflow scheduling approaches in cloud computing environments: Taxonomy and opportunities[J], future generation computer systems- the international journal of grid computing and escience, 2015, 50: 3-21.

[4] Wen Yiwen, Research on the CRM Current Situation Analysis and Prediction Based on Cloud Computing Technology [D], Tianjin University, 2013.

[5] Li Ting. Research on Big Data and Artificial Intelligence Aided Decision-Making Mechanism with the Applications on Video Website Homemade Program Innovation [J], international journal of technology management, 2016, (3): 21-23.

[6] Wang Xiuying, Zhang Ling, Zhang Congcong. On Application of Big Data Mining in Earthquake Precursor Observation[J], earthquake research in China, 2015, 29(4): 452-458.

[7] Pan Yuanyuan, Big data era mobile health care management system construction [J]. Medicine and Hygiene, 2015, 1(8): 273-274.

[8] Mather T, Kumaraswamy S, Latif S. Cloud Security and Privacy: An Enterprise Perspective on Risks and Compliance[J]. O'Reilly Media, Inc, 2009, 12(5): 18-25.

[9] Chen Xianhui, Sun Guoping, Luo Jianye. Construction of Health Management Model of Ischemic Stroke's Early Recognition Based on the Internet of Things and Cloud Computing [J]. Chinese General Practice, 2015, 18 (13): 1560-1562.

Research on Early Safety Warning Evaluation Based on Multi-hierarchy Fuzzy AHP in Civil Aviation Airport

Yu-ye LE[1], You-yuan WANG[1,*], Yun-Feng YAN[2]
[1]School of Economics & Management, Nanchang Hangkong University, Nanchang, China
[2]School of Aircraft Engineering, Nanchang Hangkong University, Nanchang, China
(*yywnc@sina.com)

Abstract: **This paper, in comprehensive consideration of the factors affecting the safety of civil aviation airport, by following the integration of theory with practice, expounds the construction of the civil aviation airport safety forewarning evaluation index system in terms of personnel, maintenance, environment and organizational management and application of multi-hierarchy comprehensive fuzzy method to carry out the early warning for assessment as well as computation to determine the safety status of civil aviation airport so as to prevent the hidden accident from occurrence.**

Keywords: **Airport safety； early warning； index system； weight； Multi-hierarchy comprehensive fuzzy evaluation method**

I.INTRODUCTION

In recent years, with the development of the science and technology, the aviation industry thrives while the aviation accidents happen frequently. Therefore, reducing the rate of the aviation accidents, enhancing the proficiency of prevention of the accidents and the safety of the airport operation are extremely urgent.

The researchers at home and abroad have related research on the early warning system for safety of the civil aviation airport. For example, the theory from Liling Yu [1] etc., based on the dissipative structure theory and synergetic principle, built a structural model of CAAC safety management system—ROSE model; Lexin Xu[2] et al analyzed current situation and problems for CAAC safety management, proposed that construction of a CAAC safety management should have "fault tolerance" mechanism; Li Jing[3] et al took the specific data of CAAC as a sample, founded resource

Li Hang[4] et al constructed the early waning model for civil aviation airport safety based on the expansibility theory and thus the subjectivity of the early safety warning model is greatly reduced.

The research activities above-mentioned mainly aim to the construction of early safety warning evaluation system and improvement approach for early safety warning proficiency, while the analytic hierarchy process method (AHP), as a common evaluation method, have been widely applied to the civil aviation airport operation. Among it the data are not commonly applied. This paper will perform specific data evaluation by the expert on the operational situation of airport safety system.

II.AIRPORT EARLY SAFETY WARING EVALUATION INDEX SYSTEM

The main construction principles of the airport early safety warning evaluation index system include: principle of accordance with the actual airport situation; principle of quantification; principle of controllability; principle of pragmatics; principle of comparability [5].

The early safety warning of civil aviation airport is a complex systematic project involved in many influence factors, indexes are quantitative or qualitative or both. Safety forewarning management system gets the reasonable data for the relevant departments. Reference source of the index system in this paper [6-7], this paper uses multi-hierarchy fuzzy comprehensive evaluation methods to evaluate airport.

TABLE I

First-Hierarchy Early Warning Indexes	Second Hierarchy Warning Indexes	Evaluation Methods

E. Qi et al. (eds.), *Proceedings of the 23rd International Conference on Industrial Engineering and Engineering Management 2016,* DOI 10.2991/978-94-6239-255-7_53

Doer's f actorsF_1	F_{11} Illegal operation rate	Number of illegal Operations / total number of operations
	F_{12} Number of the errors in airport safety	To measure the quality of t he airport safety and safety management level of the a irport
	F_{13} Number of improper emergency treatment	The analysis of the result of the emergency treatment
	F_{14} unqualified rate of the technical e xaminations	Number of the unqualified examinations/the number of executing missions
	F_{15} Occurrence rate of unsafe events due to the factor of doer	Account for the records of the unsafe events
	F_{16} Tacit understanding degree of the teams	Scoring Method
Maintena nce factor F_2	F_{21} Error rate of the maintenance in aviation	Error number of the mainte nance in aviation /the total number of the maintenance operations
	F_{22} Trouble rate of the airport equipment	Number of the equipment errors/ total number of the ground. equipment
	F_{23} Oversight rate of the technical standard	1-(the number of the illegal investigations of the technical standard)/the number of the illegal operations
	F_{24} Rate of the occurrence of the unsafe event because of maintenance	Account for the unsafe events records
	F_{25} Error rate of thousands of flying aircraft resulting from maintenance	Number of delaying and canceling flights because of maintenance /total number of the flights
environm ental fac torF_3	F_{31} Severe weather degree during the flights	Contrast weather forecast w ith minimum standard of flight weather, and abno rmal flight events et al
	F_{32} The number of instant weather change during the flights	Frequency of sudden deterioration of the weather in flight
	F_{33} Number of failed ATC equipment	Measure of ATC hardware quality and technical support level
	F_{34} Risk index of air routes	Comprehensive evaluation of ATC early warning subsystem on air route warning index
	F_{35} Rate of false weather forecasts	Number of rate of false we ather forecasts/total number of airport weather forecast s
	F_{36}Hazard degree of the airport's birds	Bird hazard degree Scoring Method
Organiza tional m anageme nt factor sF_4	F_{41}Critical brain drain rate	Number of critical brain dr ain/total number of critical brains
	F_{42} Distortion rate of information communication	Stability status for personne l in critical posts. Staff me mbers' love and dedication degree and human resource management proficiency
	F_{43} Frequency and intensity of departmental conflicts	Number of distortion of inf ormation communication /to tal number of information
	F_{44} Rationality of organizational structure	(Monitoring in four ranges (highest ,higher, low ,lower)
	F_{45} Group cohesion	Number of choosing each o ther between the members/t otal number of possible cho ices between each other
	F_{46} Oversight rate of standard management	To evaluate safety manage ment personnel and constr uction of safety manageme nt system, routine safety management work and im plementation degree for sa fety management system、
	F_{47} discipline	Number of discipline-violati

	violation rate of staff	on times staff/total number of staff

III. MULIT-HIERARCHY FUZZY COMPREHENSIVE EVALUATION OF EARLY SAFETY WARNING[8]

In the airport early safety warning evaluation index system, besides quantifiable indicators, a part of the indexes are difficult to be quantified and there is a complicated relationship between them. To process, describe and analyze this kind of fuzzy phenomena, fuzzy mathematics offers available methods-Multi-hierarchy Fuzzy Comprehensive Evaluation. The method, from low -hierarchy transition to high- hierarchy, through the expert scoring, determines the degree of membership vector of the single factor evaluation and formed membership degree matrix, then determines the weight vector, and then determines the fuzzy comprehensive evaluation assemblage, and calculates comprehensive evaluation assemblage.

A Establish factors set of dynamic monitoring

By comprehensive consideration of hierarchy division for each factor, the factor assemblage F has two hierarchies.

The first-hierarchy factor F:

$F=\{F_1,F_2,F_3,F_4\}=\{$ doer's factor, maintenan ce factor, environment factor and organizational man agement factor$\}$;

The second-hierarchy factor F_i (see 1.1-1.4) $(i=1, 2, 3, 4)$:

$F_i=\{F_{i1}, F_{i2}, F_{i3}, \dots, F_{ij}\}$

B. Establishment of the single-factor weight matrix A

Using analytic hierarchy process [9-10] to weight determination, the relative comparison between the two factors of the same hierarchy is from 1 to 9. Structuring the first hierarchy judgment matrix S and the second hierarchy of the single-factor judgment matrix S_i (i takes1,2,3,4) ,S and Si matrices are normalized and consistency checking is performed, get first hierarchy weight matrix

$$A=w^*=[a_1,a_2,a_3,a_4] \quad (1)$$

and second hierarchy weight matrix

$$A_i=w^*=[a_1,a_2\dots a_n] \quad (2)$$

C. Establishment of a single-factor fuzzy evaluation matrix multi-factor R_i

The detailed process is as follows: at first, one group of experts conduct the grade evaluation on the checked basic factors, then the statistics are made on the evaluation result and the matrix R_i is established:

$$R_i=\begin{Bmatrix} r_{i11} & r_{i12} & \cdots & r_{i1m} \\ r_{i21} & r_{i22} & \cdots & r_{i2m} \\ \cdots & \cdots & \cdots & \cdots \\ r_{in1} & r_{in2} & \cdots & r_{inm} \end{Bmatrix} \quad (3)$$

where, $i=1, 2, 3, 4$; the number m is determined by number of evaluation assemblage R_i, in this paper, it takes m=4.

$$\gamma_{ijk}=\frac{Number\cdot of\cdot evaluators\cdot who\cdot think\cdot factor\cdot F_{ij}\cdot affiliated\cdot to\cdot V_K(k=1,2,3,4)}{Total\cdot number\cdot of\cdot persons\cdot for\cdot evaluation} \quad (4)$$

D. Comprehensive fuzzy evaluation

Step 1, establish the second-hierarchy comprehensive fuzzy evaluation assemblage, which can be obtained from A and R: comprehensive fuzzy evaluation assemblage =weight matrix × comprehensive fuzzy evaluation matrix i.e.:

$$B_i=A_i\cdot R_i=(b_{i1},b_{i2},b_{i3},b_{i4}) \quad (5)$$

Where $b_i=\bigvee_{i=1}^{n}(a_{ij}\wedge r_{ij})(k=1,2,3,4)$ (6)

It means that the degree of membership of the evaluation object F *to each evaluation element V_k in case of comprehensive evaluation of the respective evaluation elements F_{ij} (j=1, 2, …, n) determining* F_i;

Step 2, establish the first-hierarchy comprehensive fuzzy evaluation assemblage, and take the matrix obtained from the evaluation in the second hierarchy as the first-hierarchy evaluation matrix R, i.e.

$$R=\begin{Bmatrix} B_1 \\ B_2 \\ B_3 \\ B_4 \end{Bmatrix}=\begin{Bmatrix} A_1\cdot R_1 \\ A_2\cdot R_2 \\ A_3\cdot R_3 \\ A_4\cdot R_4 \end{Bmatrix}=\left[R_{ij}\right]_{4\times 4} \quad (7)$$

Then the first-hierarchy comprehensive fuzzy evaluation assemblage is as follows:

$$B = A \cdot R = A \cdot \begin{Bmatrix} B_1 \\ B_2 \\ B_3 \\ B_4 \end{Bmatrix} = (b_1, b_2, b_3, b_4) \tag{8}$$

Where, $b_k = \overset{n}{\underset{j=1}{\vee}} (a_{ij} \wedge r_{ijk})$, k=1,2,3,4; (9)

Assume the weighing vector is:

μ_v= (α_1, α_2, α_3, α_4), where $\alpha_i \in [0, 1]$, (10)

$N = \mu_v \times B_k^T = \mu_v \times (b_1, b_2, b_3, b_4)$ (11)

Divide the comprehensive evaluation values into 4 intervals, which are respectively correspond to the four states of airport organization, that is:

Hazard state [1, 0.75]; Low-hierarchy hazard state [0.75, 0.5]; Almost normal state [0.5, 25]; normal management state [0.25, 0].

When the computed comprehensive evaluation value is at a certain value or interval above-mentioned, the detailed state for the airport organizational management can be visibly displayed.

IV. APPLICATION EXAMPLE

Adopt the index system above-mentioned to perform comprehensive fuzzy evaluation on a certain airport. The matrix computation can be conducted by following AHP method with the help of MATLAB. 25 experts are invited in this paper to make scoring to construct the judgment matrix. Meanwhile, the experts will make evaluation and judgment to construct the R_i *matrix*, and the number listed in the R_i matrix, is determined by the number of evaluation assemblage. The evaluation assemblage is classified as: excellent, good, just so-so and poor.

A. Establishment of single-factor weight matrix

Doer's factor judgment matrix:

$$S_1 = \begin{pmatrix} 1 & 2 & 3 & 1/2 & 1/3 & 1/2 \\ 1/2 & 1 & 1/2 & 5 & 1/3 & 2 \\ 1/3 & 2 & 1 & 3 & 1/2 & 2 \\ 2 & 1/5 & 1/3 & 1 & 1/3 & 1/2 \\ 3 & 3 & 2 & 3 & 1 & 3 \\ 3 & 1/2 & 2 & 2 & 1/2 & 1 \end{pmatrix}$$

Maintenance factor judgment matrix:

$$S_2 = \begin{pmatrix} 1 & 3 & 1/2 & 3 & 2 \\ 1/3 & 1 & 3 & 1/5 & 5 \\ 2 & 1/3 & 1 & 1/2 & 1/2 \\ 1/3 & 5 & 2 & 1 & 2 \\ 1/2 & 1/5 & 2 & 1/2 & 1 \end{pmatrix}$$

Environment basis factor judgment matrix:

$$S_3 = \begin{pmatrix} 1 & 2 & 3 & 1/3 & 2 & 1/2 \\ 1/2 & 1 & 5 & 2 & 3 & 1/2 \\ 1/3 & 1/5 & 1 & 3 & 1/2 & 1/3 \\ 3 & 1/2 & 1/3 & 1 & 3 & 1 \\ 1/2 & 1/3 & 2 & 1/3 & 1 & 1/2 \\ 2 & 2 & 3 & 1 & 2 & 1 \end{pmatrix}$$

Organizational management factor judgment matrix:

$$S_4 = \begin{pmatrix} 1 & 3 & 2 & 1/2 & 2 & 1/2 & 2 \\ 1/3 & 1 & 3 & 2 & 2 & 1 & 5 \\ 1/2 & 1/3 & 1 & 2 & 3 & 1/2 & 2 \\ 2 & 1/2 & 1/2 & 1 & 3 & 2 & 2 \\ 1/2 & 1/2 & 1/3 & 1/3 & 1 & 3 & 1/2 \\ 2 & 1 & 2 & 1/2 & 1/3 & 1 & 2 \\ 1/2 & 1/2 & 1/2 & 1/2 & 2 & 1/2 & 1 \end{pmatrix}$$

By verification, the above-mentioned judgment matrices have all passed the consistency check. After normalization treatment on the above-mentioned matrices, the formulae (2), the weight matrix for each single factor in the second hierarchy shall be obtained as follow:

A_1=w*=[0.1454,0.5101,0.1621,0.0833,0.2916,0.1674]

A_2=w*=[0.2762,0.2105,0.1553,0.2445,0.1135]

A_3=w*=[0.1708,0.2058,0.1117,0.1946,0.0873,0.2298]

A_4=w*=[0.1771,0.2041,0.1351,0.1668,0.0995,0.1448,0.07 26]

B. Establishment of single-factor fuzzy evaluation matrix R_i multi-factors

According to formulae (3) (4), construct R_i *matrix*

$$R_1 = \begin{pmatrix} 0.3600 & 0.5600 & 0.0800 & 0.0000 \\ 0.1200 & 0.5600 & 0.2800 & 0.0400 \\ 0.0800 & 0.4400 & 0.4800 & 0.0000 \\ 0.1600 & 0.2400 & 0.5200 & 0.0800 \\ 0.1200 & 0.3200 & 0.4800 & 0.0800 \\ 0.2000 & 0.4800 & 0.2000 & 0.1200 \end{pmatrix}$$

$$R_2 = \begin{pmatrix} 0.2000 & 0.6000 & 0.2000 & 0.0000 \\ 0.0400 & 0.4000 & 0.4400 & 0.1200 \\ 0.0800 & 0.4400 & 0.4800 & 0.0000 \\ 0.1600 & 0.2400 & 0.5200 & 0.0800 \\ 0.3600 & 0.5600 & 0.0800 & 0.0000 \end{pmatrix}$$

$$R_3 = \begin{pmatrix} 0.2000 & 0.5600 & 0.2400 & 0.0000 \\ 0.2800 & 0.4400 & 0.2000 & 0.0800 \\ 0.1600 & 0.6000 & 0.2400 & 0.0000 \\ 0.1600 & 0.2400 & 0.5200 & 0.0800 \\ 0.2000 & 0.4800 & 0.2800 & 0.0400 \\ 0.1600 & 0.6800 & 0.0800 & 0.0800 \end{pmatrix}$$

$$R_4=\begin{pmatrix}0.2800 & 0.5200 & 0.0800 & 0.1200\\ 0.3200 & 0.4400 & 0.2000 & 0.0400\\ 0.2400 & 0.6000 & 0.1600 & 0.0000\\ 0.1600 & 0.6400 & 0.1200 & 0.0800\\ 0.2400 & 0.2400 & 0.4800 & 0.0400\\ 0.2000 & 0.3200 & 0.4800 & 0.0000\\ 0.2800 & 0.3600 & 0.3600 & 0.0000\end{pmatrix}$$

C. Comprehensive fuzzy evaluation and judgment

Step 1, According to formulae (5) (6), establish the second-hierarchy comprehensive fuzzy valuation assemblage

$B_1 = A_1 \cdot R_1=(0.1651,0.4305,0.3483,0.0561)$

$B_2 = A_2 \cdot R_2=(0.1561,0.4405,0.3586,0.0448)$

$B_3 = A_3 \cdot R_3=(0.1950,0.4981,0.2530,0.0539)$

$B_4 = A_4 \cdot R_4 =(0.2472,0.4660,0.2400,0.0467)$

Step 2, According to formula (7); establish the First-hierarchy evaluation matrix:

$$R=\begin{pmatrix}B1\\B2\\B3\\B4\end{pmatrix}=\begin{pmatrix}0.1651,0.4305,0.3483,0.0561\\0.1561,0.4405,0.3586,0.0448\\0.1950,0.4981,0.2530,0.0539\\0.2472,0.4660,0.2400,0.0467\end{pmatrix}$$

According to AHP, construct the first-hierarchy judgment matrix:

$$S=\begin{pmatrix}1 & 2 & 0.200 & 2\\ 2 & 1 & 1/3 & 0.500\\ 5 & 3 & 1 & 2\\ 2 & 2 & 0.500 & 1\end{pmatrix}$$

Judge that the matrix *S passed normalization Inspection.*

The first-hierarchy weight matrix formula (1) is o btained after normalization treatment.

$A=(0.2030, 0.1450, 0.4326, 0.2194)$

The first-hierarchy comprehensive fuzzy evaluation assemblage is obtained based on formulae (1) (8) (9):

$B=(0.2194\ 0.4326\ 0.2530\ 0.0561)$.

According to formula (10) and characteristics of organizational management and problem research, assume this weight matrix vector $u=(0.3, 0.15, 0.3, 0.25)$, based on formula (11), $N=B^{*}\cdot u$=0.2206 is obtained.

Correspond the result to the respective four states of airport organization; it was found that airport is a *t* normal management state.

V. CONCLUTION

The experiments verified and proved that the safety situation of that airport can be judged with the help of this model, Aiming to the problem existing in different factors, adjustment can be done and relevant safety measures can be formulated in order to enhance the early safety warning proficiency of the airport.

REFERENCE

[1] Yu Liling: "Construction of CAAC safety system". Journal of China Civil Aviation University, vol. 2008, 26(6), pp. 49-51

[2] Xu Lexin, "Research on CAAC safety management system reform and innovation". Economic Control, vol. 2013, 08, pp. 226.

[3] Yao Qian, "Research on CAAC airport early safety warning evaluation system". Communications and Safety, vol. 2009, 15, pp, :198-201.

[4] Ping Yun, "Research on early safety warning managem ent system in civil airport". Wuhan University of Tech nology, 2003， pp.40-45.

[5] Li Jing, Chen Yanqiu, He Pei, "Research on CAAC safety risk monitoring and simulation". Chinese Safety Science, vol. 2009, 19(7), pp.:21-24.

[6] Li Hang and Zhao Jiahong, "Research on the expandable early warning model for airport safety". Journal of China Civil Aviation University, vol. 2009, 45(14): pp. 238-240.

[7] Wen Qiang Gao, "Evaluation index system and its application in civil aviation airport safety management". Northeastern University, 2008, pp: 29-47.

[8] Zhongbao quan, "An evaluation method based on Fuzzy Mathematics -- fuzzy comprehensive evaluation method". China Tendering, vol. 2008.4:12-24

[9] Deng Xue Li Jiaming and Zeng Haojian, "Research on AHP weight computation method and its application". Practice and Cognition of Mathematics, vol. 2012.42(7):pp. 93-99.

[10] Yang Gao, Deyi Mu, "Analytic hierarchy process in aviation safety assessment". Journal of China Safety Science, vol. 2000, 10(3):pp. 38-41.

Modeling and Analysis of the Customer Checkout Process with Flexible Servers for a Retail Store

Wen-he Yang [1,*], Soemon Takakuwa [1]
[1] Department of Industrial and Systems Engineering, Chuo University, Tokyo, Japan
(*yangwh@indsys.chuo-u.ac.jp)

Abstract – **Staff should perform several jobs beside checkout operations during working hours in a retail store. To maintain a high service level, those who are performing other jobs may switch to serve customers when there is a customer waiting queue. In this study, a simulation model of the checkout process with the staff who are cross-trained is constructed to examine the server switching policy. Different scenarios compare system performance under various customer arrival conditions and queuing modes. The simulation model provides a potential decision support tool resolving staffing and queuing arrangement problems in a retail store.**

Keywords – **Simulation modelling, checkout process**

I. INTRODUCTION

In retail, competitiveness depends heavily on the ability to reduce costs and increase customer satisfaction. Waiting in a checkout line in the store is a necessary but undesirable activity that customers must undertake to complete their purchase. The customer may be dissatisfied with a long queue or a long waiting time, and these are considered important indicators for measuring store service quality [1] [2]. In Japan, the convenience store is a common retail format that meets customers' daily needs. Products and services include food, drink, magazines, stationery items, household commodities, bill payment, even ATM and delivery services. However, because of high labour costs, staffing is limited. Therefore, staff members must perform several operations during the work hours besides serving at the checkout counter such as receiving, checking and displaying goods, preparing the Oden and deep-fried food, cleaning the floor, and others [3]. Deciding when to open a new checkout counter and switch from other work is important to the shop manager during the staff training period for maintaining service levels with a limited workforce. In this study, a simulation model is constructed to study the staff switching policy to meet different queuing problems in a retail store. Experiments with different scenarios compare system performance under various customer arrival patterns.

II. CUSTOMER SERVICE AND STAFFING PROBLEMS IN THE RETAIL INDUSTRY

Problems related to retail store operations and workforce management such as shop layout, customer flow, and staffing schedules have been studied extensively. As a powerful analyzing and design tool, several studies have adopted simulation techniques to solve the problems. Miwa and Takakuwa (2008) presented a procedure for simulation modelling in-store merchandizing to examine customer flow and goods layout in a retail store [4]. Miwa and Takakuwa (2010) also presented a simulation analysis for optimization of staffing problems at the retail store considering operation types and frequency [3]. Liu (2009) discussed a simulation IP-based approach to optimize the retail store personnel planning [5]. A discrete event modelling method was used to construct a simulation model of the store checkout process and input customer shopping information from POS data as in [3] [4] [5]. To more naturally describe and simulate system behaviour [6], recently, agent-based modelling (ABM) method has been used to examine customer flow and the checkout process of retail industries. Terano et al. (2009) developed an agent-based in-store simulator to examine the effects of changes to operations through a detailed simulation setting. A supermarket was simulated, and the effect of a layout change and promotions was evaluated in their study [7]. Yamane et al. (2012) built a model for analysing checkout layout design and discussed experimental results with the stakeholders of a specific supermarket [8]. ABM methods in [7] [8] focus on customer behaviour and movement patterns. Siebers et al. (2007) simulated management practices such as training and empowerment in regard to worker behaviour. Their study sought to determine the department where customers usually seek help before purchase [9]. The input data and parameters in [7] [8] [9] were collected from POS terminals or manager experience to reflect system characteristics.

The problem of the staff switching policy for retail services having both front and back room operations was studied in [10] [11] [12]. In these studies, the front room serves customers, and the back room focuses on restocking shelves or other less time-sensitive and interruptible work. The queuing and staffing cost model was constructed, and the optimal switching number was found by the OR method.

Register layout and operation planning of the checkout process were described in several studies using a simulation method [7] [8] [13], but few simulation studies considered the staff switching policy. In this study, simulation of the switching policy using different scenarios is used to compare service level and remaining work time

E. Qi et al. (eds.), *Proceedings of the 23rd International Conference on Industrial Engineering and Engineering Management 2016*, DOI 10.2991/978-94-6239-255-7_54

at other jobs using various customer arrival levels and queuing modes.

III. SIMULATION MODEL

A. Problem Definition

In a retail store, customers move around the shop to pick up the goods that they want to purchase. When customers finish shopping, they walk to the checkout area, and then stand in line waiting to pay the bill to complete the purchase. Meanwhile, the shop staff who are working other jobs frequently check the queue to decide whether to open a new register to shorten customer waiting time.

The outline of the staff switching problem in the study is defined as follows:

1) Number of staff and severs:

N_r =total number of registers in the shop ($N_r > 0$)

S = total number of staff working in the shop ($S > 0$),

S_r= number of staff who are fixed at the checkout counter ($N_r \geqq S_r \geqq 0$),

2) Switching points (Threshold):

The switching point to decide when to open a new register with current staff is denoted as T_0.

The number of customers waiting is used to set the threshold, which means if the waiting number of the queue is more than T_0, and there still has any available staff, he or she will open a new register to serve the customer.

3) Queuing mode:

The queuing mode: (1) customers standing in front of each register (QM_1) and (2) customers standing in a line (QM_2) are examined in the simulation model.

In the case of QM_1, the customer will try to pick a lane with the shortest wait time. We assume that once a customer picks a lane, they remain in the lane. In the case of QM_2, the customer will go to the end of the queue.

4) Output Data:

Customer average waiting time (W_a) and remaining free time for staff to do other jobs (T_f) are used as output data to evaluate system performance.

B. General Description of the Retail-Store

The store in the study is a franchise retail store located at a university campus in Japan. The shop is open from 7AM to 11PM. Sale items are goods that are commonly sold at Japanese convenience stores, such as basic grocery items, magazines, cup noodles, soft drinks, and snack foods, among others. A shop layout with areas is shown in Fig. 1. Most of the customers in the store are university students. Customer arrival rates vary depending on the timetable of the university. Peak time at the store is mid-day lunchtime (12:00 pm-1:00 pm) and during class breaks (approximately 8:30 am-8:45 am; 10:15 am-10:30 am; 2:30 pm-2:45 pm; 4:15 pm-4:30 pm; 6:00 pm-6:15 pm) [4] [5]. If the customer waiting queue is long, store staff will help work at checkout. Occasionally, a customer will take a long time at a register, to pay a bill, deliver services or buy several items. In this case, other staff will check the queue frequently, or the staff who operate the registers will ask for help. The server switching policy of the store staff at the checkout process and other operations is analysed by the simulation modeling method.

C. Model Construction

The simulation model of this study focuses on store checkout out processes, from the point when the customer finishes picking up goods. The flow reflecting customer and staff behaviour is shown in Fig. 2.

The model is developed using Simio™ software [14] that provides independently acting intelligent objects to model system behaviour. The threshold to switching the servers in the model is set through the Monitor Element and Event-driven Process [15].

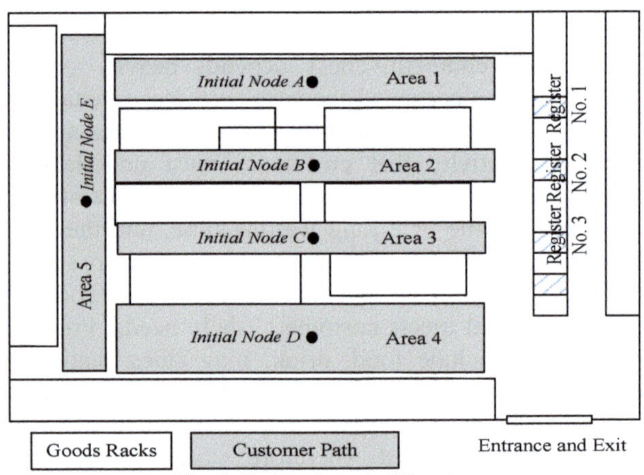

Fig. 1. Facility layout of the store.

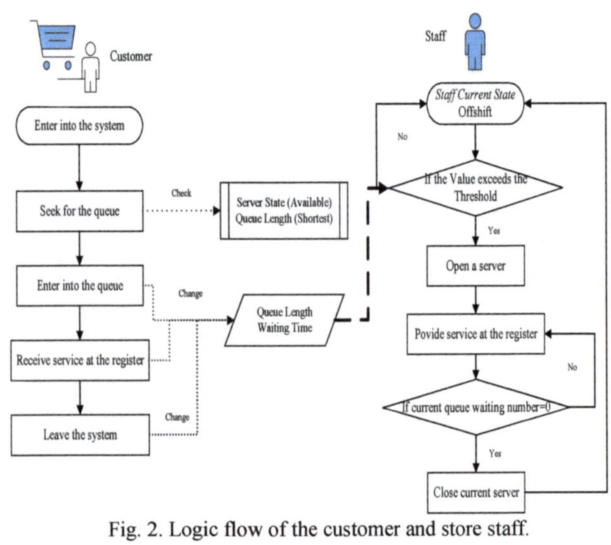

Fig. 2. Logic flow of the customer and store staff.

TABLE I. PARAMETERS FOR PERFORMING SIMULATION

Input Data	**Customer**	
	Initial Node (%)	Node (A,B,C,D,E) =(33,11,3,18,35)
	Buying Items Number (n)	Discrete(1,0.449,2,0.754, 3, 0.904, 4, 0.963, 5, 0.983, 6, 0.991, 7, 0.994, 8, 0.997, 9, 0.999, 10, 1)
	Select Time(seconds)	[0.999+EXPO(23.40)]n
	Staff	
	Nr	3
	S	3
	Sr	1 (Register No. 2)
	Priority of the register open	No.1, No.3
	Processing Time (seconds)	
	Scan	TRIA(n, 2.43*n+0.03, 2.43*n+0.03+3*3.17)
	Transaction	TRIA(7, 6.76*n+14.76, 6.76*n+3*9.31)
Output Data	**Time**	*Wa*
		Tf

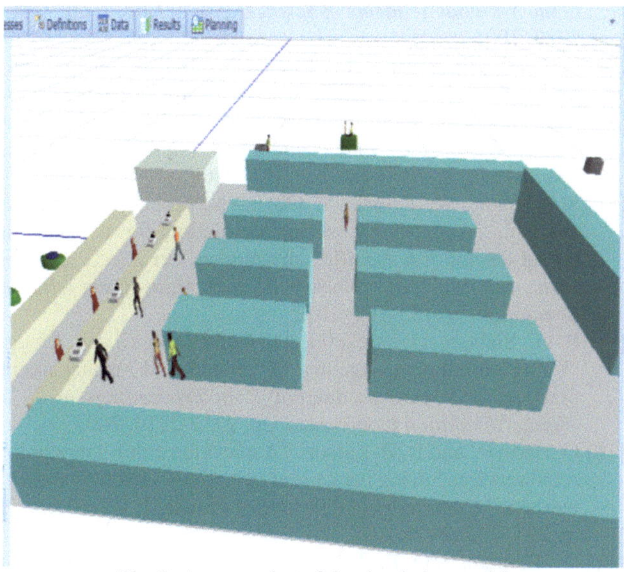

Fig. 3. A screen shot of the simulation model.

Because this is a collaborative study project between the university and the retail company, POS data are provided by the store. One month's weekday data (June, 2014) was used for analysis. After analysing the POS data, the period 1:00 pm~5:00 pm, with three staff is selected to study the server switching problem. After the lunch-time peak, customer arrival rates tend to be stable except for two 15-minute class-break peaks. Preliminary analysis of the POS data indicated that the average number of customers varied from 110 to 160 during normal time and rose to 300 during the class-break peaks. Therefore, four levels of customer arrival rates are used for the experiments: 110 (a_1), 130 (a_2), 160 (a_3), and 300 (a_4) customers per hour. The input data obtained from the POS data and selected simulation parameters are given in Table I.

The model constructed by Simio™ can be confirmed dynamically and visually. Part of the screen image for the simulation model is shown in Fig. 3.

IV. APPLICATIONS

A. Scenarios Constriction

The application examines the staff switching policy in the store. The scenarios are designed to examine the switching points and queue mode under the different arrival rates. Detail of the scenario construction is shown in Table II.

B. Results Comparison among Scenarios

Run length of the experiments is set to 30 minutes for case arrival rates a_4, and 1.5 hours for the other cases. The warm-up period is selected as fifteen minutes before output statistic collection. Simulation experiments of corresponding models are executed with 20 replications. The results of the scenarios using different customer arrival rates are shown in Fig. 4.

TABLE II. SCENARIOS

Customer Arrival Rates	T_0 (person)	QM_1 Scenario 1_#	QM_2 Scenario2_#
a_1	4	1_1	2_1
	3	1_2	2_2
	2	1_3	2_3
	1	1_4	2_4
	0	1_5	2_5
a_2	4	1_6	2_6
	3	1_7	2_7
	2	1_8	2_8
	1	1_9	2_9
	0	1_10	2_10
a_3	4	1_11	2_11
	3	1_12	2_12
	2	1_13	2_13
	1	1_14	2_14
	0	1_15	2_15
a_4	4	1_16	2_16
	3	1_17	2_17
	2	1_18	2_18
	1	1_19	2_19
	0	1_20	2_20

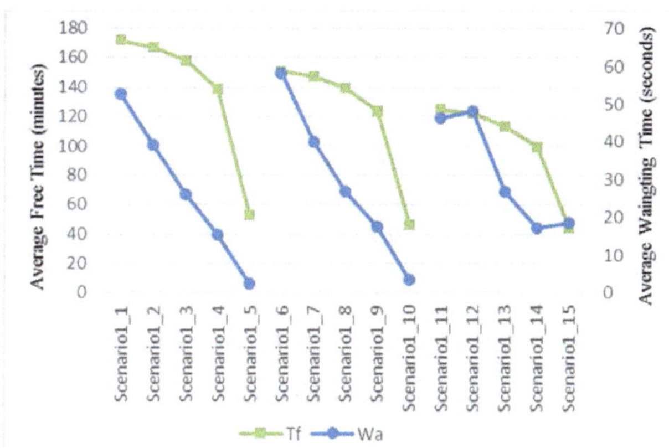

(1)

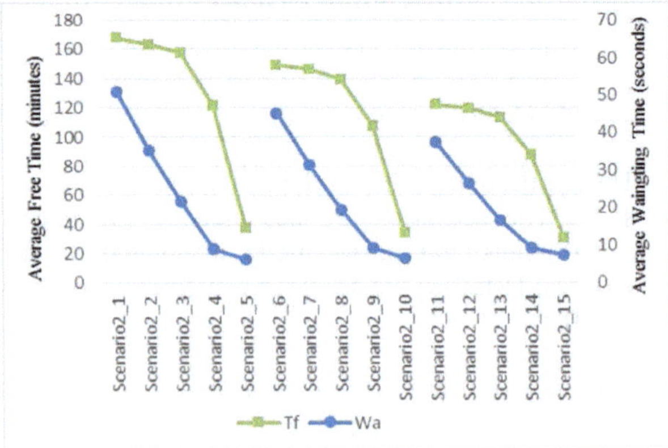

(2)

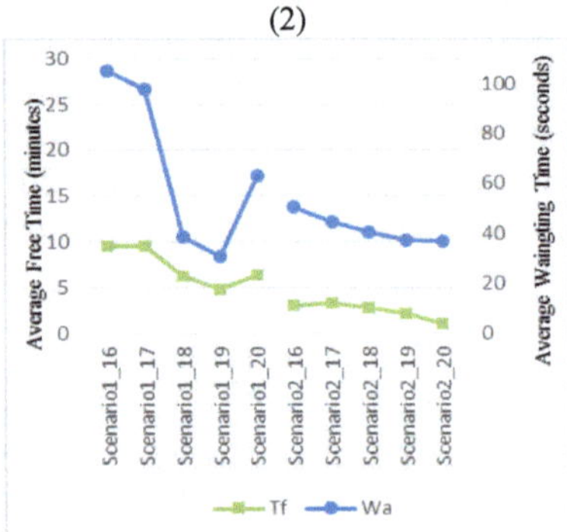

(3)

Fig. 4. Results comparison among the scenarios.

From Fig. 4, it is found that the W_a is shorten significantly with a smaller T_0 by using T_f in the case arrival rates a_1, a_2 for both two queue modes and the case a_3 for QM_2. However, in the busy time period case (arrival rates a_4), the W_a is not improved effectively when the T_0 little than two among the scenarios for QM_1 and varied little among the scenarios for QM_2. In the peak time, if a register is open, it is rarely closed until there is no one waiting for service. The switching policy should be adopted in the normal time period for effectively keeping customer service level and also ensuring enough time for staff completing other operations.

V. CONCLUSIONS

In this study, a simulation model that focused on the customer checkout process was used to examine the staff switching rules of a retail store. The model provides a potential decision support tool s on the store staffs' work assignments and customer checkout queuing patterns. Because the assumption that the customer would wait in the chosen line, some customers may wait for a long time if those ahead take too much time, further model improvements will consider the case where customers may change a queue when a new register opens and the processing time for staff performing other operations.

ACKNOWLEDGMENT

The authors wish to express their sincere gratitude to the FamilyMart Company, Ltd. for their cooperation in completing this research.

REFERENCES

[1] G. Tom, and S. Lucey, "Waiting time delays and customer satisfaction in supermarkets," *Journal of Services Marketing*, vol. 9, no. 5, pp. 20–29, 1995.

[2] P. Jones, and E. Peppiatt, "Managing perceptions of waiting times in service queues," *International Journal of Service Industry Management*, vol. 7, no. 5, pp. 47–61, 1996.

[3] K. Miwa, and S. Takakuwa, "Optimization and analysis of staffing problems at a retail store," in *Proceedings of 2010 Winter Simulation Conference*, pp. 1911-1923, 2010.

[4] K. Miwa, and S. Takakuwa, "Simulation modeling and analysis for in-store merchandizing of retail stores with enhanced information technology," in *Proceedings of 2008 Winter Simulation Conference*, pp. 1702-1710, 2008.

[5] Y. Liu., "Personnel planning of a retail store using POS data," *International Journal of Simulation Modelling*, vol. 8, no. 4, pp. 185-196, 2009.

[6] E. Bonabeau, "Agent-based modeling: methods and techniques for simulating human systems," in *Proceedings of the National Academy of Sciences*, vol. 99, no. 3, pp. 7280-7287, 2001.

[7] T. Terano, A. Kishimoto, T. Takahashi, T. Yamada, and M. Takahashi, "Agent-based in-store simulator for analyzing customer behaviors in a super-market," in *Knowledge-Based and Intelligent Information and Engineering Systems*, pp. 244-251, 2009.

[8] S. Yamane, K. Ohori, A. Obata, N. Kobayashi, and N. Yugami, "Agent-based social simulation for a checkout layout design of a specific supermarket," in *Proceedings of 13th International Workshop on Multi-Agent Based Simulation*, pp.153-164, 2012.

[9] P.-O. Siebers, U. Aickelin, H. Celia, and C. W. Clegg, "Using intelligent agents to understand management practices and retail productivity," in *Proceedings of 2007 Winter Simulation Conference*, pp. 2212-2220, 2007.

[10] O. Berman, and R. C. Larson, "A queueing control model for retail services having back room operations and cross-trained workers," *Computers & Operations Research*, vol. 31, no. 2, pp.201–222, 2004.

[11] O. Berman, and E. Ianovsky, "Optimal management of cross-trained workers, using Markov decision approach," *International Journal of Operational Research*, vol. 3, no. 1-2, pp.154–182, 2008.

[12] D. Terekhov, and J. C. Beck, "An extended queueing control model for facilities with front room and back room operations and mixed-skilled workers," European Journal of Operational Research, vol. 198, no. 1, pp. 223–231, 2009.

[13] M. D. Rossetti, and A. T. Pham, "Simulation modeling of customer checkout configurations," in *Proceedings of 2015 Winter Simulation Conference*, pp. 1151-1162, 2015.

[14] W. D. Kelton, J. S. Smith, and D. T. Sturrock, *Simio and Simulation: Modeling, Analysis, Applications,* 3rd ed. Pittsburgh, Pennsylvania: CreateSpace Independent Publishing Platform, 2013.

[15] R. M. Thiesing, and C. D. Pegden, "Introduction to SIMIO," in *Proceedings of 2013 Winter Simulation Conference*, pp. 4052-4061, 2013.

An Integrated Strategy for Information Security: Outsourcing and In-house

Nan FENG [1], Bin CHEN [1]
[1]Department of Information Management, Tianjin University, Tianjin, China
(fengnan@tju.edu.cn)

Abstract - **This paper examines a mixture of in-house efforts and fee-for-service outsourcing for firms to develop security protection. Due to the different degree of outsourcing, firms divide their information assets into two parts: one part is outsourced and the other part will be protected themselves. For the outsourced part, the optimal contract built by the firm and the MSSP, shows that the MSSP should bear full compensation when the firm was attacked by hacker successfully. Moreover, the unit price the firm should pay to the MSSP will decrease as the degree of outsourcing increases, which means that firms prefer to choose outsourcing because of MSSP's cost advantage. However, the degree of outsourcing does not have an impact on the remaining part, which is not outsourced.**

Keywords - **Outsourcing, Information Security, Cost**

I. INTRODUCTION

With the growth of information security incidents, such as cyber-attacks and security breaches, information security has become a major concern in recent times. Some firms has turned to choose outsourcing information security [1]. This has gave birth to the managed security service, as described in [2]. Schneier [3] showed the case for outsourcing security and considered what to outsource and how to choose outsourcer. In recent years, the market for managed security services is showing strong levels of growth, as in. According to a report issued by Infonetics Research, the worldwide market for managed security services will grow to 18bn dollars in 2016. The numbers show a clear trend, and there's a growing consensus that outsourcing security is a viable option for many companies. The foremost reason that a firm outsources security to a managed security provider is the lack of in-house skills. The reason for this is followed by cost-reduction and temporary need for flex force capacity. Obviously, choosing a MSSP can also release the recruiting limitations and alleviate the burden of tedious tasks. There is no need for firms to spending more efforts and investment in retraining staff.

However, some firms, especially large-scale enterprise, have still some doubts in this regard. Though outsourcing information security is a new wave of managed services, it itself is more sophisticated and full of risk [4]. Taking into account the MSSP, its clients' apparent homogeneity implies that if one client suffers some problems, other clients maybe suffer more probably throw the MSSP's client network. Especially, MSSPs do not always provide a high quality of service, because of mandatory security requirement or moral hazard problems.

From this we can know that decision-making in outsourcing for firms is still a problem. Firms can develop their information security in house, so they should be free to outsource security protection. Many studies on information security focus on two aspects: information security investment and outsourcing information security. The former mainly studies how to invest on information security, like purchasing security tools or software and training & education among staff within firms. The latter focuses on how to build an optimal contract, whether firm or MSSP can get the expected utility

In fact, many firms don't outsource all of its information assets to a MSSP. Except from outsourcing and in-house development, some firms will choose selective outsourcing or partial outsourcing, which means they will outsource some functions, such as mainframe data centers, software development. In the contrast, firms still spend some efforts and monetary investment to develop security development. An empirical investigation in IT outsourcing was showed by Lacity [5] that, the percentage of decisions perceived as successful for selective outsourcing is 85%, higher than total outsourcing or total insourcing. In other product market, Coolidge et al. [6] introduced that this strategy is also understood as a mixture of in-house efforts and fee-for-service outsourcing. For some manufacturers, Wang et al. [7] analyzed three options: full outsourcing, partial outsourcing and no outsourcing. They discovered that in addition to exclusive vertical integration and exclusive outsourcing, partial outsourcing could also be a viable strategy.

In this paper, the question we addressed is that how the amount of information assets outsourced will affect the contract built up between the firm and the MSSP. We define this as the degree of outsourcing, and we study how the firm obtains best utility. For firms, they can outsource one part of information assets to a MSSP and build up a contract. And they develop security protection on the remaining part of information assets by themselves. Due to the outsourced part, firm will build up a contract with a MSSP in order to outsource information security. We model the contracting problem as a one-shot game in which the firm offers a contract and the MSSP accept or reject the contract. Facing with the remaining part, the firm will exert its own effort to develop security protection.

E. Qi et al. (eds.), *Proceedings of the 23rd International Conference on Industrial Engineering and Engineering Management 2016,* DOI 10.2991/978-94-6239-255-7_55

And different from other researches, there are some obvious aspects that we will plan to discuss in this paper. Firstly, we remove the hypothesis that many studies has used: if the firm makes the decision to outsource, it means that the profits which the firm achieves must be higher than developing information security in house. To induce the firm to choose MSSP's service, the MSSP has to ensure that the firm is no worse off than developing in house. However, these two ways which is better or worse still difficult to determine in a nutshell though outsourcing has a cost advantage. Secondly, we'll take the degree of outsourcing into account and analyze its impact on firm's decision. For firms, they need to know how the degree of outsourcing will affect the contract built between the firm and the MSSP.

II. LITERATURE REVIEW

Researchers in prior studies on the economics of information security have examined many issues. There have been many solutions to deal with the problem about information security from different perspectives. Gordon and Loeb [8] developed an economic model to determine the optimal level of investment in information security. From the welfare-maximizing firm's perspective, security investment would lead to reduced likelihood of security breach, but too high an investment level would bring only marginal benefit. Since then, there has been a growing literature that specifically addresses information security investment. Debar et al. [9] presented an architecture for outsourcing a SIM (Security Information Management) platform and analyzed it. Zhao et al. [10] analyses interdependent information security risks in three different ways to solve information security risk: Cyberinsurance, Managed Security Services (MSS), and Risk Pooling Arrangements (RPA). In their study, cyber insurance can help firms optimize security spending but is ineffective in addressing the investment inefficiency caused by risk interdependency. Kai-Lung et al. [11] analyses how system interdependency risks interact with a mandatory security requirement to affect the equilibrium behaviors of an MSSP and its clients. Similar to this study, Ping [12] considered credence goods problem when firms choose to outsource information security to a MSSP. In our paper, our model only discuss one of these measure mentioned above, Managed Security Service and is similar to Gordon and Loeb, based on the trade-off between the probability of security breach and security protection efforts.

A small stream of research has highlighted the importance of how should the firm outsource to the MSSP about information security. In the industry, managed service can be grouped into two main categories: prevention and detection. Thus, there has been two forms of outsourcing, one is that outsourcing both prevention and detection to the same MSSP which is known as the current practice and the other is outsourcing prevention and detection to different MSSPs. Cezar et al. [13] showed that the former outsourcing contract leads to a significant disincentive to provide detection effort. The latter contract alleviates this problem but introduces misalignment of incentives between the firm and the MSSPs and eliminates the advantages offered by complementarity between prevention and detection functions, which may lead to a worse outcome than the current contract. Facing with this dilemma, they proposed a new contract that is superior to these two on various dimensions. Ding et al. [14] studies the characteristics of optimal contracts (payment to MSSPs) for security outsourcing market by setting up an economic framework that combines moral hazard problem and reputation. They [15] also considered transaction costs in outsourcing decisions and analyzed the effect of them. Different from the classification above, we adopt two ways: in-house development and outsourcing, at the same time, to protect firm's information assets.

III. MODEL

We consider a firm that chooses the degree of outsourcing in order to achieve the best utility. Thus, the firm's utility consists of the outsourced part and the in-house part. For some information assets, the firm chooses to protect information security in-house, which means that the firm will invest its effort and time to develop information security. We can get the firm's in-house utility under this part. For the remaining part, the firm chooses to outsource information security, which means that it should build up a contract with a MSSP. Because of this, the firm is faced with the problem of determining the optimal outsourcing contract.

TABLE I
MODEL NOTATION

Variable	Description
v	Value of information assets ($v > 0$)
a	Probability of attack by a hack ($0 < a < 1$)
e_m	The MSSP's effort
e_f	The firm's effort
c_m	Cost coefficient of MSSP
c_f	Cost coefficient of firm
P	Unit price paid by the firm to the MSSP
β	Ratio based on compensation from the MSSP to the firm
α	Degree of outsourcing
π_0	MSSP's reservation payoff

Tips: Subscript 'f' represents firm, and subscript 'm' represents MSSP

In this model, there is one firm and one MSSP. The firm values all of its information assets at v. Because the firm take the a mixture of in-house efforts and fee-for-service outsourcing for its information security, it should make the decision to confirm the degree of outsourcing at α, which means the value of outsourced part is αv and the in-house part is $(1-\alpha)v$. For the in-house part, the firm should determine its effort e_f to develop information security. For the outsourced part, MSSP determines its effort e_m. The cost of security protection also depends on the efforts, which is expressed like: $1/2 * c_i e_i^2$, where c_i is a cost coefficient. Obviously, the function of cost is convex and the firm's cost function is similar to the MSSP's.

Besides this, we model the contracting problem as a one-shot game in which the firm offers a contract and the MSSP accepts or rejects the contract. The firm will provide a fixed fee P to the MSSP and both sides will sign an agreement like SLAs. In the model, the SLA between the firm and the MSSP includes a compensation term ('liability'), $\beta \in [0,1]$. If the firm suffer a loss of v because of the hacker's attack, then the MSSP has to compensate her by βv.

Consistent with the one-shot game, we model security attacks during security protection period. It means a hacker may attack the firm with it, a probability $a \in [0,1]$. Above all, our model encompasses the following assumptions:

Assumption 1: $av < c_m < c_f$.

Assumption 2: c_m, c_f, a and v are public information.

Assumption 1 has two uses in this model. One ensures the analysis will not arrive at a corner solution. If $av > c_m$, then the expected loss to the firm is excessive. The other ensures that when outsourcing information security to a MSSP, the firm will have a cost advantage, which directly induces it to choose to outsource instead of in house.

Now, we can get the utility for the in-house part and outsourced part when developing security protection:

$$u_{in-house} = (1-\alpha)v\left(1-a(1-e_f)\right) - \frac{1}{2}(1-\alpha)c_f e_f{}^2. \qquad (1)$$

$$u_{outsourced} = \alpha v(1-a(1-e_m)) - \alpha vP + \alpha a v(1-e_m)\beta. \qquad (2)$$

For $u_{in-house}$, the first term is the remaining value when the firm has exerted its effort e_f but its in-house part of information assets was still attacked successfully. The second term means the firm's cost base on the effort e_f. The firm should select an optimal e_f^* to maximize its utility.

For $u_{outsourced}$, the first term is the remaining value when the MSSP has exerted its effort e_m on the firm's outsourced information assets but this outsourced part was still attacked successfully. The second term is the price which the firm should pay for the outsourced part. The last term is the compensation which the MSSP should compensate to the firm because of the MSSP's fault that it cannot prevent these attacks absolutely.

Now, we can get the payoff of the MSSP:

$$\pi_m = \alpha vP - \alpha a v(1-e_m)\beta - \frac{1}{2}c_m e_m{}^2 \qquad (3)$$

For π_m, the former two terms has explained above and the last term is the cost of MSSP based on e_m.

When the firm decides to outsource some part of information assets, the sequence is as follows:

1. The firm offers the contract $[P, \beta]$.
2. If the MSSP accepts the contract, it selects e_m to protect the information assets the firm has outsourced; otherwise, the game ends.
3. If the firm is attacked successfully, the MSSP will pay the firm a certain amount of compensation calculated by β.

We user backward induction to solve the firm's contracting problem. In the stage of game, the MSSP determines its optimum effort by maximizing π_m. The first derivative of π_m with respect to e_m:

$$\frac{\partial \pi_m}{\partial e_m} = \alpha a v\beta - c_m e_m \qquad (4)$$

And we obtain the MSSP's optimum effort: $e_m^* = \alpha a v\beta / c_m$. Anticipating how the MSSP will determine its best response in effort, the firm solves the below problem:

$$\max_{(P,\beta,e_f)} \begin{array}{l} u = \alpha v\left(1-a(1-e_m^*)\right) - \alpha vP + \alpha a v(1-e_m^*)\beta \\ +(1-\alpha)v\left(1-a(1-e_f)\right) - \dfrac{1}{2}(1-\alpha)c_f e_f{}^2 \\ s.t. \ \{\pi_m >= \pi_0. \end{array} \qquad (5)$$

IV. OPTIMAL DECISION

We can solve the previous section's problem using Lagrange multipliers on constraints. The solution is:

$$e_m^* = \frac{\alpha a v}{c_m}. \qquad (6)$$

$$e_f^* = \frac{av}{c_f}. \qquad (7)$$

$$P^* = \frac{\pi_0}{\alpha v} + a - \frac{\alpha v a^2}{2c_m}. \qquad (8)$$

$$\beta^* = 1. \qquad (9)$$

Proposition 1:

(a) The effort of the MSSP decreases as the cost coefficient of the MSSP increases, while increases as the degree of outsourcing increases.

(b) The MSSP should bear the 100% loss when the hack attacks the firm successfully.

(c) The unit price decreases as the degree of outsourcing increase and increases as the remaining payoff the MSSP wants to achieve.

For (a), it's easy to understand that when the cost the MSSP invested increases, the MSSP has to do something to reduce it and the effort the MSSP exert will decrease.

For (b), the MSSP has to bear all of loss because the firm has strong bargaining power in their contract and it wants to achieve full of compensation. The last point is that when the outsourced part is bigger, the firm will enjoy more MSSP's cost advantage. In other words, the unit price will decrease with the degree of outsourcing increasing.

From the Proposition 1, we can conclude that the degree of outsourcing has a big effect on the contract and we can describe the relationship by a numerical example:

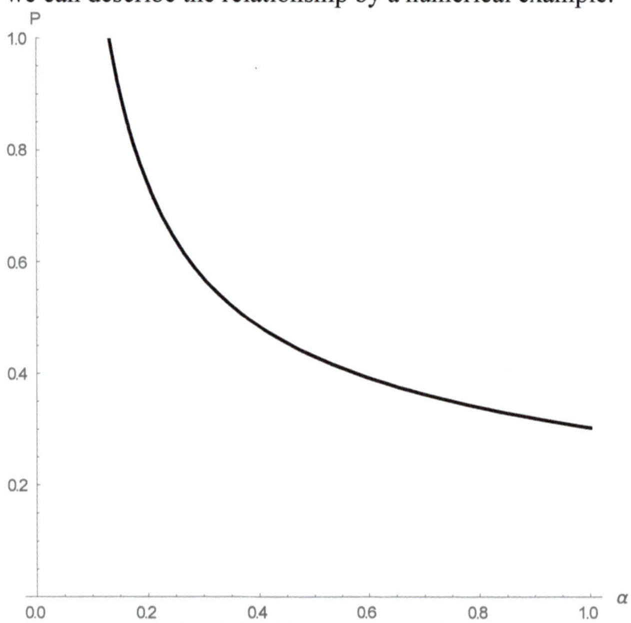

Fig. 1. The relationship between the unit price and the degree of outsourcing
($a = 0.2, c_m = 0.5 < c_f = 1.5, \pi_0 = 0.1, v = 1$)

. Obviously, the firm prefers to choose to outsource more information assets to develop security protection from the point of view of cost.

V. DISCUSSION

In order to distinguish between in-house strategy and mixed strategy, we also use a numerical example to describe the difference. We use u_{mixed} as the firm's utility when taking mixed security strategy and use $u_{in-house}$ as the firm's utility when taking in-house strategy.

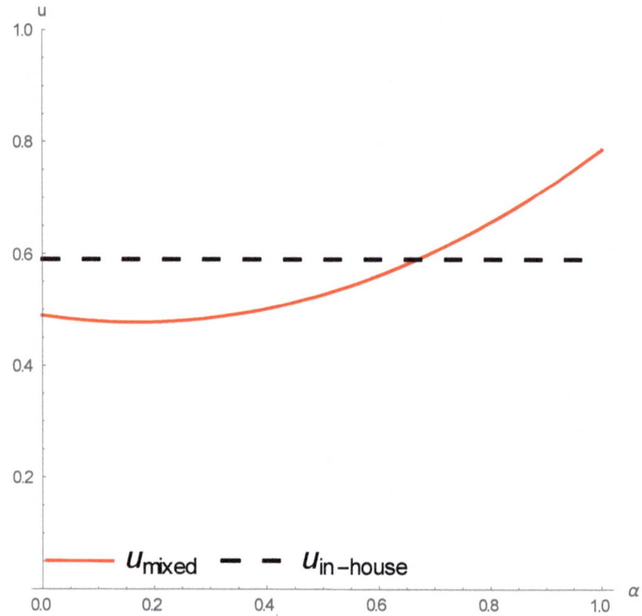

Fig. 2. The utility of mixed strategy and in-house strategy
($a = 0.6, c_m = 0.5 < c_f = 1.5, \pi_0 = 0.1, v = 1.1$)

From figure 2, the firm's utility when taking mixed strategy first decreases and then increases as the degree of outsourcing. Moreover, when the degree of outsourcing is low, the in-house strategy is better than the mixed strategy for the firm.

VI. CONCLUSION

Our analysis considers not only some factor mentioned by other researcher, like cost advantage, outsourcing contract, and security protection in-house, but also more interesting and meaningful factor. In this paper, we take the degree of outsourcing into the model. And we found that whether how many information assets the firm outsources, the loss should be full compensated by the MSSP after the firm was attacked successfully. Additionally, the unit price decreases when the degree of outsourcing increases in the contract. That means the more firms outsourced, the lower unit price t firms should pay to the MSSP. Thus, we proved that the degree of outsourcing will have a big effect on firm's decisions and the motivation for firms to choose outsourcing is cost advantage. Compare with in-house strategy, the mixed strategy may be a better decision for firms when firms choose to outsource the most of information assets. There some limitations in this paper: the first is that there are some other factors, like information leakage in outsourcing, which affect firm's degree of outsourcing. The second is that this paper is lack of comparison with more security strategies.

ACKNOWLEDGMENTS

The research was supported by the National Natural Science Foundation of China (no. 71271149). The authors are very grateful to all anonymous reviewers whose invaluable comments and suggestions substantially helped improve the quality of this paper.

REFERENCES

[1] Axelrod, C. Warren, "Outsourcing Information Security," Artech House Inc, 2004.

[2] Allen, Julia H., D. Gabbard, and C. J. May, "Outsourcing Managed Security Services," Outsourcing Managed Security Services 1, 2003.

[3] Schneier B, "The case for outsourcing security," *Computer*, vol. 35, no. 4, pp. 20-26, 2002.

[4] Navarro L, "Information Security Risks and Managed Security Service," *Information Security Technical Report*, vol. 6, no. 3, pp. 28-36, 2001.

[5] Lacity, M. C., and Willcocks, L. P., "An empirical investigation of information technology sourcing practices: lessons from experience," *Mis Quarterly*, vol. 22, no. 3, pp. 363-408, 1998.

[6] Coolidge, J., Ilic, D., and Kisunko, G., "Small businesses in south Africa: who outsources tax compliance work and why ?," *Policy Research Working Paper*, 2009.

[7] Wang, L. M., Liu, L. W., and Wang, Y. J., "Capacity decisions and supply price games under flexibility of backward integration," *International Journal of Production Economics*, vol. 110, no. 1-2, pp. 85-96, 2007.

[8] Gordon, L. A., and Loeb, M. P., "The economics of information security investment," *Acm Transactions on Information & System Security*, vol. 5, no. 4, pp. 438-457, 2002.

[9] Debar, H., and Viinikka, J., "Security information management as an outsourced service," *Information Management & Computer Security*, vol. 14, no. 14, pp. 417-435, 2013.

[10] Xia Zhao, Ling Xue, and Andrew B. Whinston., "Managing interdependent information security risks: cyberinsurance, managed security services, and risk pooling arrangements," *Journal of Management Information Systems*, vol. 30, no. 1, pp. 123-152, 2009.

[11] Kai-Lung Hui, Wendy Hui, and Wei T. Yue., "Information security outsourcing with system interdependency and mandatory security requirement," *Journal of Management Information Systems*, vol. 29, no. 3, pp. 117-155, 2012.

[12] Ke, P. F., Hui, K. L., and Yue, W. T., "Information Security as a Credence Good," *Financial Cryptography and Data Security*, 2013.

[13] Cezar, A., and Raghunathan, S., "Outsourcing information security: contracting issues and security implications," *Management Science*, vol. 60, no. 3, pp. 638-657, 2010.

[14] Ding, Wen, W. Yurcik, and X. Yin, "Outsourcing Internet Security: Economic Analysis of Incentives for Managed Security Service Providers," Internet and Network Economics, First International Workshop, WINE 2005, Hong Kong, China, December 15-17, 2005, Proceedings 2005:947-958.

[15] Ding, W., Yurcik, W., Ding, W., and Yurcik, W., "Outsourcing Internet Security: The Effect of Transaction Costs on Managed Service Providers," Prof.of Intl.conf.on Telecomm.systems (2005).

MIX
Papier aus verantwortungsvollen Quellen
Paper from responsible sources
FSC® C105338

If you have any concerns about our products,
you can contact us on
ProductSafety@springernature.com

In case Publisher is established outside the EU,
the EU authorized representative is:
Springer Nature Customer Service Center GmbH
Europaplatz 3, 69115 Heidelberg, Germany

Printed by Libri Plureos GmbH
in Hamburg, Germany